MATHEMATICS
FOR
ELECTRONICS

D0139180

MATHEMATICS
FOR
ELECTRONICS

Nancy Myers
BUNKER HILL COMMUNITY COLLEGE

West Publishing Company

Minneapolis/St. Paul   New York   Los Angeles   San Francisco

☐ CREDITS

Art: Kevin Tucker and Caroline Jumper
Composition: G&S Typesetters, Inc.
Copyediting: Caroline Jumper
Cover and Interior Design: Roslyn Stendahl, Dapper Design
Cover Photo: Andy Caulfield, The Image Bank
Index: Linda Buskus, Northwind Editorial Services

Production, Prepress, Printing and Binding by West Publishing Company.

WEST'S COMMITMENT TO THE ENVIRONMENT

In 1906, West Publishing Company began recycling materials left over from the production of books. This began a tradition of efficient and responsible use of resources. Today, up to 95 percent of our legal books and 70 percent of our college texts are printed on recycled, acid-free stock. West also recycles nearly 22 million pounds of scrap paper annually—the equivalent of 181,717 trees. Since the 1960s, West has devised ways to capture and recycle waste inks, solvents, oils, and vapors created in the printing process. We also recycle plastics of all kinds, wood, glass, corrugated cardboard, and batteries, and have eliminated the use of styrofoam book packaging. We at West are proud of the longevity and the scope of our commitment to our environment.

COPYRIGHT © 1993   By WEST PUBLISHING COMPANY
                        610 Opperman Drive
                        P.O. Box 64526
                        St. Paul, MN 55164-0526

All rights reserved
Printed in the United States of America
00 99 98 97 96 95 94 93   8 7 6 5 4 3 2 1 0

Library of Congress Cataloging-in-Publication Data

Myers, Nancy.
    Mathematics for electronics / Nancy Myers.
        p.   cm.
    Includes index.
    ISBN 0-314-01266-4 (soft)
    1. Electronics—Mathematics.   I. Title.
TK7835.M94   1993
    621.382′01′51—dc20   92-25704
                                                      CIP

Contents

Preface

Mathematics for Electronics is intended for one-semester or two-semester courses covering algebra and trigonometry topics used in electronics technology. Applications of the mathematical topics covered are drawn from areas of electronics.

The text is a mathematics book, placing emphasis on the mathematics underlying the study of electronics rather than on the electronics itself. The mathematics is developed as carefully as in many pure mathematics texts. Mathematical considerations, such as cases involving negatives, and extraneous and "no solution" cases, are discussed for appropriate topics. On the other hand, topics that are not needed at all for electronics are omitted entirely. Exceptions are a unit on quadratics, which completes the essentials of algebra, and a unit on analytical trigonometry, which completes the essentials of trigonometry. These units enable students using this book to move on to calculus if desired.

LEVEL

The text can be used by students in two-year colleges, four-year colleges, and post-secondary technical/vocational schools. For a one-semester course, students should have at least a year of high school algebra or a semester of post-secondary preparatory math. The text is self-contained, however, and can be used in a two-semester sequence by students who have never studied algebra. Additionally, Unit R is included for students who need to review pre-algebra mathematics and basic calculator functions.

On average, the mathematical difficulty is intermediate in level. The beginning units are comparable to, but perhaps somewhat more extensive than, beginning algebra texts. Later units are more difficult, perhaps approaching some college algebra and trigonometry texts. With the exception of technical terminology that must be mastered, the reading level is moderate.

CALCULATORS

The text is written with the expectation that the student will use a scientific calculator. Outdated topics such as using logarithm and trigonometry tables, and calculating by using logarithms, are not included. To help the student learn to use a calculator efficiently, sample calculator algorithms have been included. These algorithms are given in detail, including both key sequences and the expected display. Various key definitions used on common types of basic scientific calculators such as Texas Instruments, Sharp, Casio, and Radio Shack are included. An appendix covers the essentials of reverse Polish notation with algorithms for some Hewlett-Packard calculators.

Use of some calculator functions is not included or is reserved until the student has mastered the concept by hand. Calculators are not used for rounding off, writing numbers in scientific or engineering notations, or other fundamental skills. The EXP or equivalent exponent entry key is introduced only after the student has had extensive practice with powers of ten.

OTHER FEATURES

Mathematics for Electronics is unique among electronics mathematics texts in its style of presentation. The text has been developed for self-study and self-paced courses as well as for lecture courses. The essentials of the lecture are included in the text. Thus verbal explanation not only precedes worked examples, but also is provided between steps of examples. Formulas, equations, steps of worked examples, and other items that might be written on a blackboard during a lecture, are set apart in display lines rather than written into paragraphs. Definitions and rules that would be emphasized in a lecture are set off by boxes. Dialogue with the student is established by wording explanations in terms of "we" as it might be used in class, and sometimes speaking directly to the student in terms of "you," for example, when the student is using a calculator.

UNIT FORMAT

The text is also unique in its organization into units rather than chapters, with the units then subdivided into sections. Each unit has no more than five related topics, and each section discusses aspects of just one topic. This unit form of organization and its related features offer many advantages:

- *Unit form of organization*: This format allows the student to master a limited number of related topics in each unit, and related aspects of a single topic in each section. Each unit covers two to five closely related objectives. Each section within a unit covers the subject matter described by just one objective.

- *Cross-referencing*: The entire text is cross-referenced by objective numbers. At the beginning of each unit there are at least two but not more than five objectives. Each section within the unit corresponds by number to an objective. Each item in the self-test at the end of the unit is keyed to an objective number by a reference in the answer section. Thus, an incorrect answer on a self-test leads the student by number to the appropriate objective, and then by number to the appropriate section for further study.

- *Worked examples*: Worked examples show various aspects of each topic; for example, the different cases that might be encountered. As the examples progress, the amount of explanation, especially for steps done in previous examples, gradually decreases, encouraging students to fill in their own explanations.

- *Exercise sets*: Every section has an exercise set. Since the sections are generally short, exercise sets occur frequently. The student works on a new concept right away, on only one concept at a time, and with a list of problems that is not so long as to be discouraging. Odd and even exercises are paired, the first two or four matching the first example, the next couple matching the second example, and so on. Answers to all exercises are given at the end of the text.

- *Self-Tests*: Each unit ends with a self-test consisting of five items. Answers to all self-test items are included at the end of the text, and are keyed to objectives and sections as described above.

- *Cumulative Reviews*: After each four or five units there is a cumulative review unit. The self-tests in cumulative review units have ten items with answers at the end of the text. The answers are keyed to units but not to specific sections.

SEQUENCING

Choices of topics will be influenced by the program in which the text is used as well as the preparation level of the students. For programs where a concurrent course in basic electricity or electronics is required, less emphasis can be put on applications such as those included in Units 7 and 8, with more emphasis on mathematical topics such as those in Units 13 and 14.

For a one-semester course taught independently of an electronics course, students should have prior knowledge of high school algebra. Such a course might cover parts of Units 2, 5, and 7 by way of review and an introduction to mathematics applied to electronics. The main course would include Units 9, 10, 12, 15, 17, 18, and 19, with parts of Units 13, 14, 20 and 22 included if time allows.

For students who have little or no algebra, a two-semester course is recommended. Such a course would cover parts of Unit R if necessary, and then Units 1, 2, 3, 4, 5, 7, and 8. The course could then continue as described above.

Units 6, 11, 16, and 21 are not listed in the sequences above since they are review units. Some adaption of these review units might be necessary when preceding units or sections have been omitted.

SUPPLEMENTS

Two types of supplements are available to instructors using this text:

- *Solutions Manual*: The solutions manual contains worked solutions for all Exercise and Self-Test problems.

- *Test Manual*: The test manual includes four forms of quizzes for each unit. Like the Self-Tests, quizzes for content units contain 5 items and quizzes for cumulative review units contain 10 items. The test manual also includes two forms of final exams, each containing 25 items.

The test manual has also been adapted as a test bank in multiple-choice format for the WESTEST 3.0 computerized testing software for IBM and Macintosh.

ACKNOWLEDGMENTS

The text was reviewed extensively throughout the developmental process. I am grateful to those who patiently reviewed it at various stages of development and revision:

Luis Acevedo
DeVry Institute, Woodbridge

Ben Bartlett
College of Southern Idaho

Robert Campbell
Los Angeles Trade-Technical College

Richard Cotter
De Anza College

Chabel Fahed
Northern Virginia Community College

Paul Fraley
Savanna Technical Institute

Tom Hansen
Clayton State College

David Hata
Portland Community College

Donald Huskey
Bainbridge College

Wendell Johnson
University of Akron Community &
Technical College

Robert Kimball
Wake Technical Community College

John Knox
Vermont Technical College

Ronald Reis
Los Angeles Trade-Technical College

Ames Stewart
College of Southern Idaho

William Thomas
University of Toledo

Joel Turner
Blackhawk Technical Institute

James Ward
Pensacola Junior College

Terry Wright
Central Oklahoma ATVS

Judith Freedman, a former student, checked all the worked examples and exercise answers, and also constructed the problems for the test banks. In addition, she wrote out detailed solutions for the exercise and self-test problems for the solutions manual.

At West Publishing, Denise Bayko, Developmental Editor, and Ron Pullins, Acquisitions Editor, saw the manuscript through the developmental process. Laura Nelson, Production Editor, shepherded both me and the manuscript through the production process with great care and good humor.

Nancy Myers
Bunker Hill Community College

UNIT R

Review: Basic Concepts

Introduction

You have used numbers all your life, so much so that some numerical concepts have become second nature to you. You make change using quarters (a quarter is a fraction), and talk about chances of an event happening (often in terms of percents). Yet many people experience difficulty when faced with formal manipulations with numbers. To succeed in science and technology, including electronics, you need to feel comfortable with numbers and with formal arithmetic operations with numbers. In this unit, you will review whole numbers, fractions, and decimals, and their arithmetic operations: addition, subtraction, multiplication, and division. Then, you will apply some of these concepts to percents. You will learn how to do arithmetic operations by using a calculator. You will also review squares and square roots, and learn how to find squares and square roots by using typical scientific calculators.

OBJECTIVES

When you have finished this unit you should be able to:

1. Add, subtract, multiply, and divide whole numbers.
2. Reduce fractions to lowest terms, and add, subtract, multiply, and divide fractions.
3. Add, subtract, multiply, and divide decimals, and convert fractions to decimals.
4. Convert percents to fractions and to decimals, and find percentages.
5. Use a calculator to add, subtract, multiply, and divide, and find squares and square roots.

SECTION R.1

Whole Numbers

The first use of numbers, long before the beginning of recorded history, was most likely for counting. Our number system is based on the numbers we use to count. These numbers are the **counting numbers** or **natural numbers**:

$$1, 2, 3, 4, 5, 6, 7, 8, 9, 10, 11, 12, \ldots$$

The three dots, called an *ellipsis*, indicate that the list continues indefinitely.

The development of our number system can be traced to ancient times. Many types of numbers, including fractions and even square roots, were used by ancient civilizations. The Egyptians wrote numbers by using symbols like those in their picture writing, called hieroglyphics. Symbols were repeated in groups of up to ten, so the system was a base ten system. The Babylonians used wedge-shaped symbols called cuneiform characters, with a base of sixty! The ancient Greeks used the letters of their alphabet, or more often, Roman numerals. We still use Roman numerals: the 1992 football Super Bowl was Super Bowl XXVI (the Roman numeral for 26), and the year 1992 is sometimes written as MCMXCII. In these systems, new symbols must be created as larger numbers are encountered.

In tracing the development of number systems, it is important to understand that new numbers are *invented* when a need arises. None of the systems mentioned above used a number equivalent to our zero. Our concept of zero was invented about A.D. 500 by Hindu mathematicians, who also developed the symbols we use for the counting numbers. The set of counting numbers with zero is the set of **whole numbers**:

$$0, 1, 2, 3, 4, 5, 6, 7, 8, 9, 10, 11, 12, \ldots$$

The system of writing numbers devised by the Hindus uses the base ten. However, their ingenious system uses zero to allow numbers of any size to be written by using just ten symbols. The ten symbols are the **digits**:

$$0, 1, 2, 3, 4, 5, 6, 7, 8, 9.$$

This list does *not* continue indefinitely; there are exactly ten digits.

PLACE VALUE

The number system based on the ten digits is called the Hindu-Arabic system, for the Hindu mathematicians who developed it and the Arab traders who brought it to Europe. In the Hindu-Arabic system, the whole numbers are written by giving each digit a **place value** or **weight**. The place values are the base ten and products of tens:

$$10 = 10$$

$$100 = 10 \times 10$$

$$1000 = 10 \times 10 \times 10$$

and so on. Only the digit at the far right of a whole number represents the actual value of the digit. The digit to its left has place value 10, the next has place value 100, the next has the place value 1000, and so on.

EXAMPLE R.1   Write 222 in terms of place values.

SOLUTION The digit 2 at the far right has the value two. The digit 2 in the middle represents two tens, and the digit 2 at the left represents two hundreds. Thus, we interpret 222 as

$$222 = 2 \times 100 + 2 \times 10 + 2$$

or

$$222 = 200 + 20 + 2. \qquad \blacktriangle$$

EXAMPLE R.2   Write 5194 in terms of place values.

SOLUTION Starting at the far right, the digit 4 has its actual value. Then, continuing by products of ten, we have

$$5194 = 5 \times 1000 + 1 \times 100 + 9 \times 10 + 4$$

or

$$5194 = 5000 + 100 + 90 + 4. \qquad \blacktriangle$$

The absence of any amount for a place value is indicated by the digit zero.

EXAMPLE R.3   Write 202 in terms of place values.

SOLUTION The digit zero means there are no tens:

$$202 = 2 \times 100 + 0 \times 10 + 2$$

or

$$202 = 200 + 2. \qquad \blacktriangle$$

The **arithmetic operations** are addition, subtraction, multiplication, and division. Arithmetic operations for whole numbers are based on place values. A review of these operations follows; the use of a calculator for longer problems is introduced in Section R.5.

ADDITION

To add two whole numbers, we add corresponding weighted digits; that is, digits with the same place values.

EXAMPLE R.4

Add 41 + 56.

SOLUTION

We add the corresponding weighted digits:

$$
\begin{array}{rcr}
40 + 1 = & & 41 \\
+\ 50 + 6 = & & +\ 56 \\
\hline
90 + 7 = & & 97
\end{array}
$$

▲

When the sum of two weighted digits exceeds the base ten, we *carry* into the next place to the left.

EXAMPLE R.5

Add 28 + 37.

SOLUTION

We add the corresponding weighted digits:

$$
\begin{array}{rcr}
20 + 8 & & 28 \\
+\ 30 + 7 = & & +\ 37 \\
\hline
50 + 15 & &
\end{array}
$$

The result 15 exceeds the base ten; therefore, we carry into the next place to the left:

$$
\begin{array}{rcr}
50 + 15 & & \\
50 + 10 + 5 & & \\
60 + 5 = & & 65
\end{array}
$$

We can summarize the process of carrying by writing the carried digit above the column into which it is carried:

$$
\begin{array}{r}
1 \\
28 \\
+\ 37 \\
\hline
65
\end{array}
$$

▲

SUBTRACTION

To subtract one whole number from another, we subtract corresponding weighted digits.

EXAMPLE R.6

Subtract 89 − 32.

SOLUTION

We subtract the corresponding weighted digits:

$$
\begin{array}{rcr}
80 + 9 = & & 89 \\
-\ 30 - 2 = & & -\ 32 \\
\hline
50 + 7 = & & 57
\end{array}
$$

▲

When a subtraction of two digits cannot be done in whole numbers, we *borrow* from the next place in the first number.

| EXAMPLE R.7 | | Subtract 55 − 38. |

SOLUTION   We subtract the corresponding weighted digits:

$$50 + 5 = 55$$
$$-\ 30 - 8 = -\ 38$$

The subtraction $5 - 8$ cannot be done in whole numbers; therefore, we borrow from the next place in the first number:

$$50 + 5$$
$$40 + 10 + 5$$
$$40 + 15$$

Then, we have

$$40 + 15 = 55$$
$$-\ 30 - 8 = -\ 38$$
$$\overline{10 + 7 = 17}$$

We can summarize the process of borrowing by writing above the columns:

$$\begin{array}{cc} 4 & 15 \\ \cancel{5} & \cancel{5} \\ -\ 3 & 8 \\ \hline 1 & 7 \end{array}$$

    We recall that we can check subtraction by addition. To check the subtraction in Example R.7, we add $38 + 17$:

$$\begin{array}{r} 1 \\ 38 \\ +\ 17 \\ \hline 55 \end{array}$$

MULTIPLICATION

Multiplication is repeated addition. For example,

$$3 \times 2 = 2 + 2 + 2$$
$$= 6.$$

We observe that any whole number multiplied by zero is zero. For example,

$$3 \times 0 = 0 + 0 + 0$$
$$= 0.$$

Any whole number multiplied by 1 is that whole number. For example,

$$3 \times 1 = 1 + 1 + 1$$
$$= 3.$$

Multiplication of any whole number by 10 puts a zero on the end of the number. For example,

$$3 \times 10 = 10 + 10 + 10$$
$$= 30.$$

Then, another multiplication by 10 results in another zero:

$$30 \times 10 = 3 \times 10 \times 10$$
$$= 3 \times 100$$
$$= 300,$$

and so on.

To multiply a whole number by a one-digit number, we multiply each weighted digit by the one-digit number. Carried digits, when necessary, are added as in addition.

EXAMPLE R.8 Multiply 16×3.

SOLUTION We multiply each weighted digit by 3:

$$
\begin{array}{rcr}
10 + 6 = & & 16 \\
\times \qquad 3 = & \_ & \times \ 3 \\
\hline
30 + 18 & & \\
30 + 10 + 8 & & \\
40 + 8 = & & 48 \\
\end{array}
$$

We can summarize this carrying by writing the digit carried above the column:

$$
\begin{array}{r}
1 \\
16 \\
\times \ 3 \\
\hline
48 \\
\end{array}
$$

▲

To multiply two-digit numbers, we use the observation that multiplication by 10 puts a zero on the end of the number.

EXAMPLE R.9 Multiply 58×33.

SOLUTION First, we multiply each weighted digit of 58 by 3:

$$
\begin{array}{rcr}
50 + \ 8 = & & 58 \\
\times \qquad 3 = & \times & 3 \\
\hline
150 + 24 = & & 174 \\
\end{array}
$$

Then, we multiply each weighted digit of 58 by 30:

$$
\begin{array}{rcr}
50 + \ 8 = & & 58 \\
\times \qquad 30 = & \times & 30 \\
\hline
1500 + 240 = & & 1740 \\
\end{array}
$$

To find the total product, we add these results:

$$
\begin{array}{r}
58 \\
\times \ 33 \\
\hline
174 \\
+ \ 1740 \\
\hline
1914 \\
\end{array}
$$

▲

DIVISION

To divide a whole number by a one-digit number, we may divide each weighted digit by the one-digit number.

EXAMPLE R.10 Divide $365 \div 5$.

SOLUTION   We can divide each weighted digit by 5:

$$\frac{300 + 60 + 5}{5} = \frac{300}{5} + \frac{60}{5} + \frac{5}{5}$$

$$= 60 + 12 + 1$$

$$= 73. \quad \blacktriangle$$

We recall that we can check division by multiplication. To check the division in Example R.10, we multiply 73×5:

$$
\begin{array}{r}
1 \\
73 \\
\times \quad 5 \\
\hline
365
\end{array}
$$

The division in Example R.10 is often written in this form:

$$5 \overline{)\, 365}$$

We divide the digits without their weights. Since 3 divided by 5 does not result in a whole number, we divide 36 by 5. The result is 7 with a remainder 1; that is,

$$36 = 5 \times 7 + 1.$$

We express this result by writing

$$
\begin{array}{r}
7 \\
5 \overline{)\, 365} \\
35 \\
\hline
1
\end{array}
$$

Now, we bring down the next digit:

$$
\begin{array}{r}
7 \\
5 \overline{)\, 365} \\
35 \\
\hline
15
\end{array}
$$

Then, we divide 15 by 5. The result is 3 with no remainder; that is,

$$15 = 5 \times 3 + 0.$$

We express this result by writing

$$
\begin{array}{r}
73 \\
5 \overline{)\, 365} \\
35 \\
\hline
15 \\
15 \\
\hline
0
\end{array}
$$

Finally, we observe that we cannot divide by zero. For example, if we attempt to write

$$\frac{15}{0} = ?,$$

we are seeking a number such that

$$15 = 0 \times ?.$$

But there is no number to fill the place of question mark, because any number times zero is zero; that is,

$$0 \times ? = 0$$

for every whole number. We say that division by zero is *undefined*.

<table>
<tr><td>

EXERCISE

R.1

</td><td>

Write the number in terms of place values:

1. 26	**2.** 892	**3.** 4769	**4.** 85,531
5. 608	**6.** 5040	**7.** 5005	**8.** 60,018

Perform the indicated operation:

9. $38 + 51$	**10.** $814 + 72$	**11.** $43 + 29$	**12.** $87 + 77$
13. $199 + 594$	**14.** $170 + 966$	**15.** $77 - 36$	**16.** $398 - 153$
17. $75 - 66$	**18.** $95 - 49$	**19.** $206 - 185$	**20.** $805 - 98$
21. 43×2	**22.** 13×4	**23.** 28×5	**24.** 67×7
25. 43×32	**26.** 69×42	**27.** 540×20	**28.** 206×85
29. $84 \div 2$	**30.** $336 \div 6$	**31.** $195 \div 3$	**32.** $344 \div 8$
33. $155 \div 7$	**34.** $399 \div 4$	**35.** $105 \div 5$	**36.** $367 \div 6$

</td></tr>
</table>

<table>
<tr><td>

SECTION

R.2

</td><td>

Fractions

Numerical **fractions** are quotients, or divisions, of whole numbers. A fraction is written in the form

$$\frac{a}{b}$$

</td></tr>
</table>

where a and b are whole numbers but $b \neq 0$ (b is not equal to zero). A fraction in this form means $a \div b$, where a can be any whole number, but b cannot be zero because we cannot divide by zero. The whole number a is called the **numerator** and b is called the **denominator**.

Any fraction with denominator 1 is also a whole number. For example,

$$\frac{2}{1} = 2.$$

Thus, the set of all numerical fractions includes the set of whole numbers, and any whole number can be written as a fraction with denominator 1.

A fraction with the same numerator and denominator is equal to 1. For example,

$$\frac{2}{2} = 1.$$

Two whole numbers that are multiplied are called **factors**. For example,

$$6 = 2 \times 3,$$

so 6 has factors 2 and 3. Two whole numbers have a **common factor** if one factor of each is the same. For example,

$$6 = 2 \times 3$$
$$8 = 2 \times 4,$$

so 6 and 8 have the common factor 2.

If the numerator and the denominator of a fraction have a common factor then we **reduce** the fraction by dividing out the common factor.

EXAMPLE R.11 ▶   Reduce $\dfrac{6}{8}$.

SOLUTION   Since 6 and 8 have the common factor 2, we may write

$$\frac{6}{8} = \frac{2 \times 3}{2 \times 4}.$$

We know that $2 \div 2 = 1$, so we divide out the common factor 2 to obtain

$$\frac{6}{8} = \frac{\cancel{2}^{\,1} \times 3}{\cancel{2} \times 4}$$

$$= \frac{3}{4}.$$

▲

MULTIPLICATION

Suppose we have two one-quarter pieces of a pie. We can write one-quarter as a fraction. Then, we have

$$2 \times \frac{1}{4} = \frac{2}{4}.$$

We reduce this fraction by writing

$$\frac{2}{4} = \frac{2 \times 1}{2 \times 2}$$

$$= \frac{1}{2}.$$

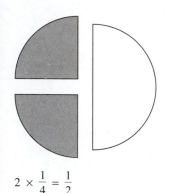

$$2 \times \frac{1}{4} = \frac{1}{2}$$

If we write the whole number 2 as a fraction, we can multiply the numerators and multiply the denominators to obtain

$$\frac{2}{1} \times \frac{1}{4} = \frac{2 \times 1}{1 \times 4}$$

$$= \frac{2}{4}$$

$$= \frac{1}{2}.$$

Now, suppose we have one-half of a one-quarter piece of the pie. Then, we have one-eighth of the pie. We can write "one-half of one-quarter" as fractions. We multiply the numerators and multiply the denominators to obtain

$$\frac{1}{2} \times \frac{1}{4} = \frac{1 \times 1}{2 \times 4}$$

$$= \frac{1}{8}.$$

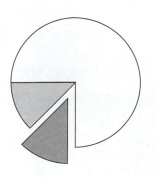

$$\frac{1}{2} \times \frac{1}{4} = \frac{1}{8}$$

In general, to multiply two fractions, we multiply the numerators and multiply the denominators.

EXAMPLE R.12 ▶ Multiply $\frac{3}{8} \times \frac{4}{9}$.

SOLUTION   We multiply the numerators and multiply the denominators:

$$\frac{3}{8} \times \frac{4}{9} = \frac{3 \times 4}{8 \times 9}$$

$$= \frac{12}{72}.$$

This result can be reduced by dividing out the common factor 12:

$$\frac{12}{72} = \frac{12 \times 1}{12 \times 6}$$

$$= \frac{1}{6}.$$

▲

It is usually more efficient to divide out any common factors *before* multiplying. Starting with

$$\frac{3}{8} \times \frac{4}{9} = \frac{3 \times 4}{8 \times 9}$$

we can write

$$\frac{3 \times 4}{8 \times 9} = \frac{\overset{1}{\cancel{3}} \times \overset{1}{\cancel{4}}}{2 \times \cancel{4} \times \cancel{3} \times 3}$$

$$= \frac{1}{2 \times 3}$$

$$= \frac{1}{6}.$$

We can multiply both the numerator and the denominator of a fraction by any number (except zero) without changing the value of the fraction. For example,

$$\frac{1}{2} = \frac{1 \times 2}{2 \times 2}$$

$$= \frac{2}{4}.$$

DIVISION

Now, suppose we have a division of fractions. Such a division is a fraction made up of fractions. For example,

$$\frac{1}{2} \div \frac{1}{4} = \frac{\frac{1}{2}}{\frac{1}{4}}.$$

If we multiply both the numerator and the denominator by 4, we obtain

$$\frac{\frac{1}{2}}{\frac{1}{4}} = \frac{\frac{1}{2} \times 4}{\frac{1}{4} \times 4}$$

$$= \frac{\frac{1}{2} \times \frac{4}{1}}{\frac{1}{4} \times \frac{4}{1}}$$

$$= \frac{\frac{1}{2} \times \frac{4}{1}}{1}$$

$$= \frac{1}{2} \times \frac{4}{1}.$$

Thus, the division

$$\frac{1}{2} \div \frac{1}{4}$$

is the same as the multiplication

$$\frac{1}{2} \times \frac{4}{1}.$$

The fraction $\frac{4}{1}$ is the *inverse* of the fraction $\frac{1}{4}$. Thus, to divide by a fraction we *invert and multiply*. The fraction inverted is the second fraction, called the divisor.

EXAMPLE R.13 ▶ Divide $\frac{3}{8} \div \frac{4}{9}$.

SOLUTION We invert the divisor and multiply:

$$\frac{3}{8} \div \frac{4}{9} = \frac{3}{8} \times \frac{9}{4}$$

$$= \frac{27}{32}. \qquad ▲$$

The **reciprocal** of a fraction is the inverse of the fraction. For example, the reciprocal of 4 is $\frac{1}{4}$, and the reciprocal of $\frac{1}{4}$ is 4.

EXAMPLE R.14 ▶ Write the reciprocal of $\frac{4}{9}$.

SOLUTION The reciprocal is

$$\frac{1}{\frac{4}{9}} = 1 \times \frac{9}{4}$$

$$= \frac{9}{4}. \qquad ▲$$

ADDITION

To add fractions, suppose we have one-quarter, and we add another one-quarter to it. Then, we have one-half. Thus, we can write

$$\frac{1}{4} + \frac{1}{4} = \frac{2}{4}$$

$$= \frac{1}{2}.$$

$\frac{1}{4} + \frac{1}{4} = \frac{1}{2}$

Observe that we add the numerators but *not* the denominators.

In general, to add two fractions with the same denominator, we add the numerators. The denominator of the fractions is called the **common denominator.** The common denominator is also the denominator of the sum.

EXAMPLE R.15   Add $\dfrac{2}{9} + \dfrac{4}{9}$.

SOLUTION We add the numerators, keeping the common denominator:

$$\frac{2}{9} + \frac{4}{9} = \frac{6}{9}.$$

Then, we can reduce the resulting fraction:

$$\frac{6}{9} = \frac{3 \times 2}{3 \times 3}$$

$$= \frac{2}{3}. \qquad \blacktriangle$$

If the two fractions do not have a common denominator, we can create a common denominator.

EXAMPLE R.16   Add $\dfrac{2}{9} + \dfrac{2}{3}$.

SOLUTION We create a common denominator by multiplying both the numerator and the denominator of $\frac{2}{3}$ by 3:

$$\frac{2}{9} + \frac{2}{3} = \frac{2}{9} + \frac{2 \times 3}{3 \times 3}$$

$$= \frac{2}{9} + \frac{6}{9}.$$

Then, we add the numerators, keeping the common denominator:

$$\frac{2}{9} + \frac{6}{9} = \frac{8}{9}. \qquad \blacktriangle$$

EXAMPLE R.17   Add $\dfrac{2}{3} + \dfrac{3}{2}$.

SOLUTION We create a common denominator by multiplying both the numerator and the denominator of $\frac{2}{3}$ by 2 and both the numerator and the denominator of $\frac{3}{2}$ by 3:

$$\frac{2}{3} + \frac{3}{2} = \frac{2 \times 2}{3 \times 2} + \frac{3 \times 3}{2 \times 3}$$

$$= \frac{4}{6} + \frac{9}{6}.$$

Then, we add the numerators, keeping the common denominator:

$$\frac{4}{6} + \frac{9}{6} = \frac{13}{6}. \qquad \blacktriangle$$

A mixed number is the sum of a whole number and a fraction. For example, the result of dividing 13 by 6 is 2 with a remainder of 1. The remainder can be written as a fraction:

$$\frac{13}{6} = 2 + \frac{1}{6}.$$

We write this mixed number as

$$\frac{13}{6} = 2\frac{1}{6}.$$

In algebra and trigonometry, the pure fraction form, sometimes called an "improper" fraction, is usually preferred.

EXAMPLE R.18 Write $3\frac{2}{5}$ as an improper fraction.

SOLUTION   The mixed number is the sum of a whole number and a fraction:

$$3\frac{2}{5} = 3 + \frac{2}{5}$$

$$= \frac{3}{1} + \frac{2}{5}$$

$$= \frac{5 \times 3}{5 \times 1} + \frac{2}{5}$$

$$= \frac{15}{5} + \frac{2}{5}$$

$$= \frac{17}{5}.$$

We can shorten this process by observing that

$$5 \times 3 + 2 = 17.$$

Then, we can write

$$3\frac{2}{5} = \frac{5 \times 3 + 2}{5}$$

$$= \frac{17}{5}.$$ ▲

SUBTRACTION

Subtraction of fractions is similar to addition. To subtract two fractions with the same denominator, we subtract the numerators, keeping the common denominator. If the fractions do not have a common denominator, we can create a common denominator.

EXAMPLE R.19 Subtract $\frac{5}{8} - \frac{1}{4}$.

SOLUTION   We create a common denominator by multiplying both the numerator and the denominator of $\frac{1}{4}$ by 2:

$$\frac{5}{8} - \frac{1}{4} = \frac{5}{8} - \frac{1 \times 2}{4 \times 2}$$

$$= \frac{5}{8} - \frac{2}{8}.$$

Then, we subtract the numerators, keeping the common denominator:

$$\frac{5}{8} - \frac{2}{8} = \frac{3}{8}.$$ ▲

EXAMPLE R.20 Subtract $3\frac{1}{3} - 2\frac{1}{2}$.

SOLUTION First, we write the mixed numbers as improper fractions:

$$3\frac{1}{3} - 2\frac{1}{2} = \frac{10}{3} - \frac{5}{2}.$$

Then, we create the common denominator 6, and subtract the resulting numerators to obtain

$$\frac{10}{3} - \frac{5}{2} = \frac{10 \times 2}{3 \times 2} - \frac{5 \times 3}{2 \times 3}$$

$$= \frac{20}{6} - \frac{15}{6}$$

$$= \frac{5}{6}.$$

▲

EXERCISE

R.2

Reduce the fraction:

1. $\frac{4}{12}$ 　　2. $\frac{6}{30}$ 　　3. $\frac{8}{36}$ 　　4. $\frac{18}{24}$

Perform the indicated multiplication or division:

5. $\frac{1}{2} \times \frac{1}{3}$ 　　6. $\frac{1}{3} \times \frac{2}{5}$ 　　7. $\frac{2}{3} \times \frac{6}{5}$ 　　8. $\frac{4}{5} \times \frac{3}{16}$

9. $\frac{3}{4} \times \frac{2}{9}$ 　　10. $\frac{6}{25} \times \frac{5}{18}$ 　　11. $\frac{4}{15} \times \frac{9}{10}$ 　　12. $\frac{10}{21} \times \frac{14}{15}$

13. $\frac{1}{3} \div \frac{1}{2}$ 　　14. $\frac{2}{3} \div \frac{1}{4}$ 　　15. $\frac{5}{6} \div \frac{1}{2}$ 　　16. $\frac{8}{9} \div \frac{2}{3}$

Write the reciprocal:

17. 10 　　18. $\frac{1}{5}$ 　　19. $\frac{5}{8}$ 　　20. $\frac{9}{2}$

Perform the indicated addition or subtraction:

21. $\frac{1}{8} + \frac{3}{8}$ 　　22. $\frac{2}{5} + \frac{3}{5}$ 　　23. $\frac{1}{4} + \frac{1}{8}$ 　　24. $\frac{1}{3} + \frac{5}{12}$

25. $\frac{1}{3} + \frac{1}{4}$ 　　26. $\frac{1}{5} + \frac{1}{8}$ 　　27. $\frac{2}{5} + \frac{3}{4}$ 　　28. $\frac{5}{6} + \frac{3}{5}$

29. $\frac{5}{6} + \frac{3}{8}$ 　　30. $\frac{5}{8} + \frac{5}{12}$ 　　31. $1\frac{1}{4} + 2\frac{1}{2}$ 　　32. $2\frac{1}{2} + 3\frac{1}{3}$

33. $\frac{5}{6} - \frac{1}{3}$ 　　34. $\frac{7}{8} - \frac{5}{6}$ 　　35. $2\frac{1}{4} - 1\frac{1}{2}$ 　　36. $3\frac{3}{5} - 2\frac{3}{4}$

SECTION

R.3

Decimals

Decimals are written by using place values that are the reciprocal of ten and reciprocals of products of tens:

$$\frac{1}{10} = \frac{1}{10}$$

$$\frac{1}{100} = \frac{1}{10 \times 10}$$

$$\frac{1}{1000} = \frac{1}{10 \times 10 \times 10}$$

and so on. These fractional place values are read "one ten*th*," "one hundred*th*," and "one thousand*th*.

A decimal has a **whole number part** and a **fractional part**. The fractional part is separated from the whole number part by a **decimal point**. The first digit to the right of the decimal point has place value $\frac{1}{10}$, the next digit to the right has place value $\frac{1}{100}$, the next has place value $\frac{1}{1000}$, and so on.

EXAMPLE R.21 Write 2.22 in terms of place values.

SOLUTION The digit 2 to the left of the decimal point has the value two. The digit 2 to the right of the decimal point represents two one-tenths, and the next digit 2 to the right represents two one-hundredths. Thus, we interpret 2.22 as

$$2.22 = 2 + 2 \times \frac{1}{10} + 2 \times \frac{1}{100}$$

or

$$2.22 = 2 + \frac{2}{10} + \frac{2}{100}.$$

By adding the fractions, we see that the fractional part is

$$\frac{2}{10} + \frac{2}{100} = \frac{20}{100} + \frac{2}{100}$$

$$= \frac{22}{100}.$$

Therefore, we can also write

$$2.22 = 2 + \frac{22}{100}. \qquad \blacktriangle$$

EXAMPLE R.22 Write 0.305 in terms of place values.

SOLUTION At the left of the decimal point, the whole number part is zero. At the right of the decimal point, the digit 3 has place value one-tenth, there are no one-hundredths, and the digit 5 has place value one-thousandth:

$$0.305 = 3 \times \frac{1}{10} + 0 \times \frac{1}{100} + 5 \times \frac{1}{1000}$$

or

$$0.305 = \frac{3}{10} + \frac{5}{1000}.$$

We can also write

$$0.305 = \frac{305}{1000}. \qquad \blacktriangle$$

ADDITION

Suppose we have two decimals, for example 0.2 and 0.3. We can write

$$0.2 + 0.3 = \frac{2}{10} + \frac{3}{10}$$

$$= \frac{5}{10}$$

$$= 0.5.$$

In general, to add two decimals, we add corresponding weighted digits; that is, digits with the same place values.

EXAMPLE R.23 Add $2.1 + 3.4$.

SOLUTION We add the corresponding weighted digits:

$$2 + \frac{1}{10} = \quad 2.1$$

$$+\ 3 + \frac{4}{10} = +\ 3.4$$

$$\overline{\qquad\qquad\qquad}$$

$$5 + \frac{5}{10} = \quad 5.5$$

▲

When the sum of two weighted digits exceeds the base ten, we carry into the next place to the left.

EXAMPLE R.24 Add $7.8 + 2.9$.

SOLUTION We add the corresponding weighted digits:

$$7 + \frac{8}{10} = \quad 7.8$$

$$+\ 2 + \frac{9}{10} = +\ 2.9$$

$$\overline{\qquad\qquad\qquad}$$

$$9 + \frac{17}{10}$$

The result 17 exceeds the base ten; therefore, we carry into the next place to the left:

$$9 + \frac{17}{10} = 9 + 1 + \frac{7}{10}$$

$$= 10 + \frac{7}{10}$$

$$= 10.7.$$

We can summarize the process of carrying by writing the digit carried above the column into which it is carried:

$$\begin{array}{r} 1 \\ 7.8 \\ +\ 2.9 \\ \hline 10.7 \end{array}$$

▲

Because digits with the same place values are added, the decimal points must be aligned for addition of decimals.

EXAMPLE R.25 Add $27.7 + 0.55$.

SOLUTION We align the decimal points to add the corresponding weighted digits:

$$
\begin{array}{r}
1 \\
27.70 \\
+\ \ 0.55 \\
\hline
28.25
\end{array}
$$

▲

SUBTRACTION

To subtract one decimal from another, we subtract corresponding weighted digits. We borrow from the next place in the first number when necessary.

EXAMPLE R.26 ▶ Subtract 9.5 − 6.7.

SOLUTION We subtract the corresponding weighted digits:

$$9 + \frac{5}{10} = \ \ 9.5$$

$$- 6 - \frac{7}{10} = -6.7$$

To subtract $\frac{5}{10} - \frac{7}{10}$ we borrow from the next place in the first number:

$$9 + \frac{5}{10}$$

$$8 + 1 + \frac{5}{10}$$

$$8 + \frac{15}{10}.$$

Then, we have

$$8 + \frac{15}{10}$$

$$- 6 - \frac{7}{10}$$

$$\overline{}$$

$$2 + \frac{8}{10} = 2.8$$

We can summarize the process of borrowing by writing above the columns:

$$
\begin{array}{r}
8\ \ \ 15 \\
\cancel{9}.\ \ \cancel{5} \\
-\ 6.\ \ 7 \\
\hline
2.\ \ 8
\end{array}
$$

To check, we add 6.7 + 2.8:

$$
\begin{array}{r}
1 \\
6.7 \\
+\ 2.8 \\
\hline
9.5
\end{array}
$$

▲

Because digits with the same place values are subtracted, the decimal points must be aligned to subtract decimals.

EXAMPLE R.27 Subtract $3.5 - 2.09$.

SOLUTION We align the decimal points to subtract the corresponding weighted digits:

$$
\begin{array}{r}
{\scriptstyle 4\ 10} \\
3.\cancel{5}\ \cancel{0} \\
-\ 2.0\ \ 9 \\
\hline
1.4\ \ 1
\end{array}
$$

To check, we add $2.09 + 1.41$:

$$
\begin{array}{r}
{\scriptstyle 1} \\
2.09 \\
+\ 1.41 \\
\hline
3.50
\end{array}
$$

▲

MULTIPLICATION

Now, suppose we have the whole number 2 and the decimal 0.3. We can write

$$2 \times 0.3 = 2 \times \frac{3}{10}$$

$$= \frac{6}{10}$$

$$= 0.6.$$

In general, to multiply a decimal by a whole number, we multiply as if both numbers were whole numbers. The resulting fractional part has the same number of digits as the original fractional part.

EXAMPLE R.28 Multiply 12×2.5.

SOLUTION We multiply as if both numbers were whole numbers:

$$
\begin{array}{r}
2.5 \\
\times\ \ \ 12 \\
\hline
5\,0 \\
+\ 25\,0 \\
\hline
30.0
\end{array}
$$

We place the decimal point with one digit in the resulting fractional part, which corresponds with the original fractional part. ▲

Multiplication of a decimal by 10 moves the decimal point one place to the right. For example,

$$10 \times 0.3 = 10 \times \frac{3}{10}$$

$$= 3.$$

Then, each further multiplication by 10 moves the decimal point another place to the right.

If we have two decimals, for example, 0.2 and 0.3, we can write:

$$0.2 \times 0.3 = \frac{2}{10} \times \frac{3}{10}$$

$$= \frac{6}{100}$$

$$= 0.06.$$

This result is six *hundredths*. To multiply two decimals, we multiply as if both numbers were whole numbers. The resulting fractional part has the same number of digits as the *total number* of digits in the original fractional parts.

EXAMPLE R.29 Multiply 2.2 × 1.6.

SOLUTION We multiply as if both numbers were whole numbers:

$$
\begin{array}{r}
1.6 \\
\times\ \ 2.2 \\
\hline
3\,2 \\
+\ \ 32\,0 \\
\hline
3.5\,2
\end{array}
$$

We place the decimal point with *two* digits in the resulting fractional part, which corresponds with the *total number* of digits in the original fractional parts. ▲

DIVISION

To divide a decimal by a whole number, we divide as if both numbers were whole numbers. The decimal point in the result corresponds with the decimal point in the original decimal.

EXAMPLE R.30 Divide 3.55 ÷ 5.

SOLUTION We divide as if both numbers were whole numbers.

$$
\begin{array}{r}
0.71 \\
5\,\overline{)\,3.55} \\
3\,5 \\
\hline
0\,5 \\
5 \\
\hline
0
\end{array}
$$

We have placed the decimal point to correspond with its place in the original decimal. ▲

Division of a decimal by 10 moves the decimal point one place to the left. For example,

$$
\begin{aligned}
0.6 \div 10 &= \frac{6}{10} \div \frac{10}{1} \\
&= \frac{6}{10} \times \frac{1}{10} \\
&= \frac{6}{100} \\
&= 0.06.
\end{aligned}
$$

Then, each further division by 10 moves the decimal point another place to the left.

To divide two decimals, we recall that multiplication by 10 moves the decimal point one place to the right. We multiply both numbers by ten until the decimal point of the second number, which is the divisor, is at the end of the number. Then, we can divide by the resulting whole number.

EXAMPLE R.31 Divide 2.16 ÷ 0.6.

SOLUTION We write the division as a fraction:

$$2.16 \div 0.6 = \frac{2.16}{0.6}.$$

Multiplying both numbers by 10 moves the decimal point to the end of the divisor:

$$\frac{2.16}{0.6} = \frac{2.16 \times 10}{0.6 \times 10}$$

$$= \frac{21.6}{6}.$$

Then, we divide by the resulting whole number, and place the decimal point to correspond with its place in the decimal:

$$
\begin{array}{r}
3.6 \\
6\,)\overline{21.6} \\
18 \\
\overline{3\,6} \\
3\,6 \\
\overline{0}
\end{array}
$$

▲

When division of decimals results in a remainder, we can continue the division indefinitely by placing zeros at the end of the fractional part of the decimal. To end the process, we **round off** the result, called the quotient, after several places.

There are many conventions for rounding off numbers. In this book we will use 4 / 5 rounding. We mark the last digit we wish to keep. Then, whenever the digit immediately following is 4 or less, we drop all the digits that follow the marked digit. Whenever the digit immediately following is 5 or more, we increase the marked digit by 1 and drop all the following digits.

EXAMPLE R.32 Divide $0.283 \div 0.12$. Round off the quotient to three digits.

SOLUTION Multiplying both numbers by 100 moves the decimal point to the end of the divisor:

$$\frac{0.283}{0.12} = \frac{0.283 \times 100}{0.12 \times 100}$$

$$= \frac{28.3}{12}.$$

Then, we divide by the resulting whole number. We place zeros at the end of the fractional part of the decimal to continue the division:

$$
\begin{array}{r}
2.358 \\
12\,)\overline{28.300} \\
24 \\
\overline{4\,3} \\
3\,6 \\
\overline{7\,0} \\
6\,0 \\
\overline{1\,0\,0} \\
9\,6 \\
\overline{4}
\end{array}
$$

Observe that we have carried the quotient to four digits. Now, we round off to three digits. We mark the third digit:

$$2.3\underline{5}8$$

The digit immediately following is an 8, which is 5 or more, so we increase the marked digit by 1 and drop all the following digits. The quotient is

$$2.36$$

when rounded off to three digits. ▲

    Any fraction can be written in decimal form. For some fractions, the decimal form is a **terminating decimal**. For example,

$$\frac{3}{4} = \frac{3 \times 25}{4 \times 25}$$

$$= \frac{75}{100}.$$

Therefore, we can write the fraction as the terminating decimal

$$\frac{3}{4} = 0.75.$$

    The decimal form of a fraction can be obtained by dividing the numerator by the denominator.

 EXAMPLE R.33 ▶ Write $\frac{3}{4}$ in decimal form.

SOLUTION We divide 3 by 4. Because 3 is not divisible by 4 as a whole number, we write 3 as 3.0, and continue placing zeros at the end of the fractional part as necessary:

$$
\begin{array}{r}
0.75 \\
4 \overline{)\ 3.00} \\
\underline{2\ 8} \\
20 \\
\underline{20} \\
0
\end{array}
$$

For a terminating decimal, the division eventually results in a zero remainder. ▲

    The decimal form of some fractions is a **nonterminating repeating** decimal.

 EXAMPLE R.34 ▶ Write $\frac{5}{11}$ in decimal form.

SOLUTION We divide 5 by 11:

$$
\begin{array}{r}
0.4545 \\
11 \overline{)\ 5.0000} \\
\underline{4\ 4} \\
60 \\
\underline{55} \\
50 \\
\underline{44} \\
60 \\
\underline{55} \\
5
\end{array}
$$

We see that this process continues indefinitely. The decimal form does not terminate, but repeats in the pattern 0.4545 . . . . Every fraction has a decimal form that is either a terminating decimal or a nonterminating repeating decimal. ▲

EXERCISE

R.3

Write the number in terms of place values:

1. 3.15   **2.** 74.91   **3.** 0.206   **4.** 0.045

5. 0.00302   **6.** 0.000001

Perform the indicated operation:

7. 3.6 + 4.2   **8.** 5.1 + 3.8   **9.** 5.4 + 3.9   **10.** 6.9 + 4.8

11. 0.23 + 9.5   **12.** 3.99 + 0.019   **13.** 5.8 − 4.6   **14.** 7.4 − 0.4

15. 9.1 − 1.6   **16.** 1.2 − 0.8   **17.** 4.2 − 1.06   **18.** 8.47 − 4.8

19. 5 × 5.2   **20.** 13 × 2.4   **21.** 1.5 × 1.8   **22.** 5.2 × 2.3

23. 2.6 × 0.14   **24.** 0.803 × 3.3   **25.** 3.14 ÷ 2   **26.** 0.393 ÷ 3

27. 3.24 ÷ 0.4   **28.** 39 ÷ 0.6

29. 4.54 ÷ 0.06 (Round off to three digits)

30. 0.0229 ÷ 0.16 (Round off to three digits)

Write in decimal form:

31. $\dfrac{1}{2}$ **32.** $\dfrac{5}{8}$ **33.** $\dfrac{4}{25}$ **34.** $\dfrac{15}{4}$

35. $\dfrac{1}{3}$ **36.** $\dfrac{2}{3}$ **37.** $\dfrac{3}{11}$ **38.** $\dfrac{5}{7}$

SECTION

R.4

Percents

Percent means "divided by 100." When we say 15% we mean "fifteen divided by 100." Thus, we can write

$$15\% = \frac{15}{100}$$

and

$$15\% = 0.15.$$

Any percent can be written as a fraction and also as a decimal.

EXAMPLE R.35

Write 11.5% as a fraction.

SOLUTION

We write the percent as a fraction with denominator 100:

$$11.5\% = \frac{11.5}{100}.$$

Then, we multiply the numerator and denominator by 10 to move the decimal point to the end:

$$\frac{11.5}{100} = \frac{11.5 \times 10}{100 \times 10}$$

$$= \frac{115}{1000}.$$

If desired, we can reduce the resulting fraction:

$$\frac{115}{1000} = \frac{23 \times 5}{200 \times 5}$$

$$= \frac{23}{200}.$$

▲

EXAMPLE R.36   Write 11.5% as a decimal.

SOLUTION   From Example R.35 we have

$$11.5\% = \frac{115}{1000}.$$

Then, decimal place values give

$$\frac{115}{1000} = 0.115.$$ ▲

Recall that each division by 10 moves the decimal point one place to the left. Since $100 = 10 \times 10$, a percent represents two divisions by 10. Thus, we can convert a percent to a decimal by moving the decimal point two places to the left:

$$11.5\% = 0.115.$$

EXAMPLE R.37   Write 36% as a decimal.

SOLUTION   The decimal point is at the end of the whole number. We can convert to decimal form by moving the decimal point two places to the left:

$$36\% = 0.36.$$ ▲

To find a percent of a number, we multiply the number by the percent in decimal form. The result is called the **percentage**.

EXAMPLE R.38   Find 36% of 250.

SOLUTION   From Example R.37 we have

$$36\% = 0.36.$$

To find the percentage, we multiply 250 by 0.36:

$$
\begin{array}{r}
250 \\
\times\ \ 0.36 \\
\hline
15\,00 \\
75\,0 \\
\hline
90.00
\end{array}
$$

Therefore, 36% of 250 is 90.   ▲

EXAMPLE R.39   Find 125% of 50.

SOLUTION   We convert the percent to decimal form by moving the decimal point two places to the left:

$$125\% = 1.25.$$

Then, to find the percentage, we multiply 50 by 1.25:

$$
\begin{array}{r}
1.25 \\
\times\ \ \ 50 \\
\hline
62.50
\end{array}
$$

Therefore, 125% of 50 is 62.5. We observe that when we use a percent that is more than 100%, the percentage is larger than the original number.   ▲

In real life, numbers are rarely exact. For example, the cover of this book is supposed to be $8\frac{1}{2}$ inches wide by 11 inches long. However, it is probably slightly less, or maybe slightly more. Similarly, components of electronic circuits cannot be manufactured to exact specifications.

A component of electronic circuits called a resistor is measured in a unit called **ohms**. However, there is no such thing as a resistor that is rated at exactly 1 ohm. A resistor rated at 1 ohm may actually be a little less than 1 ohm or a little more than 1 ohm. Manufacturers express the limits of such variation by rating resistors at ±5%, or ±10%, or ±20%. A variation of ±5%, for example, means that the resistor may be as much as 5% less than the rated value or as much as 5% more than the rated value. Thus, a resistor rated at 1 ohm ±5% may actually be as little as 0.95 ohm or as much as 1.05 ohms.

EXAMPLE R.40   Find the range of values of a resistor rated at

a. 47 ohms ±5%   **b.** 47 ohms ±10%   **c.** 47 ohms ±20%

SOLUTIONS   **a.**   We find 5% of 47:

$$
\begin{array}{r}
47 \\
\times\ 0.05 \\
\hline
2.35
\end{array}
$$

If we subtract 2.35 from 47, we obtain

$$
\begin{array}{r}
47.00 \\
-\ \ 2.35 \\
\hline
44.65
\end{array}
$$

If we add 2.35 to 47, we obtain

$$
\begin{array}{r}
47.00 \\
+\ \ 2.35 \\
\hline
49.35
\end{array}
$$

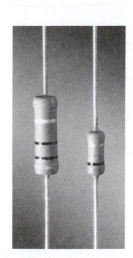

Typical resistors: The rings show the resistance in ohms and the percent variation. Courtesy of *Philips Components.*

Therefore, a resistor rated at 47 ohms ±5% may actually be as little as 44.65 ohms or as much as 49.35 ohms.

b.   We find 10% of 47:

$$
\begin{array}{r}
47 \\
\times\ 0.1 \\
\hline
4.7
\end{array}
$$

If we subtract 4.7 from 47, we obtain

$$
\begin{array}{r}
47.0 \\
-\ \ 4.7 \\
\hline
42.3
\end{array}
$$

If we add 4.7 to 47, we obtain

$$
\begin{array}{r}
47.0 \\
+\ \ 4.7 \\
\hline
51.7
\end{array}
$$

Therefore, a resistor rated at 47 ohms ±10% may actually be as little as 42.3 ohms or as much as 51.7 ohms.

c.   We find 20% of 47:

$$
\begin{array}{r}
47 \\
\times\ 0.2 \\
\hline
9.4
\end{array}
$$

If we subtract 9.4 from 47, we obtain

$$\begin{array}{r} 47.0 \\ -\ \ 9.4 \\ \hline 37.6 \end{array}$$

If we add 9.4 to 47, we obtain

$$\begin{array}{r} 47.0 \\ +\ \ 9.4 \\ \hline 56.4 \end{array}$$

Therefore, a resistor rated at 47 ohms $\pm 20\%$ may actually be as little as 37.6 ohms or as much as 56.4 ohms.   ▲

EXERCISE R.4

Write as a fraction:

1. 20%   2. 150%   3. 1.5%   4. 10.5%

Write as a decimal:

5. 33%   6. 20.2%   7. 4.5%   8. 121%

9. Find 25% of 500.   10. Find 16% of 150.

11. Find 22% of 20.   12. Find 75% of 50.

13. Find 150% of 12.   14. Find 110% of 300.

15. Find 1.5% of 20.   16. Find 5.5% of 12.5.

17. Find the range of values of a resistor rated at

a. 22 ohms $\pm 5\%$ b. 22 ohms $\pm 10\%$ c. 22 ohms $\pm 20\%$

18. Find the range of values of a resistor rated at

a. 6.8 ohms $\pm 5\%$ b. 6.8 ohms $\pm 10\%$ c. 6.8 ohms $\pm 20\%$

SECTION R.5

Calculators, Significant Digits, Squares, and Square Roots

Over the past twenty years or so, calculators have emerged as an essential tool of scientists and engineers. In this book, you will learn how to use a **scientific calculator** to do many types of calculations. You will need a scientific calculator to complete this unit, and to complete most of the other units.

Throughout this book you will find calculator **algorithms** marked with a symbol in the margin. An algorithm is a sequence of steps to solve a problem. The calculator algorithms in this book lead you through the sequence of keys you must use to do the calculations for some of the worked examples.

Calculators are changing continually, so the algorithms in this book are not written for any specific brand or model. Most scientific calculators work on the same general principles. The keys may be marked slightly differently. We have included some common variations, but be sure to have your direction book handy. Exceptions are calculators such as some Hewlett-Packard models that work on a system called *reverse Polish notation*. A few algorithms for reverse Polish notation are given in Appendix C.

Some calculators have more features than others. In this book it is assumed that you have a relatively low-priced "bare bones" calculator. It must, however, be labeled as a *scientific* calculator. One way to be sure is to look for keys marked SIN, COS, and TAN,

A typical scientific calculator.
Courtesy of *Texas Instruments*.

which are abbreviations for the trigonometric functions sine, cosine, and tangent. Also look for keys marked LOG and LN (or LNX), which are abbreviations for the logarithmic functions. If your calculator has these five keys, it should have all the other keys described in this book.

Finally, this book is written on the philosophy that, when possible, you should know how to do something yourself before you use a calculator to do it for you. We would not give calculators to first graders before they have learned how to add, subtract, multiply, and divide for themselves. Similarly, you will learn certain mathematical principles before you are shown calculator methods for them.

In this book, we will put the symbols for the calculator keys in boxes. For example, the first two keys you will use are the equals key,

$$\boxed{=} ,$$

and the addition key,

$$\boxed{+} .$$

These keys are usually at the lower right corner of the key pad. When you must press the equals key in an algorithm, we will say

Press $\boxed{=}$.

When you must press the plus key, we will say

Press $\boxed{+}$.

When we say to enter a number, use the digit keys to enter the digits in the number, from left to right, including the decimal point if there is one. We will show the number you should see on the display at the right in each step of the algorithm.

EXAMPLE R.41 Use a scientific calculator to add 23.45 + 67.89.

SOLUTION Find the addition and equals keys on your calculator, and follow these steps:

		display:
Enter 23.45		23.45
Press $\boxed{+}$		23.45
Enter 67.89		67.89
Press $\boxed{=}$		91.34 ▲

The subtraction key,

$$\boxed{-} ,$$

is usually just above the equals and addition keys. Continuing up the key pad, there are the multiplication key,

$$\boxed{\times} ,$$

and the division key,

$$\boxed{\div} .$$

EXAMPLE R.42 Use a scientific calculator to subtract 98.32 − 45.67.

SOLUTION Find the subtraction key on your calculator, and follow these steps:

		display:
Enter 98.32		98.32
Press $\boxed{-}$		98.32
Enter 45.67		45.67
Press $\boxed{=}$		52.65 ▲

EXAMPLE R.43 ▶ Use a scientific calculator to multiply 6.8 × 5.7.

SOLUTION Find the multiplication key on your calculator, and follow these steps:

	display:
Enter 6.8	6.8
Press ⊠ ×	6.8
Enter 5.7	5.7
Press ⊡ =	38.76 ▲

EXAMPLE R.44 ▶ Use a scientific calculator to divide 6.8 ÷ 5.7.

SOLUTION Find the division key on your calculator, and follow these steps:

	display:
Enter 6.8	6.8
Press ⊡ ÷	6.8
Enter 5.7	5.7
Press ⊡ =	1.192982456 ▲

We recall from Section R.3 that we can continue division of decimals indefinitely. Usually, we round off such results; for example, we might round off the preceding result to 1.19.

The results given on scientific calculators usually just stop at the end of the display. Since most calculators carry one additional digit internally, the number on the display probably is not rounded off. You can follow these steps, for example, to convert $\frac{2}{3}$ to decimal form:

	display:
Enter 2	2.
Press ⊡ ÷	2.
Enter 3	3.
Press ⊡ =	0.666666666

Using common rounding, the last 6 would be rounded up to a 7. The result is not shown with a 7 at the end on your calculator display because yet another 6 is being carried internally.

SIGNIFICANT DIGITS

We must round off results to some reasonable number of digits. The digits we choose are called **significant digits**. When we encounter a number containing many digits ("many" is often more than three), we round off the number to a few significant digits.

Digits that are significant digits are determined by these general guidelines:

1. All nonzero digits are significant.

2. All zeros between two nonzero digits are significant.

3. Zeros preceding all nonzero digits are *not* significant.

4. Zeros following all nonzero digits *might* or *might not* be significant.

We illustrate each of the guidelines:

1. The number 12.34 has four significant digits because all the digits are nonzero.

2. The number 10.04 has four significant digits because the zeros are between two non-zero digits.

3. The number 0.0034 has two significant digits because zeros preceding all of the non-zero digits are not significant digits (although they may be needed as placeholders).

4.   The number 3400 can have two, three, or four significant digits:
   a.   The number 3400 has two significant digits if the zeros are needed only as place-holders.
   b.   The number 3400 has three significant digits if the last zero is only a placeholder, but the first zero means that we can determine the tens place of 3400 exactly, not 3390 or 3410.
   c.   The number 3400 has four significant digits if the zeros mean that we can determine the number as exactly 3400, not 3399 or 3401.

Because of the complexity of the fourth guideline, we will assume throughout this book that trailing zeros are placeholders. Thus, we will assume that 3400 has two significant digits. We will also assume that 0.3400 has two significant digits. We will write numbers such as 0.3400 as 0.34, dropping the trailing zeros.

EXAMPLE R.45   Round off to four significant digits:

a.   1.414213562   b.   3.141592654

SOLUTIONS   We mark the fourth significant digit by an underscore. Then, we use 4 / 5 rounding, as in Section R.3, to round off to that digit.

a.   All digits in the number are significant. We mark the fourth digit:

$$1.41\underline{4}213562$$

The digit immediately following is a 2, so we drop all the digits that follow the marked digit. The number is

$$1.414$$

when rounded off to four significant digits.

b.   All digits in the number are significant. We mark the fourth digit:

$$3.14\underline{1}592654$$

The digit immediately following is a 5, so we increase the marked digit by 1 and drop all the following digits. The number is

$$3.142$$

when rounded off to four significant digits.   ▲

EXAMPLE R.46   Round off to three significant digits:

a.   0.707106781   b.   159154.9431

SOLUTIONS   We mark the third significant digit, and use 4 / 5 rounding to round off to that digit.
a.   The leading zero is not significant. The other zeros are between two significant digits, so they are significant. We mark the third significant digit:

$$0.70\underline{7}106781$$

The digit immediately following is a 1, which is 4 or less, so we drop all the digits that follow the marked digit. The number is

$$0.707$$

when rounded off to three significant digits.

b.   All digits in the number are significant. We mark the third digit:

$$15\underline{9}154.9431$$

The digit immediately following is a 1, so we drop all the digits that follow the marked digit. However, we must supply some zeros as placeholders. The number is

$$159,000$$

when rounded off to three significant digits.   ▲

EXAMPLE R.47   Use a scientific calculator to multiply 2.345×6.789; round off the result to three significant digits.

SOLUTION Use steps similar to those in Example R.43 to find

$$2.345 \times 6.789 = 15.920205.$$

Thus, we have

$$2.345 \times 6.789 = 15.9$$

when rounded off to three significant digits.   ▲

EXAMPLE R.48   Use a scientific calculator to divide $6.789 \div 2.345$; round off the result to three significant digits.

SOLUTION Use steps similar to those in Example R.44 to find

$$6.789 \div 2.345 = 2.895095949.$$

Since the fourth significant digit is a 5, we should increase the third significant digit by 1 and drop all the following digits. However, the third significant digit is a 9. When we increase a 9 by 1, we get 10. Therefore, we write a 0 and carry the 1 into the next place. Thus, we have

$$6.789 \div 2.345 = 2.90$$

when rounded off to three significant digits. We will write

$$6.789 \div 2.345 = 2.9,$$

dropping the trailing zero.   ▲

SQUARES

The **square** of a number is the number multiplied by itself. The square is indicated by a **superscript** 2, which is a small 2 written to the upper right of the number. For example,

$$2^2 = 2 \times 2 = 4,$$

$$3^2 = 3 \times 3 = 9,$$

$$4^2 = 4 \times 4 = 16,$$

and so on.

Although squares can be found by multiplication, scientific calculators also have a square key. The square key is usually marked

$$\boxed{x^2}.$$

On a few models of scientific calculators, this symbol is *above* a key. When this is the case, the inverse or second key, marked

$$\boxed{\text{INV}} \text{ or } \boxed{2^{\text{nd}}}$$

must be pressed first, and then the key that has the square key label above it. When we say to press

$$\boxed{x^2},$$

we mean to press the square key or, if necessary, the combination of the inverse or second key followed by the square key.

EXAMPLE R.49   Use a scientific calculator to find 9.25^2.

SOLUTION Find the square key and, if necessary, the inverse or second key on your calculator, and follow these steps:

	display:
Enter 9.25	9.25
Press x^2	85.5625

We might round off to three significant digits, and write the result as 85.6.   ▲

SQUARE ROOTS

The **square root** of a number is the opposite, or in mathematical terms, the **inverse of the square** of a number. The inverse of any operation "undoes" the operation. Thus, the square root "undoes" the square. The square root is indicated by the "check-mark" symbol $\sqrt{}$. For example,

$$\sqrt{4} = 2 \text{ because } 2^2 = 4,$$
$$\sqrt{9} = 3 \text{ because } 3^2 = 9,$$
$$\sqrt{16} = 4 \text{ because } 4^2 = 16,$$

and so on.

Scientific calculators have a square-root key marked

$$\boxed{\sqrt{x}} ,$$

or sometimes just

$$\boxed{\sqrt{}} .$$

On some calculators the square-root symbol is above the square key because the square root is the inverse of the square. In this case, the inverse or second key, marked

$$\boxed{\text{INV}} \text{ or } \boxed{2^{nd}}$$

must be pressed first, and then the key that has the square-root symbol above it. When we say to press

$$\boxed{\sqrt{x}} ,$$

we mean to press the square-root key or, if necessary, the combination of the inverse or second key followed by the square-root key.

For most models of scientific calculators, you enter the number and then press the square-root key.

EXAMPLE R.50 ► Use a scientific calculator to find $\sqrt{3}$.

 SOLUTION Find the square-root key and, if necessary, the inverse or second key on your calculator, and follow these steps:

	display:
Enter 3	3.
Press $\boxed{\sqrt{x}}$	1.732050808

or 1.73 when rounded off to three significant digits.   ▲

There are some newer models for which you press the square-root key and then enter the number. If you have such a calculator, which might be identified as a *direct algebraic logic* (DAL) calculator, you must first press the square-root key, and then enter 3. Finally, you must press the equals key to display the value of the square root.

Numerical fractions, which can be written in the form

$$\frac{a}{b}$$

where b is not zero, are called **rational numbers**. The name "rational numbers" means numbers that can be written in a form called a ratio. Whole numbers are rational numbers where $b = 1$. Some square roots are also rational numbers. For example,

$$\sqrt{4} = 2 = \frac{2}{1}$$

is a rational number. Every rational number can also be written in decimal form as a terminating decimal or a nonterminating repeating decimal, as in Section R.3.

There are also numbers whose decimal forms neither terminate nor repeat. These numbers, which are not quotients of whole numbers, are called **irrational numbers.** The name "irrational numbers" means "not rational" or "not a ratio"; that is, numbers that cannot be written in ratio form. Square roots such as

$$\sqrt{3} \approx 1.732050808$$

are irrational numbers. The symbol $\approx$ means "approximately equal to." Irrational numbers can be approximated by decimals but their actual values are nonterminating, nonrepeating decimals.

EXERCISE R.5

Use a scientific calculator to perform the indicated operation:

1. $36.54 + 13.51$
2. $1.72 + 95.78$
3. $0.459 + 1.972$
4. $0.0916 + 0.7576$
5. $81.64 - 28.58$
6. $10.23 - 5.62$
7. $3.35 - 0.214$
8. $0.289 - 0.0882$
9. 7.1×2.3
10. 0.39×1.7
11. 0.46×0.62
12. 0.035×0.66

Round off each result to two places after the decimal point:

13. $4.7 \div 1.2$
14. $9.88 \div 4.2$

Round off to four significant digits:

15. 3.162277660
16. 2.718281828
17. 0.159154943
18. 0.434294481
19. 0.301029995
20. 1.660964047

Round off to three significant digits:

21. 4.472135955
22. 1.505149978
23. 0.022360679
24. 0.000318309
25. 7389.056099
26. $98,696.04401$

Use a scientific calculator to perform the indicated operation; round off the result to three significant digits:

27. 9.68×2.78
28. 89.5×39.3
29. 0.932×0.418
30. 0.0826×0.00152
31. $4.29 \div 1.69$
32. $5.42 \div 8.47$
33. $0.434 \div 2.23$
34. $261 \div 0.78$

Use a scientific calculator to find the square; round off the result to three significant digits:

35. 1.75^2
36. 0.566^2
37. 59.3^2
38. 852^2

Use a scientific calculator to find the square root; round off the result to three significant digits:

39. $\sqrt{5}$
40. $\sqrt{10}$
41. $\sqrt{74.3}$
42. $\sqrt{188.8}$

□ **Self-Test** □

Perform the indicated operation:

1. **a.** 30×20

    b. $525 \div 5$

1a. _____

1b. _____

2. **a.** $\dfrac{3}{8} \div \dfrac{3}{2}$

    b. $1\dfrac{1}{2} - \dfrac{2}{3}$

2a. _____

2b. _____

3. **a.** 3.6×0.05

    b. $1.61 \div 0.7$.

3a. _____

3b. _____

4. Find the range of values of a resistor rated at 1.2 ohms $\pm 20\%$.

4. _____

5. Use a scientific calculator to find the square or square root; round off the result to three significant digits:

    a. 0.959^2

    b. $\sqrt{0.355}$

5a. _____

5b. _____

UNIT 1

The Real Numbers

Introduction

Algebra and trigonometry, and other branches of mathematics used in science and technology, have their basis in the concepts and properties of numbers. After some review in Unit R, if necessary, you should feel comfortable with the concepts and arithmetic of whole numbers, fractions, and decimals. In this unit, you will extend these concepts to the larger system of real numbers. You will learn about the number line, the properties of order, and the meaning of the absolute value of real numbers. Then, you will learn the rules for arithmetic operations with real numbers. Finally, you will use the arithmetic operations to combine real numbers in expressions.

OBJECTIVES

When you have finished this unit you should be able to:

1. Plot real numbers on the number line, and determine the order of two real numbers.
2. Find the absolute value of any real number, and evaluate expressions containing absolute values.
3. Add two real numbers, and subtract two real numbers.
4. Multiply two real numbers, and divide two real numbers.
5. Apply the rules for order of operations to evaluate expressions consisting of real numbers and arithmetic operations for real numbers.

SECTION 1.1

The Number Line

In Unit R, we saw that whole numbers, fractions, and even square roots have been used since ancient times, while the concept of zero is somewhat more recent. At about the same time they invented zero, the Hindu mathematicians also began to use a concept of numbers to represent debts. Such numbers were not generally used in mathematics, however, until about the sixteenth century, and mathematicians seem not to have been quite comfortable with them until as recently as the nineteenth century.

The numbers we used in Unit R are called **positive numbers**. The numbers that were invented for the concept of representing a debt are called **negative numbers**. The number zero is neither a positive number nor a negative number. The **real numbers** consist of the positive numbers, zero, and the negative numbers.

The real numbers can be represented by means of the **number line**. The number line associates every real number—positive, negative, and zero—with a point on a line. The number associated with a specific point is called the **coordinate** of the point.

To construct the number line, we first choose a **direction** for the line. The number line is usually drawn horizontally, with its direction to the right:

The arrow indicates the direction.

Next we choose the **origin** for the number line. The origin is the point with the coordinate 0:

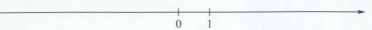

The origin may be placed anywhere we like on the line.

Finally, we choose a **unit** for the number line. The unit establishes the distance from the origin to the point with coordinate 1:

The unit may be any length we like. We must place 1 to the right of the origin, however, because the line is directed to the right. Then, the point with coordinate 2 is placed one unit to the right of 1, the point with coordinate 3 is placed one unit to the right of 2, and so on:

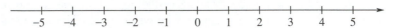

Now, we place the negative numbers to correspond with the positive numbers. The point with coordinate −1, read "negative one," is placed one unit to the left of the origin. Then, the point with coordinate −2, read "negative two," is placed one unit to the left of −1; the point with coordinate −3, read "negative three," is placed one unit to the left of −2; and so on:

The numbers we have placed so far, the whole numbers and their negatives, are called the **integers**.

Once we have placed the integers on the number line, we can place fractions by using fractional parts of the unit. For example, the point with coordinate

$$\frac{1}{2} = 0.5$$

is one-half of a unit to the right of the origin. We can place decimals with an accuracy of about one-tenth of a unit by estimating 0.1, 0.2, . . . , 0.9 as fractional parts of the unit. Then, we place irrational numbers by using their decimal approximations. For example, we place the point with coordinate

$$\sqrt{2} \approx 1.414213562$$

at approximately 1.4:

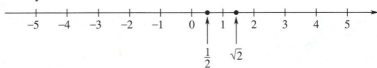

We observe that, due to the inexact nature of making measurements, the width of pencil marks, and so forth, all point placements on the number line are approximate. Approximation of coordinates to one-tenth of a unit is usually sufficient. (The method for approximating square roots by using a scientific calculator is given in Section R.5.)

EXAMPLE 1.1 ▶ Place the point with the given coordinate on the number line:

a. $\sqrt{3}$ **b.** $-\sqrt{3}$

SOLUTIONS We approximate each coordinate to one-tenth of a unit.

a. Use a scientific calculator to find that

$$\sqrt{3} \approx 1.732050808$$

Therefore, we place the point at approximately 1.7:

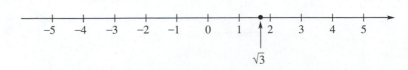

b. The number $-\sqrt{3}$ is the negative of $\sqrt{3}$, and its decimal approximation is

$$-\sqrt{3} \approx -1.732050808$$

Therefore, we place the point at approximately -1.7, or 1.7 units to the left of zero:

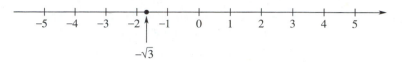

We observe that numbers such as

$$\sqrt{-3} \quad \text{and} \quad -\sqrt{-3}$$

are not real numbers. You can use the square-root key on your calculator to verify that $\sqrt{-3}$ is not a real number. First, find the change-sign key, marked

$$\boxed{+/-} \ .$$

Then, enter 3, press the change-sign key to obtain -3, and attempt to use the square root key. (For a DAL calculator, press the square-root key, enter 3, press the change-sign key, and press the equals key.) The resulting display is some kind of error symbol, possibly the word "error," the letter "E," or "E" with 0, or on older calculators, possibly just 0. Numbers that are not real numbers are not coordinates of points on the number line.

ORDER

The **order relations** compare the sizes of two real numbers. There are three order relations:

1. The first number is equal to the second; or

2. The first number is less than the second; or

3. The first number is greater than the second.

Each of the order relations has a symbol. We are familiar with the symbol =, the equal sign. The symbol < means "is less than" or "is smaller than." The symbol > means "is greater than" or "is larger than." We can summarize the order relations by using these symbols.

If A and B are two real numbers, then

      1. $A = B$, or

      2. $A < B$, or

      3. $A > B$.

For any two real numbers, we can determine which order relation is true by using the number line. First, we observe that $A = B$ if A and B are coordinates of the same point on the number line.

To determine the second case, we observe that $A < B$ if the point that has coordinate A is to the left of the point that has coordinate B. For example, we know that

$$2 < 3,$$

and the point that has coordinate 2 is to the left of the point that has coordinate 3.

To determine the third case, we observe that $A > B$ if the point that has coordinate A is to the right of the point that has coordinate B. For example, we know that

$$4 > 3,$$

and the point that has coordinate 4 is to the right of the point that has coordinate 3.

EXAMPLE 1.2 Use the number line to determine the order of the given numbers:

a. $\sqrt{5}, \ 2$ **b.** $\sqrt{5}, \ \dfrac{5}{2}$

SOLUTIONS We place the points with the given coordinates on the number line.

a. Use a scientific calculator to find that

$$\sqrt{5} \approx 2.236067978.$$

Figure for Example 1.2**a.**

Therefore, we place the point at approximately 2.2 on the number line. We see that the point with coordinate $\sqrt{5}$ is to the right of the point with coordinate 2. Therefore,

$$\sqrt{5} > 2.$$

b. We write the fraction $\frac{5}{2}$ as

$$\frac{5}{2} = 2\frac{1}{2}$$

or

$$\frac{5}{2} = 2.5,$$

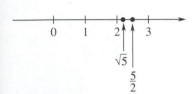

Figure for Example 1.2**b.**

and place the point at approximately one-half of a unit between 2 and 3. We see that the point with coordinate $\sqrt{5}$ is to the left of the point with coordinate $\frac{5}{2}$. Therefore,

$$\sqrt{5} < \frac{5}{2}. \qquad \blacktriangle$$

All points whose coordinates are positive numbers are to the right of the origin, which has coordinate zero. Therefore, all positive numbers are greater than zero. Also, all points whose coordinates are negative numbers are to the left of the origin. Therefore, all negative numbers are less than zero.

EXAMPLE 1.3 ▶ Use the number line to determine the order of the given numbers:

a. $\sqrt{3}$, 0 **b.** $-\sqrt{3}$, 0

SOLUTIONS We place the points with the given coordinates on the number line.

a. We recall that

$$\sqrt{3} \approx 1.732050808,$$

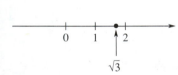

Figure for Example 1.3**a.**

and place the point at approximately 1.7 on the number line. Since the point is to the right of the origin, we see that

$$\sqrt{3} > 0.$$

b. We recall that

$$-\sqrt{3} \approx -1.732050808,$$

Figure for Example 1.3**b.**

and place the point at approximately -1.7 on the number line. Since the point is to the left of the origin, we see that

$$-\sqrt{3} < 0. \qquad \blacktriangle$$

The number line is especially useful in determining the order of two negative numbers.

EXAMPLE 1.4 ▶ Use the number line to determine the order of the given numbers:

a. -2, -3 **b.** $-\sqrt{5}$, $-\frac{5}{2}$

SOLUTIONS We place the points with the given coordinates on the number line.

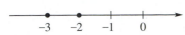

Figure for Example 1.4**a.**

a. We place the points with coordinates -2 and -3 on the number line. We see that the point with the coordinate -2 is to the right of the point with coordinate -3. Therefore,

$$-2 > -3.$$

b. We recall that

$$\sqrt{5} \approx 2.236067978,$$

so we know that

$$-\sqrt{5} \approx -2.236067978.$$

Therefore, we place the point at approximately -2.2 on the number line. Similarly, we know that

$$\frac{5}{2} = 2\frac{1}{2}$$

or

$$\frac{5}{2} = 2.5.$$

Thus, we know that

$$-\frac{5}{2} = -2\frac{1}{2}$$

or

$$-\frac{5}{2} = -2.5.$$

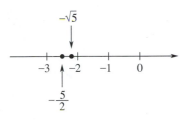

Figure for Example 1.4**b.**

We place the point at approximately one-half of a unit between -2 and -3. We see that the point with coordinate $-\sqrt{5}$ is to the right of the point with coordinate $-\frac{5}{2}$. Therefore,

$$-\sqrt{5} > -\frac{5}{2}.$$ ▲

EXERCISE
1.1

Place the point with the given coordinate on the number line:

1. **a.** $\dfrac{5}{4}$ **b.** $-\dfrac{5}{4}$ **c.** $\sqrt{6}$ **d.** $-\sqrt{6}$ **e.** $\sqrt{-6}$

2. **a.** $\dfrac{5}{3}$ **b.** $-\dfrac{5}{3}$ **c.** $\sqrt{10}$ **d.** $-\sqrt{10}$ **e.** $\sqrt{-10}$

Use the number line to determine the order of the given numbers:

3. $\sqrt{17},\ 4$ 4. $\dfrac{17}{4},\ 4$ 5. $\sqrt{8},\ 0$ 6. $-\sqrt{8},\ 0$

7. $-4,\ -5$ 8. $-\dfrac{7}{2},\ -\sqrt{7}$ 9. $-\dfrac{10}{3},\ -\dfrac{11}{4}$ 10. $-\dfrac{4}{3},\ -\dfrac{7}{5}$

SECTION
1.2

Absolute Value

The **distance** of any point from the origin is the number of units between the point and the origin, without regard to whether the point is to the right or left of the origin. For example, the point with coordinate 2 and the point with coordinate -2 are the same distance from the origin:

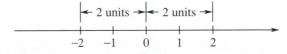

Distance is an example of an unsigned quantity; that is, we do not usually distinguish between positive and negative distance.

Any real number can be indicated as an unsigned number by using the **absolute value** symbol. The absolute value of any positive real number A is written

$$|A|,$$

read "the absolute value of A." The absolute value of any negative real number $-A$ is written

$$|-A|,$$

read "the absolute value of negative A."

Absolute Value: If A is positive or 0, then
$$|A| = A.$$
If A is positive, then $-A$ is negative and
$$|-A| = A.$$

The absolute value of any positive number is positive; the absolute value of zero is zero; and the absolute value of any negative number is the corresponding positive number.

EXAMPLE 1.5   Find the absolute value:

a. $|2|$ b. $|-2|$ c. $|0|$

SOLUTIONS   We use the definition of absolute value.

a. Since 2 is positive,
$$|2| = 2.$$

b. Since -2 is the negative of 2,
$$|-2| = 2.$$

c. For zero,
$$|0| = 0.$$ ▲

When absolute values are to be combined by arithmetic operations, we must find the absolute values before combining the numbers.

EXAMPLE 1.6   Find the value of $|-7| + |5|$.

SOLUTION   We find the absolute values first and then add. Since
$$|-7| = 7 \quad \text{and} \quad |5| = 5,$$
we write
$$|-7| + |5| = 7 + 5$$
$$= 12.$$ ▲

EXAMPLE 1.7   Find the value of $|-3| \times |-5|$.

SOLUTION   We find the absolute values first and then multiply. Since
$$|-3| = 3 \quad \text{and} \quad |-5| = 5,$$
we write
$$|-3| \times |-5| = 3 \times 5$$
$$= 15.$$ ▲

EXAMPLE 1.8 Find the value of $|-10| - |-6|$.

SOLUTION We find the absolute values first, and then subtract. Since

$$|-10| = 10 \quad \text{and} \quad |-6| = 6,$$

we write

$$|-10| - |-6| = 10 - 6$$
$$= 4. \qquad \blacktriangle$$

There may be arithmetic operations inside absolute value symbols. We first do any calculations within absolute value symbols. Then, we find the absolute values and complete the calculation.

EXAMPLE 1.9 Find the value of $|15 - 5| + |-6|$.

SOLUTION First, we do the subtraction inside the first pair of absolute value symbols:

$$|15 - 5| + |-6| = |10| + |-6|.$$

Then, we find each absolute value and complete the calculation:

$$|10| + |-6| = 10 + 6$$
$$= 16. \qquad \blacktriangle$$

So far, we have used only integers. We can use absolute values with any real numbers.

EXAMPLE 1.10 Find the value of $3 \times \left|-\dfrac{1}{6}\right|$.

SOLUTION We know that $\frac{1}{6}$ is positive, so $-\frac{1}{6}$ is negative. Therefore, we find

$$\left|-\frac{1}{6}\right| = \frac{1}{6}.$$

Thus, we write

$$3 \times \left|-\frac{1}{6}\right| = 3 \times \frac{1}{6}$$
$$= \frac{1}{2}. \qquad \blacktriangle$$

When absolute values contain square roots, we can find decimal approximations to a given number of significant digits. (Significant digits are discussed in Section R.5).

EXAMPLE 1.11 Find the value of $|-\sqrt{3}| - |-\sqrt{2}|$ to three significant digits.

SOLUTION We know that $\sqrt{3}$ is positive, so $-\sqrt{3}$ is negative. Therefore, we find

$$|-\sqrt{3}| = \sqrt{3}.$$

Similarly, $\sqrt{2}$ is positive, so $-\sqrt{2}$ is negative. Therefore, we also find

$$|-\sqrt{2}| = \sqrt{2}.$$

Thus, we write

$$|-\sqrt{3}| - |-\sqrt{2}| = \sqrt{3} - \sqrt{2}.$$

You can use a calculator to complete the calculation, using this combination of square root calculations:

		display:
Enter	3	3.
Press	$\sqrt{x}$	1.732050808
Press	$-$	1.732050808
Enter	2	2.
Press	$\sqrt{x}$	1.414213562
Press	$=$	0.317837245

Therefore, we have

$$\left| -\sqrt{3} \right| - \left| -\sqrt{2} \right| = 0.318$$

to three significant digits. (Remember that, for a DAL calculator, you must reverse the first two steps and also reverse the fourth and fifth steps.)   ▲

EXERCISE 1.2

Find the value:

1. $|6|$
2. $|-6|$
3. $-|6|$
4. $-|-6|$

5. $|-1| + |8|$
6. $|-6| + |-6|$
7. $|-2| \times |-7|$
8. $|-4| \times |-4|$

9. $|-1| \times |0|$
10. $|0| \times |-2|$
11. $|3| \times |-6|$
12. $|-8| \times |10|$

13. $|-12| - |-2|$
14. $|-6| - |-5|$
15. $|4| - |-3|$
16. $|-15| - |6|$

17. $\dfrac{|-8|}{|4|}$

18. $\dfrac{|-18|}{|-12|}$

19. $|6 - 5| + |-7|$

20. $|14 - 6| + |-9|$

21. $|10 + 2| - |6 - 3|$

22. $|8 - 6| - |10 - 8|$

23. $2 \times \left| -\dfrac{1}{10} \right|$

24. $\left| \dfrac{2}{3} \right| \times |-6|$

25. $|-\sqrt{5}| - |-\sqrt{3}|$

26. $|-\sqrt{3}| - |-\sqrt{3}|$

27. $|\sqrt{3}| + |-\sqrt{6}|$

28. $|\sqrt{3}| + |-\sqrt{3}|$

SECTION 1.3

Addition and Subtraction

We can picture operations with real numbers by using the number line. We picture addition of positive numbers by measuring distances in the positive direction on the number line. We measure a distance from the origin to the point whose coordinate is the first number. Then we measure a distance equal to the second number from that point. The coordinate of the final point is the sum. For example, to add 2 and 1, we measure the distance 2 from the origin to the point with coordinate 2, and then we measure the distance 1 from that point. We finish at the point with coordinate 3. Therefore,

$$2 + 1 = 3.$$

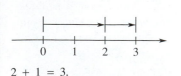

$2 + 1 = 3.$

ADDITION

Now, we consider the sum of two negative numbers. We picture addition of negative numbers by measuring distances in the negative direction. Since distances are unsigned numbers, we can represent them by absolute values. For example, to add −2 and −1, we write their absolute values:

$$|-2| = 2 \quad \text{and} \quad |-1| = 1.$$

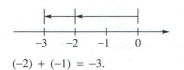

$$(-2) + (-1) = -3.$$

We measure the distance 2 from the origin in the negative direction, to the point with coordinate −2. Then we measure the distance 1, also in the negative direction, from that point. We finish at the point with coordinate −3. Therefore,

$$(-2) + (-1) = -3.$$

In the preceding process, we added the absolute values of the numbers, using the negative direction to indicate the negative numbers. We state this process as a rule for adding two negative numbers.

Addition of Two Negative Numbers: Add the absolute values, and write the negative of the result.

EXAMPLE 1.12 ▶

Add the numbers:

a. $(-4) + (-7)$

b. $-14 + (-22)$

SOLUTIONS

We add the absolute values, and write the negative of the result.

a. The absolute values are

$$|-4| = 4 \quad \text{and} \quad |-7| = 7.$$

Adding the absolute values, we have

$$4 + 7 = 11.$$

Then, we write the negative of the result to obtain

$$(-4) + (-7) = -11.$$

b. The absolute values are

$$|-14| = 14 \quad \text{and} \quad |-22| = 22.$$

Adding the absolute values, we have

$$14 + 22 = 36.$$

Then, we write the negative of the result to obtain

$$-14 + (-22) = -36. \qquad \blacktriangle$$

We note that, in part **a** of Example 1.12, each negative number is written in parentheses. These parentheses are grouping symbols that group the negative sign with the number to indicate a negative number. The negative sign in this sense is different from the minus sign of subtraction. The parentheses around the first negative number can be omitted, as in part **b**, because a negative sign with the first number could only indicate a negative number. The parentheses around the second negative number cannot be omitted.

Now, we consider the sum of two numbers where one is positive and one is negative. We picture this addition by measuring the distance of the positive number in the positive direction and the distance of the negative number in the negative direction. For example, to add 2 and −1, we write their absolute values:

$$|2| = 2 \quad \text{and} \quad |-1| = 1.$$

We measure the distance 2 from the origin in the positive direction, to the point with coordinate 2. Then we measure the distance 1 in the negative direction from that point. We have finished at the point with coordinate 1. Therefore,

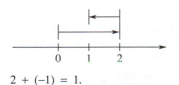

$$2 + (-1) = 1.$$

$$2 + (-1) = 1.$$

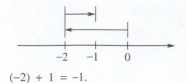

$(-2) + 1 = -1.$

Similarly, to add −2 and 1, we write their absolute values:

$$|-2| = 2 \quad \text{and} \quad |1| = 1.$$

We measure the distance 2 from the origin in the negative direction, to the point with coordinate −2. Then we measure the distance 1 in the positive direction from that point. We have finished at the point with coordinate −1. Therefore,

$$-2 + 1 = -1.$$

In the two preceding cases, we have subtracted the absolute values of the numbers. In the first case, the absolute value of the positive number is larger, and the result is positive. In the second case, the absolute value of the negative number is larger, and the result is negative. We state this process as a rule for adding two numbers with opposite signs.

Addition of Two Numbers with Opposite Signs: Subtract the absolute values, and write the negative of the result if the absolute value of the negative number is larger.

EXAMPLE 1.13 ▶ Add the numbers:

a. $-4 + 7$ **b.** $14 + (-22)$

SOLUTIONS   We subtract the absolute values, and write the negative if the absolute value of the negative number is larger.

a.   The absolute values are

$$|-4| = 4 \quad \text{and} \quad |7| = 7.$$

Subtracting the absolute values, we have

$$7 - 4 = 3.$$

The absolute value of the positive number is larger. Therefore, we write

$$-4 + 7 = 3.$$

b.   The absolute values are

$$|14| = 14 \quad \text{and} \quad |-22| = 22.$$

Subtracting the absolute values, we have

$$22 - 14 = 8.$$

Since the absolute value of the negative number is larger, we write the negative of the result to obtain

$$14 + (-22) = -8. \qquad ▲$$

It is a common error simply to say "the negative number is larger" rather than *the absolute value of* the negative number is larger. We cannot say, for example, that −22 is larger than 14. We can say, however, that $|-22| > |14|$.

The sum of zero and any number equals the number. We can measure a distance from the origin to the point whose coordinate is the number, and then measure a distance equal to zero. For example,

$$-2 + 0 = -2.$$

The sum of any number and its corresponding negative number is zero. For example, to add 2 and −2, we first write their absolute values:

$$|2| = 2 \quad \text{and} \quad |-2| = 2.$$

We measure the distance 2 from the origin in the positive direction, to the point with coordinate 2. Then we measure the distance 2 in the negative direction from that point. We have finished at the origin. Therefore,

$$2 + (-2) = 0.$$

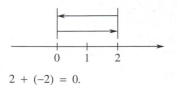

$2 + (-2) = 0.$

SUBTRACTION

Subtraction is the opposite, or in mathematical terms the inverse, of addition. Subtracting a number is the same as adding the number with the opposite sign. For example,

$$2 - 1 = 1$$

because $1 + 1 = 2$. But we also know that

$$2 + (-1) = 1,$$

and, therefore,

$$2 - 1 = 2 + (-1)$$
$$= 1.$$

The operation on the left of the equals sign is a subtraction. The operation on the right is the addition of a negative number. We say that 2 *minus* 1 is equal to 2 *plus negative* 1.

Similarly, we can write

$$1 - 2 = 1 + (-2).$$

We say that 1 *minus* 2 is equal to 1 *plus negative* 2. But we know how to add real numbers, so we can say

$$1 - 2 = 1 + (-2)$$
$$= -1.$$

We state this process as a rule for subtracting two numbers.

Subtraction of Two Numbers: Write the second number with the opposite sign and add.

EXAMPLE 1.14   Subtract the numbers:

a. $-4 - 7$ **b.** $14 - (-22)$

SOLUTIONS   We write the second number with the opposite sign and add.

a.   The subtraction is read "negative 4 minus 7." The second number 7 is positive, so we write the corresponding negative number -7, and add:

$$-4 - 7 = -4 + (-7)$$
$$= -11.$$

b.   The subtraction is read "14 minus negative 22." The second number -22 is negative, so we write the corresponding positive number 22, and add:

$$14 - (-22) = 14 + 22$$
$$= 36.$$ ▲

EXAMPLE 1.15   Subtract $-12 - (-16)$.

SOLUTION   The second number -16 is negative, so we write the corresponding positive number 16 and add:

$$-28 - (-16) = -28 + 16.$$

We recall that, to add numbers with opposite signs, we subtract the absolute values:

$$|-28| = 28 \quad \text{and} \quad |16| = 16.$$

Therefore, we have

$$28 - 16 = 12.$$

Then, since the absolute value of the negative number is larger, we write the negative of the result to obtain

$$-28 + 16 = -12. \qquad \blacktriangle$$

Zero subtracted from any number is the number. For example,

$$-2 - 0 = -2,$$

because $-2 + 0 = -2$. A real number subtracted from zero, however, is the negative of that number. For example,

$$0 - 2 = -2$$

because $-2 + 2 = 0$.

EXAMPLE 1.16 ▶ Find the value of $-(-2)$.

SOLUTION We can think of $-(-2)$ as $0 - (-2)$. Then, we write the corresponding positive number, and add to obtain

$$0 - (-2) = 0 + 2$$
$$= 2.$$

Therefore, we have

$$-(-2) = 2. \qquad \blacktriangle$$

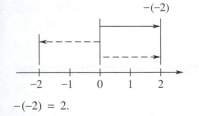

$-(-2) = 2.$

Example 1.16 is an example of a "double negative." The negative inside the parentheses reverses the direction from positive to negative. Then, the negative sign outside the parentheses reverses the direction again, from negative back to positive. Thus we have the familiar expression, "a double negative makes a positive."

EXERCISE 1.3

Add or subtract, as indicated:

1. $(-6) + (-5)$ 2. $(-10) + (-25)$ 3. $-12 + (-14)$ 4. $-29 + (-19)$

5. $-6 + 8$ 6. $-18 + 24$ 7. $10 + (-4)$ 8. $23 + (-7)$

9. $8 + (-15)$ 10. $13 + (-27)$ 11. $-18 + 33$ 12. $-37 + 26$

13. $(-5) + (-5)$ 14. $-16 + (-16)$ 15. $5 + (-5)$ 16. $(-12) + 12$

17. $-9 - 12$ 18. $-26 - 24$ 19. $18 - (-8)$ 20. $19 - (-23)$

21. $-11 - (-14)$ 22. $-6 - (-18)$ 23. $-20 - (-11)$ 24. $-36 - (-18)$

25. $5 - (-5)$ 26. $(-10) - 10$ 27. $-5 - (-5)$ 28. $(-15) - (-15)$

29. $0 - 8$ 30. $0 - (-9)$ 31. $-12 - 0$ 32. $0 - 5$

33. $-(-5)$ 34. $-(-1)$ 35. $-(-4) + 4$ 36. $-(-4) - 4$

37. $-(-(-1))$ 38. $-(-(-(-1)))$

SECTION 1.4

Multiplication and Division

Multiplication can be defined as repeated addition. For example, to multiply 3 times 1, we add three ones:

$$3 \times 1 = 1 + 1 + 1$$
$$= 3.$$

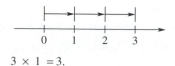

$3 \times 1 = 3.$

To picture this multiplication on the number line, we measure the distance 1 from the origin three times. We finish at the point with coordinate 3. Therefore,

$$3 \times 1 = 3.$$

MULTIPLICATION

Now, we consider multiplication of a negative number. For example, to multiply 3 times -1, we add three negative ones:

$$3 \times (-1) = (-1) + (-1) + (-1)$$
$$= -3.$$

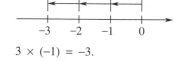

$3 \times (-1) = -3.$

To picture this multiplication on the number line, we measure the distance 1 from the origin in the negative direction three times. We finish at the point with coordinate -3. Therefore,

$$3 \times (-1) = -3.$$

Similarly, to multiply -3 times 1, we measure the distance 3 from the origin in the negative direction one time, also finishing at -3. Therefore,

$$(-3) \times 1 = -3.$$

In the preceding cases, we have multiplied the absolute values of the numbers, using the negative direction to indicate the negative numbers. We state this process as a rule for multiplying two numbers with opposite signs.

Multiplication of Two Numbers with Opposite Signs: Multiply the absolute values and write the negative of the result.

EXAMPLE 1.17 ▶

Multiply the numbers:

a. $(-8) \times 3$ **b.** $7 \times (-5)$

SOLUTIONS

We multiply the absolute values, and write the negative of the result.

a.   The absolute values are

$$|-8| = 8 \quad \text{and} \quad |3| = 3.$$

Multiplying the absolute values, we have

$$8 \times 3 = 24.$$

Then, we write the negative of the result to obtain

$$(-8) \times 3 = -24.$$

b.   The absolute values are

$$|7| = 7 \quad \text{and} \quad |-5| = 5.$$

Multiplying the absolute values, we have

$$7 \times 5 = 35.$$

Then, we write the negative of the result to obtain

$$7 \times (-5) = -35.$$

▲

We can think of a negative as reversing direction on the number line. Multiplying by a negative number reverses the direction of the corresponding multiplication by a positive number. For example, to multiply −3 times 1, we reverse the direction of 3 times 1:

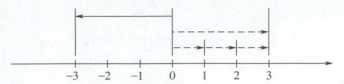

Thus, we can write

$$(-3) \times 1 = -(3 \times 1)$$

by reversing the direction of 3×1. But we know that

$$3 \times 1 = 3,$$

and, therefore,

$$(-3) \times 1 = -3.$$

Now, we consider the product of two negative numbers. For example, to multiply −3 times −1, we reverse the direction of 3 times −1:

Thus, we can write

$$(-3) \times (-1) = -[3 \times (-1)]$$

by reversing the direction of $3 \times (-1)$. But we know that

$$3 \times (-1) = -3,$$

and, therefore,

$$(-3) \times (-1) = -(-3)$$
$$= 3.$$

We state this result as a rule for multiplying two negative numbers.

Multiplication of Two Negative Numbers: Multiply the absolute values.

EXAMPLE 1.18   Multiply $(-9) \times (-6)$.

SOLUTION   The absolute values are

$$\left|-9\right| = 9 \quad \text{and} \quad \left|-6\right| = 6.$$

Multiplying the absolute values, we have

$$9 \times 6 = 54.$$

Therefore, we write

$$(-9) \times (-6) = 54. \qquad \blacktriangle$$

DIVISION

Division is the opposite, or in mathematical terms the inverse, of multiplication. Dividing by a number is the same as multiplying by the **reciprocal** of the number. The reciprocal of any number A is the fraction

$$\frac{1}{A}.$$

For example, the reciprocal of 2 is

$$\frac{1}{2}.$$

Then, for example,

$$6 \div 2 = 3$$

because $6 = 2 \times 3$. But we also know that

$$6 \times \frac{1}{2} = 3,$$

and, therefore,

$$6 \div 2 = 6 \times \frac{1}{2}.$$

If a number is positive, its reciprocal is also positive, and if the number is negative, its reciprocal is also negative. Thus, the rules for division of two numbers are the same as the rules for multiplication.

Division of Two Numbers with Opposite Signs: Divide the absolute values, and write the negative of the result.

Division of Two Negative Numbers: Divide the absolute values.

EXAMPLE 1.19 ▶

Divide the numbers:

a. $18 \div (-6)$ **b.** $(-11) \div 22$

SOLUTIONS We divide the absolute values, and write the negative of the result.

a. The absolute values are

$$|18| = 18 \quad \text{and} \quad |-6| = 6.$$

Dividing the absolute values, we have

$$18 \div 6 = 3.$$

Then, we write the negative of the result to obtain

$$18 \div (-6) = -3.$$

b. The absolute values are

$$|-11| = 11 \quad \text{and} \quad |22| = 22.$$

Dividing the absolute values, we have

$$11 \div 22 = 0.5.$$

Then, we write the negative of the result to obtain

$$(-11) \div 22 = -0.5.$$

▲

EXAMPLE 1.20 ▶

Divide $(-9) \div (-6)$.

SOLUTION The absolute values are

$$|-9| = 9 \quad \text{and} \quad |-6| = 6.$$

Dividing the absolute values, we have

$$9 \div 6 = 1.5.$$

Therefore, we write

$$(-9) \div (-6) = 1.5. \qquad \blacktriangle$$

EXAMPLE 1.21 ▶ Divide $(-2) \div (-\sqrt{3})$.

SOLUTION The absolute values are

$$|-2| = 2 \quad \text{and} \quad |-\sqrt{3}| = \sqrt{3}.$$

You can use a calculator to divide the absolute values:

	display:
Enter 2	2.
Press $\div$	2.
Enter 3	3.
Press $\sqrt{x}$	1.732050808
Press $=$	1.154700538

Therefore, we write

$$(-2) \div (-\sqrt{3}) = 1.15,$$

rounded off to three significant digits.   ▲

Multiplication of any real number by zero is zero. For example,

$$0 \times (-3) = 0.$$

Division of zero by any real number except zero is also zero. For example,

$$0 \div (-3) = 0,$$

because $0 \times (-3) = 0$. However, division of any real number by zero is *undefined*. For example,

$$(-3) \div 0 \text{ is undefined}$$

because there is no number that, when multiplied by zero, gives -3. Also, zero has no reciprocal because

$$1 \div 0 \text{ is undefined}$$

EXERCISE 1.4

Multiply or divide, as indicated:

1. $(-10) \times 4$ **2.** $(-12) \times 11$ **3.** $5 \times (-11)$ **4.** $8 \times (-12)$

5. $(-4) \times (-12)$ **6.** $\left(-\dfrac{1}{2}\right) \times (-4)$ **7.** $48 \div (-8)$ **8.** $80 \div (-4)$

9. $(-100) \div 10$ **10.** $(-13) \div 52$ **11.** $(-54) \div (-9)$ **12.** $(-50) \div (-20)$

Multiply or divide; round off to three significant digits:

13. $(-2) \div 3$ **14.** $4 \div (-9)$ **15.** $(-15) \div (-11)$ **16.** $(-7) \div (-6)$

17. $(-4) \times (-\sqrt{3})$ **18.** $(-1) \times \sqrt{3}$ **19.** $5 \div (-\sqrt{2})$ **20.** $-\sqrt{10} \div (-2)$

Multiply or divide, or write "undefined":

21. $0 \times (-15)$ **22.** $(-20) \times 0$ **23.** $\left(-\dfrac{1}{2}\right) \times 0$ **24.** $0 \times (-\sqrt{3})$

25. $0 \div (-15)$ **26.** $(-20) \div 0$ **27.** $(-0.5) \div 0$ **28.** $0 \div (-\sqrt{10})$

SECTION
1.5

Order of Operations

An **expression** consists of numbers combined by arithmetic operations. Each example in Sections 1.3 and 1.4 is an expression with just one arithmetic operation that combines two real numbers. Expressions can contain more than one operation, combining several numbers.

Every expression represents a number. When we **evaluate** an expression, we find the number. To evaluate an expression, we perform the operations with the numbers in the expression. When an expression has more than one operation, we follow rules for the **order of operations**. Three rules express the priority order for operations.

Priority Order for Operations:

Priority 1: Powers and Roots

Priority 2: Multiplication and Division

Priority 3: Addition and Subtraction

We do operations in order of their priority: first, all powers and roots; then, all multiplications and divisions; and, finally, all additions and subtractions.

EXAMPLE 1.22 Evaluate $2 - 3 \times 4$.

SOLUTION There are no powers or roots, so we do the multiplication first, and then the subtraction:

$$2 - 3 \times 4 = 2 - 12$$
$$= -10. \quad \blacktriangle$$

Squares are an example of powers. (Squares are reviewed with square roots in Section R.5).

EXAMPLE 1.23 Evaluate $5 - 2^2$.

SOLUTION We do the power first, and then the subtraction:

$$5 - 2^2 = 5 - 4$$
$$= 1. \quad \blacktriangle$$

When we have more than one operation of the same priority, we proceed from left to right.

EXAMPLE 1.24 Evaluate $6 - 9 + 7$.

SOLUTION Addition and subtraction have the same priority, so we proceed from left to right:

$$6 - 9 + 7 = -3 + 7$$
$$= 4. \quad \blacktriangle$$

EXAMPLE 1.25 Evaluate $10 \div 2 \times 4$.

SOLUTION Multiplication and division have the same priority, so we proceed from left to right:

$$10 \div 2 \times 4 = 5 \times 4$$
$$= 20. \quad \blacktriangle$$

We use grouping parentheses to change the order of operations. We do operations grouped in parentheses first, regardless of their priority in the order of operations.

EXAMPLE 1.26 Evaluate each expression:

a. $(2 - 3) \times 4$ **b.** $(5 - 2)^2$ **c.** $6 - (9 + 7)$ **d.** $10 \div (2 \times 4)$

SOLUTIONS In each case, we do the operation inside the parentheses first.

a. We do the subtraction first because it is in the parentheses:

$$(2 - 3) \times 4 = -1 \times 4$$
$$= -4.$$

b. We do the subtraction first because it is in the parentheses:

$$(5 - 2)^2 = 3^2$$
$$= 9.$$

c. We do the addition first because it is in the parentheses:

$$6 - (9 + 7) = 6 - 16$$
$$= -10.$$

d. We do the multiplication first because it is in the parentheses:

$$10 \div (2 \times 4) = 10 \div 8$$
$$= 1.25. \qquad \blacktriangle$$

Several other symbols can act as grouping symbols. We recall the absolute value symbol

$$|A|$$

from Section 1.2. The absolute value symbol is a grouping symbol. We do any operations inside the symbol first.

EXAMPLE 1.27 Evaluate $|4 - 6| - |-4|$.

SOLUTION We do the subtraction inside the absolute value symbol first:

$$|4 - 6| - |-4| = |-2| - |-4|.$$

Then, we find each absolute value:

$$|-2| = 2 \quad \text{and} \quad |-4| = 4.$$

Therefore, we have

$$|-2| - |-4| = 2 - 4$$
$$= -2. \qquad \blacktriangle$$

We observe that the negative of an absolute value is *not a double negative*; it is just *negative*. For example, we know that

$$|-4| = 4,$$

and, therefore,

$$-|-4| = -4.$$

A division line separating parts of a fraction is a grouping symbol. The operations above the line are grouped and the operations below the line are grouped.

EXAMPLE 1.28 Evaluate $\dfrac{3 \times 4^2}{2 \times 10^2}$.

SOLUTION We evaluate the expressions above and below the division line, and then do the division:

$$\frac{3 \times 4^2}{2 \times 10^2} = \frac{3 \times 16}{2 \times 100}$$

$$= \frac{48}{200}$$

$$= 0.24. \quad \blacktriangle$$

The square root symbol is also a grouping symbol.

EXAMPLE 1.29 ▶ Evaluate $1 - 2 \times \sqrt{4 \times 10^2}$.

SOLUTION First, we do the operations in the square root symbol:

$$1 - 2 \times \sqrt{4 \times 10^2} = 1 - 2 \times \sqrt{400}.$$

Then, following the priority order of operations, we find the square root, then multiply, and finally subtract:

$$1 - 2 \times \sqrt{400} = 1 - 2 \times 20$$

$$= 1 - 40$$

$$= -39. \quad \blacktriangle$$

You might have noticed that, in some cases, you can get the same result by other methods. For example, in Example 1.29, you will get the same result if you first find $\sqrt{4}$ and $\sqrt{10^2}$ and then multiply the results. The rules for powers and roots will be covered in Unit 2. In the absence of another rule, you should always follow the priority order.

EXAMPLE 1.30 ▶ Evaluate $\sqrt{3^2 + 4^2}$.

SOLUTION The square root is a grouping symbol; therefore, we do all operations inside the square root first:

$$\sqrt{3^2 + 4^2} = \sqrt{9 + 16}$$

$$= \sqrt{25}$$

$$= 5.$$

Observe that, in this example, you *must* treat the square root symbol as a grouping symbol, doing the addition in the square root *before* you take any square roots. If you take the square roots first, you will get $3 + 4 = 7$, which is clearly not the same as 5. ▲

EXAMPLE 1.31 ▶ Evaluate $\dfrac{3.3 \times 8.2}{3.3 + 8.2}$.

SOLUTION We must do the operations in the numerator and the denominator, and then divide:

$$\frac{3.3 \times 8.2}{3.3 + 8.2} = \frac{27.06}{11.5}$$

$$= 2.35$$

to three significant digits. ▲

You can use a scientific calculator efficiently for the preceding calculation, but you must be sure to indicate that the division line is a grouping symbol. On a scientific calculator, grouping is indicated by the parentheses keys, marked

$$\boxed{(} \quad \text{and} \quad \boxed{)} .$$

Calculators vary with respect to the number shown on the display when you press the left parentheses key. Many display a zero, as we show in the algorithm that follows. Find the parentheses keys on your calculator, and then follow these steps:

	display:
Enter 3.3	3.3
Press ×	3.3
Enter 8.2	8.2
Press ÷	27.06
Press (	0.
Enter 3.3	3.3
Press +	3.3
Enter 8.2	8.2
Press)	11.5
Press =	2.353043478

or 2.35 to three significant digits.

There are many other ways to use a calculator for this example. However, in some way you must take into account the division line as a grouping symbol. If you do not account for grouping, you will find you have calculated

$$\frac{3.3 \times 8.2}{3.3} + 8.2.$$

The result is 8.2 + 8.2, or 16.4, which is clearly not the same as the result in Example 1.31.

EXERCISE 1.5

Evaluate the expression:

1. $10 - 6 \times 4$ **2.** $3 + 4 \div 2$ **3.** $2 + 3^2$ **4.** $6 - 4^2$

5. $5 - 8 + 9$ **6.** $10 - 8 - 6$ **7.** $20 \div 40 \times 2$ **8.** $16 \div 4 \div 2$

9. $(10 - 6) \times 4$ **10.** $(3 + 4) \div 2$ **11.** $(2 + 3)^2$ **12.** $(6 - 4)^2$

13. $5 - (8 + 9)$ **14.** $10 - (8 - 6)$ **15.** $20 \div (40 \times 2)$ **16.** $16 \div (4 \div 2)$

17. $30 - 15 \times 2^2 \div 5$ **18.** $3 \times 6 - 10^2 \div 5$

19. $20 \div (3^2 - 11) \times (2^2 + 1)$ **20.** $16 \times (2^2 - 6) \div 16 \times (21 - 3^2)$

21. $|5 - 7| - |-3|$ **22.** $|8 - 2^2| - |2 - 3^2|$

23. $\dfrac{3 \times 10^2}{8 \times 5^2}$ **24.** $\dfrac{8 \times 3^2}{15 \times 20^2}$

25. $20 - 3\sqrt{16 \times 10^2}$ **26.** $6 + 2\sqrt{3^2 \times 5^2}$

27. $\sqrt{5^2 + 12^2}$ **28.** $\sqrt{17^2 - 15^2}$

29. $\dfrac{12 \times 56}{12 + 56}$ **30.** $\dfrac{2.7 \times 10^2 \times 4.7 \times 10^2}{2.7 \times 10^2 + 4.7 \times 10^2}$

31. $\dfrac{1}{2 \times 3.142 \times 0.5 \times 6}$ **32.** $\dfrac{1}{2 \times 3.142 \times \sqrt{3} \times 5}$

Self-Test

1. Use the number line to determine the order of $-\sqrt{11}$ and $-\dfrac{11}{3}$.

    1. _____

2. Find the value:

    a. $|-3| \times |-4|$

    b. $|-16| - |-4|$

    2a. _____

    2b. _____

3. Find the value:

    a. $(-8) + (-12)$

    b. $-15 - (-6)$

    3a. _____

    3b. _____

4. Find the value:

    a. $22 \times \left(-\dfrac{1}{2}\right)$

    b. $(-3) \div (-\sqrt{6}\,)$

    4a. _____

    4b. _____

5. Evaluate $(5^2 - 1) \div 6 \times 2$.

    5. _____

UNIT 2

Powers of Ten

Introduction

In science in general, and electronics in particular, you will often encounter numbers that are very large and numbers that are very close to zero. For example, this book was written on a computer that has 2,000,000 bytes of memory with an access time of 0.00000008 second. Clearly, there are better ways to write these numbers, ways in which you use fewer zeros. These ways are scientific notation and engineering notation. Both of these notations use powers of ten to represent zeros. In this unit, you will learn about powers of ten and basic rules for exponents. Then, you will use powers of ten in scientific notation and in engineering notation.

OBJECTIVES

When you have finished this unit you should be able to:

1. Simplify expressions containing powers of ten that have positive integer exponents.

2. Simplify expressions containing powers of ten that have zero or negative exponents, and simplify expressions containing square roots of powers of ten.

3. Write numbers in scientific notation, and evaluate expressions containing numbers in scientific notation.

4. Write numbers in engineering notation, change prefixes of numbers in engineering notation, and evaluate expressions containing numbers in engineering notation.

SECTION 2.1

Positive Integer Exponents

In Units R and 1, we used squares of numbers. For example, the square of ten is

$$10^2 = 10 \times 10 = 100.$$

The square of ten is an example of a **power of ten**. In this square, 10 is called the **base** of the power, and 2 is called the **exponent**.

We can write a power of ten with any positive integer as the exponent. The first power of ten, or ten to the first power, is simply ten:

$$10^1 = 10.$$

The second power of ten, or ten to the second power, is two factors of ten:

$$10^2 = 10 \times 10,$$

also called "ten squared." The third power of ten, or ten to the third power, is three factors of ten:

$$10^3 = 10 \times 10 \times 10,$$

also called "ten cubed." The fourth power of ten, or ten to the fourth power, is four factors of ten:

$$10^4 = 10 \times 10 \times 10 \times 10,$$

and so on. (Only ten squared and ten cubed have special names.)

In general, the power of ten with exponent n,

$$10^n,$$

is n factors of ten; that is, n tens multiplied, where n is any positive integer.

The nth power of ten, or ten to the nth power, is n factors of ten:

$$10^n = 10 \times 10 \times 10 \times \ldots \times 10.$$

$$\longleftarrow \qquad n \text{ factors} \qquad \longrightarrow$$

Basic rules for exponents are used to combine powers. In this unit, we will show these rules by examples with powers of ten.

To multiply powers of ten we add the exponents.

EXAMPLE 2.1 ▶ Multiply $10^2 \times 10^3$.

SOLUTION   We add the exponents to obtain

$$10^2 \times 10^3 = 10^{2+3}$$
$$= 10^5.$$

Observe that

$$10^2 \times 10^3 = (10 \times 10) \times (10 \times 10 \times 10),$$

or five factors of 10. Although such an example is not a proof of the rule, it helps us to see how the rule works.   ▲

To divide powers of ten we subtract the exponents. We will write divisions of powers as fractions, by using division lines.

EXAMPLE 2.2 ▶ Divide each power of ten:

a. $\dfrac{10^3}{10^2}$ **b.** $\dfrac{10^2}{10^3}$ **c.** $\dfrac{10^3}{10^3}$

SOLUTIONS   We subtract the exponents. Since we have defined powers of ten only for positive exponents, we will subtract the smaller exponent from the larger. Thus, the exponent will remain positive.

a. To divide

$$\frac{10^3}{10^2}$$

we subtract the smaller exponent from the larger:

$$\frac{10^3}{10^2} = 10^{3-2}$$

$$= 10^1$$

$$= 10.$$

Recall that we can reduce fractions by dividing out common factors from the numerator and the denominator. We observe that

$$\frac{10^3}{10^2} = \frac{\cancel{10} \times \cancel{10} \times 10}{\cancel{10} \times \cancel{10}}.$$

Dividing out factors of ten leaves one 10 in the numerator. We do not need to write the denominator when it is 1. (Reducing fractions is reviewed in Section R.2.)

b. To divide

$$\frac{10^2}{10^3}$$

we subtract the smaller exponent from the larger:

$$\frac{10^2}{10^3} = \frac{1}{10^{3-2}}$$

$$= \frac{1}{10^1}$$

$$= \frac{1}{10}.$$

We observe that

$$\frac{10^2}{10^3} = \frac{\cancel{10} \times \cancel{10}}{\cancel{10} \times \cancel{10} \times 10}.$$

Dividing out factors of ten leaves one 10 in the denominator, but no factors in the numerator. When there are no factors in the numerator, we write the numerator 1.

c. To divide

$$\frac{10^3}{10^3}$$

we observe that the exponents are the same. We divide out all the factors of ten to obtain

$$\frac{10^3}{10^3} = \frac{\cancel{10} \times \cancel{10} \times \cancel{10}}{\cancel{10} \times \cancel{10} \times \cancel{10}}$$

$$= \frac{1}{1}$$

$$= 1. \qquad \blacktriangle$$

A power of ten can be **raised** to another power, for example,

$$(10^2)^3.$$

To raise one power of ten to another power, we multiply the exponents.

| EXAMPLE 2.3 | | Simplify $(10^2)^3$.

| SOLUTION | We multiply the exponents to obtain

$$(10^2)^3 = 10^{2 \times 3}$$

$$= 10^6.$$

Observe that

$$(10^2)^3 = (10 \times 10) \times (10 \times 10) \times (10 \times 10),$$

or six factors of ten. $\qquad \blacktriangle$

We summarize the preceding examples as basic rules for exponents.

Basic Rules for Exponents:
1. To multiply powers of ten, add the exponents.
2. To divide powers of ten, subtract the smaller exponent from the larger.
3. To raise one power of ten to another, multiply the exponents.

The basic rules for exponents are true for any base except zero, not just the base 10.

You should be careful not to confuse the first rule with the third rule. To *multiply* powers of ten, we *add* the exponents. To *raise* one power of ten to another, we *multiply* the exponents.

We may use more than one rule to simplify an expression.

EXAMPLE 2.4   Simplify $(10^3 \times 10^3)^2$.

SOLUTION   We start inside the parentheses by multiplying the powers of ten. Recall that, to multiply powers of ten, we add the exponents:

$$(10^3 \times 10^3)^2 = (10^{3+3})^2.$$
$$= (10^6)^2.$$

Then, to raise one power of ten to another, we multiply the exponents:

$$(10^6)^2 = 10^{6 \times 2}$$
$$= 10^{12}. \qquad \blacktriangle$$

EXAMPLE 2.5   Simplify $\left(\dfrac{10^6}{10^2}\right)^3$.

SOLUTION   We start inside the parentheses by dividing the powers of ten. Recall that, to divide powers of ten, we subtract the exponents:

$$\left(\frac{10^6}{10^2}\right)^3 = (10^{6-2})^3$$
$$= (10^4)^3.$$

Then, to raise one power of ten to another, we multiply the exponents:

$$(10^4)^3 = 10^{4 \times 3}$$
$$= 10^{12}. \qquad \blacktriangle$$

When the exponent in the denominator is larger, as in Example 2.2**b**, we divide powers of ten and write the result in the denominator. To raise a power of ten in a denominator to another power, we multiply the exponents as in the preceding examples.

EXAMPLE 2.6   Simplify $\left(\dfrac{1}{10^2}\right)^3$.

SOLUTION   We multiply the exponents to obtain

$$\left(\frac{1}{10^2}\right)^3 = \frac{1}{10^{2 \times 3}}$$
$$= \frac{1}{10^6}.$$

Observe that

$$\left(\frac{1}{10^2}\right)^3 = \frac{1}{10 \times 10} \times \frac{1}{10 \times 10} \times \frac{1}{10 \times 10}$$

which has six factors of ten in the denominator. $\qquad \blacktriangle$

EXAMPLE 2.7 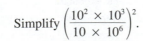   Simplify $\left(\dfrac{10^2 \times 10^3}{10 \times 10^6}\right)^2$.

SOLUTION · We start inside the parentheses by multiplying the powers of ten, in both the numerator and the denominator. To simplify the denominator, we observe that $10 = 10^1$:

$$\left(\frac{10^2 \times 10^3}{10 \times 10^6}\right)^2 = \left(\frac{10^2 \times 10^3}{10^1 \times 10^6}\right)^2$$

$$= \left(\frac{10^5}{10^7}\right)^2.$$

Then, to divide the resulting powers of ten, we subtract the smaller exponent from the larger:

$$\left(\frac{10^5}{10^7}\right)^2 = \left(\frac{1}{10^2}\right)^2.$$

Finally, we raise the power of ten in the denominator to another power by multiplying the exponents:

$$\left(\frac{1}{10^2}\right)^2 = \frac{1}{10^{2 \times 2}}$$

$$= \frac{1}{10^4}.$$

▲

EXERCISE 2.1

Simplify:

1. $10^5 \times 10^4$
2. 10×10^5
3. $\dfrac{10^8}{10^4}$
4. $\dfrac{10^5}{10^5}$

5. $\dfrac{10^5}{10^7}$
6. $\dfrac{10}{10^4}$
7. $(10^3)^4$
8. $(10^5)^2$

9. $(10^2 \times 10^2)^3$
10. $(10^6 \times 10)^2$
11. $\left(\dfrac{10^4}{10^2}\right)^3$
12. $\left(\dfrac{10}{10^5}\right)^2$

13. $\left(\dfrac{10 \times 10^8}{10^4 \times 10^3}\right)^2$
14. $\left(\dfrac{10^5 \times 10^2}{10^4 \times 10^6}\right)^3$
15. $\dfrac{(10^2 \times 10^3)^5}{(10^2 \times 10^3)^4}$
16. $\dfrac{(10 \times 10^8)^2}{(10^7 \times 10)^3}$

SECTION 2.2

Zero, Negative, and Rational Exponents

The definition of powers of ten given in Section 2.1, in terms of numbers of factors, makes sense only for powers with positive integer exponents. However, the rules for simplifying powers of ten can be used to extend the concepts to other real number exponents.

THE ZERO EXPONENT

To find a value for the power of ten with exponent zero,

$$10^0,$$

we start by dividing any two powers with the same exponent. For example, we recall that

$$\frac{10^3}{10^3} = \frac{\cancel{10} \times \cancel{10} \times \cancel{10}}{\cancel{10} \times \cancel{10} \times \cancel{10}}$$

$$= \frac{1}{1}$$

$$= 1.$$

But, if we subtract the exponents as we did in other divisions of powers of ten, we have

$$\frac{10^3}{10^3} = 10^{3-3}$$

$$= 10^0.$$

Therefore, we may conclude that

$$10^0 = 1.$$

We state this conclusion in the form of a rule.

Zero Exponent Rule: The zero power of ten equals one.

The zero exponent rule is true for any base except zero, not just the base 10.

EXAMPLE 2.8 ▶ Simplify $\frac{10^0 \times 10^3}{10^2}$.

SOLUTION   We use the zero exponent rule:

$$\frac{10^0 \times 10^3}{10^2} = \frac{1 \times 10^3}{10^2}$$

$$= \frac{10^3}{10^2}.$$

Then, we divide by subtracting the exponents to obtain

$$\frac{10^3}{10^2} = 10.$$ ▲

EXAMPLE 2.9 ▶ Simplify $\left(\frac{10^3 \times 10^9}{10^6}\right)^0$.

SOLUTION   Because the zero exponent rule is true for any base, we may simply write

$$\left(\frac{10^3 \times 10^9}{10^6}\right)^0 = 1.$$ ▲

NEGATIVE EXPONENTS

We can find a meaning for negative exponents by using the zero exponent rule. For example, suppose we multiply

$$10^3 \times 10^{-3}.$$

Multiplying by adding the exponents, we have

$$10^3 \times 10^{-3} = 10^{3+(-3)}$$

$$= 10^{3-3}$$

$$= 10^0$$

$$= 1.$$

Therefore, we can write

$$\frac{10^3 \times 10^{-3}}{10^3} = \frac{1}{10^3}.$$

But also, we can write

$$\frac{10^3 \times 10^{-3}}{10^3} = 10^{-3}.$$

Therefore, we may conclude that

$$10^{-3} = \frac{1}{10^3}.$$

We also can show that

$$10^3 = \frac{1}{10^{-3}}.$$

We can use any positive integer n as the exponent. We state these conclusions as a rule for any negative exponent $-n$ and the corresponding positive integer n.

Negative Exponent Rule:
1. The $-n$th power of ten in a numerator can be replaced with the nth power of ten in the denominator.
2. The $-n$th power of ten in a denominator can be replaced with the nth power of ten in the numerator.

The negative exponent rule is true for any base except zero.

EXAMPLE 2.10 Simplify $\frac{10^{-3}}{10^2}$.

SOLUTION Using Part 1 of the negative exponent rule, we replace 10^{-3} in the numerator with 10^3 in the denominator:

$$\frac{10^{-3}}{10^2} = \frac{1}{10^3 \times 10^2}$$

$$= \frac{1}{10^5}.$$ ▲

We may also simplify the division of powers in Example 2.10 by subtracting the exponents. Since the positive exponent is larger, we write

$$\frac{10^{-3}}{10^2} = \frac{1}{10^{2-(-3)}}$$

$$= \frac{1}{10^{2+3}}$$

$$= \frac{1}{10^5}.$$

We also know from the negative exponent rule that

$$10^{-5} = \frac{1}{10^5}.$$

Thus, we may subtract in the opposite order to obtain

$$\frac{10^{-3}}{10^2} = 10^{-3-2}$$

$$= 10^{-5}.$$

EXAMPLE 2.11 Simplify $\frac{10^3}{10^{-2}}$.

SOLUTION Using Part 2 of the negative exponent rule, we replace 10^{-2} in the denominator with 10^2 in the numerator:

$$\frac{10^3}{10^{-2}} = \frac{10^3 \times 10^2}{1}$$

$$= 10^5.$$ ▲

We may also simplify the division of powers in Example 2.11 by subtracting the exponents:

$$\frac{10^3}{10^{-2}} = 10^{3 - (-2)}$$

$$= 10^{3 + 2}$$

$$= 10^5.$$

By using negative exponents, we may have several ways to simplify expressions containing powers of ten.

EXAMPLE 2.12 Simplify $\dfrac{10^{-2} \times 10^4}{10^{-3} \times 10^{-1} \times 10^2}$.

SOLUTION   We may proceed in either of two ways.

Method 1.   We may make all negative exponents positive by using the negative exponent rule:

$$10^{-2} = \frac{1}{10^2},$$

$$\frac{1}{10^{-3}} = 10^3,$$

and

$$\frac{1}{10^{-1}} = 10^1.$$

Therefore, we have

$$\frac{10^{-2} \times 10^4}{10^{-3} \times 10^{-1} \times 10^2} = \frac{10^3 \times 10^1 \times 10^4}{10^2 \times 10^2}.$$

Then, we multiply by adding exponents, and finally divide by subtracting the resulting exponents to obtain

$$\frac{10^3 \times 10^1 \times 10^4}{10^2 \times 10^2} = \frac{10^8}{10^4}.$$

$$= 10^4.$$

Method 2.   We may multiply first, by adding the exponents in the numerator, and adding all the exponents in the denominator:

$$\frac{10^{-2} \times 10^4}{10^{-3} \times 10^{-1} \times 10^2} = \frac{10^{-2 + 4}}{10^{-3 - 1 + 2}}$$

$$= \frac{10^2}{10^{-2}}.$$

Then, we divide by subtracting the resulting exponents to obtain

$$\frac{10^2}{10^{-2}} = 10^{2 - (-2)}$$

$$= 10^{2 + 2}$$

$$= 10^4.$$ ▲

EXAMPLE 2.13  Simplify $\left(\dfrac{10^2 \times 10^{-4}}{10^{-3}}\right)^{-1}$.

SOLUTION We may proceed in any of several ways.

Method 1. We may simplify inside the parentheses first:

$$\left(\frac{10^2 \times 10^{-4}}{10^{-3}}\right)^{-1} = \left(\frac{10^{-2}}{10^{-3}}\right)^{-1}$$
$$= (10^1)^{-1}.$$

Then, we raise the resulting power to another power by multiplying the exponents:

$$(10^1)^{-1} = 10^{1 \times (-1)}$$
$$= 10^{-1}$$
$$= \frac{1}{10}.$$

Method 2. When there are negative exponents both inside and outside parentheses, it is often easiest to start by raising each power of ten to the outside power by multiplying the exponents:

$$\left(\frac{10^2 \times 10^{-4}}{10^{-3}}\right)^{-1} = \frac{10^{2 \times (-1)} \times 10^{(-4) \times (-1)}}{10^{(-3) \times (-1)}}$$
$$= \frac{10^{-2} \times 10^4}{10^3}.$$

Then, we multiply by adding exponents, and finally divide by subtracting the resulting exponents to obtain

$$\frac{10^{-2} \times 10^4}{10^3} = \frac{10^2}{10^3}$$
$$= \frac{1}{10}. \qquad \blacktriangle$$

FRACTIONAL EXPONENTS

We can also find a meaning for exponents that are fractions. For example, suppose we multiply

$$10^{\frac{1}{2}} \times 10^{\frac{1}{2}}.$$

Multiplying by adding the exponents, we have

$$10^{\frac{1}{2}} \times 10^{\frac{1}{2}} = 10^{\frac{1}{2} + \frac{1}{2}}$$
$$= 10^1$$
$$= 10.$$

But the definition of the square root of 10 is given by

$$\sqrt{10} \times \sqrt{10} = 10.$$

Therefore, we may conclude that

$$\sqrt{10} = 10^{\frac{1}{2}}.$$

More generally,

$$\sqrt{10^n} = (10^n)^{\frac{1}{2}}.$$

We state this conclusion as a rule for the fractional exponent one-half.

Fractional Exponent Rule: The square root of any power of ten can be replaced with the power of ten raised to the fractional exponent one-half.

The fractional exponent rule is true for any base for which the square root is defined. Similar fractional exponent rules are true for cube roots, fourth roots, and so on.

EXAMPLE 2.14   Simplify $\sqrt{10^4}$.

SOLUTION   Using the fractional exponent rule, we write

$$\sqrt{10^4} = (10^4)^{\frac{1}{2}}.$$

Then, we raise the power of ten to another power by multiplying the exponents:

$$(10^4)^{\frac{1}{2}} = 10^{4 \times \frac{1}{2}}$$

$$= 10^2. \qquad \blacktriangle$$

EXAMPLE 2.15   Simplify $\sqrt{10^2 \times 10^4}$.

SOLUTION   We multiply the powers of ten by adding the exponents:

$$\sqrt{10^2 \times 10^4} = \sqrt{10^{2 + 4}}$$

$$= \sqrt{10^6}.$$

Then, using the fractional exponent rule, we write

$$\sqrt{10^6} = (10^6)^{\frac{1}{2}}$$

$$= 10^{6 \times \frac{1}{2}}$$

$$= 10^3. \qquad \blacktriangle$$

SQUARE ROOTS

We can use the fractional exponent rule to derive rules for square roots. For example, we know from Example 2.15 that

$$\sqrt{10^2 \times 10^4} = 10^3.$$

We can also use the fractional exponent rule to write

$$\sqrt{10^2} \times \sqrt{10^4} = (10^2)^{\frac{1}{2}} \times (10^4)^{\frac{1}{2}}$$

$$= 10^{2 \times \frac{1}{2}} \times 10^{4 \times \frac{1}{2}}$$

$$= 10^1 \times 10^2$$

$$= 10^3.$$

Therefore, we may conclude that

$$\sqrt{10^2 \times 10^4} = \sqrt{10^2} \times \sqrt{10^4}.$$

Similarly, we can show, for example, that

$$\sqrt{\frac{10^4}{10^2}} = \frac{\sqrt{10^4}}{\sqrt{10^2}}.$$

We state these conclusions as a rule for square roots.

> **Rule for Square Roots:**
> 1. The square root of a product of powers of ten is equal to the product of the square roots of the powers of ten.
> 2. The square root of a quotient of powers of ten is equal to the quotient of the square roots of the powers of ten.

The rule for square roots is true for any base for which the square roots are defined and is also true for higher order roots.

EXAMPLE 2.16   Simplify $\sqrt{10^4 \times 10^3}$.

SOLUTION We can add the exponents to obtain

$$\sqrt{10^4 \times 10^3} = \sqrt{10^7}.$$

However, the fractional exponent rule will result in an integer exponent only if the power of ten has an *even* exponent. Therefore, we rewrite this result as

$$\sqrt{10^7} = \sqrt{10^6 \times 10^1}.$$

Then, using Part 1 of the rule for square roots, we have

$$\sqrt{10^6 \times 10^1} = \sqrt{10^6} \times \sqrt{10^1}$$
$$= 10^3 \times \sqrt{10}. \quad \blacktriangle$$

EXAMPLE 2.17   Simplify $\sqrt{16 \times 10^6}$.

SOLUTION Because the rule for square roots is true for any base, we may write

$$\sqrt{16 \times 10^6} = \sqrt{16} \times \sqrt{10^6}$$
$$= 4 \times 10^3. \quad \blacktriangle$$

EXAMPLE 2.18   Simplify $\sqrt{14.6 \times 10^9}$.

SOLUTION To use the fractional exponent rule, we need an even exponent. Therefore, we adjust the power of ten by writing

$$\sqrt{14.6 \times 10^9} = \sqrt{14.6 \times 10 \times 10^8}.$$

Then, we multiply 14.6 by 10 to obtain

$$\sqrt{14.6 \times 10 \times 10^8} = \sqrt{146 \times 10^8}.$$

Now, we can apply the rule for square roots, and use a calculator to find the square root of 146:

$$\sqrt{146 \times 10^8} = \sqrt{146} \times \sqrt{10^8}$$
$$= 12.1 \times 10^4$$

to three significant digits. $\quad \blacktriangle$

EXERCISE 2.2

Simplify:

1. $\dfrac{10^6 \times 10^0}{10^4}$

2. $\dfrac{10}{10^4 \times 10^0}$

3. $\left(\dfrac{10^4 \times 10^5}{10^6}\right)^0$

4. $(10^2 \times 10^3 \times 10)^0$

5. $\dfrac{10^5}{10^{-4}}$

6. $\dfrac{10}{10^{-5}}$

7. $\dfrac{10^{-1}}{10^3}$ 8. $\dfrac{10^{-3}}{10^3}$

9. $\dfrac{10^{-3}}{10^{-2}}$ 10. $\dfrac{10^0}{10^{-6}}$

11. $\dfrac{10^{-1} \times 10^2}{10^4 \times 10^{-3} \times 10^{-2}}$ 12. $\dfrac{10^{-3} \times 10^3}{10^{-6} \times 10 \times 10^{-4}}$

13. $\dfrac{10^4 \times 10^0 \times 10^{-4}}{10^{-8} \times 10^{-2}}$ 14. $\dfrac{10^{-6} \times 10^3 \times 10^6}{10^3 \times 10^0}$

15. $\left(\dfrac{10^3 \times 10^{-6}}{10^{-9}}\right)^{-1}$ 16. $\left(\dfrac{10^{-2} \times 10^{-4}}{10^2}\right)^{-1}$

17. $\sqrt{10^6}$ 18. $\sqrt{10^{-12}}$

19. $\sqrt{10^4 \times 10^6}$ 20. $\sqrt{10^{-4} \times 10^6}$

21. $\sqrt{10^3 \times 10^9}$ 22. $\sqrt{10^{-3} \times 10^9}$

23. $\sqrt{10^5 \times 10^6}$ 24. $\sqrt{10^{12} \times 10^3}$

25. $\sqrt{10^{-12} \times 10^{-3}}$ 26. $\sqrt{10^{-12} \times 10^3}$

27. $\sqrt{25 \times 10^8}$ 28. $\sqrt{144 \times 10^{-10}}$

29. $\sqrt{12.1 \times 10^9}$ 30. $\sqrt{0.65 \times 10^7}$

31. $\sqrt{25 \times 10^{-7}}$ 32. $\sqrt{2.5 \times 10^{-9}}$

SECTION	# Scientific Notation
2.3	

Powers of ten offer convenient ways to write numbers, especially numbers that are very large and numbers that are very close to zero. One way to write numbers by using powers of ten is called **scientific notation**. A positive number is written in scientific notation as the product of a number between 1 and 10 in decimal form and a power of ten. These are examples of numbers written in scientific notation:

$$4.39 \times 10^2, \quad 6.05 \times 10^{12}, \quad 7.34 \times 10^{-2}, \quad 1.92 \times 10^{-7}, \quad 5.03 \times 10^0.$$

Observe that the decimal point in the first part of the number is always placed after the first nonzero digit. The exponent of the power of ten is an integer that can be positive, negative, or zero.

Any number written in scientific notation can be rewritten in ordinary notation. If the exponent is a positive integer, we multiply by the number of factors of ten indicated by the exponent. Each multiplication by ten moves the decimal point one place to the right.

EXAMPLE 2.19   Write the number in ordinary notation:

a. 4.39×10^2 **b.** 6.05×10^{12}

SOLUTIONS   We multiply by the number of factors of ten indicated by the exponent.

a. To rewrite 4.39×10^2, we multiply by two factors of ten. This process moves the decimal point two places to the right:

$$4.39 \times 10^2 = 439.$$

b. To rewrite 6.05×10^{12}, we multiply by twelve factors of ten. This process moves the decimal point twelve places to the right. We count the second digit 0, and the third digit 5, and we must supply ten zeros:

$$6.05 \times 10^{12} = 6{,}050{,}000{,}000{,}000. \qquad \blacktriangle$$

When the exponent is a negative integer, we use the negative exponent rule given in Section 2.2. Multiplying by the $-n$th power of ten is equivalent to dividing by the corresponding nth power. For example,

$$7.34 \times 10^{-2} = \frac{7.34}{10^2}.$$

Therefore, we divide by the number of factors of ten indicated by the absolute value of the exponent. Each division by ten moves the decimal point one place to the left.

EXAMPLE 2.20 Write the number in ordinary notation:

a. 7.34×10^{-2} **b.** 1.92×10^{-7}

SOLUTIONS We divide by the number of factors of ten indicated by the absolute value of the exponent.

a. To rewrite 7.34×10^{-2}, we divide by two factors of ten. This process moves the decimal point two places to the left. We count the first digit 7, and we must supply a zero for the additional place:

$$7.34 \times 10^{-2} = 0.0734.$$

We have also placed a zero in front of the decimal point for clarity.

b. To rewrite 1.92×10^{-7}, we divide by seven factors of ten. This process moves the decimal point seven places to the left. We count the first digit 1, and we must supply zeros for the six additional places:

$$1.92 \times 10^{-7} = 0.000000192.$$

We have also placed a zero in front of the decimal point. ▲

EXAMPLE 2.21 Write 5.03×10^0 in ordinary notation.

SOLUTION The exponent is 0. But we know from the zero exponent rule given in Section 2.2 that

$$10^0 = 1.$$

Therefore, the decimal point remains after the first digit:

$$5.03 \times 10^0 = 5.03 \times 1$$
$$= 5.03. \qquad ▲$$

Any number written in ordinary notation can be rewritten in scientific notation. We use this mark $_\wedge$, called a caret, to mark the new location of the decimal point after the first nonzero digit. Then, we count the number of places from the caret to the location of the decimal point in ordinary notation. This number is the absolute value of the exponent of the power of ten. We can determine the sign of the exponent by following this reasoning:

If we have counted *to the right from the caret*, then the exponent is a *positive* integer because we would *multiply* to get to the position of the decimal point in ordinary notation.

If we have counted *to the left from the caret*, then the exponent is a *negative* integer because we would *divide* to get to the position of the decimal point in ordinary notation.

EXAMPLE 2.22 Write the number in scientific notation:

a. 2970 **b.** 50,400,000

SOLUTIONS We mark the new location of the decimal point with a caret, and then count *to the right from the caret*.

a. To rewrite 2970, we place a caret after the 2:

$$2_\wedge 970.$$

Because the decimal point does not appear in the ordinary notation, we assume its position is at the end of the number. We count three places from the caret to the end of the number. Therefore, the exponent is 3:

$$2970 = 2.97 \times 10^3.$$

We have dropped the trailing zero in the decimal part of the number.

b. To rewrite 50,400,000, we place a caret after the 5:

$$5_\wedge0,400,000.$$

We count seven places from the caret to the end of the number. Therefore, the exponent is 7:

$$50,400,000 = 5.04 \times 10^7.$$

We have dropped the trailing zeros.

EXAMPLE 2.23   Write the number in scientific notation:

a. 0.0323   **b.** 0.0000000905

SOLUTIONS   We mark the new location of the decimal point with a caret, and then count *to the left from the caret*.

a. To rewrite 0.0323 we place a caret after the first 3:

$$0.03_\wedge23.$$

We count two places to the left from the caret to the decimal point. Therefore, the exponent is −2:

$$0.0323 = 3.23 \times 10^{-2}.$$

We have dropped the preceding zeros.

b. To rewrite 0.0000000905 we place a caret after the 9:

$$0.00000009_\wedge05.$$

We count eight places to the left from the caret to the decimal point. Observe that we count the 9, but we do not count the zero following the 9 or the zero preceding the decimal point. The exponent is −8:

$$0.0000000905 = 9.05 \times 10^{-8}.$$

We have dropped the preceding zeros.   ▲

    To multiply two numbers written in scientific notation, we use the rule for multiplication of powers of ten given in Section 2.1. We multiply the decimal parts of the numbers, but we add the exponents of the powers of ten.

EXAMPLE 2.24   Multiply $(6.43 \times 10^8)(4.91 \times 10^9)$, and write the result in scientific notation.

SOLUTION   We multiply the decimal parts of the numbers to obtain

$$(6.43)(4.91) = 31.6$$

to three significant digits. Then, we find the power of ten by adding the exponents:

$$(10^8)(10^9) = 10^{17}.$$

Therefore,

$$(6.43 \times 10^8)(4.91 \times 10^9) = 31.6 \times 10^{17}.$$

This result is not written in scientific notation. We write 31.6 in scientific notation, and combine the resulting powers of ten:

$$31.6 \times 10^{17} = 3_\wedge1.6 \times 10^{17}$$

$$= 3.16 \times 10^1 \times 10^{17}$$

$$= 3.16 \times 10^{18}$$

in scientific notation.   ▲

    To divide two numbers written in scientific notation, we use the rule for division of powers of ten given in Section 2.1. We divide the decimal parts of the numbers, but we subtract the exponents of the powers of ten.

EXAMPLE 2.25 Divide $\dfrac{1.54 \times 10^2}{5 \times 10^{-5}}$, and write the result in scientific notation.

SOLUTION We divide the decimal parts of the numbers to obtain

$$\frac{1.54}{5} = 0.308.$$

Then, we find the power of ten by subtracting the exponents:

$$\frac{10^2}{10^{-5}} = 10^7.$$

Therefore,

$$\frac{1.54 \times 10^2}{5 \times 10^{-5}} = 0.308 \times 10^7.$$

This result is not written in scientific notation. We write 0.308 in scientific notation, and combine the resulting powers of ten:

$$0.308 \times 10^7 = 0.3{_\wedge}08 \times 10^7$$
$$= 3.08 \times 10^{-1} \times 10^7$$
$$= 3.08 \times 10^6$$

in scientific notation. ▲

EXAMPLE 2.26 The mass of a proton is 1836 times the mass of an electron. If the mass of an electron is 9.11×10^{-28} grams, what is the mass of a proton in scientific notation?

SOLUTION We observe that, in scientific notation,

$$1836 = 1.836 \times 10^3.$$

We multiply the numbers in scientific notation:

$$(1.836 \times 10^3)(9.11 \times 10^{-28}) = 16.7 \times 10^{-25}.$$

Then, we write the result in scientific notation:

$$16.7 \times 10^{-25} = 1.67 \times 10^1 \times 10^{-25}$$
$$= 1.67 \times 10^{-24}.$$

Therefore, the mass of a proton is 1.67×10^{-24} grams. ▲

**EXERCISE
2.3**

Write the number in ordinary notation:

1. 2.54×10^1 **2.** 6.19×10^3 **3.** 8.07×10^8 **4.** 1.05×10^{10}

5. 4.64×10^{-1} **6.** 7.04×10^{-3} **7.** 3.02×10^{-5} **8.** 4.16×10^{-8}

9. 1.29×10^0 **10.** 5.02×10^0

Write the number in scientific notation:

11. 575   **12.** 87,400   **13.** 2,010,000   **14.** 603,000,000,000

15. 0.643   **16.** 0.000802   **17.** 0.00000404   **18.** 0.000000000269

19. 2.34   **20.** 1.00

Multiply or divide as indicated, and write the result in scientific notation:

21. $(5.12 \times 10^3)(3.73 \times 10^4)$ **22.** $(6.09 \times 10^8)(4.14 \times 10^6)$

23. $(2.19 \times 10^{-2})(1.86 \times 10^6)$ **24.** $(6.68 \times 10^8)(3.05 \times 10^{-4})$

25. $(3.93 \times 10^{-17})(4.07 \times 10^4)$

26. $(8.52 \times 10^{-8})(7.05 \times 10^{-9})$

27. $\dfrac{7.56 \times 10^{12}}{8.58 \times 10^3}$

28. $\dfrac{3.87 \times 10^{20}}{9.12 \times 10^8}$

29. $\dfrac{4.39 \times 10^6}{1.06 \times 10^{-6}}$

30. $\dfrac{2.03 \times 10^4}{6.90 \times 10^4}$

31. $\dfrac{1.40 \times 10^3}{3.94 \times 10^{12}}$

32. $\dfrac{3.79 \times 10^{-20}}{7.18 \times 10^{-10}}$

Write the answers to Exercises 33–36 in scientific notation.

33. The circumference of the sun is approximately 109 times the circumference of the earth. If the circumference of the earth is approximately 24,800 miles, what is the circumference of the sun?

34. The mass of the sun is approximately 332,000 times the mass of the earth. If the mass of the earth is approximately 5.97×10^{24} kilograms, what is the mass of the sun?

35. A unit of atomic mass is approximately 1.66×10^{-24} gram. If there are approximately 238 units of atomic mass in a uranium atom, what is the mass in grams of a uranium atom?

36. The electric charge of an electron is approximately 1.60×10^{-19} coulomb. If there are 92 electrons in a uranium atom, what is the total charge in coulombs of the electrons in a uranium atom?

Write the answers to Exercises 37 and 38 in ordinary notation.

37. It is approximately 9.30×10^7 miles from the sun to the earth, and light travels at a rate of approximately 186,000 miles per second. How many seconds does it take for light from the sun to reach the earth? (Divide the distance by the rate.)

38. The total charge in coulombs of the electrons in a radium atom is approximately 1.41×10^{-17} coulomb. The charge per electron is approximately 1.60×10^{-19} coulomb. How many electrons are there in a radium atom? (Divide the total charge by the charge per electron.)

SECTION 2.4

2.4 Engineering Notation

Any quantity that is measured in any way, is measured in terms of some **unit**. If we measured the width of this page, we might use inches. If we measured the width of a room, we might use feet. If we measured the distance to the next town, we might use miles. Inches, feet, and miles are examples of units.

You should be familiar with some units of the metric system. For example, meters (m), grams (g), and seconds (s), are used to measure length, weight, and time, respectively. The letters in parentheses are the symbols used for the units. You are familiar with many other units. For example, the **watt** (W) is a unit of power that you have seen used to rate light bulbs. The watt is named for James Watt (1736–1819), the Scottish engineer who invented the steam engine. The **hertz** (Hz) is the unit of frequency that you use to tune in stations on your radio. The hertz is named for Heinrich Rudolf Hertz (1857–1894), a German physicist. In Exercise 2.3, you saw the **coulomb** (C), which is named for Charles Augustin de Coulomb (1736–1806), a French physicist, and is used to measure electrical charge.

Using the metric system, we would measure the width of this page either in millimeters or in centimeters. We would measure the width of the room in meters, and we would measure the distance to the next town in kilometers. In each of these metric units, the basic unit is *meters*. Each of the modifiers, milli-, centi-, and kilo-, is a **prefix** that changes the basic unit to a smaller or a larger unit.

Engineering notation provides a way to write quantities by using certain prefixes. The prefixes used in engineering notation represent powers of ten where the exponents are multiples of three. Thus, milli-, which represents 10^{-3}, and kilo-, which represents 10^3, are used in engineering notation. However, centi-, which represents 10^{-2}, is not used in engineering notation. Prefixes commonly used in engineering notation are listed in the following chart:

Prefix	Symbol	Power of Ten
giga	G	10^{9}
mega	M	10^{6}
kilo	k	10^{3}
milli	m	10^{-3}
micro	μ	10^{-6}
nano	n	10^{-9}
pico	p	10^{-12}

(The symbol μ for micro- is the Greek letter mu.)

In engineering notation, a prefix is always used with a unit. We say 10 millimeters, or milligrams, or milliseconds, or even milliunits, but not 10 millis. ("Kilo" is sometimes used in common language to mean kilogram, but not in engineering notation.) We may, however, have a unit without a prefix.

EXAMPLE 2.27

Write the quantity without a prefix:

a. 85 kHz

b. 1.05 MW

SOLUTIONS We replace each prefix symbol by the power of ten given in the chart.

a. The prefix k-, for kilo-, means 10^{3}. Therefore,

$$85 \ \text{kHz} = 85 \times 10^{3} \ \text{Hz}$$

$$= 85{,}000 \ \text{Hz.}$$

b. The prefix M-, for mega-, means 10^{6}. Therefore,

$$1.05 \ \text{MW} = 1.05 \times 10^{6} \ \text{W}$$

$$= 1{,}050{,}000 \ \text{W.}$$ ▲

EXAMPLE 2.28

Write the quantity without a prefix:

a. 115 mg

b. 0.965 ns

SOLUTIONS We replace each prefix symbol by the power of ten given in the chart.

a. The prefix m-, for milli-, means 10^{-3}. Therefore,

$$115 \ \text{mg} = 115 \times 10^{-3} \ \text{g}$$

$$= 0.115 \ \text{g.}$$

b. The prefix n-, for nano-, means 10^{-9}. Therefore,

$$0.965 \ \text{ns} = 0.965 \times 10^{-9} \ \text{s}$$

$$= 0.000000000965 \ \text{s.}$$ ▲

In scientific notation, the decimal point is always placed after the first nonzero digit. In engineering notation, we cannot always place the decimal point directly after the first nonzero digit. We can, however, always place the decimal point after the first nonzero digit or one of the two following digits. We place the decimal point so that the power of ten will be a multiple of three.

EXAMPLE 2.29

Write the quantity in engineering notation with the decimal point after the first nonzero digit or one of the two following digits:

a. 55,200 g

b. 302,000,000,000 Hz

SOLUTIONS We mark the new location of the decimal point with a caret, and then count to the right from the caret.

a. To rewrite 55,200 g, we place a caret after the second 5, which is three places from the end of the number:

$$55{_\land}200 \text{ g}.$$

Therefore, the exponent is 3 and the prefix is kilo-:

$$55,200 \text{ g} = 55.2 \times 10^3 \text{ g}$$
$$= 55.2 \text{ kg}.$$

b. To rewrite 302,000,000,000 Hz, we place a carat after the 2, which is nine places from the end of the number:

$$302{_\land}000,000,000 \text{ Hz}.$$

Therefore, the exponent is 9 and the prefix is giga-:

$$302,000,000,000 \text{ Hz} = 302 \times 10^9 \text{ Hz}$$
$$= 302 \text{ GHz}. \qquad \blacktriangle$$

EXAMPLE 2.30   Write the quantity in engineering notation with the decimal point after the first nonzero digit or one of the two following digits:

a. 0.11 W   **b.** 0.00000125 s

SOLUTIONS We mark the new location of the decimal point with a caret, and then count to the left from the caret.

a. To rewrite 0.11 W, we must place a caret three places after the decimal point. Because there are less than three digits after the decimal point, we write a trailing zero. Then, we place the caret after the zero:

$$0.110{_\land} \text{ W}.$$

There are three places to the left from the caret to the decimal point. Therefore, the exponent is −3 and the prefix is milli-:

$$0.11 \text{ W} = 110 \times 10^{-3} \text{ W}$$
$$= 110 \text{ mW}.$$

b. To rewrite 0.00000125 s, we place the caret after the 1, which is six places from the decimal point:

$$0.000001{_\land}25 \text{ s}.$$

There are six places to the left from the caret to the decimal point. Therefore, the exponent is −6 and the prefix is micro-:

$$0.00000125 \text{ s} = 1.25 \times 10^{-6} \text{ s}$$
$$= 1.25 \text{ µs}. \qquad \blacktriangle$$

We can change numbers given in scientific notation to engineering notation by adjusting the exponent to a multiple of three, and adjusting the location of the decimal point to correspond.

EXAMPLE 2.31   Write the quantity in engineering notation with the decimal point after the first nonzero digit or one of the two following digits:

a. 9.2×10^7 W   **b.** 5.7×10^{11} Hz

SOLUTIONS We adjust the exponent to a multiple of three.

a. To rewrite 9.2×10^7 W, we adjust the exponent to 6 by replacing 10^7 with 10×10^6:

$$9.2 \times 10^7 \text{ W} = 9.2 \times 10 \times 10^6 \text{ W}$$

$$= 92 \times 10^6 \text{ W}$$

$$= 92 \text{ MW}.$$

b. To rewrite 5.7×10^{11} Hz, we adjust the exponent to 9 by replacing 10^{11} with $10^2 \times 10^9$:

$$5.7 \times 10^{11} \text{ Hz} = 5.7 \times 10^2 \times 10^9 \text{ Hz}$$

$$= 570 \times 10^9 \text{ Hz}$$

$$= 570 \text{ GHz}. \qquad \blacktriangle$$

EXAMPLE 2.32 ▶ Write the quantity in engineering notation with the decimal point after the first nonzero digit or one of the two following digits:

a. 4.66×10^{-5} W
b. 3.25×10^{-7} s

SOLUTIONS We adjust the exponent to a multiple of three.

a. To rewrite 4.66×10^{-5} W, we adjust the exponent to -6 by replacing 10^{-5} with 10×10^{-6}:

$$4.66 \times 10^{-5} \text{ W} = 4.66 \times 10 \times 10^{-6} \text{ W}$$

$$= 46.6 \times 10^{-6} \text{ W}$$

$$= 46.6 \text{ } \mu\text{W}.$$

Observe that we always adjust by using a power of ten with a positive exponent so the decimal point moves to the right.

b. To rewrite 3.25×10^{-7} s, we adjust the exponent to -9 by replacing 10^{-7} with $10^2 \times 10^{-9}$:

$$3.25 \times 10^{-7} \text{ s} = 3.25 \times 10^2 \times 10^{-9} \text{ s}$$

$$= 325 \times 10^{-9} \text{ s}$$

$$= 325 \text{ ns}. \qquad \blacktriangle$$

When we change from one unit to another, we change the number of items measured. For example, if we have measured 36 inches, we also can say that we have measured 3 feet or 1 yard. We observe that, as the unit increases in size, the number of items decreases. When we change a prefix in engineering notation, we may determine the new number by following this reasoning:

When the unit is changed by a prefix to a *larger unit*, the *number is decreased*.
When the unit is changed by a prefix to a *smaller unit*, the *number is increased*.

EXAMPLE 2.33 ▶ Convert the quantity to the given prefix:

a. 282 kHz to MHz
b. 5.67 μs to ms

The new prefix changes the unit to a *larger unit* so the *number is decreased.*

SOLUTIONS **a.** To change from 282 kHz to MHz, the exponent is increased by 3. Therefore, we decrease the number by three places:

$$_\wedge 282 \text{ kHz}.$$

We replace 282 with 0.282×10^3:

$$282 \text{ kHz} = 0.282 \times 10^3 \text{ kHz}$$
$$= 0.282 \times 10^3 \times 10^3 \text{ Hz}$$
$$= 0.282 \times 10^6 \text{ Hz}$$
$$= 0.282 \text{ MHz}.$$

b. We recall that -3 is *larger* than -6. Therefore, to change from 5.67 μs to ms, the exponent is *increased* by 3. Thus, we decrease the number by three places:

$$_\wedge 005.67 \text{ μs}.$$

We replace 5.67 with 0.00567×10^3:

$$5.67 \text{ μs} = 0.00567 \times 10^3 \text{ μs}$$
$$= 0.00567 \times 10^3 \times 10^{-6} \text{ s}$$
$$= 0.00567 \times 10^{-3} \text{ s}$$
$$= 0.00567 \text{ ms}.$$

We observe that to change to a larger unit when the prefix has a negative exponent, we can simply split the exponent into two parts:

$$5.67 \text{ μs} = 5.67 \times 10^{-6} \text{ s}$$
$$= 5.67 \times 10^{-3} \times 10^{-3} \text{ s}.$$

Then, we replace 5.67×10^{-3} with 0.00567:

$$5.67 \text{ μs} = 0.00567 \times 10^{-3} \text{ s}$$
$$= 0.00567 \text{ ms}. \qquad \blacktriangle$$

EXAMPLE 2.34 Convert the quantity to the given prefix:

a. 762 GHz to MHz **b.** 0.012 ms to ns

SOLUTIONS The new prefix changes the unit to a *smaller unit*, so the *number is increased*.

a. To change from 762 GHz to MHz, the exponent is decreased by 3. Therefore, we increase the number by three places:

$$762.000_\wedge \text{ GHz}.$$

We replace 762 with $762{,}000 \times 10^{-3}$:

$$762 \text{ GHz} = 762{,}000 \times 10^{-3} \text{ GHz}$$
$$= 762{,}000 \times 10^{-3} \times 10^9 \text{ Hz}$$
$$= 762{,}000 \times 10^6 \text{ Hz}$$
$$= 762{,}000 \text{ MHz}.$$

We observe that to change to a smaller unit when the prefix has a positive exponent, we can simply split the exponent into two parts. Thus,

$$762 \text{ GHz} = 762 \times 10^9 \text{ Hz}$$
$$= 762 \times 10^3 \times 10^6 \text{ Hz}.$$

Then, we replace 762×10^3 with 762,000:

$$762 \text{ GHz} = 762{,}000 \times 10^6 \text{ Hz}$$
$$= 762{,}000 \text{ MHz}.$$

b. To change from 0.012 ms to ns, the exponent is decreased by 6. Therefore, we increase the number by six places:

$$0.012000_\wedge \text{ ms.}$$

We replace 0.012 by $12,000 \times 10^{-6}$:

$$0.012 \text{ ms} = 12,000 \times 10^{-6} \text{ ms}$$
$$= 12,000 \times 10^{-6} \times 10^{-3} \text{ s}$$
$$= 12,000 \times 10^{-9} \text{ s}$$
$$= 12,000 \text{ ns.} \qquad \blacktriangle$$

EXAMPLE 2.35 ▶ The speed of light is approximately 3×10^8 meters per second. It takes approximately 2.59×10^3 seconds for light from the planet Jupiter to reach the earth. What is the distance from Jupiter to the earth in kilometers? in gigameters?

SOLUTION We multiply the speed of light by the time the light travels:

$$\left(3 \times 10^8 \ \frac{\text{m}}{\text{s}}\right)\left(2.59 \times 10^3 \text{ s}\right) = 7.77 \times 10^{11} \text{ m.}$$

Converting the unit to kilometers, we have

$$7.77 \times 10^{11} \text{ m} = 7.77 \times 10^8 \times 10^3 \text{ m}$$
$$= 777,000,000 \times 10^3 \text{ m}$$
$$= 777,000,000 \text{ km.}$$

To convert to gigameters, we write

$$7.77 \times 10^{11} \text{ m} = 7.77 \times 10^2 \times 10^9 \text{ m}$$
$$= 777 \times 10^9 \text{ m}$$
$$= 777 \text{ Gm.} \qquad \blacktriangle$$

EXERCISE 2.4

Write the quantity without a prefix:

1. 16 MHz	**2.** 2.5 kHz	**3.** 180 GHz	**4.** 0.702 MHz
5. 150 ms	**6.** 693 ns	**7.** 12.2 μs	**8.** 1020 ps
9. 23.8 kW	**10.** 74.5 mg	**11.** 1.73 pg	**12.** 6.05 GW

Write the quantity in engineering notation with the decimal point after the first nonzero digit or one of the two following digits:

13. 272,000 Hz	**14.** 9420 Hz
15. 69,000,000,000 Hz	**16.** 350,000 Hz
17. 0.492 s	**18.** 0.000189 s
19. 0.0000622 s	**20.** 0.00000000082 s
21. 0.00000000585 s	**22.** 4,650,000,000 Hz
23. 7,960,000 W	**24.** 0.00188 W
25. 6.85×10^5 Hz	**26.** 7.67×10^{10} Hz
27. 2.5×10^{-4} s	**28.** 1.55×10^{-8} s
29. 3.8×10^4 W	**30.** 8.93×10^8 W
31. 5.44×10^{-11} s	**32.** 9.08×10^{-2} s

Convert the quantity to the given prefix:

33. 466 kW to MW   34. 7300 MW to GW

35. 7.08 μs to ms   36. 30.4 ns to μs

37. 965 MHz to kHz   38. 7.71 GHz to MHz

39. 0.0474 ms to μs   40. 0.00808 μs to ns

41. 0.785 μs to ps   42. 0.018 GW to kW

43. 722 ns to ms   44. 6280 kHz to GHz

45. The speed of light is approximately 3×10^8 meters per second. It takes approximately 500 seconds for light from the sun to reach the earth. What is the distance from the sun to the earth in kilometers? in gigameters?

46. The prefix tera- (T) represents 10^{12}. If it takes 250×10^5 seconds for light from a certain star to reach the earth, what is the distance from the star to the earth in terameters? If the prefix peta- (P) represents 10^{15}, what is the distance from the star to the earth in petameters? (Use the speed of light given in Exercise 45.)

47. The electric charge of an electron is approximately 1.60×10^{-19} coulomb (C). There are 29 electrons in a copper atom. What is the total charge of the electrons in a copper atom in picocoulombs? If the prefix femto- (f) represents 10^{-15}, what is the total charge of the electrons in a copper atom in femtocoulombs? If the prefix atto- (a) represents 10^{-18}, what is the total charge of the electrons in a copper atom in attocoulombs?

48. There are 99 electrons in an einsteinium atom. What is the total charge of the electrons in an einsteinium atom in femtocoulombs? in attocoulombs? (Use the definitions of the prefixes femto- and atto- and the charge of an electron given in Exercise 47.)

☐ **Self-Test** ☐

1. Convert the quantity to the measure in engineering notation with the given prefix:

   a. 0.0000787 s to μs

   b. 150×10^8 Hz to GHz

   1a. _____

   1b. _____

2. Write the number in scientific notation:

   a. 29,700,000

   b. 0.0000599

   2a. _____

   2b. _____

3. Simplify $\dfrac{(10 \times 10^4)^3}{(10 \times 10^3)^4}$.

   3. _____

4. Simplify $\dfrac{10^3 \times 10^{-3} \times 10^3}{10^{-3} \times 10^{-1}}$.

   4. _____

5. A computer can do a certain kind of operation in 455 ns. How much time does it take to do 600 such operations? Write the result in appropriate engineering notation.

   5. _____

UNIT 3

Algebraic Expressions

Introduction

Algebra and arithmetic are closely related. Algebra might be said to be a generalization of arithmetic. In the preceding units, you did arithmetic using numbers and arithmetic operations. In this unit, you will learn how letters can replace numbers. You will learn how numbers, letters, and operations are combined to form algebraic expressions. Then you will learn how algebraic expressions are combined and simplified. You will learn an important property called the distributive property, and its role in simplifying algebraic expressions. You will use the distributive property to add and subtract algebraic expressions, and to multiply and divide some basic types of algebraic expressions. You will also extend the rules for exponents and roots to simplify other basic types of algebraic expressions.

OBJECTIVES

When you have finished this unit you should be able to:

1. Evaluate algebraic expressions for given values of the variables.

2. Multiply algebraic expressions by constants, factor out constants from algebraic expressions, and combine like terms of algebraic expressions.

3. Simplify general linear expressions.

4. Simplify algebraic expressions multiplied or divided by a monomial, and algebraic expressions involving square roots.

SECTION 3.1

Evaluating Algebraic Expressions

The basic component of algebra is the **algebraic expression**. Algebraic expressions use combinations of numbers, letters, and arithmetic operations to represent numbers. Letters used in algebraic expressions are called **variables** because they may vary among numerical values. Numbers used in algebraic expressions are called **constants**. These are some examples of algebraic expressions:

$$x + 7, \quad x + y, \quad 2x - y, \quad x + xy - y, \quad x^2 + y^2, \quad x^2 y^{-2}.$$

The variables are letters such as x and y. The constants are numbers such as 7 and 2.

Multiplication involving variables is commonly indicated by two variables, or a constant and a variable, written next to each other. For example,

$$2x \text{ means } 2 \text{ times } x$$

and

$$xy \text{ means } x \text{ times } y.$$

Of course we cannot use this notation for two numbers because, for example, 23 means twenty-three, not 2 times 3. We use parentheses to indicate the product of two numbers. For example,

$$(2)(3) \text{ and } 2(3) \text{ mean } 2 \text{ times } 3.$$

We may also use a dot to indicate multiplication. For example,

$$2 \cdot 3 \text{ means } 2 \text{ times } 3.$$

This notation must be used with caution because it is easily confused with 2.3, where the dot is a decimal point.

A **term** of an algebraic expression consists of a product of constants and variables. These expressions have just one term:

$$7, \quad x, \quad 2x, \quad xy, \quad x^2, \quad x^2y^{-2}.$$

A constant by itself, or a variable by itself, is an expression with one term. These expressions have two terms:

$$x + 7, \quad x + y, \quad 2x - y, \quad x^2 + y^2.$$

This expression has three terms:

$$x + xy - y.$$

When all of the variables in an algebraic expression have positive integer exponents, the expression is a **polynomial**. If the expression has just one term, it is a **monomial**. All the expressions in the first group of examples above are monomials, except the expression x^2y^{-2}, because of the negative exponent. A polynomial with two terms is a **binomial**. The expressions in the second group are binomials. A polynomial with three terms is a **trinomial**. The three-term expression above is a trinomial.

Algebraic expressions represent numbers. When specific numerical values are assigned to the variables, we can calculate the number represented by the algebraic expression. The phrase **evaluate the expression** means this process of finding the number.

EXAMPLE 3.1 Evaluate $2x - 3$ for $x = 4$.

SOLUTION We replace x by 4 in the expression:

$$2x - 3 = 2(4) - 3.$$

Then, we calculate the number represented by the expression:

$$2(4) - 3 = 8 - 3$$
$$= 5. \qquad \blacktriangle$$

When we use a given number to replace a variable, we **substitute** the number for the variable.

EXAMPLE 3.2 Evaluate $5x + 5y$ for $x = -3$ and $y = 5$.

SOLUTION We substitute -3 for x and 5 for y in the expression:

$$5x + 5y = 5(-3) + 5(5).$$

Then, we calculate the value of the expression to obtain

$$5(-3) + 5(5) = -15 + 25$$
$$= 10. \qquad \blacktriangle$$

EXAMPLE 3.3 Evaluate $5xy^2$ for $x = 0.5$ and $y = 3$.

SOLUTION We substitute 0.5 for x and 3 for y in the expression:

$$5xy^2 = 5(0.5)(3)^2.$$

Then, we calculate the value of the expression to obtain

$$5(0.5)(3)^2 = 5(0.5)(9)$$
$$= 22.5. \qquad \blacktriangle$$

We must be careful to recognize the difference between $-y^2$ and $(-y)^2$. For example, if $y = 3$, then for $-y^2$, we have

$$-y^2 = -3^2.$$

The exponent applies only to 3, and so we square only the 3:

$$-3^2 = -(3)(3)$$
$$= -9.$$

However, if $y = 3$, then for $(-y)^2$, we have

$$(-y)^2 = (-3)^2.$$

The exponent applies to the entire number, -3 with its sign, and so we square the entire number:

$$(-3)^2 = (-3)(-3)$$
$$= 9.$$

EXAMPLE 3.4 ▶ Evaluate $2xy^2$ for $x = 0.25$ and $y = -4$.

SOLUTION We substitute 0.25 for x and -4 for y in the expression:

$$2xy^2 = 2(0.25)(-4)^2.$$

We observe that the number replacing y is the entire number, -4 with its sign. As the parentheses show, we square the entire number, -4 with its sign:

$$(-4)^2 = 16.$$

Therefore, we calculate the value of the expression to obtain:

$$2(0.25)(-4)^2 = 2(0.25)(16)$$
$$= 8. \qquad \blacktriangle$$

When an expression describes a specific application, we often choose variables that describe the application. We can use both lowercase (small) and uppercase (capital) letters. For example, we might use the variables D and T in an expression about distance and time. The constants substituted for the variables might be quantities that include units.

EXAMPLE 3.5 ▶ Evaluate $\dfrac{D}{T}$ for $D = 20$ mi (miles) and $T = 0.44$ h (hour).

SOLUTION We substitute 20 mi for D and 0.44 h for T, and calculate the value of the expression:

$$\frac{D}{T} = \frac{20 \text{ mi}}{0.44 \text{ h}}$$

$$= 45.5 \frac{\text{mi}}{\text{h}},$$

or 45.5 miles per hour, to three significant digits. ▲

EXAMPLE 3.6 ▶ Evaluate $\dfrac{E}{R}$ for $E = 6$ V (volts) and $R = 2.2$ kΩ (kilohms: Ω is the Greek capital letter omega).

SOLUTION We substitute 6 V for E and 2.2 kΩ for R:

$$\frac{E}{R} = \frac{6 \text{ V}}{2.2 \text{ k}\Omega}.$$

Recalling from Section 2.4 that k-, for kilo-, represents 10^3, we can write

$$\frac{6 \text{ V}}{2.2 \text{ k}\Omega} = \frac{6 \text{ V}}{2.2 \times 10^3 \ \Omega}.$$

We use the negative exponent rule, and then perform the division:

$$\frac{6 \text{ V}}{2.2 \times 10^3 \ \Omega} = \frac{6 \times 10^{-3} \text{ V}}{2.2 \ \Omega}$$

$$= 2.73 \times 10^{-3} \frac{\text{V}}{\Omega},$$

or 2.73×10^{-3} volts per ohm. ▲

Like numbers, units can be manipulated by arithmetic operations. For example, we can write

$$\frac{\text{mi}}{\text{h}} \times \text{h} = \frac{\text{mi} \times \text{h}}{\text{h}}$$

$$= \text{mi},$$

by dividing out the unit h from the numerator and the denominator.

EXAMPLE 3.7 ▶ Evaluate RT for $R = 120 \dfrac{\text{mi}}{\text{h}}$ and $T = 0.75$ h.

SOLUTION We substitute $120 \dfrac{\text{mi}}{\text{h}}$ for R and 0.75 h for T, and calculate the value of the expression:

$$RT = \left(120 \frac{\text{mi}}{\text{h}}\right)(0.75 \text{ h})$$

$$= (120)(0.75) \frac{\text{mi} \times \text{h}}{\text{h}}$$

$$= 90 \text{ mi}.$$ ▲

EXAMPLE 3.8 ▶ Evaluate VI for $V = 6$ V and $I = 2$ mA (milliamperes).

SOLUTION We substitute 6 V for V and 2 mA for I, and calculate the value of the expression:

$$VI = (6 \text{ V})(2 \text{ mA}).$$

Then, we recall that m-, for milli-, represents 10^{-3}:

$$(6 \text{ V})(2 \text{ mA}) = (6 \text{ V})(2 \times 10^{-3} \text{ A}).$$

Performing the multiplication, we have

$$(6 \text{ V})(2 \times 10^{-3} \text{ A}) = 12 \times 10^{-3} \text{ V} \cdot \text{A}.$$

Since we do not know the unit for the resulting quantity, we have written the original units as V·A, read *volt-amperes*. ▲

When an expression contains squares or square roots, we use the rules for exponents to simplify powers of ten resulting from prefixes.

EXAMPLE 3.9 ▶ Evaluate I^2R for $I = 5$ mA and $R = 2.2$ kΩ.

SOLUTION We substitute 5 mA for I and 2.2 kΩ for R:

$$I^2R = (5 \text{ mA})^2(2.2 \text{ k}\Omega).$$

Replacing the prefix symbols by the powers of ten they represent, we have

$$(5 \text{ mA})^2(2.2 \text{ k}\Omega) = (5 \times 10^{-3} \text{ A})^2(2.2 \times 10^3 \ \Omega).$$

We apply rules for exponents to simplify the square, and then perform the multiplications:

$$(5 \times 10^{-3} \text{ A})^2 (2.2 \times 10^3 \text{ }\Omega) = [5^2 \times (10^{-3})^2 \text{ A}^2](2.2 \times 10^3 \text{ }\Omega)$$

$$= (25 \times 10^{-6} \text{ A}^2)(2.2 \times 10^3 \text{ }\Omega)$$

$$= (25)(2.2) \times (10^{-6})(10^3) \text{ A}^2 \cdot \Omega$$

$$= 55 \times 10^{-3} \text{ A}^2 \cdot \Omega. \qquad \blacktriangle$$

EXAMPLE 3.10 ▶ Evaluate $\sqrt{\dfrac{P}{R}}$ for $P = 500$ mW and $R = 15$ MΩ.

SOLUTION We substitute 500 mW for P and 15 MΩ for R:

$$\sqrt{\frac{P}{R}} = \sqrt{\frac{500 \text{ mW}}{15 \text{ M}\Omega}}.$$

Replacing the prefix symbols by the powers of ten they represent, we have

$$\sqrt{\frac{500 \text{ mW}}{15 \text{ M}\Omega}} = \sqrt{\frac{500 \times 10^{-3} \text{ W}}{15 \times 10^6 \text{ }\Omega}}.$$

We perform the division, using the rule for exponents to divide the powers of ten:

$$\sqrt{\frac{500 \times 10^{-3} \text{ W}}{15 \times 10^6 \text{ }\Omega}} = \sqrt{33.3 \times 10^{-9}} \sqrt{\frac{\text{W}}{\Omega}}.$$

Then, using the rule for square roots, we write

$$\sqrt{33.3 \times 10^{-9}} \sqrt{\frac{\text{W}}{\Omega}} = \sqrt{33.3 \times 10^{-1} \times 10^{-8}} \sqrt{\frac{\text{W}}{\Omega}}$$

$$= \sqrt{3.33 \times 10^{-8}} \sqrt{\frac{\text{W}}{\Omega}}$$

$$= \sqrt{3.33} \times \sqrt{10^{-8}} \sqrt{\frac{\text{W}}{\Omega}}$$

$$= 1.82 \times 10^{-4} \sqrt{\frac{\text{W}}{\Omega}}.$$

The quantities were given in engineering notation, so we use the method in Section 2.4 to write the exponent as a multiple of three. Replacing 10^{-4} by $10^2 \times 10^{-6}$, we have

$$1.82 \times 10^{-4} \sqrt{\frac{\text{W}}{\Omega}} = 1.82 \times 10^2 \times 10^{-6} \sqrt{\frac{\text{W}}{\Omega}}$$

$$= 182 \times 10^{-6} \sqrt{\frac{\text{W}}{\Omega}}. \qquad \blacktriangle$$

EXERCISE 3.1

Evaluate the expression for the given values of the variables:

1. $6x + 1$ for $x = 2$ 　　　　　　　　　　**2.** $4y - 8$ for $y = 1$

3. $2x + 3y$ for $x = -3$ and $y = 4$ 　　　　**4.** $5x + 10y$ for $x = 5$ and $y = -1$

5. $9x - 10y + z$ for $x = -1, y = -2$, and $z = -3$

6. $3x - 3y - 2z$ for $x = 2, y = 3$, and $z = 4$

7. $3x^2y$ for $x = 2$ and $y = 0.3$ 　　　　　**8.** $4x^2y^2$ for $x = 1.5$ and $y = 0.5$

9. $2xy^2$ for $x = 3$ and $y = -3$ 　　　　　**10.** $5x^2y^2$ for $x = 0.2$ and $y = -0.4$

11. $3xy^2$ for $x = -3$ and $y = 1.2$ 　　　　**12.** $5x^2 - 3y^2$ for $x = 1.5$ and $y = 1.8$

Evaluate $\dfrac{D}{T}$ for the given values of the variables:

13. $D = 60$ mi and $T = 1.88$ h

14. $D = 2.5$ mi and $T = 50$ min (minutes)

Evaluate $\dfrac{E}{R}$ for the given values of the variables:

15. $E = 90$ V and $R = 27$ kΩ

16. $E = 120$ V and $R = 18$ MΩ

Evaluate RT for the given values of the variables:

17. $R = 50\ \dfrac{\text{mi}}{\text{h}}$ and $T = 0.17$ h

18. $R = 44\ \dfrac{\text{ft}}{\text{s}}$ and $T = 3$ s (seconds)

Evaluate VI for the given values of the variables:

19. $V = 12$ V and $I = 2$ mA

20. $V = 15$ V and $I = 450$ μA

Evaluate $\dfrac{V}{I}$ for the given values of the variables:

21. $V = 20$ V and $I = 800$ μA

22. $V = 1$ mV and $I = 2.5$ mA

Evaluate IR for the given values of the variables:

23. $I = 3$ mA and $R = 4.7$ kΩ

24. $I = 330$ μA and $R = 150$ kΩ

Evaluate I^2R for the given values of the variables:

25. $I = 100$ mA and $R = 5.6$ MΩ

26. $I = 12$ mA and $R = 8.2$ kΩ

Evaluate $\dfrac{V^2}{R}$ for the given values of the variables:

27. $V = 120$ V and $R = 75$ kΩ

28. $V = 60$ mV and $R = 0.5$ Ω

Evaluate $\sqrt{\dfrac{P}{R}}$ for the given values of the variables:

29. $P = 100$ mW and $R = 33$ MΩ

30. $P = 200$ W and $R = 24$ kΩ

Evaluate $\sqrt{PR}$ for the given values of the variables:

31. $P = 25$ W and $R = 2.7$ kΩ

32. $P = 1.1$ kW and $R = 3.6$ MΩ

SECTION
3.2

Simplifying Algebraic Expressions

We recall the priority order of operations from Section 1.5. In particular, recall that multiplication has priority before addition. For example, to evaluate $4(3) + 4(5)$, we multiply first, writing

$$4(3) + 4(5) = 12 + 20$$

$$= 32.$$

The parentheses in this expression indicate multiplication, and do not affect the order of operations. We recall, however, that grouping parentheses take precedence over any operation. Thus, to evaluate $4(3 + 5)$, we add first, writing

$$4(3 + 5) = 4(8)$$

$$= 32.$$

Observe that we get the same result from each expression. Thus,

$$4(3 + 5) = 4(3) + 4(5).$$

This example illustrates a general rule called the **distributive property**.

> **Distributive Property for Multiplication over Addition:** For any three algebraic terms A, B, and C,
>
> $$A(B + C) = AB + AC$$

EXAMPLE 3.11   Use the distributive property to remove the parentheses from $5(x + y)$.

SOLUTION   Using the distributive property for multiplication over addition, we write

$$5(x + y) = 5x + 5y.$$ ▲

The distributive property is also true for multiplication over subtraction, that is,

$$A(B - C) = AB - AC.$$

EXAMPLE 3.12   Use the distributive property to remove the parentheses from $3(x - y)$.

SOLUTION   Using the distributive property for multiplication over subtraction, we write

$$3(x - y) = 3x - 3y.$$ ▲

We can extend the distributive property to any number of terms, where the terms are added or subtracted.

EXAMPLE 3.13   Use the distributive property to remove the parentheses from $10(x - y + z)$.

SOLUTION   Since the distributive property can be used for addition and subtraction, and for any number of terms, we write

$$10(x - y + z) = 10x - 10y + 10z.$$ ▲

Each term in an expression is the product of a constant and one or more variables. The constant is called the numerical **coefficient** of the variables. In the expression

$$10x - 10y + 10z,$$

the coefficient of x is 10 and the coefficient of z is also 10. Because the second term is subtracted, the coefficient of y is taken to be -10.

When no constant appears in a term, the coefficient is taken to be 1. In the expression

$$x - y + z,$$

the coefficient of x is 1 and the coefficient of z is also 1, but the coefficient of y is taken to be -1.

EXAMPLE 3.14   Use the distributive property to remove the parentheses from $-(x - y + z)$.

SOLUTION   We can think of the negative in front of the parentheses as -1:

$$-(x - y + z) = -1(x - y + z).$$

Then, using the distributive property, we write

$$-1(x - y + z) = (-1)(x) - (-1)y + (-1)z.$$

Recalling the rules of signs for multiplication and subtraction from Unit 1, we have

$$(-1)(x) - (-1)y + (-1)z = -1x + 1y - 1z$$

$$= -x + y - z.$$

Observe that the negative in front of the algebraic expression changes the sign of each term of the expression. ▲

| **EXAMPLE 3.15** | | Use the distributive property to remove the parentheses from $-3(4x - 5y - z)$. |

SOLUTION   Using the distributive property, we multiply each term by -3:

$$-3(4x - 5y - z) = (-3)(4x) - (-3)(5y) - (-3)z.$$

Recalling the rules of signs for multiplication and subtraction, we have

$$(-3)(4x) - (-3)(5y) - (-3)z = -12x - (-15y) - (-3)z$$
$$= -12x + 15y + 3z.$$

Observe that the negative number in front of the expression changes the sign of each term of the expression as well as multiplying each coefficient. ▲

COMMON FACTORS

The two parts of a product are called **factors** of the product. In the distributive properties

$$A(B + C) = AB + AC$$

and

$$A(B - C) = AB - AC,$$

A and B are factors of the term AB, and A and C are factors of the term AC. Because A is common to both AB and AC, we say that A is a **common factor** of the expression $AB + AC$.

If we read the distributive property from right to left, we have

$$AB + AC = A(B + C)$$

and

$$AB - AC = A(B - C).$$

We **factor out** the common factor A from the expression $AB + AC$ or $AB - AC$ when we use the distributive property in this way.

| **EXAMPLE 3.16** | | Factor out a common factor from $5x + 5y$. |

SOLUTION   There is a common factor 5. Therefore, we write

$$5x + 5y = 5(x + y).$$ ▲

LIKE TERMS

The terms $5x$ and $5y$ have the same numerical coefficient but different variables. Terms such as $5x$ and $3x$ have different numerical coefficients but the same variable. If all of the literal parts of two terms—the variables and their exponents—are identical, the terms are **like terms**. For example, $5x$ and $3x$ are like terms.

When two or more terms are like terms, we can combine their coefficients. We can use the process of factoring out the common factor to **combine like terms** of algebraic expressions.

| **EXAMPLE 3.17** | | Combine like terms of $5x + 3x$. |

SOLUTION   We factor out the common factor x:

$$5x + 3x = (5 + 3)x$$
$$= (8)x.$$

Therefore, we have

$$5x + 3x = 8x.$$ ▲

In actual practice, we usually skip the middle steps and add or subtract the numerical coefficients directly.

EXAMPLE 3.18   Combine like terms of $9x + 16x$.

SOLUTION   We add the numerical coefficients to obtain

$$9x + 16x = 25x. \qquad \blacktriangle$$

EXAMPLE 3.19   Combine like terms of $5x - 6x$.

SOLUTION   We subtract the numerical coefficients to obtain

$$5x - 6x = (-1)x$$
$$= -x. \qquad \blacktriangle$$

EXAMPLE 3.20   Combine like terms of $7x - 5x - 2x$.

SOLUTION   We proceed from left to right, combining $7x - 5x$ first:

$$7x - 5x - 2x = 2x - 2x$$
$$= 0x.$$

Since any number multiplied by zero is zero, $0x = 0$. Thus, we have

$$7x - 5x - 2x = 0. \qquad \blacktriangle$$

ASSOCIATIVE AND COMMUTATIVE PROPERTIES

Two other basic properties are used in arithmetic. If we proceed from left to right to evaluate, for example, $4 + 3 + 5$, we group the first two numbers:

$$(4 + 3) + 5 = 7 + 5$$
$$= 12.$$

If we group the second two numbers, however, we must add them first:

$$4 + (3 + 5) = 4 + 8$$
$$= 12.$$

Observe that we get the same result from each expression. Thus,

$$(4 + 3) + 5 = 4 + (3 + 5).$$

This example illustrates the **associative property for addition**. There is a corresponding associative property for multiplication.

Associative Property for Addition: For any three algebraic terms A, B, and C,

$$(A + B) + C = A + (B + C).$$

Associative Property for Multiplication: For any three algebraic terms A, B, and C,

$$(AB)C = A(BC).$$

We observe, however, that

$$(4 - 3) - 5 = 1 - 5 = -4,$$

whereas

$$4 - (3 - 5) = 4 - (-2) = 6.$$

Therefore, $(4 - 3) - 5$ is *not the same* as $4 - (3 - 5)$. Similarly, $(16 \div 8) \div 2$ is *not the same* as $16 \div (8 \div 2)$. There are *no* associative properties for subtraction or division.

When we add or multiply numbers, we can add or multiply in any order. For example,

$$2 + 3 = 3 + 2.$$

This example illustrates the **commutative property for addition**. There is a corresponding commutative property for multiplication.

Commutative Property for Addition: For any two algebraic terms A and B,

$$A + B = B + A.$$

Commutative Property for Multiplication: For any two algebraic terms A and B,

$$AB = BA.$$

We observe that $2 - 3$ is *not the same* as $3 - 2$, and similarly $2 \div 4$ is *not the same* as $4 \div 2$. There are *no* commutative properties for subtraction or division.

EXAMPLE 3.21 ▶   Combine like terms of $(x - 4y) - (3x - 6y)$.

SOLUTION   We remove the parentheses, recalling that the minus in front of the second expression changes the signs of that expression:

$$(x - 4y) - (3x - 6y) = x - 4y - 3x + 6y.$$

Since there is no associative property for subtraction, we rewrite the expression using addition of the negative:

$$x - 4y - 3x + 6y = x + (-4y) + (-3x) + 6y.$$

Now, we can use the associative and commutative properties for addition to rearrange the terms, and then combine like terms:

$$x + (-4y) + (-3x) + 6y = x + (-3x) + (-4y) + 6y$$
$$= -2x + 2y. \quad ▲$$

In actual practice, we usually skip some steps by using addition of the negative without actually writing it out. When we rearrange terms, however, we must always be sure to take each sign with the term.

EXAMPLE 3.22 ▶   Combine like terms of $(2x - 4y + z) - (2x + 3y - 5z)$.

SOLUTION   We remove the parentheses as before:

$$(2x - 4y + z) - (2x + 3y - 5z) = 2x - 4y + z - 2x - 3y + 5z.$$

We rearrange the terms, being sure to take each sign with the term, and then combine like terms:

$$2x - 4y + z - 2x - 3y + 5z = 2x - 2x - 4y - 3y + z + 5z$$
$$= 0x - 7y + 6z$$
$$= -7y + 6z. \quad ▲$$

EXAMPLE 3.23 ▶   Combine like terms of $3x + 4xy - 3yx - 5y$.

SOLUTION   The terms $3x$ and $5y$ are not like any other terms. Therefore, $3x$ and $5y$ cannot be combined with any other terms. However, we observe that, using the commutative property for multiplication,

$$3yx = 3xy.$$

Therefore, we can write

$$3x + 4xy - 3yx - 5y = 3x + 4xy - 3xy - 5y$$
$$= 3x + 1xy - 5y$$
$$= 3x + xy - 5y.$$

▲

**EXERCISE
3.2**

Use the distributive property to remove the parentheses:

1. $10(x + y)$ **2.** $0.5(x + y)$ **3.** $5(x - y)$ **4.** $3.3(y - z)$

5. $9(x - y - z)$ **6.** $1.2(z - y - x)$ **7.** $-(x - y)$ **8.** $-(x + y - z)$

9. $-2(x - y)$ **10.** $-3(y - x)$

11. $-5(2x - 3y - 2z)$ **12.** $-0.5(4x + 6y - 8z)$

Factor out the common factor:

13. $10x + 10y$ **14.** $5x - 5y$ **15.** $4x + 8y$ **16.** $10x - 15y$

Combine like terms:

17. $2x + 4x$ **18.** $4x + 8x$ **19.** $7x - 6x$ **20.** $9x - 12x$

21. $15x - 16x$ **22.** $0.5x - 2.5x$ **23.** $10x - 8x - 4x$ **24.** $3x - 8x + x$

25. $(2x - 3y) - (3x - 2y)$ **26.** $(4x - y) - (x - 6y)$

27. $(3x + 2y) - (2x + 3y)$ **28.** $(6x - 6y) - (4x + 3y)$

29. $(5x - 3y + z) - (4x - y + z)$ **30.** $(4x - 6y + 8z) - (6x - 6y + 2z)$

31. $(x - 2y + 3z) - (2x - 3y - 3z)$ **32.** $(5x + 5y - 4z) - (6x + 6y - 3z)$

33. $2x + xy + yx + 8y$ **34.** $2x + 4xy - 5yx - 3y$

35. $xy - 2x - 6yx + 7y$ **36.** $y + yx - 5xy - 3x$

**SECTION
3.3**

General Linear Expressions

Linear expressions are algebraic expressions containing only terms of the form a, x, and ax, where a is a constant and x is a variable. Thus, each term of a linear expression is a constant, a variable, or a product of a constant and a variable. These are examples of linear expressions:

$$x + 7, \quad x + y, \quad 2x - y, \quad x + y - z.$$

Expressions including terms such as x^2, xy, and $\dfrac{1}{x}$ are not linear expressions.

To simplify general linear expressions, we may use one or more of the methods in Section 3.2. First, we combine any like terms that appear within parentheses. Then, we remove all parentheses by using the distributive property. Finally, we combine any resulting like terms. The phrase **simplify the expression** means this or similar processes.

EXAMPLE 3.24

Simplify $-5(8x - 6x - 5)$.

SOLUTION

First, we combine the like terms within the parentheses:

$$-5(8x - 6x - 5) = -5(2x - 5).$$

Then, we remove the parentheses by using the distributive property:

$$-5(2x - 5) = -10x + 25.$$

There are no like terms in this result, so the process is complete. ▲

| EXAMPLE 3.25 | | Simplify $-3x - 2(4x - 9y - 4)$. |

SOLUTION   There are no like terms within the parentheses. Therefore, we remove the parentheses by using the distributive property:

$$-3x - 2(4x - 9y - 4) = -3x - 8x + 18y + 8.$$

Then, we combine the resulting like terms to obtain

$$-3x - 8x + 18y + 8 = -11x + 18y + 8. \qquad \blacktriangle$$

| EXAMPLE 3.26 | | Simplify $5 - (6x - 8x - 5)$. |

SOLUTION   We combine the like terms within the parentheses:

$$5 - (6x - 8x - 5) = 5 - (-2x - 5).$$

Observe that the quantity within the parentheses is *subtracted* from 5, not multiplied by 5. Therefore, we remove the parentheses by using the distributive property to change the signs of the terms in the parentheses:

$$5 - (-2x - 5) = 5 + 2x + 5.$$

Then, we combine the resulting like terms to obtain

$$5 + 2x + 5 = 2x + 10.$$

It is conventional to write the constant term last. $\qquad \blacktriangle$

| EXAMPLE 3.27 | | Simplify $5(x - 3y) + 3(3x - 5y)$. |

SOLUTION   There are no like terms within either set of parentheses. We remove each set of parentheses by using the distributive property:

$$5(x - 3y) + 3(3x - 5y) = 5x - 15y + 9x - 15y.$$

Then, we combine both sets of resulting like terms to obtain

$$5x - 15y + 9x - 15y = 14x - 30y.$$

It is conventional to write terms including variables in the alphabetical order of the variables. $\qquad \blacktriangle$

| EXAMPLE 3.28 | | Simplify $2(4x - 2) - 4(2x - 7)$. |

SOLUTION   Since there are no like terms within either set of parentheses, we remove each set of parentheses, being careful to change the signs of the terms in the second set:

$$2(4x - 2) - 4(2x - 7) = 8x - 4 - 8x + 28.$$

Then, we combine the resulting like terms:

$$8x - 4 - 8x + 28 = 0x + 24$$

$$= 24. \qquad \blacktriangle$$

**EXERCISE
3.3**

Simplify:

1. $-3(x + 5x - 3)$

2. $-(2x - 5x - 6)$

3. $-x - (3x - 8y - 1)$

4. $-2y - (2x + 7y - 3)$

5. $-2x - (9x - 2x + 3)$

6. $8y - (6y + 6y - 4)$

7. $8 - (2x - 5x - 5)$ **8.** $1 - (9x + 2x - 1)$

9. $3(x - y) + 4(2x + y)$ **10.** $-(5x - 7y) + 6(2x - 5y)$

11. $-3(7x - 8) + 3(7x - 6)$ **12.** $5(2x - 6) + 6(3x + 5)$

13. $2(8x - y) - 4(7x - y)$ **14.** $-2(9x - 3y) + 3(7x - 5y)$

15. $-(8x - 7) - (2x - 8)$ **16.** $2(10x - 4) - (5x - 8)$

SECTION
3.4

Nonlinear Expressions

In Unit 2, we used the powers of ten:

$$10^1 = 10$$

$$10^2 = 10 \times 10$$

$$10^3 = 10 \times 10 \times 10$$

$$10^4 = 10 \times 10 \times 10 \times 10$$

and so on. Similarly, for any variable x we can write positive integer powers of x. The first power of x, or x to the first power is

$$x^1 = x.$$

The second power of x, or x to the second power, or x squared is

$$x^2 = x \cdot x.$$

The third power of x, or x to the third power, or x cubed is

$$x^3 = x \cdot x \cdot x.$$

The fourth power of x, or x to the fourth power, is

$$x^4 = x \cdot x \cdot x \cdot x,$$

and so on.

The **nth power of x**, or x to the nth power, is n factors of x:

$$x^n = x \cdot x \cdot x \cdot \ldots \cdot x$$

$$\longleftarrow \quad n \text{ factors} \quad \longrightarrow$$

We can use the rules for exponents that we developed in Unit 2 when the base is a variable such as x. We repeat the basic rules for reference, with variables m and n as the exponents.

Basic Rules for Exponents:

1. $x^m \cdot x^n = x^{m+n}$.

2. $\dfrac{x^m}{x^n} = x^{m-n}$ if $x \neq 0$.

3. $(x^m)^n = x^{m \cdot n}$.

The zero exponent and negative exponent rules are also true for powers with any base x except zero.

> **Zero Exponent Rule:**
>
> $$x^0 = 1 \text{ if } x \neq 0.$$
>
> **Negative Exponent Rule:**
>
> 1. $\dfrac{1}{x^{-n}} = x^n$ if $x \neq 0.$
>
> 2. $x^{-n} = \dfrac{1}{x^n}$ if $x \neq 0.$

EXAMPLE 3.29 ▶ Simplify $(2x^2)(3x^4)$.

SOLUTION We multiply the constants, and multiply the powers of x by adding the exponents:

$$(2x^2)(3x^4) = 6x^{2+4}$$
$$= 6x^6. \quad ▲$$

EXAMPLE 3.30 ▶ Simplify $4x(2x + x^2)$.

SOLUTION We cannot combine the terms inside the parentheses because they are not like terms. We use the distributive property to write

$$4x(2x + x^2) = 4x(2x) + 4x(x^2).$$

Since $x = x^1$, we can write

$$4x(2x) + 4x(x^2) = 4x^1(2x^1) + 4x^1(x^2).$$

Then, we multiply the constants, and multiply the powers of x by adding the exponents, as in the preceding example:

$$4x^1(2x^1) + 4x^1(x^2) = 8x^{1+1} + 4x^{1+2}$$
$$= 8x^2 + 4x^3.$$

The two terms in this result are not like terms and cannot be combined. ▲

When the rules for exponents involve division or negative exponents, the variables are restricted to *allowable* values by the rules of arithmetic. For example, when x is in a denominator, zero is not an allowable value of x. When we use the rules for exponents with variables, *we will assume implicitly that all values of variables are allowable values.*

EXAMPLE 3.31 ▶ Simplify $\dfrac{4x^2y^3}{2xy^2}$.

SOLUTION Since $x = x^1$, we can write

$$\frac{4x^2y^3}{2xy^2} = \frac{4x^2y^3}{2x^1y^2}.$$

Then, we divide the constants and divide the powers of x by subtracting the exponents:

$$\frac{4x^2y^3}{2x^1y^2} = 2x^{2-1}y^{3-2}$$

$$= 2x^1y^1$$

$$= 2xy. \quad ▲$$

EXAMPLE 3.32 ▶ Simplify $\dfrac{3x^{-4}y^2}{2y^{-2}}$.

SOLUTION We can subtract the exponents, or we can use the negative exponent rule and add the exponents, to combine the powers with base y. Using the negative exponent rule and adding, we write

$$\frac{3x^{-4}y^2}{2y^{-2}} = \frac{3}{2}x^{-4}y^2(y^2)$$

$$= \frac{3}{2}x^{-4}y^4.$$ ▲

We can write the result in Example 3.32 in several forms. If we prefer to have only positive exponents, we can use the negative exponent rule to write the result in this form:

$$\frac{3}{2}x^{-4}y^4 = \frac{3y^4}{2x^4}.$$

If we prefer to keep the negative exponent, we can divide the constants to write the result in this form:

$$\frac{3}{2}x^{-4}y^4 = 1.5x^{-4}y^4.$$

For the remainder of this section, we will write results in this second form, eliminating the denominator.

DIVISION BY ONE TERM

When we add or subtract fractions by combining the numerators over a common denominator, we use a property related to the distributive property:

$$\frac{A}{C} + \frac{B}{C} = \frac{A + B}{C}.$$

If we read this property from right to left, we have

$$\frac{A + B}{C} = \frac{A}{C} + \frac{B}{C}.$$

This property can be extended to subtraction, and to expressions with more than two terms. We use this property to divide expressions with two or more terms by expressions with just one term.

EXAMPLE 3.33 ▶ Simplify $\dfrac{2x^3 + 4xy}{2x^{-3}}$.

SOLUTION We use the preceding property to separate the terms:

$$\frac{2x^3 + 4xy}{2x^{-3}} = \frac{2x^3}{2x^{-3}} + \frac{4xy}{2x^{-3}}.$$

Then, we divide the constants, use the negative exponent rule, and add the exponents, to obtain

$$\frac{2x^3}{2x^{-3}} + \frac{4xy}{2x^{-3}} = \frac{x^3}{x^{-3}} + \frac{2xy}{x^{-3}}$$

$$= x^3(x^3) + 2xy(x^3)$$

$$= x^6 + 2x^4y.$$ ▲

EXAMPLE 3.34 ▶ Simplify $\dfrac{x^2 - x^2y^2 + y^2}{xy}$.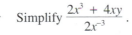

SOLUTION Extending the preceding property to subtraction, and to expressions with more than two terms, we can separate the terms:

$$\frac{x^2 - x^2y^2 + y^2}{xy} = \frac{x^2}{xy} - \frac{x^2y^2}{xy} + \frac{y^2}{xy}.$$

Then, subtracting the exponents, we have

$$\frac{x^2}{xy} - \frac{x^2y^2}{xy} + \frac{y^2}{xy} = \frac{x}{y} - xy + \frac{y}{x}.$$

If we wish, we can eliminate the denominators by using the negative exponent rule:

$$\frac{x}{y} - xy + \frac{y}{x} = xy^{-1} - xy + x^{-1}y.$$ ▲

When we use variable bases, we have a rule for powers with different bases.

Rule for Different Bases:

    1. $(x \cdot y)^n = x^n \cdot y^n.$

    2. $\left(\dfrac{x}{y}\right)^n = \dfrac{x^n}{y^n}$ if $y \neq 0.$

EXAMPLE 3.35 Simplify $(x^3y^{-2})^2$.

SOLUTION We use Part 1 of the rule for different bases to write

$$(x^3y^{-2})^2 = (x^3)^2(y^{-2})^2.$$

Then, we can multiply the exponents to obtain

$$(x^3)^2(y^{-2})^2 = x^6y^{-4}.$$ ▲

SQUARE ROOTS

The fractional exponent and square roots rules are also true for powers with any base x that is not negative.

Fractional Exponent Rule:

$$\sqrt{x^n} = (x^n)^{\frac{1}{2}} \text{ if } x \geq 0.$$

Rule for Square Roots:

    1. $\sqrt{x^m \cdot y^n} = \sqrt{x^m} \cdot \sqrt{y^n}$ if $x \geq 0$ and $y \geq 0.$

    2. $\sqrt{\dfrac{x^m}{y^n}} = \dfrac{\sqrt{x^m}}{\sqrt{y^n}}$ if $x \geq 0$ and $y > 0.$

We must be careful when we use rules for exponents in conjunction with the fractional exponent rule. For example, if $x = 3$, by using the rules for exponents we have

$$\sqrt{3^2} = (3^2)^{\frac{1}{2}}$$

$$= 3^{2 \cdot \frac{1}{2}}$$

$$= 3^1$$

$$= 3.$$

However, if $x = -3$, by using rules for exponents we might try to write $\sqrt{(-3)^2}$ as $(-3)^{2 \cdot \frac{1}{2}} = -3$. But, using ordinary arithmetic, we know that

$$\sqrt{(-3)^2} = \sqrt{9}$$

$$= 3.$$

Thus, $\sqrt{(-3)^2}$ is *not the same* as -3. In general, for any base x,

$$\sqrt{x^2} = x \ \ only \ \ if \ \ x \geq 0.$$

As before, *we will assume implicitly that all values of variables are allowable values.*

| EXAMPLE 3.36 | | Simplify $\sqrt{x^2 y^{-6}}$.

| SOLUTION | We use Part 1 of the rule for square roots to write

$$\sqrt{x^2 y^{-6}} = \sqrt{x^2}\sqrt{y^{-6}}.$$

Then, we use the fractional exponent rule to write

$$\sqrt{x^2}\sqrt{y^{-6}} = (x^2)^{\frac{1}{2}}(y^{-6})^{\frac{1}{2}}$$
$$= x^{2 \cdot \frac{1}{2}} y^{(-6) \cdot \frac{1}{2}}$$
$$= x^1 y^{-3}$$
$$= xy^{-3}. \qquad \blacktriangle$$

| EXAMPLE 3.37 | | Simplify $\sqrt{\dfrac{4x^3}{2y^2}}$.

| SOLUTION | We divide the constants, and use Part 2 of the rule for square roots to write

$$\sqrt{\frac{4x^3}{2y^2}} = \sqrt{\frac{2x^3}{y^2}}$$
$$= \frac{\sqrt{2x^3}}{\sqrt{y^2}}.$$

Then, we can replace x^3 by $x^2 \cdot x$, and use Part 1 of the rule for square roots to write

$$\frac{\sqrt{2x^3}}{\sqrt{y^2}} = \frac{\sqrt{2 \cdot x^2 \cdot x}}{\sqrt{y^2}}$$
$$= \frac{\sqrt{2} \cdot \sqrt{x^2} \cdot \sqrt{x}}{\sqrt{y^2}}$$
$$= \frac{1.41x\sqrt{x}}{y}.$$

If we wish, we can eliminate the denominator by using the negative exponent rule:

$$\frac{1.41x\sqrt{x}}{y} = 1.41xy^{-1}\sqrt{x}. \qquad \blacktriangle$$

| EXERCISE
3.4 | Assume that values of the variables are restricted as necessary.
Simplify:

1. $(3x^5)(5x^3)$ **2.** $(7x^{-1})(8x^8)$ **3.** $5x(x + 3x^2)$ **4.** $3x(x^3 + x^2)$

5. $2x(3x^2 + 8x^{-2})$ **6.** $7x^2(5x^3 - 4x^{-3})$ **7.** $x^{-1}(x^{-3} - x)$ **8.** $5x^2(2x^2 + x^{-2})$

9. $\dfrac{10x^4 y^3}{5x^2 y^2}$ **10.** $\dfrac{12x^6 y^9}{3x^3 y^3}$ **11.** $\dfrac{9x^2 y^{-2}}{2x^{-3}}$ **12.** $\dfrac{2x^{-3}}{6x^3 y^{-4}}$

13. $\dfrac{9x^2 + 6x}{3x^{-2}}$ **14.** $\dfrac{2x^2 y^2 + 2xy}{2xy^{-1}}$ **15.** $\dfrac{8x^3 y + 4y^3}{2xy^2}$ **16.** $\dfrac{7x^3 y + 9xy^3}{2x^2 y^2}$

17. $\dfrac{x^3 + x^2 - x}{x^3}$

18. $\dfrac{x^3 - x^2y^2 + y^3}{x^2y^2}$

19. $(xy^3)^2$

20. $(x^2y^{-3})^{-2}$

21. $\left(\dfrac{x^3}{y^{-2}}\right)^2$

22. $\left(\dfrac{x}{y^2}\right)^4$

23. $\left(\dfrac{xy^{-2}}{z^{-3}}\right)^3$

24. $\left(\dfrac{x^2}{y^{-1}z^3}\right)^4$

25. $\sqrt{x^4y^{-2}}$

26. $\sqrt{x^{-2}y^2}$

27. $\sqrt{x^3y^2}$

28. $\sqrt{x^5y^3}$

29. $\sqrt{\dfrac{9x^9}{3y^4}}$

30. $\sqrt{\dfrac{25xy^4}{5xy}}$

☐ **Self-Test** ☐

1. Combine like terms of $8x - 3x + 4x$.

1. _____

2. Simplify $-3(2x + 3y) - 2(3x - 2y)$.

2. _____

3. Simplify $\dfrac{5x^2y + 3xy^2}{4xy^{-1}}$.

3. _____

4. Simplify $\sqrt{6x^3y^5}$.

4. _____

5. Evaluate $\sqrt{\dfrac{P}{R}}$ for $P = 50$ W and $R = 2.2$ kΩ.

5. _____

UNIT 4

Equations

Introduction

The equation is one of the primary tools of problem solving. When an applied problem can be reduced to one or more equations, and the equations can be solved, then the problem is solved. In this unit, you will learn the fundamental rules for solving equations. You will use these rules to solve some basic types of equations. First, you will learn how to solve linear equations, which involve only linear expressions. Then, you will learn how to solve a basic type of equation where variables are in denominators, and then basic types of equations where variables are squared or in square roots. Finally, you will apply these types of equations to fundamental laws of electronics.

OBJECTIVES

When you have finished this unit you should be able to:

1. Solve equations of the form $x + a = b$, $x - a = b$, $ax = b$, and $\frac{x}{a} = b$.
2. Solve general linear equations.
3. Solve elementary rational equations, where the variable is in denominators.
4. Solve elementary quadratic equations, where the variable is squared, and related radical equations where the variable is in a square root.
5. Use Ohm's law to find current, voltage, or resistance, given two of the quantities, and use related equations involving power.

SECTION 4.1

Solving Basic Equations

Historically, the growth of the study of algebra, from prehistoric times to about the sixteenth century, can be attributed to attempts to solve different types of equations. The word *algebra* derives from the word *al-jabr* in the title of a book written by an Arabian mathematician, Mohammed ibn Musa Al-Khowârizmî, in the ninth century. The full title of the book is *Hisâb al-jabr w'al muqâbala*. It was translated into Latin and introduced into Europe in the twelfth century. The word *al-jabr* meant restoring; that is, balancing the terms on the sides of an equation by rules we will show in this unit.

An **equation** consists of two equal algebraic expressions. These are some examples of equations:

$$x + 7 = 10, \quad x + y = 6, \quad 2x - (x - 1) = 3x,$$

$$\frac{1}{x} + 1 = 3, \quad x^2 - 1 = 3, \quad I = \frac{V}{R}.$$

The algebraic expressions on the two sides of the equal sign are the **sides** of the equation. The expression to the left of the equal sign is the **left-hand side** of the equation. The expression to the right of the equal sign is the **right-hand side** of the equation.

An equation is **in one variable** if it contains only one variable. These are examples of equations in one variable:

$$x + 7 = 10, \quad 2x - (x - 1) = 3x, \quad \frac{1}{x} + 1 = 3, \quad x^2 - 1 = 3.$$

The variable can appear more than once in an equation in one variable.

A **solution** of an equation in one variable is a number that, when substituted for the variable, causes the two sides to be numerically equal. For example, the equation

$$x + 7 = 10$$

has the solution

$$x = 3.$$

To check the solution, we substitute 3 for x in the expression on the left-hand side of the equation. We may write

$$x + 7 = 10$$
$$3 + 7 \overset{?}{=} 10$$
$$10 = 10.$$

We have used the symbol $\overset{?}{=}$ in the second line to ask the question, "is the left-hand side equal to the right-hand side?" Since the third line shows that the two sides are numerically equal, $x = 3$ is a solution of the equation $x + 7 = 10$.

If we were to try $x = 4$ as a proposed solution of the equation $x + 7 = 10$, we would substitute 4 for x:

$$x + 7 = 10$$
$$4 + 7 \overset{?}{=} 10.$$

Then, we would have

$$11 \neq 10,$$

where the symbol $\neq$ means "is not equal to." Thus, $x = 4$ is *not* a solution of the equation $x + 7 = 10$.

To find solutions of equations, we must isolate the variable on one side of the equation, with a single number on the other side. This process is called **solving an equation**. To solve an equation, we use a series of **equivalent equations**. Two equations are equivalent equations if they have the same solutions. For example, the equations

$$x + 7 = 10$$

and

$$x = 10 - 7$$

both have the solution $x = 3$, so they are equivalent equations.

There are four fundamental ways to derive an equivalent equation.

Fundamental Methods to Derive an Equivalent Equation:
1. Add the same expression to both sides of the equation.
2. Subtract the same expression from both sides of the equation.
3. Multiply both sides of the equation by the same nonzero expression.
4. Divide both sides of the equation by the same nonzero expression.

In Methods 3 and 4, the expression must be nonzero. It is important that we *cannot* multiply or divide both sides of an equation by an expression that is equal to zero.

If an equation has the form

$$x - a = b,$$

we solve the equation by *adding* the same expression to both sides of the equation. The expression we add is the constant a.

> **Fundamental Rule 1:**   For an equation of the form
> $$x - a = b,$$
> *add a* to both sides.

EXAMPLE 4.1   Solve the equation $x - 5 = 4$ and check the solution.

SOLUTION   We add 5 to both sides of the equation:

$$x - 5 = 4$$
$$x - 5 + 5 = 4 + 5$$
$$x + 0 = 9$$
$$x = 9.$$

We check the solution by substituting 9 for x in the original equation:

$$x - 5 = 4$$
$$9 - 5 \stackrel{?}{=} 4$$
$$4 = 4.$$

Since the two sides are numerically equal, $x = 9$ is the solution of the equation.   ▲

If an equation has the form

$$x + a = b,$$

we solve the equation by *subtracting* the same expression from both sides of the equation. The expression we subtract is the constant a.

> **Fundamental Rule 2:**   For an equation of the form
> $$x + a = b,$$
> *subtract a* from both sides.

EXAMPLE 4.2   Solve the equation $x + 7 = 10$ and check the solution.

SOLUTION   We subtract 7 from both sides of the equation:

$$x + 7 = 10$$
$$x + 7 - 7 = 10 - 7$$
$$x + 0 = 3$$
$$x = 3.$$

We check the solution by substituting 3 for x in the original equation. The check for this solution was done earlier in this section.   ▲

If an equation has the form

$$\frac{x}{a} = b,$$

we solve the equation by *multiplying* both sides of the equation by the same nonzero expression. The expression we multiply by is the constant a, which will not be zero because the fraction $\frac{x}{0}$ is undefined.

Fundamental Rule 3: For an equation of the form
$$\frac{x}{a} = b,$$
multiply by a on both sides.

EXAMPLE 4.3 Solve the equation $\frac{x}{8} = 2$ and check the solution.

SOLUTION We multiply both sides of the equation by 8:

$$\frac{x}{8} = 2$$

$$8\left(\frac{x}{8}\right) = 8(2)$$

$$1x = 16$$

$$x = 16.$$

To check, we substitute 16 for x in the original equation:

$$\frac{x}{8} = 2$$

$$\frac{16}{8} \overset{?}{=} 2$$

$$2 = 2.$$

Therefore, $x = 16$ is the solution of the equation. ▲

If an equation has the form

$$ax = b,$$

we solve the equation by *dividing* both sides of the equation by the same nonzero expression. The expression we divide by is the constant a, which will not be zero because the product $0x = 0$.

Fundamental Rule 4: For an equation of the form
$$ax = b,$$
divide by a on both sides.

EXAMPLE 4.4 Solve the equation $3x = 18$ and check the solution.

SOLUTION We divide both sides of the equation by 3:

$$3x = 18$$

$$\frac{3x}{3} = \frac{18}{3}$$

$$1x = 6$$

$$x = 6.$$

To check, we substitute 6 for x in the original equation:

$$3x = 18$$

$$3(6) \overset{?}{=} 18$$

$$18 = 18.$$

Therefore, $x = 6$ is the solution of the equation. ▲

EXAMPLE 4.5 ▶ Solve the equation $\frac{3}{4}x = \frac{1}{2}$ and check the solution.

SOLUTION This equation has the form $ax = b$, where both a and b are fractions. We can avoid division of fractions, however, by *multiplying* both sides of the equation by $\frac{4}{3}$:

$$\frac{4}{3}\left(\frac{3}{4}x\right) = \frac{4}{3}\left(\frac{1}{2}\right)$$

$$1x = \frac{2}{3}$$

$$x = \frac{2}{3}.$$

To check, we substitute $\frac{2}{3}$ for x in the original equation:

$$\frac{3}{4}x = \frac{1}{2}$$

$$\left(\frac{3}{4}\right)\left(\frac{2}{3}\right) \overset{?}{=} \frac{1}{2}$$

$$\frac{1}{2} = \frac{1}{2}.$$

Therefore, $x = \frac{2}{3}$ is the solution of the equation. ▲

EXERCISE 4.1

Solve the equation and check the solution:

1. $x + 1 = 8$ 2. $x + 5 = 4$ 3. $x + 4.9 = 3$ 4. $x + \frac{1}{2} = 2$

5. $x - 11 = 7$ 6. $x - 6.2 = 3.9$ 7. $4x = 60$ 8. $4.8x = 5.6$

9. $-3x = 18$ 10. $5x = \frac{1}{2}$ 11. $\frac{x}{9} = 4$ 12. $\frac{x}{7.6} = 2.5$

13. $\frac{1}{4}x = \frac{3}{2}$ 14. $\frac{4}{3}x = \frac{5}{6}$ 15. $\frac{2}{3}x = \frac{1}{4}$ 16. $\frac{1}{8}x = \frac{7}{12}$

SECTION 4.2

General Linear Equations

An equation is a **linear equation** if the two algebraic expressions it contains are linear expressions; that is, expressions with terms of the form a, x, and ax. These are examples of linear equations in one variable:

$$x + 7 = 10, \quad 2x - 3 = 5, \quad 2x - (x - 1) = 3x.$$

The first of these equations can be solved by using one of the fundamental methods, as we showed in Section 4.1. The second equation must be reduced to an equivalent equation by using one of the fundamental methods, and then solved by using another. The third equation must first be reduced to an equivalent equation by simplifying the expression on the left-hand side, and then solved by using the fundamental methods. Any linear equation in one variable can be solved by simplifying expressions if necessary, and then applying one or more of the four fundamental methods.

EXAMPLE 4.6

Solve the equation $2x - 3 = 5$ and check the solution.

SOLUTION

We begin as if the equation were of the basic form $x - a = b$. Thus, we isolate the term $2x$ by adding 3 to both sides of the equation:

$$2x - 3 = 5$$
$$2x - 3 + 3 = 5 + 3$$
$$2x = 8.$$

We now have an equivalent equation of the form $ax = b$. We isolate x by dividing both sides of this equivalent equation by 2:

$$2x = 8$$
$$\frac{2x}{2} = \frac{8}{2}$$
$$x = 4.$$

We check the solution by substituting 4 for x in the original equation:

$$2x - 3 = 5$$
$$2(4) - 3 \overset{?}{=} 5$$
$$8 - 3 \overset{?}{=} 5$$
$$5 = 5.$$

Since the two sides are numerically equal, $x = 4$ is the solution of the equation.   ▲

EXAMPLE 4.7

Solve the equation $3x + 20 = 5$ and check the solution.

SOLUTION

First, we isolate the term $3x$ by subtracting 20 from both sides of the equation:

$$3x + 20 = 5$$
$$3x + 20 - 20 = 5 - 20$$
$$3x = -15.$$

Then, we isolate x by dividing both sides by 3:

$$\frac{3x}{3} = \frac{-15}{3}$$
$$x = -5.$$

To check, we substitute -5 for x in the original equation:

$$3x + 20 = 5$$
$$3(-5) + 20 \overset{?}{=} 5$$
$$-15 + 20 \overset{?}{=} 5$$
$$5 = 5.$$

Therefore, $x = -5$ is the solution of the equation.   ▲

EXAMPLE 4.8   Solve the equation $\frac{x}{5} - 6 = -2$ and check the solution.

SOLUTION   We isolate the term $\frac{x}{5}$ by adding 6 to both sides:

$$\frac{x}{5} - 6 = -2$$

$$\frac{x}{5} - 6 + 6 = -2 + 6$$

$$\frac{x}{5} = 4.$$

Then, we isolate x by multiplying both sides by 5:

$$5\left(\frac{x}{5}\right) = 5(4)$$

$$x = 20.$$

You should check this solution by substituting 20 for x in the original equation.   ▲

When the variable appears on both sides of the equation, we must add or subtract terms containing variables as well as constant terms.

EXAMPLE 4.9   ▶   Solve the equation $4x - 5 = x + 10$ and check the solution.

SOLUTION   First, we collect the x-terms by subtracting x from both sides:

$$4x - 5 = x + 10$$

$$4x - 5 - x = x + 10 - x$$

$$3x - 5 = 10.$$

Now the equation is in the form of the preceding examples. We add 5 to both sides and then divide both sides by 3:

$$3x - 5 + 5 = 10 + 5$$

$$3x = 15$$

$$\frac{3x}{3} = \frac{15}{3}$$

$$x = 5.$$

To check, we substitute 5 for x, being sure to substitute for x in both x-terms in the original equation:

$$4x - 5 = x + 10$$

$$4(5) - 5 \overset{?}{=} 5 + 10$$

$$20 - 5 \overset{?}{=} 15$$

$$15 = 15.$$ ▲

It is sometimes convenient to collect the x-terms on the right-hand side of the equation, and the constants on the left-hand side.

EXAMPLE 4.10   ▶   Solve the equation $4x + 4 = 5x + 8$ and check the solution.

SOLUTION   We can avoid a negative x-term by collecting the x-terms on the right. We subtract $4x$ from both sides:

$$4x + 4 = 5x + 8$$

$$4x + 4 - 4x = 5x + 8 - 4x$$

$$4 = x + 8.$$

Now, we can subtract 8 from both sides:

$$4 - 8 = x + 8 - 8$$

$$-4 = x.$$

It is conventional to write this result with variable on the left:

$$x = -4.$$

You should check this solution by substituting -4 for x, being sure to substitute for x in both x-terms in the original equation.   ▲

Often, an equation must be reduced to an equivalent equation by simplifying the expressions on one or both sides of the equation. Any parentheses should be removed by using the distributive property, and any like terms that are on the same side should be combined.

EXAMPLE 4.11   Solve the equation $3x + 8 - 4x = x + 6$ and check the solution.

SOLUTION   First, we simplify the expression on the left-hand side by combining like terms:

$$3x + 8 - 4x = x + 6$$

$$-x + 8 = x + 6.$$

Now, we add x to both sides, eliminating the negative x-term:

$$-x + 8 + x = x + 6 + x$$

$$8 = 2x + 6.$$

We complete the solution by subtracting 6 from both sides and then dividing by 2:

$$8 - 6 = 2x + 6 - 6$$

$$2 = 2x$$

$$\frac{2}{2} = \frac{2x}{2}$$

$$1 = x$$

or

$$x = 1.$$

You should check this solution, being sure to substitute 1 for x in each of the three x-terms in the original equation.   ▲

EXAMPLE 4.12   Solve the equation $2x - (x - 1) = 3x$ and check the solution.

SOLUTION   We simplify the expression on the left-hand side by using the distributive property to remove the parentheses and then combining like terms:

$$2x - (x - 1) = 3x$$

$$2x - x + 1 = 3x$$

$$x + 1 = 3x.$$

Now, we solve the resulting equation by subtracting x from both sides and then dividing by the resulting coefficient:

$$x + 1 - x = 3x - x$$

$$1 = 2x$$

$$\frac{1}{2} = \frac{2x}{2}$$

$$\frac{1}{2} = x$$

or

$$x = \frac{1}{2}.$$

To check this solution, we substitute $\frac{1}{2}$ for x in each x-term of the original equation:

$$2x - (x - 1) = 3x$$

$$2\left(\frac{1}{2}\right) - \left(\frac{1}{2} - 1\right) \stackrel{?}{=} 3\left(\frac{1}{2}\right)$$

$$1 - \left(-\frac{1}{2}\right) \stackrel{?}{=} \frac{3}{2}$$

$$1 + \frac{1}{2} \stackrel{?}{=} \frac{3}{2}$$

$$\frac{3}{2} = \frac{3}{2}.$$

▲

EXAMPLE 4.13   Solve the equation $4x = 2(x - 3) - 2(2x - 5)$ and check the solution.

SOLUTION We simplify the expression on the right-hand side by using the distributive property to remove the parentheses and then combining like terms:

$$4x = 2(x - 3) - 2(2x - 5)$$

$$4x = 2x - 6 - 4x + 10$$

$$4x = -2x + 4.$$

Now, we solve the resulting equation by adding $2x$ to both sides and then dividing by the resulting coefficient:

$$4x + 2x = -2x + 4 + 2x$$

$$6x = 4$$

$$\frac{6x}{6} = \frac{4}{6}$$

$$x = \frac{2}{3}.$$

To check this solution, you can substitute the fraction $\frac{2}{3}$ for x in the original equation. Alternatively, you might approximate $\frac{2}{3}$ by 0.667, and check by using a calculator:

$$4x = 2(x - 3) - 2(2x - 5)$$

$$4(0.667) \stackrel{?}{=} 2(0.667 - 3) - 2[2(0.667) - 5]$$

$$2.67 \stackrel{?}{=} -4.67 + 7.33$$

$$2.67 \approx 2.66.$$

The symbol $\approx$ means "approximately equal to." Because we used an approximation for the solution, the two sides of the result are not exactly equal. A slight difference in the third digit due to rounding off is considered to be insignificant. ▲

**EXERCISE
4.2**

Solve the equation and check the solution:

1. $4x + 8 = 20$ 2. $3x - 5 = 13$ 3. $2x - 8 = 1$ 4. $6x + 1 = 10$

5. $7x + 30 = 9$ 6. $6x + 25 = 5$ 7. $\dfrac{x}{2} + 4 = 9$ 8. $\dfrac{x}{5} - 8 = -2$

9. $5x - 4 = 2x + 5$ 10. $7x + 2 = 5x + 10$

11. $3x + 10 = 4x + 9$ 12. $4x - 3 = 6x + 5$

13. $4x - 3 - x = 2x + 3$ 14. $x + 8 - 3x = 5x - 4 - x$

15. $4x - (x - 4) = 5x + 2$ 16. $5(2x - 1) - (x - 2) = 15$

17. $2(3x + 4) + 3(2x - 3) = 5$ 18. $2(4 - 5x) + 3(3 - 2x) = 5$

19. $5x - 2 = 4(x - 2) - 2(x - 4)$ 20. $4(3x - 1) = 6 - 2(4 - 3x)$

**SECTION
4.3**

Basic Rational Equations

The two preceding sections include some equations that contain a fraction with a variable in the numerator and a constant in the denominator. To solve such equations, we multiply both sides by the constant in the denominator. Recall, for example, that to solve the equation

$$\frac{x}{2} = 4,$$

we multiply both sides by 2:

$$2\left(\frac{x}{2}\right) = 2(4)$$

$$x = 8.$$

This equation is a linear equation because the term $\frac{x}{2}$ is of the form ax where $a = \frac{1}{2}$.

An equation such as

$$\frac{2}{x} = 4,$$

which has a variable in the denominator, is not a linear equation. For an equation to be linear, its expressions must contain only linear terms. We recall that the only forms for linear terms are a, x, or ax. Terms of the form $\frac{a}{x}$ are not linear terms, and thus the equation is not a linear equation.

An equation with a variable in a denominator is called a **rational equation**. We can solve such equations by multiplying both sides by the variable, or an expression involving the variable, *if the expression is not zero*.

EXAMPLE 4.14   Solve the equation $\dfrac{2}{x} = 4$, and check the solution.

SOLUTION   We assume x will not be zero, and multiply both sides by x:

$$\frac{2}{x} = 4$$

$$x\left(\frac{2}{x}\right) = x(4)$$

$$\frac{x(2)}{x} = 4x.$$

If x is not zero, then $\frac{x}{x} = 1$. Therefore, we can divide out x to reduce the equation to an equivalent linear equation:

$$\frac{x(2)}{x} = 4x$$

$$2 = 4x.$$

We divide both sides by 4 to complete the solution:

$$\frac{2}{4} = \frac{4x}{4}$$

$$\frac{1}{2} = x,$$

or

$$x = \frac{1}{2}.$$

We observe that x is not zero, therefore the assumption we made to multiply by x is valid. To check this solution we substitute $\frac{1}{2}$ for x in the original equation:

$$\frac{2}{x} = 4$$

$$\frac{2}{\frac{1}{2}} \overset{?}{=} 4.$$

Recall that, to divide by a fraction, we "invert and multiply." Thus, we have

$$2\left(\frac{2}{1}\right) \overset{?}{=} 4$$

$$4 = 4.$$

Alternatively, you can write $\frac{1}{2}$ as 0.5, and check by using a calculator.   ▲

EXAMPLE 4.15 ▶ Solve the equation $\dfrac{3}{x} + 1 = 3$, and check the solution.

SOLUTION First, we isolate the term with x in the denominator by subtracting 1 from both sides:

$$\frac{3}{x} + 1 = 3$$

$$\frac{3}{x} + 1 - 1 = 3 - 1$$

$$\frac{3}{x} = 2.$$

Now, we assume x will not be zero, and multiply both sides by x:

$$x\left(\frac{3}{x}\right) = x(2).$$

If x is not zero, we can divide out x to obtain the linear equation

$$3 = 2x.$$

We divide both sides by 2 to complete the solution:

$$\frac{3}{2} = \frac{2x}{2}$$

$$\frac{3}{2} = x,$$

or

$$x = \frac{3}{2}.$$

We observe that x is not zero, therefore the assumption we made to multiply by x is valid. You should check this solution by substituting $\frac{3}{2}$ for x in the original equation.   ▲

EXAMPLE 4.16   Solve the equation $\frac{1}{x} + 2 = \frac{5}{x}$ and check the solution.

SOLUTION   It is possible to combine the two rational terms on one side of the equation by using x as the common denominator. It is often easier, however, to eliminate the denominators by multiplying both sides by the common denominator. Thus, we assume x will not be zero, and multiply both sides by x:

$$\frac{1}{x} + 2 = \frac{5}{x}$$

$$x\left(\frac{1}{x} + 2\right) = x\left(\frac{5}{x}\right).$$

We apply the distributive property to the expression on the left-hand side to obtain

$$x\left(\frac{1}{x}\right) + x(2) = x\left(\frac{5}{x}\right).$$

Then, if x is not zero, we can divide out x from both rational terms and solve the resulting linear equation to obtain

$$1 + 2x = 5$$
$$2x = 4$$
$$x = 2.$$

We observe that x is not zero, therefore the assumption we made to multiply by x is valid. To check this solution we substitute 2 for x in both rational terms in the original equation:

$$\frac{1}{x} + 2 = \frac{5}{x}$$

$$\frac{1}{2} + 2 \stackrel{?}{=} \frac{5}{2}$$

$$\frac{1}{2} + \frac{4}{2} \stackrel{?}{=} \frac{5}{2}$$

$$\frac{5}{2} = \frac{5}{2}.$$ ▲

COMMON DENOMINATORS

When two or more terms in an equation have different denominators, we multiply by a common denominator. We choose a common denominator that will eliminate all of the denominators in the equation.

EXAMPLE 4.17   Solve the equation $\frac{1}{2} - \frac{2}{x} = \frac{1}{x}$ and check the solution.

SOLUTION   The common denominator $2x$ will eliminate all of the denominators. We assume that the expression $2x$ will not be zero, and multiply both sides by $2x$:

$$\frac{1}{2} - \frac{2}{x} = \frac{1}{x}$$

$$2x\left(\frac{1}{2} - \frac{2}{x}\right) = 2x\left(\frac{1}{x}\right).$$

Applying the distributive property to the expression on the left-hand side, we have

$$2x\left(\frac{1}{2}\right) - 2x\left(\frac{2}{x}\right) = 2x\left(\frac{1}{x}\right).$$

We can divide out 2 from the first term and, if x is not zero, we can divide out x from the other terms, to obtain

$$x(1) - 2(2) = 2(1)$$
$$x - 4 = 2$$
$$x = 6.$$

We observe that x is not zero, therefore $2x$ is not zero and the assumption we made to multiply by $2x$ is valid. You should check this solution, being sure to substitute for x in both rational terms in the original equation.   ▲

EXAMPLE 4.18   Solve the equation $\frac{1}{2} + \frac{1}{x} = \frac{5}{6}$ and check the solution.

SOLUTION

Because 6 is divisible by both 2 and 6, the common denominator $6x$ will eliminate all of the denominators. We assume that the expression $6x$ will not be zero, and multiply both sides by $6x$:

$$\frac{1}{2} + \frac{1}{x} = \frac{5}{6}$$

$$6x\left(\frac{1}{2} + \frac{1}{x}\right) = 6x\left(\frac{5}{6}\right)$$

$$6x\left(\frac{1}{2}\right) + 6x\left(\frac{1}{x}\right) = 6x\left(\frac{5}{6}\right).$$

We can divide out 2 from the first term, 6 from the last term, and if x is not zero, we can divide out x from the second term to obtain

$$3x(1) + 6(1) = x(5)$$
$$3x + 6 = 5x.$$

Then, subtracting $3x$ from both sides, we have

$$6 = 2x$$
$$3 = x,$$

or

$$x = 3.$$

We observe that x is not zero, therefore $6x$ is not zero and the assumption we made to multiply by $6x$ is valid. You should check this solution in the original equation.   ▲

EXAMPLE 4.19   Solve the equation $\frac{5}{x-1} = \frac{2}{x-1} + \frac{3}{2}$ and check the solution.

SOLUTION

The common denominator $2(x - 1)$ will eliminate all of the denominators. We assume that the expression $2(x - 1)$ will not be zero; that is, x will not be 1. Multiplying both sides by $2(x - 1)$, we have

$$\frac{5}{x-1} = \frac{2}{x-1} + \frac{3}{2}$$

$$2(x - 1)\left(\frac{5}{x-1}\right) = 2(x - 1)\left(\frac{2}{x-1} + \frac{3}{2}\right).$$

We apply the distributive property to the expression on the right-hand side, multiplying each term in the large parentheses by the common denominator $2(x - 1)$:

$$2(x - 1)\left(\frac{5}{x - 1}\right) = 2(x - 1)\left(\frac{2}{x - 1}\right) + 2(x - 1)\left(\frac{3}{2}\right).$$

We can divide out 2 from the last term and, if $x - 1$ is not zero, we can divide out $x - 1$ from the first and second terms, and solve the resulting linear equation:

$$2(x - 1)\left(\frac{5}{x - 1}\right) = 2(x - 1)\left(\frac{2}{x - 1}\right) + 2(x - 1)\left(\frac{3}{2}\right)$$

$$2(5) = 2(2) + (x - 1)(3)$$

$$10 = 4 + 3x - 3$$

$$10 = 1 + 3x$$

$$9 = 3x$$

$$3 = x$$

or

$$x = 3.$$

We observe that x is not 1, therefore $2(x - 1)$ is not zero, and the assumption we made to multiply by $2(x - 1)$ is valid. To check this solution we substitute x = 3 in both rational terms in the original equation:

$$\frac{5}{x - 1} = \frac{2}{x - 1} + \frac{3}{2}$$

$$\frac{5}{3 - 1} \stackrel{?}{=} \frac{2}{3 - 1} + \frac{3}{2}$$

$$\frac{5}{2} \stackrel{?}{=} \frac{2}{2} + \frac{3}{2}$$

$$\frac{5}{2} = \frac{5}{2}.$$

▲

| EXAMPLE 4.20 | ► | Solve the equation $\dfrac{2}{x - 1} + 1 = \dfrac{2x}{x - 1}$ and check the solution.

SOLUTION   We assume that the expression $x - 1$ will not be zero; that is, x will not be 1. Multiplying both sides by $x - 1$, we have

$$(x - 1)\left(\frac{2}{x - 1} + 1\right) = (x - 1)\left(\frac{2x}{x - 1}\right).$$

Applying the distributive property to the left-hand side, we have

$$(x - 1)\left(\frac{2}{x - 1}\right) + (x - 1)(1) = (x - 1)\left(\frac{2x}{x - 1}\right).$$

If $x - 1$ is not zero, we can divide out $x - 1$ from the first and last terms, and solve the resulting linear equation:

$$(x - 1)\left(\frac{2}{x - 1}\right) + (x - 1) = (x - 1)\left(\frac{2x}{x - 1}\right)$$

$$2 + x - 1 = 2x$$

$$1 + x = 2x$$

$$1 = x$$

or

$$x = 1.$$

We observe that x *is* 1, so the expression $x - 1$ by which we multiplied *is zero*. Therefore, our assumption in multiplying by $x - 1$ *is not valid*, and our result $x = 1$ is not a solution. If we attempt to check by substituting $x = 1$ in the original equation, we have

$$\frac{2}{x - 1} + 1 = \frac{2x}{x - 1}$$

$$\frac{2}{1 - 1} + 1 \stackrel{?}{=} \frac{2(1)}{1 - 1} .$$

This attempt results in division by zero. But division by zero is not possible with real numbers; therefore, $x = 1$ is not a solution of the equation. A solution such as this, which is derived by correct methods but does not fit the equation, is called an **extraneous solution**. Since we have derived no other possible solutions, we say that the equation has **no solution**. ▲

EXERCISE 4.3

Solve the equation and check the solution:

1. $\dfrac{2}{x} = 3$

2. $\dfrac{1}{2x} = 2$

3. $\dfrac{3}{x} + 4 = 7$

4. $\dfrac{1}{x} + 3 = 2$

5. $\dfrac{2}{x} + 1 = 7$

6. $\dfrac{3}{2x} - 1 = 1$

7. $\dfrac{1}{x} + 1 = \dfrac{5}{x}$

8. $\dfrac{4}{x} - 1 = \dfrac{1}{x}$

9. $\dfrac{1}{3} - \dfrac{2}{x} = \dfrac{1}{x}$

10. $\dfrac{5}{x} - \dfrac{1}{2} = \dfrac{3}{x}$

11. $\dfrac{3}{x} - \dfrac{1}{5} = \dfrac{1}{x}$

12. $\dfrac{1}{4} + \dfrac{3}{x} = \dfrac{4}{x}$

13. $\dfrac{1}{4} + \dfrac{1}{x} = \dfrac{1}{3}$

14. $\dfrac{2}{5} - \dfrac{3}{x} = \dfrac{1}{10}$

15. $\dfrac{4}{x - 2} = \dfrac{1}{x - 2} + \dfrac{3}{2}$

16. $\dfrac{3}{x - 3} - \dfrac{1}{2} = \dfrac{2}{x - 3}$

17. $\dfrac{3}{x + 1} - 1 = \dfrac{x}{x + 1}$

18. $\dfrac{2x}{x + 2} - \dfrac{1}{5} = \dfrac{5}{x + 2}$

19. $\dfrac{x}{x + 1} = 2 - \dfrac{1}{x + 1}$

20. $\dfrac{1}{2x - 1} - \dfrac{1}{2} = \dfrac{2x}{2x - 1}$

SECTION 4.4

Basic Quadratic and Radical Equations

Equations such as

$$x^2 = 4,$$

that include the square of a variable, are not linear equations. To solve such equations, we take the square root of each side of the equation. For example, if

$$x^2 = 4,$$

then

$$\sqrt{x^2} = \sqrt{4}.$$

We recall from Section 3.4 that $\sqrt{x^2} = x$ if $x \geq 0$. Then, for $x \geq 0$, we have

$$x = 2.$$

To check, if we substitute 2 for x, we have

$$x^2 = 4$$

$$2^2 \stackrel{?}{=} 4$$

$$4 = 4.$$

But, if we substitute -2 for x, we have

$$(-2)^2 \stackrel{?}{=} 4$$

$$(-2)(-2) \stackrel{?}{=} 4$$

$$4 = 4.$$

Therefore, -2 is also a solution. Thus, the equation

$$x^2 = 4$$

has two solutions

$$x = 2 \text{ and } x = -2.$$

We summarize these two solutions by writing

$$x = \pm 2.$$

An equation that includes the square of a variable is called a **quadratic equation**. We can solve such equations by solving for the square of the variable and then taking the square root of each side of the equation. We must *include both the positive and negative roots of the constant.*

EXAMPLE 4.21   Solve the equation $x^2 = 10$ and check the solutions.

SOLUTION   The equation is solved for x^2, so we take the square root of each side. We must include both the positive and negative roots of the constant:

$$x^2 = 10$$

$$\sqrt{x^2} = \pm\sqrt{10}$$

$$x = \pm\sqrt{10}.$$

You can use a calculator to find that $x = 3.16$ and $x = -3.16$ to three significant digits, and check each solution in the original equation. The slight difference you will have in the third digit due to rounding off is insignificant.   ▲

EXAMPLE 4.22   Solve the equation $x^2 - 1 = 3$ and check the solutions.

SOLUTION   First, we solve for x^2 by adding 1 to both sides:

$$x^2 - 1 = 3$$

$$x^2 = 4.$$

Then, we take the square root of each side, including both the positive and negative roots of the constant:

$$\sqrt{x^2} = \pm\sqrt{4}$$

$$x = \pm 2.$$

Thus, there are two possible solutions, $x = 2$ and $x = -2$. We check each possible solution in the original equation. Substituting 2 for x, we have

$$x^2 - 1 = 3$$

$$2^2 - 1 \stackrel{?}{=} 3$$

$$4 - 1 \stackrel{?}{=} 3$$

$$3 = 3.$$

Substituting -2 for x, we have

$$x^2 - 1 = 3$$
$$(-2)^2 - 1 \stackrel{?}{=} 3$$
$$4 - 1 \stackrel{?}{=} 3$$
$$3 = 3.$$

Therefore, $x = 2$ and $x = -2$ both are solutions of the equation.   ▲

EXAMPLE 4.23   Solve the equation $4x^2 + 4 = 19$ and check the solutions.

SOLUTION   First, we solve for x^2:

$$4x^2 + 4 = 19$$
$$4x^2 + 4 - 4 = 19 - 4$$
$$4x^2 = 15$$
$$\frac{4x^2}{4} = \frac{15}{4}$$
$$x^2 = \frac{15}{4}.$$

Then, we take the square root of each side:

$$\sqrt{x^2} = \pm\sqrt{\frac{15}{4}}$$
$$x = \pm\sqrt{\frac{15}{4}}.$$

You can use a calculator to find that $x = 1.94$ and $x = -1.94$ to three significant digits. You also can check each possible solution in the original equation by using a calculator. For example, replacing x by 1.94, you will find

$$4x^2 + 4 = 19$$
$$4(1.94)^2 + 4 \stackrel{?}{=} 19$$
$$19.1 \approx 19.$$

Again, the slight difference in the third digit due to rounding off is insignificant. You should check that $x = -1.94$ is also a solution.   ▲

EXAMPLE 4.24   Solve the equation $\dfrac{x^2}{5} + 6 = 1$ and check the solutions.

SOLUTION   First, we solve for x^2:

$$\frac{x^2}{5} + 6 = 1$$
$$\frac{x^2}{5} + 6 - 6 = 1 - 6$$
$$\frac{x^2}{5} = -5$$
$$5\left(\frac{x^2}{5}\right) = 5(-5)$$
$$x^2 = -25.$$

If we take the square root of each side, we have

$$\sqrt{x^2} = \pm\sqrt{-25}$$

$$x = \pm\sqrt{-25}.$$

But the square root of a negative is not defined for real numbers; therefore, $x = \sqrt{-25}$ and $x = -\sqrt{-25}$ do not give solutions of the equation. Since we have derived no other possible solutions, we say that the equation has no solution.   ▲

RADICAL EQUATIONS

Equations such as

$$\sqrt{x} = 2,$$

that include the square root of a variable, also are not linear equations. To solve such equations, we square each side of the equation. For example, if

$$\sqrt{x} = 2$$

then

$$(\sqrt{x})^2 = 2^2.$$

By using fractional exponents, it can be shown that $(\sqrt{x})^2 = x$ if $x \geq 0$. Therefore, we have

$$x = 4.$$

To check, if we substitute 4 for x, we have

$$\sqrt{x} = 2$$

$$\sqrt{4} \overset{?}{=} 2$$

$$2 = 2.$$

But, if we start with

$$\sqrt{x} = -2,$$

and square both sides, we can also derive

$$(\sqrt{x})^2 = (-2)^2$$

$$(\sqrt{x})^2 = (-2)(-2)$$

$$x = 4.$$

To check, if we substitute 4 for x, we have

$$\sqrt{x} = -2$$

$$\sqrt{4} \overset{?}{=} -2$$

$$2 \neq -2.$$

(Recall that the symbol $\neq$ means "is not equal to.") The result $x = 4$ is not a solution of the second equation. Such a result, derived by correct methods, is an extraneous solution.

We call an equation that includes the square root of a variable a **radical equation**. We can solve such equations by isolating the square root of the variable and then squaring each side of the equation. We must *check all results for possible extraneous solutions*.

EXAMPLE 4.25   Solve the equation $2\sqrt{x} - 3 = 5$ and check the solution.

SOLUTION   First, we solve for $\sqrt{x}$:

$$2\sqrt{x} - 3 = 5$$

$$2\sqrt{x} - 3 + 3 = 5 + 3$$

$$2\sqrt{x} = 8$$

$$\frac{2\sqrt{x}}{2} = \frac{8}{2}$$

$$\sqrt{x} = 4.$$

Then, we square each side:

$$(\sqrt{x})^2 = 4^2$$

$$x = 16.$$

We check this possible solution in the original equation:

$$2\sqrt{x} - 3 = 5$$

$$2\sqrt{16} - 3 \overset{?}{=} 5$$

$$2(4) - 3 \overset{?}{=} 5$$

$$5 = 5.$$

Therefore, $x = 16$ is the solution of the equation.

EXAMPLE 4.26 ▶ Solve the equation $1 - \dfrac{1}{\sqrt{x}} = 2$ and check the solution.

SOLUTION First, we solve for $\sqrt{x}$:

$$1 - \frac{1}{\sqrt{x}} = 2$$

$$1 - \frac{1}{\sqrt{x}} - 1 = 2 - 1$$

$$-\frac{1}{\sqrt{x}} = 1$$

$$\sqrt{x}\left(-\frac{1}{\sqrt{x}}\right) = \sqrt{x}(1)$$

$$-1 = \sqrt{x}.$$

If we square each side, we have

$$(-1)^2 = (\sqrt{x})^2$$

$$1 = x$$

or

$$x = 1.$$

We check this possible solution in the original equation:

$$1 - \frac{1}{\sqrt{x}} = 2$$

$$1 - \frac{1}{\sqrt{1}} \overset{?}{=} 2$$

$$1 - \frac{1}{1} \overset{?}{=} 2$$

$$0 \neq 2.$$

Therefore, the result $x = 1$ is an extraneous solution. Since we have derived no other possible solutions, we say that the equation has no solution. ▲

EXERCISE 4.4

Solve the equation and check the solution:

1. $x^2 = 25$ 2. $x^2 = 121$ 3. $x^2 = 13$ 4. $x^2 = 40$

5. $x^2 - 8 = 8$ 6. $x^2 + 1 = 10$ 7. $3x^2 + 5 = 32$ 8. $\frac{1}{2}x^2 - 30 = 20$

9. $2x^2 - 6 = 5$ 10. $5x^2 + 16 = 20$ 11. $5x^2 + 36 = 16$ 12. $1 - \frac{1}{4}x^2 = 5$

13. $3 + \frac{x^2}{4} = 5$ 14. $\frac{x^2}{5} - 5 = 20$ 15. $x - \frac{1}{x} = \frac{x}{2}$ 16. $\frac{x}{2} + \frac{x}{3} = \frac{4}{x}$

17. $3\sqrt{x} - 1 = 5$ 18. $2\sqrt{x} - 2 = 3$ 19. $\frac{\sqrt{x}}{2} + \frac{3}{2} = 3$ 20. $1 - \frac{\sqrt{x}}{2} = \frac{1}{2}$

21. $\frac{\sqrt{x}}{2} + 3 = \frac{5}{2}$ 22. $2 - \frac{\sqrt{x}}{3} = 4$ 23. $\frac{2}{5}\sqrt{x} - 1 = \frac{1}{2}$ 24. $\frac{1}{\sqrt{x}} - \frac{1}{2} = \frac{1}{3}$

SECTION 4.5

Ohm's Law and Power

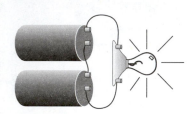

A simple electrical circuit.

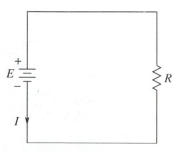

A schematic diagram for the circuit.

A simple electrical circuit consists of an electrical energy source, a resistor, and conductors. For example, such a circuit could consist of battery cells connected with a light bulb by wires. The battery cells are a source that provides the electrical energy. The light bulb is a resistor that uses the energy supplied. The wires are conductors that carry electricity between the source and the resistor.

In electronics, circuits are represented by schematic diagrams in which each component has a symbol. The figure at the left below is a schematic diagram for the circuit above.

The symbol on the left of the schematic diagram represents the battery. In this book, we will use this symbol for all electrical energy sources that provide **direct current** (DC). The symbol is labeled E for **electromotive force** (EMF), and may also be labeled V for applied **voltage**. The unit of measure for electromotive force and voltage is the **volt** (V), named for Count Alessandro Volta (1745–1827), an Italian physicist.

The symbol on the right represents the resistor. We will use this symbol for all types of resistors. The symbol is labeled R for the **resistor** or for the **resistance** it provides. The unit of measure for resistance is the **ohm** (Ω, the Greek capital letter omega), named for Georg Simon Ohm (1787–1854), a German physicist.

The straight lines connecting the source and the load represent the wires, which carry the **current**. The current is labeled I (C means something else in electronics). The arrow shows the direction of the current. The unit of measure for current is the **ampere** (A), named for André Marie Ampère (1775–1836), a French physicist.

In this book, we will take the direction of current flow from the negative terminal of E, represented by the shorter line of the battery symbol, to the positive terminal. This is called **electron flow** current. The direction of current flow may also be taken from the positive terminal to the negative terminal, which was the original convention and is called **conventional** current.

Georg Ohm, for whom the ohm is named, discovered a fundamental law that relates voltage, resistance, and current.

Ohm's law: In a simple circuit with applied voltage V and resistance R, the current I is given by

$$I = \frac{V}{R}.$$

EXAMPLE 4.27 Find I for $V = 12$ V and $R = 2$ Ω.

SOLUTION   We use Ohm's law, substituting 12 V for V and 2 Ω for R:

$$I = \frac{V}{R}$$

$$I = \frac{12 \text{ V}}{2 \text{ Ω}}.$$

We perform the division, and write the unit amperes:

$$I = 6 \text{ A}.$$ ▲

EXAMPLE 4.28 Find I for $V = 9$ V and $R = 1.8$ kΩ.

SOLUTION   We use Ohm's law, substituting 9 V for V and 1.8 kΩ for R:

$$I = \frac{V}{R}$$

$$I = \frac{9 \text{ V}}{1.8 \text{ kΩ}}.$$

Replacing k- by the power of ten it represents, and writing the unit amperes, we have

$$I = \frac{9}{1.8 \times 10^3} \text{ A}.$$

Now, we perform the division, applying the negative exponent rule to the power of ten:

$$I = 5 \times 10^{-3} \text{ A}$$

$$I = 5 \text{ mA}.$$ ▲

By using the fundamental methods for solving equations, we can derive two other forms of Ohm's law. To solve Ohm's law for V, we multiply both sides of the original form by R:

$$I = \frac{V}{R}$$

$$RI = R\left(\frac{V}{R}\right)$$

$$RI = V.$$

This form of Ohm's law is usually written

$$V = IR.$$

EXAMPLE 4.29 Find V for $I = 0.5$ μA and $R = 2.2$ kΩ.

SOLUTION   We solve Ohm's law for V, and then substitute 0.5 μA for I and 2.2 kΩ for R:

$$V = IR$$

$$V = (0.5 \text{ μA})(2.2 \text{ kΩ}).$$

We replace μ- and k- by the powers of ten they represent, write the unit volts, and multiply by adding the exponents:

$$V = (0.5 \times 10^{-6})(2.2 \times 10^3) \text{ V}$$

$$= 1.1 \times 10^{-3} \text{ V}$$

$$= 1.1 \text{ mV}.$$ ▲

To solve Ohm's law for R, we divide both sides of the previous version by I:

$$V = IR$$

$$\frac{V}{I} = \frac{IR}{I}$$

$$\frac{V}{I} = R$$

or

$$R = \frac{V}{I}.$$

EXAMPLE 4.30 ▶ Find R for $V = 30$ V and $I = 40$ mA.

SOLUTION We solve Ohm's law for R and then substitute 30 V for V and 40 mA for I:

$$R = \frac{V}{I}$$

$$R = \frac{30 \text{ V}}{40 \text{ mA}}.$$

We replace m- by the power of ten it represents, write the unit ohms, and divide by using the negative exponent rule:

$$R = \frac{30}{40 \times 10^{-3}} \; \Omega$$

$$= 0.75 \times 10^3 \; \Omega$$

$$= 0.75 \text{ k}\Omega.$$

We can write this result in the form

$$R = 750 \; \Omega. \qquad\qquad ▲$$

POWER

A resistor such as a light bulb dissipates power. The unit of measure for power is the watt (W), which we used in examples of engineering notation in Section 2.4. A basic definition relates voltage, current, and power.

> **Power:** In a simple circuit with applied voltage V and current I, the power P is given by
>
> $$P = VI.$$

EXAMPLE 4.31 ▶ Find P for $V = 12$ V and $I = 100$ μA.

SOLUTION We use the definition of power, substituting 12 V for V and 100 μA for I:

$$P = VI$$

$$P = (12 \text{ V})(100 \text{ μA}).$$

We replace μ- by the power of ten it represents, write the unit watts, and multiply:

$$P = (12)(100 \times 10^{-6}) \text{ W}$$

$$= 1200 \times 10^{-6} \text{ W}.$$

We get the best form of this result by using mW:

$$P = 1.2 \times 10^{-3} \text{ W}$$

$$= 1.2 \text{ mW.} \qquad \blacktriangle$$

We can derive two other formulas from the power definition by using Ohm's law. We can write the power formula in terms of I and R by solving Ohm's law for V:

$$V = IR.$$

Then, we replace V by IR in the power formula:

$$P = VI$$

$$P = (IR)I$$

$$P = I^2R.$$

EXAMPLE 4.32   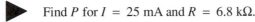   Find P for $I = 25$ mA and $R = 6.8$ kΩ.

SOLUTION   We use the power formula written in terms of I and R, and substitute 25 mA for I and 6.8 kΩ for R:

$$P = I^2R$$

$$P = (25 \text{ mA})^2(6.8 \text{ k}\Omega).$$

We replace the powers of ten by the prefixes they represent, and write the unit watts:

$$P = (25 \times 10^{-3})^2(6.8 \times 10^3) \text{ W.}$$

We must evaluate the square before we can multiply or combine the powers of ten:

$$P = [25^2 \times (10^{-3})^2](6.8 \times 10^3) \text{ W}$$

$$= (25^2 \times 10^{-6})(6.8 \times 10^3) \text{ W}$$

$$= 4250 \times 10^{-3} \text{ W}$$

$$= 4.25 \text{ W.} \qquad \blacktriangle$$

We can also write the power formula in terms of V and R by using Ohm's law solved for I:

$$I = \frac{V}{R}.$$

Then, we replace I by $\frac{V}{R}$ in the power formula:

$$P = VI$$

$$P = V\left(\frac{V}{R}\right)$$

$$P = \frac{V^2}{R}.$$

EXAMPLE 4.33   Find P for $V = 120$ mV and $R = 22$ kΩ.

SOLUTION   We use the power formula written in terms of V and R, and substitute 120 mV for V and 22 kΩ for R:

$$P = \frac{V^2}{R}$$

$$= \frac{(120 \text{ mV})^2}{22 \text{ k}\Omega}$$

$$= \frac{(120 \times 10^{-3})^2}{22 \times 10^3} \text{ W.}$$

Remembering to square before dividing or combining the powers of ten, we have

$$P = \frac{120^2 \times (10^{-3})^2}{22 \times 10^3} \text{ W}$$

$$= \frac{120^2 \times 10^{-6}}{22 \times 10^3} \text{ W}$$

$$= 655 \times 10^{-9} \text{ W}$$

$$= 655 \text{ nW}$$

to three significant digits.   ▲

There are two other useful equations that can be derived from the power formulas. To solve for I in terms of P and R, we begin with the power formula written in terms of I and R, and solve for I^2:

$$P = I^2 R$$

$$\frac{P}{R} = \frac{I^2 R}{R}$$

$$\frac{P}{R} = I^2$$

or

$$I^2 = \frac{P}{R}.$$

Now, we can take the square root of each side:

$$\sqrt{I^2} = \sqrt{\frac{P}{R}}.$$

In this section, we will always take I to be positive. Therefore, we have

$$I = \sqrt{\frac{P}{R}}.$$

EXAMPLE 4.34 ▶ Find I for $P = 220 \text{ μW}$ and $R = 20 \text{ kΩ}$.

SOLUTION We write the power formula in terms of I and R, and solve for I. Then, we can substitute 220 μW for P and 20 kΩ for R:

$$I = \sqrt{\frac{P}{R}}$$

$$= \sqrt{\frac{220 \text{ μW}}{20 \text{ kΩ}}}$$

$$= \sqrt{\frac{220 \times 10^{-6}}{20 \times 10^3}} \text{ A}$$

$$= \sqrt{11 \times 10^{-9}} \text{ A}.$$

We can write 10^{-9} as $10^{-1} \times 10^{-8}$ to obtain

$$I = \sqrt{11 \times 10^{-1} \times 10^{-8}} \text{ A}$$

$$= \sqrt{1.1 \times 10^{-8}} \text{ A}$$

$$= \sqrt{1.1} \times \sqrt{10^{-8}} \text{ A}$$

$$= 1.05 \times 10^{-4} \text{ A}.$$

Returning this result to engineering notation, we have

$$I = 105 \times 10^{-6} \text{ A}$$

$$= 105 \text{ μA}$$

to three significant digits.   ▲

    To solve for V in terms of P and R, we begin with the power formula written in terms of V and R, and solve for V^2:

$$P = \frac{V^2}{R}$$

$$PR = \left(\frac{V^2}{R}\right)R$$

$$PR = V^2$$

or

$$V^2 = PR.$$

Now, we can take the square root of each side:

$$\sqrt{V^2} = \sqrt{PR}.$$

In this section, we will always take V to be positive. Therefore, we have

$$V = \sqrt{PR}.$$

EXAMPLE 4.35 ▶ Find V for $P = 600 \text{ μW}$ and $R = 3.9 \text{ kΩ}$.

SOLUTION We write the power formula in terms of V and R, and solve for V. Then, we can substitute 600 μW for P and 3.9 kΩ for R:

$$V = \sqrt{PR}$$

$$= \sqrt{(600 \text{ μW})(3.9 \text{ kΩ})}$$

$$= \sqrt{(600 \times 10^{-6})(3.9 \times 10^{3})} \text{ V}$$

$$= \sqrt{2340 \times 10^{-3}} \text{ V}.$$

We can write the number in ordinary notation to find

$$V = \sqrt{2.34} \text{ V}$$

$$= 1.53 \text{ V}$$

to three significant digits.   ▲

EXERCISE 4.5

Use Ohm's law to find I for:

1. $V = 66 \text{ V}$ and $R = 3.3 \text{ Ω}$ 2. $V = 10 \text{ V}$ and $R = 2.2 \text{ Ω}$

3. $V = 500 \text{ V}$ and $R = 30 \text{ kΩ}$ 4. $V = 6 \text{ V}$ and $R = 1.5 \text{ MΩ}$

Solve Ohm's law for V to find V for:

5. $I = 2 \text{ mA}$ and $R = 2.7 \text{ kΩ}$ 6. $I = 0.1 \text{ mA}$ and $R = 56 \text{ kΩ}$

7. $I = 200 \text{ μA}$ and $R = 75 \text{ kΩ}$ 8. $I = 50 \text{ μA}$ and $R = 1.2 \text{ kΩ}$

Solve Ohm's law for R to find R for:

9. $V = 30 \text{ V}$ and $I = 3 \text{ mA}$ 10. $V = 120 \text{ V}$ and $I = 25 \text{ μA}$

11. $V = 40 \text{ mV}$ and $I = 16 \text{ μA}$ 12. $V = 250 \text{ mV}$ and $I = 0.5 \text{ mA}$

Use the definition of power to find P for:

13. $V = 50$ V and $I = 20$ mA

14. $V = 9$ V and $I = 160$ μA

15. $V = 750$ mV and $I = 2.4$ mA

16. $V = 900$ mV and $I = 660$ μA

Write the power formula in terms of I and R to find P for:

17. $I = 2.5$ mA and $R = 120$ kΩ

18. $I = 400$ μA and $R = 27$ MΩ

Write the power formula in terms of V and R to find P for:

19. $V = 12$ V and $R = 2.2$ kΩ

20. $V = 165$ mV and $R = 3.9$ Ω

Write the power formula in terms of I and R, and solve for I, to find I for:

21. $P = 1.6$ W and $R = 6.8$ kΩ

22. $P = 220$ mW and $R = 82$ MΩ

Write the power formula in terms of V and R, and solve for V, to find V for:

23. $P = 0.4$ W and $R = 33$ kΩ

24. $P = 5.5$ mW and $R = 1.8$ MΩ

☐ Self-Test ☐

Solve the equation and check the solution:

1. $3(5x - 4) - 2(3x - 3) = 6$

1. _____

2. $\dfrac{2}{15}x = \dfrac{8}{5}$

2. _____

3. $\dfrac{2}{x} - \dfrac{2}{3} = \dfrac{4}{x}$

3. _____

4. $3x - \dfrac{3}{x} = \dfrac{x}{3}$

4. _____

5. Write the power formula in terms of V and R to find P for $V = 250$ mV and $R = 56$ kΩ.

5. _____

UNIT 5

Formulas

Introduction

In the preceding units, you learned how to evaluate and simplify expressions, and how to solve equations in one variable. You applied these techniques to Ohm's law, the power definition, and related formulas. A formula is a type of equation that contains two or more variables. In this unit you will learn how to evaluate formulas for specified values of variables, and how to solve formulas for a specified variable. Then, you will apply these techniques to some specific types of formulas. You will write and solve variation formulas, which are used in many applications. Also, you will write and solve proportions, and apply proportions to an electrical device called a transformer.

OBJECTIVES

When you have finished this unit you should be able to:

1. Evaluate formulas given numerical values for the independent variables.

2. Solve formulas for a specified variable.

3. Write direct and inverse variation formulas, find the constant of variation, and solve for a specified variable.

4. Write proportions, and solve for a specified variable.

5. Use proportions relating the primary and secondary coils of transformers to find numbers of turns, voltages, currents, and impedances.

SECTION 5.1

Evaluating Formulas

A **formula** is an equation containing two or more variables. Generally, the left-hand side of a formula consists of a single variable, called the **dependent variable**. The right-hand side is an expression containing one or more different variables, called **independent variables**. For example, Ohm's law, which we used in Section 4.5, is a formula containing three variables. When Ohm's law is written in the form

$$I = \frac{V}{R},$$

I is the dependent variable, and V and R are independent variables. The power definition and related equations in Section 4.5 are also examples of formulas. These are some other examples of formulas:

$$y = mx + b, \quad I = PRT, \quad X_L = 2\pi fL.$$

Recall from Section 3.1 that, to evaluate an expression, we substitute given quantities for the variables in the expression and then calculate the number it represents. To **evaluate a formula**, we substitute given quantities for the independent variables and evaluate the expression on the right-hand side of the formula.

EXAMPLE 5.1   Evaluate $I = \dfrac{V}{R}$ for $V = 800$ mV and $R = 91$ kΩ.

SOLUTION   We substitute 800 mV for V and 91 kΩ for R in the expression on the right-hand side of the formula:

$$I = \frac{V}{R}$$

$$I = \frac{800 \text{ mV}}{91 \text{ kΩ}}.$$

Then, we evaluate the expression on the right-hand side to find

$$I = \frac{800 \times 10^{-3}}{91 \times 10^{3}} \text{ A}$$

$$= 8.79 \times 10^{-6} \text{ A}$$

$$= 8.79 \text{ μA}. \qquad \blacktriangle$$

STRAIGHT LINE FORMULAS

Formulas of the form

$$y = mx + b$$

represent straight lines. Usually, y is the dependent variable, x is the independent variable, and m and b are replaced by constants. We will study such formulas in Unit 10.

EXAMPLE 5.2   Evaluate $y = 2x + 5$ for $x = -1$.

SOLUTION   We substitute -1 for x in the expression on the right-hand side of the formula, and evaluate the expression:

$$y = 2x + 5$$

$$= 2(-1) + 5$$

$$= -2 + 5$$

$$= 3. \qquad \blacktriangle$$

EXAMPLE 5.3   Evaluate the formula $y = mx + b$ for $m = -2, b = -5$, and $x = -3$.

SOLUTION   We substitute the given numbers for the variables on the right-hand side, being careful to substitute each number for the correct variable, and evaluate the expression:

$$y = mx + b$$

$$= (-2)(-3) + (-5)$$

$$= 6 - 5$$

$$= 1. \qquad \blacktriangle$$

SIMPLE INTEREST FORMULA

The formula

$$I = PRT$$

is the simple interest formula. In this formula, the dependent variable is I, which represents the amount of interest earned on an investment or owed on a loan. The independent variables are P, which is the principal; R, which is the interest rate written as a decimal; and T, which is the amount of time in years.

EXAMPLE 5.4

Evaluate the simple interest formula $I = PRT$ for $P = \$500$, $R = 0.12$ per year (12% annually), and $T = \frac{1}{4}$ year (3 months).

SOLUTION   We substitute the quantities for the variables on the right-hand side:

$$I = PRT$$

$$= (\$500)\left(\frac{0.12}{\text{yr}}\right)\left(\frac{1}{4}\,\text{yr}\right)$$

$$= \$15. \qquad \blacktriangle$$

EXAMPLE 5.5

Evaluate the simple interest formula $I = PRT$ for $P = \$1200$, $R = 5\frac{1}{2}\%$ annually, and $T = 6$ months.

SOLUTION   We must write R as a decimal,

$$5\frac{1}{2}\% = 0.055,$$

and T in years,

$$6 \text{ months} = \frac{1}{2} \text{ year.}$$

Then, substituting the given quantities for the variables on the right-hand side and evaluating the expression, we have

$$I = PRT$$

$$= (\$1200)\left(\frac{0.055}{\text{yr}}\right)\left(\frac{1}{2}\,\text{yr}\right)$$

$$= \$33. \qquad \blacktriangle$$

RESISTANCE FORMULAS

Often, two or more different variables in a formula refer to the same type of object. We can use one letter with **subscripts**, which are small numbers or letters written to the lower right of the main letter. For example, to use the letter R for two resistances, we use the variable R_1 for the first resistance, and the variable R_2 for the second resistance. Moreover, we might use the variable R_T for the total resistance.

When two resistors are in a circuit called a series circuit, their total resistance is given by the formula

$$R_T = R_1 + R_2.$$

We will study series circuits in Units 7 and 8.

EXAMPLE 5.6

Evaluate the formula $R_T = R_1 + R_2$ for $R_1 = 680 \text{ k}\Omega$ and $R_2 = 820 \text{ k}\Omega$.

SOLUTION   Substituting the given quantities for the variables on the right-hand side and evaluating the expression, we have

$$R_T = R_1 + R_2$$

$$= 680 \text{ k}\Omega + 820 \text{ k}\Omega$$

$$= 1500 \text{ k}\Omega$$

or, writing this result in MΩ,

$$R_T = 1.5 \text{ M}\Omega. \qquad \blacktriangle$$

A formula may occasionally be stated in a form in which the variable is not isolated on the left-hand side. When two resistors are in a circuit called a parallel circuit, their total resistance is given by the formula

$$\frac{1}{R_T} = \frac{1}{R_1} + \frac{1}{R_2}.$$

We will study parallel circuits also in Units 7 and 8.

A scientific calculator has a reciprocal key marked

$$\boxed{1/x}\ ,$$

or possibly marked with the negative exponent form,

$$\boxed{x^{-1}}\ .$$

On some calculators, the reciprocal key is an inverse or second function. You might need to press

$$\boxed{\text{INV}}\ \text{or}\ \boxed{2^{\text{nd}}}$$

and then some other key when we say to press

$$\boxed{1/x}\ .$$

(In these cases, the reciprocal is the second function of the key but it is not an actual mathematical inverse of the primary function; a reciprocal is its own mathematical inverse. The inverse or second function key is discussed in Section R.5.)

EXAMPLE 5.7 ▶ Evaluate the formula $\dfrac{1}{R_T} = \dfrac{1}{R_1} + \dfrac{1}{R_2}$ for $R_1 = 1.2\ \Omega$ and $R_2 = 1.8\ \Omega$.

SOLUTION   Substituting the numerical quantities for the variables on the right-hand side, we have

$$\frac{1}{R_T} = \frac{1}{R_1} + \frac{1}{R_2}$$

$$\frac{1}{R_T} = \frac{1}{1.2\ \Omega} + \frac{1}{1.8\ \Omega}.$$

You can use a calculator to find

		display:
Enter	1.2	1.2
Press	$\boxed{1/x}$	0.833333333
Press	$\boxed{+}$	0.833333333
Enter	1.8	1.8
Press	$\boxed{1/x}$	0.555555555
Press	$\boxed{=}$	1.388888889

You have evaluated the formula

$$\frac{1}{R_T} = \frac{1}{R_1} + \frac{1}{R_2}.$$

To find R_T, you must press the reciprocal key one final time:

		display:
(Currently displayed)		1.388888889
Press	$\boxed{1/x}$	0.72

Thus, $R_T = 0.72\ \Omega$. The formula written with R_T isolated is

$$R_T = \cfrac{1}{\cfrac{1}{R_1} + \cfrac{1}{R_2}}.$$

▲

AC FORMULAS

In the preceding units, we have used examples involving DC circuits, in which the current flows in only one direction. In later units we will study **alternating current** (AC) circuits, in which the current reverses direction. Many formulas for AC circuits use the constant π (the Greek letter pi). The constant π occurs very often in mathematics and science. For example, in basic geometry we use π to calculate circumferences and areas of circles. Although we can approximate π to several places, π is an irrational number that has a nonterminating decimal representation.

A scientific calculator has a π key that will approximate π to several decimal places. The π key is often marked as a second function. On many calculators, π is above a key marked EXP. For such calculators, we mean to press the combination of

$$\boxed{\text{INV}} \text{ or } \boxed{2^{\text{nd}}}$$

and then

$$\boxed{\text{EXP}}$$

when we say to press

$$\boxed{\pi}.$$

The formula

$$X_L = 2\pi f L$$

is used in AC circuits to find a quantity called inductive reactance. In this formula, the dependent variable is X_L, and the independent variables are f and L (remember that 2 and π are constants). In later units, we will learn about these quantities and the units in which they are measured.

EXAMPLE 5.8 ▶ Evaluate the formula $X_L = 2\pi f L$ (X_L in ohms) for $f = 60$ MHz (hertz) and $L = 4\ \mu$H (henrys).

SOLUTION We substitute the given quantities for the variables on the right-hand side:

$$X_L = 2\pi f L$$
$$= 2\pi (60\ \text{MHz})(4\ \mu\text{H})$$
$$= 2\pi (60 \times 10^6)(4 \times 10^{-6})\ \Omega$$
$$= 2\pi (60)(4)\ \Omega.$$

Now, you can use a calculator to obtain π and multiply:

		display:
Enter 2		2.
Press $\boxed{\times}$		2.
Press $\boxed{\pi}$		3.141592654
Press $\boxed{\times}$		6.283185307
Enter 60		60.
Press $\boxed{\times}$		376.9911184
Enter 4		4.
Press $\boxed{=}$		1507.964474

Thus, $X_L = 1508\ \Omega$, or 1.51 kΩ, to three significant digits. ▲

The exponent entry key on a scientific calculator is the key marked

$$\boxed{\text{EXP}},$$

or a similar marking (sometimes just an E, but *not* EXC). The exponent entry key can be used to calculate powers of ten along with numbers. For example, to enter a number such as 60×10^6, you can follow these steps:

display:

Enter 60		60.
Press $\boxed{\text{EXP}}$		60. 00
Enter 6		60. 06

For negative exponents, recall the change sign key

$$\boxed{+/-},$$

introduced in Section 1.1. To enter a number such as 4×10^{-6}, follow these steps:

display:

Enter 4		4.
Press $\boxed{\text{EXP}}$		4. 00
Enter 6		4. 06
Press $\boxed{+/-}$		4. −06

The number in the two rightmost places, with its sign when negative, is the exponent of the power of ten. Thus, the display 60. 06 means 60×10^6, and the display 4. −06 means 4×10^{-6}.

To use the exponent entry key for the calculation in Example 5.8, you may follow these steps:

display:

Enter 2		2.
Press $\boxed{\times}$		2.
Press $\boxed{\pi}$		3.141592654
Press $\boxed{\times}$		6.283185307
Enter 60		60.
Press $\boxed{\text{EXP}}$		60. 00
Enter 6		60. 06
Press $\boxed{\times}$		376.9911184
Enter 4		4.
Press $\boxed{\text{EXP}}$		4. 00
Enter 6		4. 06
Press $\boxed{+/-}$		4. −06
Press $\boxed{=}$		1507.964474

Using the exponent entry key, we find that $X_L = 1508$ Ω, or 1.51 kΩ as before.

The square-root key might be the second function of the square key. In this case, you press

$$\boxed{\text{INV}} \text{ or } \boxed{2^{\text{nd}}}$$

and then

$$\boxed{x^2}$$

when we say to press

$$\boxed{\sqrt{x}}.$$

(The square-root key is discussed in Section R.5.)

The formula

$$f_r = \frac{1}{2\pi\sqrt{LC}}$$

is used in AC circuits to find a quantity called resonant frequency. In this formula, the dependent variable is f_r, and the independent variables are L and C.

EXAMPLE 5.9

Evaluate the formula $f_r = \dfrac{1}{2\pi\sqrt{LC}}$ (f_r in hertz) for $L = 40$ mH (henrys) and $C = 2\ \mu F$ (farads).

SOLUTION

We substitute the given quantities for the variables on the right-hand side:

$$f_r = \frac{1}{2\pi\sqrt{LC}}$$

$$= \frac{1}{2\pi\sqrt{(40\ \text{mH})(2\ \mu\text{F})}}$$

$$= \frac{1}{2\pi\sqrt{(40 \times 10^{-3})(2 \times 10^{-6})}}\ \text{Hz.}$$

As in the preceding example, we can work out the exponents by using the rules for exponents or by using the exponent entry key. We can simply multiply the quantities under the square root:

$$f_r = \frac{1}{2\pi\sqrt{80 \times 10^{-9}}}\ \text{Hz.}$$

Then, we can rewrite 10^{-9} as $10^{-1} \times 10^{-8}$:

$$f_r = \frac{1}{2\pi\sqrt{80 \times 10^{-1} \times 10^{-8}}}\ \text{Hz}$$

$$= \frac{1}{2\pi\sqrt{8 \times 10^{-8}}}\ \text{Hz}$$

$$= \frac{1}{2\pi\sqrt{8} \times 10^{-4}}\ \text{Hz}$$

$$= \frac{1}{2\pi\sqrt{8}} \times 10^{4}\ \text{Hz.}$$

Now, you can use a calculator, following these steps:

		display:
Enter 2		2.
Press $\times$		2.
Press π		3.141592654
Press $\times$		6.283185307
Enter 8		8.
Press $\sqrt{x}$		2.828427125
Press $=$		17.77153175
Press $1/x$		0.056269769

Thus, we have $f_r = 0.0563 \times 10^4$ Hz, or $f_r = 563$ Hz. ▲

You may also do the preceding calculation by using the exponent entry key. You must remember to take the square root of the entire product,

$$(40 \times 10^{-3} \times 2 \times 10^{-6}).$$

One way to indicate this product is to recall the parentheses keys,

(and)

introduced in Section 1.5. Using the parentheses keys to indicate the product, you can follow these steps:

		display:
Enter	2	2.
Press	×	2.
Press	π	3.141592654
Press	×	6.283185307
Press	(	0.
Enter	40	40.
Press	EXP	40. 00
Enter	3	40. 03
Press	+/−	40. −03
Press	×	0.04
Enter	2	2.
Press	EXP	2. 00
Enter	6	2. 06
Press	+/−	2. −06
Press	)	0.00000008
Press	√x	0.000282842
Press	=	0.001777153
Press	1/x	562.6976976

You can also start with the product, then take the square root, multiply the result by π and 2, and then press the equals and reciprocal keys.

EXERCISE 5.1

Evaluate the formula for the given quantities:

1. $I = \dfrac{V}{R}$ for $V = 14$ mV and $R = 8.2$ kΩ.

2. $I = \dfrac{V}{R}$ for $V = 7.5$ V and $R = 68$ kΩ.

3. $y = 4x + 3$ for $x = -2$

4. $v = -32t + 16$ for $t = \frac{1}{4}$

5. $z = 3x - 2y$ for $x = -2$ and $y = -4$

6. $v = -32t + v_0$ for $t = \frac{1}{2}$ and $v_0 = 8$

7. $y = mx + b$ for $m = -\frac{1}{2}$, $x = 2$, and $b = 3$

8. $y = mx + b$ for $b = -4$, $m = -\frac{1}{2}$, and $x = -4$

9. $I = PRT$ for $P = \$1000$, $R = 0.05$, and $T = 1$

10. $I = PRT$ for $P = \$5000$, $R = 0.12$, and $T = \frac{1}{4}$

11. $I = PRT$ for $P = \$100$, $R = 6\%$ annually, and $T = 6$ months

12. $I = PRT$ for $P = \$2000$, $R = 10\frac{1}{2}\%$ annually, and $T = 3$ months

13. $A = P(1 + RT)$, the total amount accumulated by simple interest, for $P = \$300$, $R = 15\%$ annually, and $T = 6$ months

14. $A = P(1 + RT)$ for $P = \$550$, $R = 6\frac{1}{4}\%$ annually, and $T = 2$ years

15. $R_T = R_1 + R_2$ for $R_1 = 270\ \Omega$ and $R_2 = 330\ \Omega$

16. $R_T = R_1 + R_2$ for $R_1 = 560\ \text{k}\Omega$ and $R_2 = 820\ \text{k}\Omega$

17. $R_t = R_0(1 + \alpha\,\Delta t)$, resistance at temperature t, for $\alpha = 0.004$ per °C, $\Delta t = 25$°C, and $R_0 = 10\ \Omega$ (α is the Greek lowercase letter alpha and Δ is the Greek capital letter delta; Δt is all one variable)

18. $R_T = R_0(1 + \alpha\,\Delta t)$ for $\alpha = -0.0003$ per °C, $\Delta t = 50$°C, and $R_0 = 5.5\ \Omega$

19. $\dfrac{1}{R_T} = \dfrac{1}{R_1} + \dfrac{1}{R_2}$ for $R_1 = 22\ \Omega$ and $R_2 = 56\ \Omega$ (find R_T)

20. $\dfrac{1}{R_T} = \dfrac{1}{R_1} + \dfrac{1}{R_2}$ for $R_1 = 4.7\ \text{k}\Omega$ and $R_2 = 7.5\ \text{k}\Omega$ (find R_T)

21. $\dfrac{1}{\alpha} = 1 + \dfrac{1}{\beta}$, amplification factor of a transistor, for $\beta = 25$ (find α; β is the Greek lowercase letter beta; α and β have no unit)

22. $\dfrac{1}{\alpha} = 1 + \dfrac{1}{\beta}$ for $\beta = 199$

23. $X_L = 2\pi f L$ for $f = 85\ \text{kHz}$ and $L = 5\ \text{mH}$

24. $X_L = 2\pi f L$ for $f = 500\ \text{kHz}$ and $L = 25\ \mu\text{H}$

25. $X_C = \dfrac{1}{2\pi f C}$, capacitive reactance, for $f = 500\ \text{kHz}$ and $C = 1.2\ \text{nF}$

26. $X_C = \dfrac{1}{2\pi f C}$ for $f = 500\ \text{kHz}$ and $C = 180\ \text{pF}$

27. $f_r = \dfrac{1}{2\pi\sqrt{LC}}$ for $L = 6\ \text{mH}$ and $C = 0.04\ \mu\text{F}$

28. $f_r = \dfrac{1}{2\pi\sqrt{LC}}$ for $L = 18\ \mu\text{H}$ and $C = 240\ \text{pF}$

29. $Q = \dfrac{1}{R}\sqrt{\dfrac{L}{C}}$, quality of an inductor, for $L = 100\ \mu\text{H}$, $C = 25\ \text{pF}$, and $R = 10\ \Omega$ (Q has no unit)

30. $Q = \dfrac{1}{R}\sqrt{\dfrac{L}{C}}$ for $L = 150\ \text{mH}$, $C = 25\ \text{nF}$, and $R = 20\ \Omega$

SECTION

5.2

Solving Formulas

To **solve a formula** for a specified variable means to isolate the specified variable on one side of the equal sign, with all the constants and other variables on the other side. We use the same methods as we use to isolate the variable in solving equations.

EXAMPLE 5.10 ▶ Solve the formula $P = VI$ for I.

SOLUTION To isolate I, we divide both sides of the formula by V:

$$P = VI$$

$$\frac{P}{V} = \frac{VI}{V}$$

$$\frac{P}{V} = I$$

or

$$I = \frac{P}{V}.$$

▲

EXAMPLE 5.11 Solve the formula $I = \dfrac{V}{R}$ for R.

SOLUTION First, we eliminate the denominator by multiplying both sides of the formula by R:

$$I = \frac{V}{R}$$

$$RI = R\left(\frac{V}{R}\right)$$

$$RI = V.$$

Then, we isolate R by dividing both sides of the formula by I:

$$RI = V$$

$$\frac{RI}{I} = \frac{V}{I}$$

$$R = \frac{V}{I}.$$

Recall that these are the forms of Ohm's law that we derived in Section 4.5.   ▲

EXAMPLE 5.12 Solve the formula $y = mx + b$ for x.

SOLUTION First, we subtract b from both sides of the formula:

$$y = mx + b$$

$$y - b = mx + b - b$$

$$y - b = mx.$$

Now, we divide both sides by m:

$$\frac{y - b}{m} = \frac{mx}{m}$$

$$\frac{y - b}{m} = x$$

or

$$x = \frac{y - b}{m}.$$ ▲

Many formulas in electronics contain variables that have double or even triple subscripts. For example, the formula

$$V_C = V_{CC} - I_C R_C$$

relates two voltages with a current and a resistance for a circuit component called a transistor.

EXAMPLE 5.13 Solve the formula $V_C = V_{CC} - I_C R_C$ for I_C.

SOLUTION When the term containing the specified variable is subtracted, it is often best to add that term to both sides of the formula. Thus we begin by adding $I_C R_C$ to both sides of the formula:

$$V_C = V_{CC} - I_C R_C$$

$$V_C + I_C R_C = V_{CC} - I_C R_C + I_C R_C$$

$$V_C + I_C R_C = V_{CC}.$$

A typical transistor.
Courtesy of *International Rectifier*.

Then, we isolate the term containing I_C by subtracting V_C from both sides:

$$V_C + I_C R_C - V_C = V_{CC} - V_C$$

$$I_C R_C = V_{CC} - V_C.$$

Finally, we divide both sides by R_C:

$$\frac{I_C R_C}{R_C} = \frac{V_{CC} - V_C}{R_C}$$

$$I_C = \frac{V_{CC} - V_C}{R_C}.$$

▲

In many formulas there is a factor containing two or more terms grouped in parentheses. If the specified variable is not in this factor, we can divide both sides by the entire factor. The formula for a transistor saturation state has been simplified for use in the next two examples.

EXAMPLE 5.14 Solve the formula $V_{CC} = I_C(R_C + R_E)$ for I_C.

SOLUTION We divide both sides by the entire factor $R_C + R_E$:

$$V_{CC} = I_C(R_C + R_E)$$

$$\frac{V_{CC}}{R_C + R_E} = \frac{I_C(R_C + R_E)}{R_C + R_E}$$

$$\frac{V_{CC}}{R_C + R_E} = I_C$$

or

$$I_C = \frac{V_{CC}}{R_C + R_E}.$$

▲

If the specified variable is in a factor in parentheses, we remove the parentheses before attempting to solve the formula.

EXAMPLE 5.15 Solve the formula $V_{CC} = I_C(R_C + R_E)$ for R_C.

SOLUTION Since R_C is in the parentheses, we remove the parentheses by using the distributive property:

$$V_{CC} = I_C(R_C + R_E)$$

$$V_{CC} = I_C R_C + I_C R_E.$$

Now, we can subtract $I_C R_E$ from both sides:

$$V_{CC} - I_C R_E = I_C R_C + I_C R_E - I_C R_E$$

$$V_{CC} - I_C R_E = I_C R_C.$$

Then, we divide both sides by I_C:

$$\frac{V_{CC} - I_C R_E}{I_C} = \frac{I_C R_C}{I_C}$$

$$\frac{V_{CC} - I_C R_E}{I_C} = R_C$$

or

$$R_C = \frac{V_{CC} - I_C R_E}{I_C}.$$

▲

Medium and high-power
transistors. Courtesy of
International Rectifier.

Recall from Section 4.4 that we can take the square root of each side of an equation, and we can square each side of an equation. We use one of these methods if we are to solve for a variable that is squared or is in a square root. For example, many formulas in mathematics and electronics are similar to the formula

$$X_L = \sqrt{Z^2 - R^2},$$

which relates a resistance R with two other types of opposition to current.

EXAMPLE 5.16 Solve the formula $X_L = \sqrt{Z^2 - R^2}$ for R.

SOLUTION   Since the square root containing R is isolated, we can square each side of the formula:

$$X_L = \sqrt{Z^2 - R^2}$$

$$X_L^2 = (\sqrt{Z^2 - R^2})^2$$

$$X_L^2 = Z^2 - R^2.$$

Since R^2 is subtracted, we add R^2 to both sides and then subtract X_L^2 from both sides to isolate R^2:

$$X_L^2 + R^2 = Z^2$$

$$R^2 = Z^2 - X_L^2.$$

Now, we can take the square root of each side of the formula to obtain

$$\sqrt{R^2} = \sqrt{Z^2 - X_L^2}.$$

We will assume $R > 0$, and so $\sqrt{R^2} = R$. Therefore, we have

$$R = \sqrt{Z^2 - X_L^2}.$$

▲

EXAMPLE 5.17 Solve the formula $f_r = \dfrac{1}{2\pi\sqrt{LC}}$ for L.

SOLUTION   We must isolate the square root containing L. To remove the square root from the denominator, we multiply both sides of the formula by $\sqrt{LC}$:

$$f_r = \frac{1}{2\pi\sqrt{LC}}$$

$$f_r(\sqrt{LC}) = \frac{1}{2\pi\sqrt{LC}}(\sqrt{LC})$$

$$f_r(\sqrt{LC}) = \frac{1}{2\pi}.$$

Then, we divide by f_r to isolate the square root:

$$\frac{f_r\sqrt{LC}}{f_r} = \frac{1}{2\pi f_r}$$

$$\sqrt{LC} = \frac{1}{2\pi f_r}.$$

Now, we can square each side of the formula:

$$(\sqrt{LC})^2 = \left(\frac{1}{2\pi f_r}\right)^2$$

$$LC = \left(\frac{1}{2\pi f_r}\right)^2$$

$$LC = \frac{1^2}{(2\pi f_r)^2}$$

$$LC = \frac{1}{(2\pi f_r)^2}.$$

Finally, we divide both sides by C:

$$\frac{LC}{C} = \frac{1}{(2\pi f_r)^2 C}$$

$$L = \frac{1}{(2\pi f_r)^2 C}.$$

EVALUATING FOR AN INDEPENDENT VARIABLE

In Section 5.1, we evaluated formulas, finding the value of the dependent variable by substituting given quantities for the independent variables in the expression on the right-hand side. Often, however, we must evaluate a formula for an independent variable that is in the expression on the right-hand side. In this case, we first solve the formula for the required variable.

EXAMPLE 5.18   Find L (in henrys) in the formula $X_L = 2\pi f L$ for $f = 1$ MHz and $X_L = 1$ kΩ.

SOLUTION Since we are to find L, we first solve the formula for L. We divide both sides by 2, and π, and f:

$$X_L = 2\pi f L$$

$$\frac{X_L}{2\pi f} = \frac{2\pi f L}{2\pi f}$$

$$\frac{X_L}{2\pi f} = L$$

or

$$L = \frac{X_L}{2\pi f}.$$

Then, we substitute the given quantities for the variables on the right-hand side:

$$L = \frac{1\ \text{k}\Omega}{2\pi(1\ \text{MHz})}$$

$$= \frac{1 \times 10^3}{2\pi(1 \times 10^6)}\ \text{H}$$

$$= 0.159 \times 10^{-3}\ \text{H}$$

$$= 0.159\ \text{mH}$$

or $L = 159\ \mu\text{H}$.

| EXAMPLE 5.19 | | Find X_L in the formula $Z = \sqrt{R^2 + X_L^2}$ for $Z = 4.2\ \Omega$ and $R = 3.6\ \Omega$. |

SOLUTION   First, we square each side of the formula:

$$Z = \sqrt{R^2 + X_L^2}$$

$$Z^2 = (\sqrt{R^2 + X_L^2})^2$$

$$Z^2 = R^2 + X_L^2.$$

Then, we solve for X_L to obtain

$$Z^2 - R^2 = X_L^2$$

$$\sqrt{Z^2 - R^2} = \sqrt{X_L^2}$$

$$\sqrt{Z^2 - R^2} = X_L$$

or

$$X_L = \sqrt{Z^2 - R^2}.$$

We substitute the given quantities for the variables on the right-hand side:

$$X_L = \sqrt{(4.2\ \Omega)^2 - (3.6\ \Omega)^2}$$

$$= \sqrt{4.2^2 - 3.6^2}\ \Omega.$$

Now, you can use a calculator to find X_L:

		display:
Enter 4.2		4.2
Press	x^2	17.64
Press	$-$	17.64
Enter 3.6		3.6
Press	x^2	12.96
Press	$=$	4.68
Press	$\sqrt{x}$	2.163330765

Thus, $X_L = 2.16\ \Omega$ to three significant digits.

▲

| EXERCISE 5.2 | Solve for the specified variable: |

1. $y = mx$ for m

2. $I = PRT$ for T

3. $I = \dfrac{Q}{t}$, the definition of current, for t

4. $a = \dfrac{F}{m}$, acceleration due to a force, for m

5. $s = \dfrac{1}{2}at^2$, distance in uniformly accelerated motion, for a

6. $F = \dfrac{kQ_1Q_2}{r^2}$, force between two charges, for Q_2

7. $I_C = \alpha I_E + I_{CO}$, a transistor formula, for I_{CO}

8. $I_C = \alpha I_E + I_{CO}$ for I_E

9. $V_{CE} = V_C - I_E R_E$ for I_E

10. $V_{CE} = V_{CC} - I_C R_C - I_E R_E$ for I_C

11. $E = I(R_1 + R_2)$, source voltage in a series circuit, for I

12. $V_{CE} = V_{CC} - I_C(R_C + R_E)$ for I_C

13. $E = I(R_1 + R_2)$ for R_1

14. $V_{CE} = V_{CC} - I_C(R_C + R_E)$ for R_C

15. $s = \dfrac{1}{2}at^2$ for t

16. $F = \dfrac{kQ_1 Q_2}{r^2}$ for r

17. $v = \sqrt{2as}$, velocity in uniformly accelerated motion, for s

18. $v = \sqrt{\dfrac{2E}{m}}$, velocity due to kinetic energy, for m

19. $Z = \sqrt{R^2 + X_C^2}$ for X_C

20. $V_C = \sqrt{E^2 - V_R^2}$ for V_R

21. $f_r = \dfrac{1}{2\pi\sqrt{LC}}$ for C

22. $Q = \dfrac{1}{R}\sqrt{\dfrac{L}{C}}$ for L

23. $Q = \dfrac{1}{R}\sqrt{\dfrac{L}{C}}$ for C

24. $T = 2\pi\sqrt{\dfrac{l}{g}}$, period of a pendulum, for g.

Find the specified variable in the formula for the given quantities:

25. $X_L = 2\pi f L$ for L if $X_L = 5$ kΩ and $f = 55$ MHz

26. $X_L = 2\pi f L$ for f if $L = 2.5$ μH and $X_L = 550$ Ω

27. $X_C = \dfrac{1}{2\pi f C}$ for C if $X_C = 600$ Ω and $f = 12$ kHz

28. $X_C = \dfrac{1}{2\pi f C}$ for f if $C = 1.2$ pF and $X_C = 7.5$ kΩ

29. $R_t = R_0(1 + \alpha \, \Delta t)$ for R_0 if $R_t = 28$ Ω, $\alpha = 0.004$ per °C, and $\Delta t = 30$°C

30. $R_t = R_0(1 + \alpha \, \Delta t)$ for α if $R_t = 4.02$ Ω, $R_0 = 3.66$ Ω, and $\Delta t = 25$°C

31. $Z = \sqrt{R^2 + X_L^2}$ for X_L if $Z = 12.1$ Ω and $R = 12$ Ω

32. $E = \sqrt{V_R^2 + V_L^2}$ for V_L if $E = 1$ V and $V_R = 750$ mV

33. $f_r = \dfrac{1}{2\pi\sqrt{LC}}$ for L if $f_r = 1.5$ MHz and $C = 750$ pF

34. $f_r = \dfrac{1}{2\pi\sqrt{LC}}$ for C if $f_r = 2$ MHz and $L = 15$ μH

35. $Q = \dfrac{1}{R}\sqrt{\dfrac{L}{C}}$ for L if $Q = 200$, $R = 500$ Ω, and $C = 350$ pF

36. $Q = \dfrac{1}{R}\sqrt{\dfrac{L}{C}}$ for C if $Q = 75$, $R = 40$ Ω, and $L = 27$ μH

Variation Formulas

Two or more variables are sometimes related by formulas called **variation** formulas. In **direct variation** formulas, when the independent variable increases, the dependent variable also increases. For example, as the speed of an automobile increases, the distance it can cover in a given time increases.

Direct Variation: The relationship "y varies directly as x" is represented by the formula

$$y = kx$$

where k is a constant.

The variable y is the dependent variable and x is the independent variable. The constant k is called the **constant of variation**. The dependent variable is equal to the constant of variation multiplied by the independent variable.

EXAMPLE 5.20   Write the variation formula for: Distance D varies directly as rate of speed R.

SOLUTION   For direct variation, the dependent variable is equal to the constant of variation multiplied by the independent variable. Therefore, we write

$$D = kR.$$ ▲

In **inverse variation** formulas, when the independent variable *increases* the dependent variable *decreases*. For example, as the speed of an automobile increases, the time it takes to cover a given distance decreases.

Inverse Variation: The relationship "y varies inversely as x" is represented by the formula

$$y = \frac{k}{x}$$

where k is a constant.

Again, y is the dependent variable, x is the independent variable, and k is the constant of variation. The dependent variable is equal to the constant of variation *divided* by the independent variable.

EXAMPLE 5.21   Write the variation formula for: Time T varies inversely as rate of speed R.

SOLUTION   In the inverse variation formula, the dependent variable is equal to the constant of variation divided by the independent variable. Therefore, we write

$$T = \frac{k}{R}.$$

Variation formulas can contain more than one independent variable.

EXAMPLE 5.22   Write the variation formula for: Inductive reactance X_L varies directly as frequency f and inductance L.

SOLUTION   Using direct variation for both f and L, we multiply the constant of variation by both f and L. Thus, the formula is

$$X_L = kfL.$$

We use just one constant of variation k. This type of variation formula is also called joint variation, and we may say "X_L varies jointly as f and L."   ▲

EXAMPLE 5.23 Capacitance is the ability of plates in a circuit component called a capacitor to store charge. Write the variation formula for: Capacitance C varies directly as the plate area A and inversely as the distance d between the plates.

SOLUTION Using direct variation, we multiply the constant of variation by A. Then, using inverse variation, we divide by d. The formula is

$$C = \frac{kA}{d}.$$

Again, we use just one constant of variation k. ▲

FINDING THE CONSTANT OF VARIATION

If we are given values for the variables in a variation formula, we can find the constant of variation k. Then, we can write the variation formula in terms of the variables only.

EXAMPLE 5.24 Write the variation formula for: Voltage V varies directly as resistance R. Find k for $V = 25$ V and $R = 10 \ \Omega$.

SOLUTION First, we write the variation formula

$$V = kR.$$

Then, we solve the formula for k:

$$k = \frac{V}{R}.$$

We substitute the given quantities for V and R:

$$k = \frac{25 \text{ V}}{10 \ \Omega}$$

$$= 2.5 \frac{\text{V}}{\Omega}.$$

Therefore, the variation formula is

$$V = \left(2.5 \frac{\text{V}}{\Omega}\right) R.$$

(Of course, we know from Ohm's law that $\frac{\text{V}}{\Omega}$ results in amperes (A), but this fact cannot be determined from the variation formula alone.) ▲

EXAMPLE 5.25 Write the variation formula for: Current I varies inversely as resistance R. Find k for $I = 0.5$ A and $R = 7.2 \ \Omega$.

SOLUTION First, we write the variation formula

$$I = \frac{k}{R}.$$

Then, we solve the formula for k:

$$k = IR.$$

We substitute the given quantities for I and R:

$$k = (0.5 \text{ A})(7.2 \ \Omega)$$

$$= 3.6 \text{ A} \cdot \Omega.$$

Therefore, the variation formula is

$$I = \frac{3.6 \text{ A} \cdot \Omega}{R}.$$

▲

EXAMPLE 5.26   Write the variation formula for: Inductive reactance X_L varies directly as frequency f and inductance L. Find k for $X_L = 201 \text{ } \Omega, f = 160 \text{ kHz}$, and $L = 2 \text{ } \mu\text{H}$.

SOLUTION   From Example 5.22, we have

$$X_L = kfL.$$

Then, solving the formula for k and substituting the given quantities, we find

$$k = \frac{X_L}{fL}$$

$$= \frac{201 \text{ } \Omega}{(160 \text{ kHz})(2 \text{ } \mu\text{H})}$$

$$= \frac{201 \text{ } \Omega}{(160 \times 10^3 \text{ Hz})(2 \times 10^{-6} \text{ H})}$$

$$= \frac{201}{320 \times 10^{-3}} \frac{\Omega}{\text{Hz} \cdot \text{H}}$$

$$= 0.628 \times 10^3 \frac{\Omega}{\text{Hz} \cdot \text{H}}$$

$$= 628 \frac{\Omega}{\text{Hz} \cdot \text{H}}.$$

Therefore, the variation formula is

$$X_L = \left(628 \frac{\Omega}{\text{Hz} \cdot \text{H}}\right) fL.$$

(It seems likely that $628 = 6.28 \times 10^2$ comes from an approximation for 2π, but this fact cannot be determined from the variation formula alone.)   ▲

EVALUATING FOR THE DEPENDENT VARIABLE

If we are given a second set of values, we can evaluate the variation formula. First, to find k, we must have a set of values for all the variables. Then, to evaluate the formula, we must have a new set of values for the independent variables.

EXAMPLE 5.27   Write the variation formula for: Voltage V varies directly as resistance R. Find k for $V = 25 \text{ V}$ and $R = 10 \text{ } \Omega$; and then evaluate the formula for $R = 5 \text{ } \Omega$.

SOLUTION   From Example 5.24, we have

$$V = \left(2.5 \frac{\text{V}}{\Omega}\right) R$$

for $V = 25 \text{ V}$ and $R = 10 \text{ } \Omega$. Now, we substitute $5 \text{ } \Omega$ for R:

$$V = \left(2.5 \frac{\text{V}}{\Omega}\right) (5 \text{ } \Omega)$$

We can divide out the unit Ω to obtain

$$V = 12.5 \text{ V}.$$

Observe that V is halved when R is halved. ▲

EXAMPLE 5.28 Write the variation formula for: Current I varies inversely as resistance R. Find k for $I = 0.5$ A and $R = 7.2\ \Omega$; and then evaluate the formula for $R = 1.8\ \Omega$.

SOLUTION From Example 5.25, we have

$$I = \frac{3.6 \text{ A} \cdot \Omega}{R},$$

for $I = 0.5$ A and $R = 7.2\ \Omega$. Now, we substitute $1.8\ \Omega$ for R:

$$I = \frac{3.6 \text{ A} \cdot \Omega}{1.8\ \Omega}$$

We can divide out the unit Ω to obtain

$$I = 2\text{A}.$$

Observe that I is quadrupled (multiplied by four) when R is divided by four. ▲

EXAMPLE 5.29 Write the variation formula for: Inductive reactance X_L varies directly as frequency f and inductance L. Find k for $X_L = 201\ \Omega, f = 160$ kHz, and $L = 2\mu$H; and then evaluate the formula for $f = 2.4$ MHz and $L = 1.5\ \mu$H.

SOLUTION From Example 5.26, we have

$$X_L = \left(628\ \frac{\Omega}{\text{Hz} \cdot \text{H}}\right)fL,$$

for $X_L = 201\ \Omega, f = 160$ kHz, and $L = 2\ \mu$H. Now, we substitute 2.4 MHz for f and $1.5\ \mu$H for L:

$$X_L = \left(628\ \frac{\Omega}{\text{Hz} \cdot \text{H}}\right)(2.4 \text{ MHz})(1.5\ \mu\text{H}).$$

We replace the prefixes by powers of ten, and divide out the units Hz and H, to obtain

$$X_L = (628)(2.4 \times 10^6)(1.5 \times 10^{-6})\ \Omega$$

$$= 2260\ \Omega$$

or $X_L = 2.26$ kΩ. ▲

EXAMPLE 5.30 The inductor in Examples 5.26 and 5.29 is made from a wire coil. Its inductance L varies directly as the square of the number of turns N in the coil, directly as its cross-sectional area A, and inversely as its length l. Write the variation formula and find k for $L = 5.04$ H, $N = 200$ turns, $A = 10$ mm^2, and $l = 100$ mm. Then, evaluate the formula for $N = 100$ turns, $A = 5$ mm^2, and $l = 50$ mm.

SOLUTION First, we write the variation formula

$$L = \frac{kN^2A}{l}.$$

Then, we solve the formula for k, and substitute the first set of given quantities to find k:

$$k = \frac{L\,l}{N^2 A}$$

$$= \frac{(5.04\ \text{H})(100\ \text{mm})}{(200\ \text{turns})^2(10\ \text{mm}^2)}$$

$$= \frac{(5.04)(10^2)}{(2 \times 10^2)^2(10)}\ \frac{\text{H}}{\text{turns}^2 \cdot \text{mm}}$$

$$= \frac{(5.04)(10^2)}{(4 \times 10^4)(10)}\ \frac{\text{H}}{\text{turns}^2 \cdot \text{mm}}$$

$$= 1.26 \times 10^{-3}\ \frac{\text{H}}{\text{turns}^2 \cdot \text{mm}}.$$

The variation formula is

$$L = \frac{\left(1.26 \times 10^{-3}\ \dfrac{\text{H}}{\text{turns}^2 \cdot \text{mm}}\right) N^2 A}{l}.$$

Then, we use the second set of given quantities to evaluate the formula for L:

$$L = \frac{\left(1.26 \times 10^{-3}\ \dfrac{\text{H}}{\text{turns}^2 \cdot \text{mm}}\right) (100\ \text{turns})^2(5\ \text{mm}^2)}{50\ \text{mm}}$$

$$= \frac{(1.26 \times 10^{-3})(10^4)(5)}{50}\ \text{H}$$

$$= 1.26\ \text{H}. \qquad \blacktriangle$$

EXERCISE

5.3

Write the variation formula:

1. Power P varies directly as voltage V.
2. Angular velocity ω (the Greek lowercase letter omega) varies directly as frequency f.
3. Power P varies inversely as resistance R.
4. Angular velocity ω varies inversely as period T.
5. Work W varies directly as force F and distance d.
6. Force of attraction F between two charged bodies varies directly as their charges Q_1 and Q_2.
7. Power P varies directly as force F and inversely time t.
8. Magnetic field intensity H of a wire varies directly as current I and inversely as distance r from the wire.

Write the variation formula and find k:

9. Power P varies directly as voltage V. Find k for $P = 330$ W and $V = 60$ V.
10. Velocity v of a falling object varies directly as time t. Find k for $v = 49\ \frac{\text{mm}}{\text{s}}$ (millimeters per second) and $t = 5$ ms.
11. Power P varies inversely as resistance R. Find k for $P = 18.5$ W and $R = 12\ \Omega$.
12. Wavelength λ (the Greek letter lambda) varies inversely as frequency f. Find k for $\lambda = 3$ m and $f = 100$ MHz.
13. Voltage V_1 across a resistor varies directly as applied voltage E and resistance R_1. Find k for $V_1 = 1.87$ V, $E = 12$ V, and $R_1 = 1.2\ \Omega$.

14. Angle measure θ (the Greek letter theta) varies directly as frequency f and time t. Find k for $\theta = 0.157$ radians, $f = 5$ kHz, and $t = 5$ μs.

15. Resistance R in a wire varies directly as length l and inversely as the cross-sectional area A. Find k for $R = 0.325$ Ω, $l = 10$ ft, and $A = 320$ c.mils (circular mils).

16. Capacitive reactance X_C varies inversely as frequency f and inversely as capacitance C. Find k for $X_C = 636$ kΩ, $f = 2.5$ kHz, and $C = 0.1$ μF.

Write the variation formula, find k, and then evaluate the formula:

17. Power P varies directly as voltage V. Find k for $P = 4.4$ W and $V = 220$ V. Evaluate for $V = 50$ V.

18. Power P varies inversely as resistance R. Find k for $P = 50$ mW and $R = 33$ Ω. Evaluate for $R = 27$ Ω.

19. Angle measure θ varies directly as frequency f and time t. Find k for $\theta = 3.77$ radians, $f = 60$ Hz, and $t = 10$ ms. Evaluate for $f = 30$ Hz and $t = 4$ ms.

20. Resistance R varies directly as length l and inversely as cross-sectional area A. Find k for $R = 10.5$ Ω, $l = 100$ ft, and $A = 101$ c.mils. Evaluate for $l = 50$ ft and $A = 404$ c.mils.

21. Power P varies directly as the square of current I. Find k for $P = 14.2$ kW and $I = 16$ A. Evaluate for $I = 10$ A.

22. Stopping distance s of a car varies directly as the square of its velocity v. Find k for $s = 50$ ft and $v = 30$ miles per hour, or 44 $\frac{\text{ft}}{\text{s}}$ (feet per second). Evaluate for $v = 60$ miles per hour, or 88 $\frac{\text{ft}}{\text{s}}$.

23. The centripetal force F toward the center of a circle varies directly as the square of the velocity v and inversely as the radius r of the circle. Find k for $F = 612.5$ lb (pounds), $v = 28$ $\frac{\text{ft}}{\text{s}}$, and $r = 80$ ft. Evaluate for $v = 36$ $\frac{\text{ft}}{\text{s}}$ and $r = 100$ ft.

24. The force of attraction F between two charged bodies varies directly as their charges q_1 and q_2 and inversely as the square of the distance r between them. Find k for $F = 0.36$ mN (millinewtons), $q_1 = 4$ μC, $q_2 = 4$ μC, and $r = 20$ mm. Evaluate for $q_1 = 5$ μC, $q_2 = 5$ μC, and $r = 15$ mm.

25. The velocity v of a falling object at the end of its fall varies directly as the square root of the height h from which it falls. Find k for $v = 62$ $\frac{\text{ft}}{\text{s}}$ and $h = 60$ ft. Evaluate for $h = 120$ ft.

26. The mutual inductance L_M of two coils varies directly as the square root of the product of their inductances L_1 and L_2. Find k for $L_M = 0.55$ mH, $L_1 = 4$ mH, and $L_2 = 6$ mH. Evaluate for $L_1 = 0.8$ mH and $L_2 = 1.2$ mH.

SECTION
5.4

Ratio and Proportion

A **ratio** compares two quantities in the form of a fraction. For example, when we say "miles per hour," we compare miles to hours as a ratio. If we travel 60 miles in two hours, we have traveled according to the ratio

$$\frac{60 \text{ miles}}{2 \text{ hours}}$$

or

$$30 \frac{\text{miles}}{\text{hour}},$$

which means 30 miles per hour.

In general, ratios are written as fractions, in the form

$$\frac{a}{b}.$$

This ratio is sometimes written

$$a : b,$$

and in either case is read "the ratio of a to b," or just "a to b."

Two equal ratios form a **proportion**. For example, if we travel 60 miles in two hours, then we could travel 120 miles in four hours, and we could travel 30 miles in one hour. The relationship between these ratios may be expressed by the proportions

$$\frac{60 \text{ miles}}{2 \text{ hours}} = \frac{120 \text{ miles}}{4 \text{ hours}}$$

and

$$\frac{60 \text{ miles}}{2 \text{ hours}} = \frac{30 \text{ miles}}{1 \text{ hour}}.$$

If the rate of travel is constant, then the relationship between two distances and their corresponding times can be expressed by the proportion

$$\frac{D_1}{T_1} = \frac{D_2}{T_2}.$$

When the time is increased, the distance also increases. When the time is decreased, the distance also decreases. The quantities in the ratios are said to be **proportional**; that is, distance is proportional to time.

We observe that proportions are related to variation. If the rate of travel is constant, then distance varies directly as time:

$$D = kT$$

or

$$\frac{D}{T} = k.$$

In general, proportions are written as two equal ratios

$$\frac{a}{b} = \frac{c}{d}$$

or, less commonly,

$$a : b = c : d.$$

Proportions are read "a is to b as c is to d," or "a to b equals c to d."

SOLVING PROPORTIONS

We **solve a proportion** by multiplying both sides of the proportion by the product of the denominators. For the proportion

$$\frac{a}{b} = \frac{c}{d},$$

we multiply both sides by bd to obtain

$$bd\left(\frac{a}{b}\right) = bd\left(\frac{c}{d}\right)$$

$$da = bc$$

or

$$ad = bc.$$

This process is summarized by the following pattern, sometimes called "cross-multiplying":

$$\frac{a}{b} \times \frac{c}{d}$$

$$ad = bc.$$

It is important to remember that "cross-multiplying" applies *only to solving proportions*. You must not try to adapt this pattern to addition or subtraction of fractions, or to solving equations containing additions and subtractions of fractions such as those in Section 4.3.

EXAMPLE 5.31 A number is to 2.2 as 10 is to 3.3. Find the number.

SOLUTION We write the proportion

$$\frac{x}{2.2} = \frac{10}{3.3}.$$

Then, we can multiply:

$$\frac{x}{2.2} = \frac{10}{3.3}$$

$$x(3.3) = (2.2)(10).$$

We solve this equation to obtain

$$x = \frac{(2.2)(10)}{3.3}$$

$$= 6.67$$

to three significant digits. ▲

We can also solve the proportion in Example 5.31 by multiplying both sides by 2.2:

$$\frac{x}{2.2} = \frac{10}{3.3}$$

$$2.2\left(\frac{x}{2.2}\right) = 2.2\left(\frac{10}{3.3}\right)$$

$$x = \frac{(2.2)(10)}{3.3}$$

$$x = 6.67.$$

EXAMPLE 5.32 Sixteen is to 39 as 24 is to a number. Find the number.

SOLUTION We write the proportion

$$\frac{16}{39} = \frac{24}{x}.$$

Then, we can multiply:

$$\frac{16}{39} = \frac{24}{x}$$

$$16x = (39)(24).$$

We solve this equation to obtain

$$x = \frac{(39)(24)}{16}$$

$$x = 58.5.$$ ▲

We can also solve the proportion in Example 5.32 by taking the reciprocal of each side. Recall that the reciprocal of a fraction "inverts" the fraction, exchanging the numerator and the denominator. Thus, we can write the proportion

$$\frac{16}{39} = \frac{24}{x}$$

as the proportion

$$\frac{39}{16} = \frac{x}{24}.$$

Then, we can multiply both sides by 24 to obtain

$$24\left(\frac{39}{16}\right) = 24\left(\frac{x}{24}\right)$$

$$\frac{(24)(39)}{16} = x$$

$$58.5 = x$$

or

$$x = 58.5.$$

It is important to remember that taking the reciprocal of each side applies *only to solving proportions*. You must not try to adapt this technique to addition or subtraction of fractions, or to solving equations containing additions and subtractions of fractions such as those in Section 4.3.

Proportions must be written so that the units of quantities in the numerators match, and the units of quantities in the denominators match.

EXAMPLE 5.33   The ratio of inches to centimeters is 1 : 2.54. How many centimeters are there in 3 feet?

SOLUTION   There are 36 inches in three feet. Therefore, we can write the proportion

$$\frac{1\ \text{in}}{2.54\ \text{cm}} = \frac{36\ \text{in}}{x\ \text{cm}}.$$

Solving by multiplying or by reciprocals, we have

$$x = \frac{(2.54\ \text{cm})(36\ \text{in})}{1\ \text{in}}$$

$$= 91.44\ \text{cm.} \qquad \blacktriangle$$

We have used Ohm's law as a formula in the form

$$I = \frac{V}{R}.$$

The right-hand side of this formula is the ratio of voltage to resistance. For example, we can write the ratio of 20 V to 10 kΩ as

$$\frac{20\ \text{V}}{10\ \text{k}\Omega}$$

or

$$2\,\frac{\text{V}}{\text{k}\Omega},$$

which means 2 volts per kilohm.

If the current in a circuit is constant, we can write the proportion

$$\frac{V_1}{R_1} = \frac{V_2}{R_2}.$$

When the resistance is increased, the voltage also increases. When the resistance is decreased, the voltage also decreases. Thus, voltage is proportional to resistance or, if the current is constant, voltage varies directly as resistance.

EXAMPLE 5.34   If $V_1 = 80\,\text{mV}$ when $R_1 = 56\,\Omega$, find the value of R_2 that causes the voltage to increase to $V_2 = 110\,\text{mV}$.

SOLUTION We can solve the proportion that relates voltage and resistance for R_2 by taking the reciprocal of each side:

$$\frac{V_1}{R_1} = \frac{V_2}{R_2}$$

$$\frac{R_1}{V_1} = \frac{R_2}{V_2}.$$

Multiplying both sides by V_2, we have

$$\frac{V_2 R_1}{V_1} = R_2$$

or

$$R_2 = \frac{V_2 R_1}{V_1}.$$

Then, we substitute the given values to obtain

$$R_2 = \frac{(110\,\text{mV})(56\,\Omega)}{80\,\text{mV}}$$

$$= 77\,\Omega.$$

We observe that the voltage increased and the resistance has also increased. ▲

The proportion relating voltage and resistance can be rewritten as a proportion that compares the ratio of the two voltages with the ratio of the two resistances. We write

$$\frac{V_1}{R_1} = \frac{V_2}{R_2}$$

$$V_1 R_2 = R_1 V_2.$$

Then, we divide both sides by $V_2 R_2$ to obtain the proportion

$$\frac{V_1 R_2}{V_2 R_2} = \frac{R_1 V_2}{V_2 R_2}$$

$$\frac{V_1}{V_2} = \frac{R_1}{R_2}.$$

INVERSE PROPORTION

Ohm's law can also be written in the form

$$V = IR.$$

Then, if the voltage is constant, we can write

$$I_1 R_1 = I_2 R_2.$$

Dividing both sides by I_2R_1, we obtain the proportion

$$\frac{I_1R_1}{I_2R_1} = \frac{I_2R_2}{I_2R_1}$$

$$\frac{I_1}{I_2} = \frac{R_2}{R_1}.$$

When the resistance is *increased*, the current *decreases*, and when the resistance is *decreased*, the current *increases*. We say that current is *inversely* proportional to resistance. In terms of variation, if the voltage is constant, then current varies *inversely* as resistance.

EXAMPLE 5.35 ▶ For the proportion

$$\frac{I_1}{I_2} = \frac{R_2}{R_1},$$

if $I_1 = 25$ mA when $R_1 = 56$ Ω, find the value of R_2 that causes the current to decrease to $I_2 = 20$ mA.

SOLUTION   To solve for R_2, we can simply multiply both sides by R_1:

$$\frac{I_1}{I_2} = \frac{R_2}{R_1}$$

$$\frac{R_1I_1}{I_2} = R_2$$

or

$$R_2 = \frac{R_1I_1}{I_2}.$$

Then, we substitute the given values to obtain

$$R_2 = \frac{(56\ \Omega)(25\ \text{mA})}{20\ \text{mA}}$$

$$= 70\ \Omega.$$

We observe that the current decreased but the resistance has increased.   ▲

PROPORTIONS INVOLVING SQUARES

We recall from Section 4.5 that power is defined by

$$P = VI.$$

By using the power definition and Ohm's law, we obtained the formula

$$P = I^2R.$$

We can solve for R to obtain

$$R = \frac{P}{I^2}.$$

Thus, if the resistance is constant, we have the proportion

$$\frac{P_1}{I_1^2} = \frac{P_2}{I_2^2}.$$

We can also write this proportion in the form

$$\frac{P_1}{P_2} = \frac{I_1^2}{I_2^2}.$$

Power is proportional to the *square* of current or, if the resistance is constant, power varies directly as the *square* of current.

In Section 4.5, we also obtained the formula

$$P = \frac{V^2}{R}.$$

We can solve for R to obtain

$$PR = V^2$$

$$R = \frac{V^2}{P}.$$

Thus, if the resistance is constant, we have the proportion

$$\frac{V_1^2}{P_1} = \frac{V_2^2}{P_2}.$$

We can also write this proportion in the form

$$\frac{P_1}{P_2} = \frac{V_1^2}{V_2^2}.$$

Power is proportional to the *square* of voltage or, if the resistance is constant, power varies directly as the *square* of voltage.

EXAMPLE 5.36 For the proportion

$$\frac{P_1}{P_2} = \frac{V_1^2}{V_2^2},$$

if $P_1 = 5$ W when $V_1 = 12$ V, find the value of V_2 that causes the power to increase to $P_2 = 10$ W.

SOLUTION To solve for V_2^2, we take the reciprocal of each side and multiply by V_1^2:

$$\frac{P_2}{P_1} = \frac{V_2^2}{V_1^2}$$

$$\frac{V_1^2 P_2}{P_1} = V_2^2$$

or

$$V_2^2 = \frac{P_2 V_1^2}{P_1}.$$

Then, taking the square root of each side, we have

$$V_2 = \sqrt{\frac{P_2 V_1^2}{P_1}}$$

$$V_2 = V_1 \sqrt{\frac{P_2}{P_1}}.$$

We substitute the given values to obtain

$$V_2 = (12 \text{ V})\sqrt{\frac{10 \text{ W}}{5 \text{ W}}}$$

$$= 12\sqrt{2} \text{ V}$$

$$= 17.0 \text{ V}$$

to three significant digits.   ▲

**EXERCISE
5.4**

1. What number is to 56 as 5 is to 100?

2. The ratio $1.2 : 1.5$ equals the ratio of what number to 5?

3. Fifty is to what number as 220 is to 300?

4. The ratio $330 : 390$ equals 8.5 to what number?

5. A car drives 15 miles in 20 minutes. How long will it take to go 125 miles at the same rate?

6. An airplane flies 105 miles in $\frac{1}{2}$ hour. How far can it fly in $1\frac{1}{4}$ hours at the same rate?

7. The ratio of inches to millimeters is $1 : 25.4$. How many millimeters are there in a foot?

8. The ratio of miles to kilometers is $0.6 : 1$. How many miles are there in 90 kilometers?

9. A computer can do 5 operations of a certain type in 2 µs. How many such operations can it do in 1 ms?

10. A computer is rated at a speed of 2.5 Gflops (billions of floating point operations per second). How long does it take to do one trillion (10^{12}) floating point operations?

11. For the proportion

$$\frac{V_1}{R_1} = \frac{V_2}{R_2},$$

if the voltage increases to $V_2 = 12$ V when the resistance is increased from $R_1 = 1\ \Omega$ to $R_2 = 1.5\ \Omega$, find V_1.

12. For the proportion

$$\frac{V_1}{I_1} = \frac{V_2}{I_2},$$

if $V_1 = 10$ V and the current is increased from $I_1 = 20$ mA to $I_2 = 25$ mA, find V_2.

13. For the proportion

$$\frac{V_1}{V_2} = \frac{R_1}{R_2},$$

if $V_1 = 125$ V when $R_1 = 1\ k\Omega$, find the value of R_2 that causes the voltage to decrease to $V_2 = 100$ V.

14. For the proportion

$$\frac{V_1}{V_2} = \frac{I_1}{I_2},$$

if $I_1 = 75$ A when $V_1 = 30$ V, find the value of V_2 that causes the current to decrease to 60 A.

15. For the proportion

$$\frac{I_1}{I_2} = \frac{R_2}{R_1},$$

if $I_1 = 15$ mA when $R_1 = 1\ k\Omega$, and the resistance is increased to $R_2 = 1.5\ k\Omega$, find I_2.

16. For the proportion

$$\frac{I_1}{I_2} = \frac{R_2}{R_1},$$

if changing the resistance to $R_2 = 22\ k\Omega$ causes the current to increase from $I_1 = 120$ µA to $I_2 = 150$ µA, find R_1.

17. For the proportion

$$\frac{P_1}{P_2} = \frac{V_1^{\,2}}{V_2^{\,2}},$$

if changing the voltage to $V_2 = 10$ V causes the power to increase from $P_1 = 25$ mW to $P_2 = 100$ mW, find V_1.

18. For the proportion

$$\frac{P_1}{P_2} = \frac{I_1^{\,2}}{I_2^{\,2}},$$

if $P_1 = 10$ W when $I_1 = 2$ A, find the value of I_2 that causes the power to decrease to $P_2 = 5$ W.

19. For the proportion

$$\frac{V_1}{V_2} = \sqrt{\frac{P_1}{P_2}},$$

if changing the power to $P_2 = 16.7$ W causes the voltage to increase from $V_1 = 9$ V to $V_2 = 10$ V, find P_1.

20. For the proportion

$$\frac{I_1}{I_2} = \sqrt{\frac{P_1}{P_2}},$$

if $I_1 = 135$ mA when $P_1 = 27$ W, find the value of P_2 that causes the current to decrease to $I_2 = 45$ mA.

Transformers

A **transformer** is an electronic device used to step up or step down AC voltage. For example, house current in the United States is transmitted from generating plants at very high voltages. The voltage is stepped down to household voltage by using transformers.

A typical transformer consists of an iron core wound with two coils of wire:

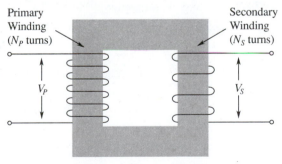

One coil is called the **primary winding** and the other is called the **secondary winding**. A voltage is applied to the primary winding, which creates a magnetic field. This magnetic field induces a voltage in the secondary winding. The primary and secondary voltages are proportional to the numbers of turns in the coils. That is,

$$\frac{N_P}{N_S} = \frac{V_P}{V_S},$$

where N_P and N_S are the numbers of turns in the primary and secondary windings and V_P and V_S are the primary and secondary voltages. If there are more turns in the primary winding than in the secondary winding, the voltage is stepped down. If there are more turns in the secondary winding, the voltage is stepped up.

EXAMPLE 5.37 ▶

A transformer has 200 turns in the primary winding and 100 turns in the secondary winding. If the voltage applied to the primary is 120 V, what voltage is induced in the secondary?

SOLUTION   We use the proportion

$$\frac{N_P}{N_S} = \frac{V_P}{V_S},$$

where $N_P = 200$ turns, $N_S = 100$ turns, and $V_P = 120$ V. Taking the reciprocals, and solving for V_S, we have

$$\frac{N_S}{N_P} = \frac{V_S}{V_P}$$

$$\frac{V_P N_S}{N_P} = V_S$$

or

$$V_S = \frac{V_P N_S}{N_P}.$$

We substitute the given values to obtain

$$V_S = \frac{(120 \text{ V})(100 \text{ turns})}{200 \text{ turns}}$$

$$= 60 \text{ V}.$$

We observe that there are more turns in the primary winding than in the secondary winding, and the voltage is stepped down.   ▲

EXAMPLE 5.38 ▶

A transformer has 10 turns in the primary winding and 150 turns in the secondary winding. If the voltage applied to the primary is 120 V, what voltage is induced in the secondary?

SOLUTION   Using the proportion

$$\frac{N_P}{N_S} = \frac{V_P}{V_S},$$

where $N_P = 10$ turns, $N_S = 150$ turns, and $V_P = 120$ V, we have

$$V_S = \frac{V_P N_S}{N_P}$$

$$= \frac{(120 \text{ V})(150 \text{ turns})}{10 \text{ turns}}$$

$$= 1800 \text{ V}.$$

We observe that there are more turns in the secondary winding than in the primary winding, and the voltage is stepped up.   ▲

TURNS RATIO

The **turns ratio** of a transformer is the ratio of the number of turns in the primary winding to the number of turns in the secondary winding

$$N_P : N_S.$$

For example, if $N_P = 200$ turns and $N_S = 100$ turns, the turns ratio is

$$N_P : N_S = 200 : 100$$

$$N_P : N_S = 2 : 1.$$

The turns ratio is always written with a 1 for either N_P or N_S. If N_S is 1, there are more turns in the primary and the voltage is stepped down. If N_P is 1, there are more turns in the secondary and the voltage is stepped up.

EXAMPLE 5.39   A transformer has turns ratio 4.5 : 1. If a voltage of 270 V is applied to the primary, what voltage is induced in the secondary?

SOLUTION The turns ratio is

$$N_P : N_S = 4.5 : 1$$

or

$$\frac{N_P}{N_S} = \frac{4.5}{1}.$$

Therefore, using

$$\frac{N_P}{N_S} = \frac{V_P}{V_S},$$

we have

$$\frac{4.5}{1} = \frac{V_P}{V_S}$$

$$4.5V_S = V_P$$

$$V_S = \frac{V_P}{4.5}$$

$$= \frac{270 \text{ V}}{4.5}$$

$$= 60 \text{ V}.$$

We observe that N_S is 1, so the voltage is stepped down. ▲

POWER AND CURRENT IN TRANSFORMERS

The power supplied by the primary of a transformer is P_i, the "power in," and the power dissipated in the secondary is P_o, the "power out." The efficiency of a transformer is the ratio

$$\frac{P_o}{P_i},$$

expressed as a percent. For example, if $P_i = 100$ W and $P_o = 70$ W, then is

$$\frac{P_o}{P_i} = \frac{70 \text{ W}}{100 \text{ W}}$$

$$= 0.70,$$

so the efficiency of the transformer is 70%.

For an **ideal** transformer, the efficiency is 100%; that is, in an ideal transformer

$$P_i = P_o.$$

We know from the definition of power in Section 4.5 that

$$P = VI.$$

Therefore, P_i is given by

$$P_i = V_P I_P,$$

and P_o is given by

$$P_o = V_S I_S.$$

Thus, for an ideal transformer,

$$P_i = P_o$$

$$V_P I_P = V_S I_S.$$

Dividing both sides by $V_S I_P$, we have

$$\frac{V_P I_P}{V_S I_P} = \frac{V_S I_S}{V_S I_P}$$

$$\frac{V_P}{V_S} = \frac{I_S}{I_P}.$$

We observe that the currents are *inversely* proportional to the voltages. Since we know that

$$\frac{N_P}{N_S} = \frac{V_P}{V_S},$$

we also have

$$\frac{N_P}{N_S} = \frac{I_S}{I_P}.$$

EXAMPLE 5.40 A voltage of 30 V is applied to the primary of a transformer, inducing 10 V in the secondary. The resulting current in the secondary is found to be 10 mA. What is the current in the primary?

SOLUTION We use the proportion

$$\frac{V_P}{V_S} = \frac{I_S}{I_P},$$

where $V_P = 30$ V, $V_S = 10$ V, and $I_S = 10$ mA. Taking the reciprocals, and solving for I_P, we have

$$\frac{V_S}{V_P} = \frac{I_P}{I_S}$$

$$\frac{I_S V_S}{V_P} = I_P.$$

Therefore,

$$I_P = \frac{I_S V_S}{V_P}$$

$$= \frac{(10 \text{ mA})(10 \text{ V})}{30 \text{ V}}$$

$$= 3.33 \text{ mA}.$$

We observe that the voltage is stepped down, but the current in the secondary is greater than the current in the primary.

IMPEDANCE IN TRANSFORMERS

Opposition to current in a transformer is called the **impedance**, Z. The impedance in the primary is Z_P and the impedance in the secondary is Z_S. Impedance is related to resistance in that it is also measured in ohms and obeys Ohm's law:

$$I = \frac{V}{Z}.$$

Thus, for the primary, we have

$$I_P = \frac{V_P}{Z_P}.$$

Then, since $P_i = V_P I_P$, we can replace I_P by $\frac{V_P}{Z_P}$ to obtain

$$P_i = V_P \left(\frac{V_P}{Z_P}\right)$$

$$P_i = \frac{V_P{}^2}{Z_P}.$$

Similarly, for the secondary,

$$P_o = \frac{V_S{}^2}{Z_S}.$$

Therefore, for an ideal transformer,

$$P_i = P_o$$

$$\frac{V_P{}^2}{Z_P} = \frac{V_S{}^2}{Z_S}.$$

We can write this proportion in the form

$$\frac{Z_P}{Z_S} = \frac{V_P{}^2}{V_S{}^2}.$$

Thus, the impedances vary as the *squares* of the voltages. Since we know that

$$\frac{N_P}{N_S} = \frac{V_P}{V_S},$$

we also have

$$\frac{Z_P}{Z_S} = \frac{N_P{}^2}{N_S{}^2}.$$

EXAMPLE 5.41 A transformer has 800 turns in the primary winding and 250 turns in the secondary winding. If the secondary impedance is 50 Ω, what is the primary impedance?

SOLUTION We use the proportion

$$\frac{Z_P}{Z_S} = \frac{N_P{}^2}{N_S{}^2},$$

where $N_P = 800$ turns, $N_S = 250$ turns, and $Z_S = 50\ \Omega$. Solving for Z_P, we have

$$Z_P = \frac{Z_S N_P{}^2}{N_S{}^2}$$

or

$$Z_P = Z_S \left(\frac{N_P}{N_S}\right)^2.$$

We substitute the given values to obtain

$$Z_P = 50\ \Omega \left(\frac{800 \text{ turns}}{250 \text{ turns}}\right)^2$$

$$Z_P = 512\ \Omega.$$ ▲

Finally, since we know that

$$\frac{V_P}{V_S} = \frac{I_S}{I_P},$$

we have

$$\frac{Z_P}{Z_S} = \frac{I_S^2}{I_P^2}.$$

Thus, the impedances vary *inversely* as the *squares* of the currents.

EXAMPLE 5.42 ▶

The current in the primary winding is 10 mA and the current in the secondary winding is 1.5 mA. If the secondary impedance is 50 Ω, what is the primary impedance?

SOLUTION

We use the proportion

$$\frac{Z_P}{Z_S} = \frac{I_S^2}{I_P^2},$$

where $I_P = 10$ mA, $I_S = 1.5$ mA, and $Z_S = 50\ \Omega$. Solving for Z_P, we have

$$Z_P = \frac{Z_S I_S^2}{I_P^2}$$

or

$$Z_P = Z_S \left(\frac{I_S}{I_P}\right)^2.$$

We substitute the given values to obtain

$$Z_P = 50\ \Omega \left(\frac{1.5 \text{ mA}}{10 \text{ mA}}\right)^2$$

$$Z_P = 1.13\ \Omega.$$ ▲

**EXERCISE
5.5**

1. A transformer has 1050 turns in the primary winding and 250 turns in the secondary winding. If the voltage applied to the primary is 25 V, what voltage is induced in the secondary?

2. A transformer has 800 turns in the primary winding. A voltage of 10 V is applied to the primary, and 2.25 V is induced in the secondary. How many turns are in the secondary winding?

3. A transformer has 100 turns in the primary winding and 600 turns in the secondary winding. What voltage applied to the primary induces 12.2 V in the secondary?

4. A transformer has 650 turns in the secondary winding. A voltage of 14.5 V is induced in the secondary when 5 V is applied to the primary. How many turns are in the primary winding?

5. A transformer has turns ratio 3 : 1. If a voltage of 40 V is applied to the primary, what voltage is induced in the secondary?

6. A transformer has turns ratio 1 : 4.5. What voltage applied to the primary induces 5.8 V in the secondary?

7. A voltage of 12 V is applied to the primary of a transformer, inducing 102 V in the secondary. What is its turns ratio?

8. A voltage of 17.7 V is measured in the primary of a transformer, and 5.71 V in the secondary. What is its turns ratio?

9. A voltage of 10 V is applied to the primary of a transformer, inducing 60 V in the secondary. The resulting current in the secondary is found to be 15 mA. What is the current in the primary?

10. A transformer has turns ratio 4.5 : 1. If the current in the primary is 2.4 mA, what is the current in the secondary?

11. A transformer has 100 turns in the primary winding and 1000 turns in the secondary winding. If the secondary impedance is 33 kΩ, what is the primary impedance?

12. A transformer has 600 turns in the secondary winding. The primary impedance is 100 Ω, and the secondary impedance is 25 Ω. How many turns are in the primary winding?

13. A voltage of 25 V is applied to the primary of a transformer, and 5 V is induced in the secondary. If the secondary impedance is 12 kΩ, what is the primary impedance?

14. A voltage of 10 V is applied to the primary of a transformer. If the primary impedance is 2.5 Ω, and the secondary impedance is 20 Ω, what is the voltage across the secondary?

15. The current in the secondary of a transformer is 2 mA. If the primary impedance is 150 Ω, and the secondary impedance is 50 Ω, what is the current in the primary?

16. The current in the secondary of a transformer is 55 mA. If the primary impedance is 15 kΩ, and the secondary impedance is 300 Ω, what is the current in the primary?

Self-Test

1. Solve the formula $V_{CC} = I_C(R_C + R_E)$ for R_E.

1. _____

2. Evaluate the formula $f_r = \dfrac{1}{2\pi\sqrt{LC}}$ for $L = 5\ \mu\text{H}$ and $C = 80\ \text{nF}$.

2. _____

3. A computer can do 400 operations of a certain type in 1 ms. How long does it take to do one million such operations?

3. _____

4. A transformer has 100 turns in the primary winding and 250 turns in the secondary winding. If the primary impedance is 2.5 kΩ, what is the secondary impedance?

4. _____

5. Resistance R of a wire varies directly as length l and inversely as cross-sectional area A. Write the variation formula and find k for $R = 0.655\ \Omega$, $l = 25$ ft, and $A = 404$ c.mils (circular mils). Then, evaluate the formula for $l = 1000$ ft and $A = 160$ c.mils.

5. _____

UNIT 6

Cumulative Review

Introduction

In the preceding units, you have learned about real numbers. In particular, you learned how to do arithmetic operations with signed numbers. You learned about powers of ten, and how to use them in scientific notation and in engineering notation. You simplified and evaluated linear expressions and some nonlinear algebraic expressions. Then, you learned one of the main tools of algebra, equation solving. You solved linear equations and some types of quadratic and related equations. You used equations to solve problems involving two basic concepts of electricity, Ohm's law and power. Finally, you used these skills to evaluate and solve formulas. You learned how to write and use some specific types of formulas such as variation formulas and proportions. Now, you should review the material in all of the preceding units.

OBJECTIVE

When you have finished this unit you should be able to fulfill every objective of each of the preceding units.

SECTION 6.1

Review Method

For this unit, you should review the Self-Tests for Units 1 through 5. Do each problem in each Self-Test over again. If you cannot do a problem, or if you have the slightest doubt or difficulty, find the appropriate material to review:

1. You will find an objective number next to the answer to every Self-Test problem. Find the objective number of the problem you are working on (this is not necessarily the problem number). Go to the first page of the unit and reread the objective. For example, if you have difficulty with a Self-Test problem in Unit 2, and the Objective number next to the answer is Objective 3, reread Objective 3 in the objectives list at the beginning of Unit 2 to find out what concept you need to review.

2. The material you should study to review the concept is in the section that has the same number as the objective. Reread the material in that section, rework the examples, and redo some of the exercises for the section. For example, if you are reviewing Unit 2, Objective 3, refer to Unit 2, Section 3; that is, Section 2.3. Reread Section 2.3, rework examples in that section, and redo problems in Exercise 2.3.

3. Try the Self-Test for the unit again to find out if there is any other objective you need to review in the unit. For example, if you have been reviewing Section 2.3 and Exercise 2.3, try the Self-Test for Unit 2 again.

Review each unit in this way. For some units, you may find that you can still do the Self-Test easily. For other units, you might need to review one or more of the objectives in more detail. As a final check, and only after you have reviewed each preceding unit, try the Self-Test for this Cumulative Review Unit 6.

□ **Self-Test** □

1. Find the value of $2 - 3\sqrt{3^2 + 4^2}$.

1. _____

2. Divide and write the result in scientific notation to three significant digits:

   $$\frac{2.92 \times 10^3}{6.48 \times 10^{-4}}$$

2. _____

3. Convert the quantity to engineering notation with the given prefix:

   a.   1.22 ns to ps

   b.   360 kHz to GHz

3a. _____

3b. _____

4. Simplify $3(2x - 3) - (2 - 3x)$.

4. _____

5. Simplify $\sqrt{\dfrac{16x^4y^8}{4x^2y^4}}$.

5. _____

6. Solve $3x - (x - 6) = 4(x - 3)$ and check the solution.

6. _____

7. Solve $\dfrac{x^2}{2} - \dfrac{2}{3} = \dfrac{4}{3}$ and check the solution.

7. _____

8. Use Ohm's law to find R if $I = 54.5\ \mu A$ and $V = 120$ mV.

8. _____

9. For the formula $X_C = \dfrac{1}{2\pi f C}$, find C if $X_C = 3.3$ kΩ and $f = 5$ kHz.

9. _____

10. The volume V of a pyramid varies directly as its base area A and directly as its height h. If $V = 33\frac{1}{3}$ cu ft (cubic feet) when $A = 20$ sq ft (square feet) and $h = 5$ ft, find V when $A = 9$ sq ft and $h = 10$ ft.

10. _____

UNIT 7

Kirchhoff's Laws

Introduction

In preceding units, you have used Ohm's law as an example of an equation and as an example of the special type of equation called a formula. Ohm's law is one of the basic principles of electronics. Two other basic laws of electronics are Kirchhoff's laws: Kirchhoff's voltage law and Kirchhoff's current law. In this unit, you will learn about these two laws. You will use Kirchhoff's voltage law to analyze voltage in closed loops of circuits, and Kirchhoff's current law to analyze currents at nodes of circuits.

OBJECTIVES

When you have finished this unit you should be able to:

1. Use Kirchhoff's voltage law to find voltages in closed loops of series and series-parallel circuits.

2. Use Kirchhoff's current law to find currents at nodes of parallel and series-parallel circuits.

SECTION 7.1

Kirchhoff's Voltage Law

To understand the behavior of voltage in electrical circuits, we must look more closely at the concept of voltage. In order to understand voltage, we might picture a book on a table. Suppose we raise the book above the table. To raise the book requires energy. The higher we raise the book, the more energy we must use. Once the book is raised, however, it has the potential to create energy by falling. The higher it is raised, the greater its potential energy.

We recall the simple circuit introduced in Section 4.5, consisting of a voltage source and a resistor connected by wires that carry a current. The voltage source E, an ordinary battery for example, has a negative terminal and a positive terminal. Like raising the book from the table, the separation of electrons to create the positive and negative terminals creates potential energy. The greater the difference in electrical charge between the positive and negative terminals, the greater the potential energy. This difference is called the **potential difference** of the source. We refer to this potential difference as a **voltage rise**.

The current I in the circuit consists of a flow of electrons, which are negatively charged, from the negative terminal to the positive terminal of the voltage source. When the current flows through the resistor R, energy is used by the resistor. If the resistor is a light bulb, for example, it lights up. We refer to the loss of energy through a resistor as a **voltage drop**.

When we start from one point in a circuit, travel through a series of voltage rises and drops, and return to our starting point, we have a **closed loop**. Gustav Robert Kirchhoff (1824–1887), a German physicist, stated the relationship between voltage rises and voltage drops in a closed loop.

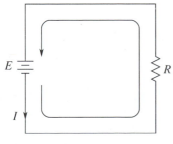

A simple circuit with a closed loop.

Kirchhoff's Voltage Law: In any closed loop, the sum of the voltage rises is equal to the sum of the voltage drops.

The simple circuit at the beginning of this section is a closed loop. This closed loop contains one voltage rise E and one voltage drop V_R across R. Therefore, the voltage rise E is equal to the voltage drop V_R. Thus, for example, if $E = 10$ V then $V_R = 10$ V.

SERIES CIRCUITS

When a voltage source and two or more resistors are contained in a closed loop, we have a **series circuit**. This series circuit has one voltage rise E, and two voltage drops, V_1 across R_1, and V_2 across R_2. According to Kirchhoff's voltage law, the voltage rise is equal to the sum of the two voltage drops; that is,

$$E = V_1 + V_2.$$

A series circuit with two resistors.

EXAMPLE 7.1

In a series circuit with two resistors, the voltage drops across the resistors are $V_1 = 4$ V and $V_2 = 8$ V. Find the voltage rise E.

SOLUTION

Using Kirchhoff's voltage law, we have

$$E = V_1 + V_2$$
$$= 4\text{ V} + 8\text{ V}$$
$$= 12\text{ V}.$$
▲

EXAMPLE 7.2

For a series circuit with two resistors, if $E = 8.5$ V and $V_1 = 3.9$ V, find V_2.

SOLUTION

Using Kirchhoff's voltage law, we have

$$E = V_1 + V_2.$$

Thus, we solve for V_2 to obtain

$$V_2 = E - V_1$$
$$= 8.5\text{ V} - 3.9\text{ V}$$
$$= 4.6\text{ V}.$$
▲

EXAMPLE 7.3

For this series circuit with three resistors, if $V_1 = 12.3$ V, $V_2 = 7.5$ V, and $V_3 = 14.4$ V, find E:

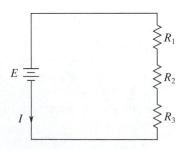

SOLUTION

Using Kirchhoff's voltage law, we have

$$E = V_1 + V_2 + V_3$$
$$= 12.3\text{ V} + 7.5\text{ V} + 14.4\text{ V}$$
$$= 34.2\text{ V}.$$
▲

EXAMPLE 7.4   For the series circuit with three resistors in Example 7.3, if $E = 4.19$ V, $V_1 = 1.27$ V, and $V_3 = 670$ mV, find V_2.

SOLUTION   Using Kirchhoff's voltage law, we have

$$E = V_1 + V_2 + V_3.$$

Thus, we solve for V_2 to obtain

$$V_2 = E - V_1 - V_3$$

$$= 4.19 \text{ V} - 1.27 \text{ V} - 670 \text{ mV}.$$

Since 670 mV $= 0.67$ V, we have

$$V_2 = 4.19 \text{ V} - 1.27 \text{ V} - 0.67 \text{ V}$$

$$= 2.25 \text{ V}.$$ ▲

PARALLEL CIRCUITS

Two resistors may be placed so that they do not form a closed loop with the voltage source. In this case, we have a **parallel circuit**. The parallel circuit has two closed loops that contain E:

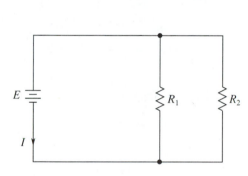

A parallel circuit with two resistors.

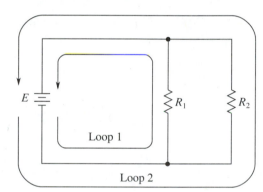

Two closed loops in the parallel circuit.

Loop 1 contains E and R_1, and Loop 2 contains E and R_2. In Loop 1, the sum of the voltage rises is equal to the sum of the voltage drops, and so

$$E = V_1.$$

Similarly, in Loop 2, the sum of the voltage rises is equal to the sum of the voltage drops, and so

$$E = V_2.$$

Thus, we have

$$V_1 = V_2.$$

The parts of the circuit containing the resistors are called **branches**. The voltage drops in each branch of a parallel circuit are equal, and are also equal to the voltage rise. Thus, for example, if $E = 10$ V, then $V_1 = 10$ V and $V_2 = 10$ V.

SERIES-PARALLEL CIRCUITS

Many circuits are **series-parallel** circuits; that is, they have components in series and also components in parallel. We can analyze voltages in series-parallel circuits by finding closed loops that include the voltage source E.

EXAMPLE 7.5 For the series-parallel circuit shown below, if $E = 15$ V and $V_1 = 6$ V, find V_2 and V_3.

SOLUTION   The series-parallel circuit has two closed loops that contain E:

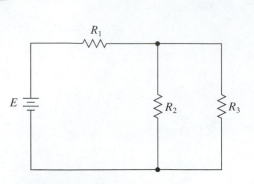

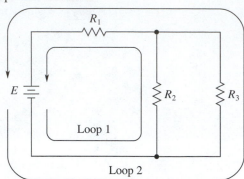

Circuit for Example 7.5.   Two closed loops in the circuit.

Loop 1 contains E, R_2, and R_1, and Loop 2 contains E, R_3, and R_1. Using Kirchhoff's voltage law for Loop 1, we have

$$E = V_2 + V_1.$$

Solving for V_2, we obtain

$$V_2 = E - V_1$$
$$= 15\text{ V} - 6\text{ V}$$
$$= 9\text{ V}.$$

Using Kirchhoff's voltage law for Loop 2, we have

$$E = V_3 + V_1.$$

Solving for V_3, we obtain

$$V_3 = E - V_1$$
$$= 15\text{ V} - 6\text{ V}$$
$$= 9\text{ V}.$$

We observe that R_2 and R_3 are in parallel branches, and so the voltage drops V_2 and V_3 are equal. These voltage drops are not equal to the voltage rise E, however, because of the presence of R_1 in series with E.   ▲

EXAMPLE 7.6 For this series-parallel circuit, if $E = 11.5$ V, $V_1 = 6.2$ V, and $V_4 = 4.2$ V, find V_2, V_3, and V_5:

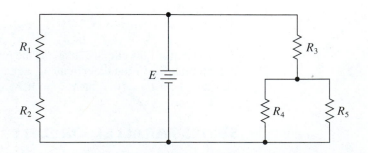

SOLUTION   This series-parallel circuit has three closed loops that contain E. Two are shown here:

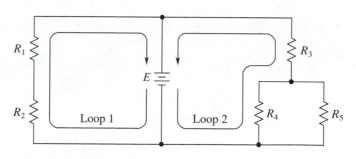

Using Kirchhoff's voltage law for Loop 1, we have

$$E = V_2 + V_1.$$

Solving for V_2, we obtain

$$V_2 = E - V_1$$

$$= 11.5 \text{ V} - 6.2 \text{ V}$$

$$= 5.3 \text{ V}.$$

Using Kirchhoff's voltage law for Loop 2, we have

$$E = V_4 + V_3.$$

Solving for V_3, we obtain

$$V_3 = E - V_4$$

$$= 11.5 \text{ V} - 4.2 \text{ V}$$

$$= 7.3 \text{ V}.$$

A third loop contains E, R_5, and R_3. This loop is identical to Loop 2 but with R_5 in place of R_4, and so

$$E = V_5 + V_3.$$

Thus, we solve for V_5 to obtain

$$V_5 = E - V_3$$

$$= 11.5 \text{ V} - 7.3 \text{ V}$$

$$= 4.2 \text{ V}.$$

We observe that R_4 and R_5 are in parallel branches. Therefore, the voltage drops V_4 and V_5 are equal. We could find V_5 simply by writing

$$V_5 = V_4$$

$$= 4.2 \text{ V}. \qquad \blacktriangle$$

EXAMPLE 7.7 ▶ The series-parallel circuit shown is a loaded voltage-divider circuit. If $E = 30$ V, $V_1 = 5$ V, and $V_2 = 10$ V, find V_3, V_4, V_5, and V_6:

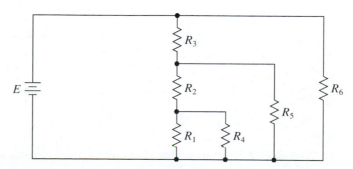

SOLUTION   This series-parallel circuit has several closed loops that contain E. Three are shown here:

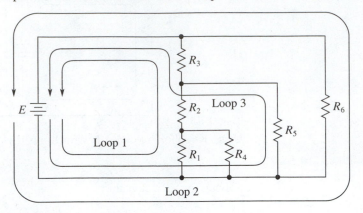

Loop 1, at the left, contains E, R_1, R_2, and R_3, and so

$$E = V_1 + V_2 + V_3.$$

We solve for V_3 to obtain

$$V_3 = E - V_1 - V_2$$
$$= 30\text{ V} - 5\text{ V} - 10\text{ V}$$
$$= 15\text{ V}.$$

Loop 2, around the outside, contains just E and R_6, and so

$$V_6 = E$$
$$= 30\text{ V}.$$

Loop 3 contains E, R_5, and R_3; therefore, we find that

$$E = V_3 + V_5.$$

We solve for V_5 to obtain

$$V_5 = E - V_3$$
$$= 30\text{ V} - 15\text{ V}$$
$$= 15\text{ V}.$$

We can draw a similar loop that contains E, R_4, R_2, and R_3. Using this loop, we find that

$$E = V_4 + V_2 + V_3.$$

We solve for V_4 to obtain

$$V_4 = E - V_2 - V_3$$
$$= 30\text{ V} - 10\text{ V} - 15\text{ V}$$
$$= 5\text{ V}.$$

We may also observe that R_4 is in a branch parallel to R_1. Therefore, we could find V_4 by writing

$$V_4 = V_1$$
$$= 5\text{ V}.$$

Similarly, R_5 is in a branch parallel to R_1 and R_2 combined. Therefore, we could find V_5 by writing

$$V_5 = V_1 + V_2$$

$$= 5\text{ V} + 10\text{ V}$$

$$= 15\text{ V}.$$

▲

EXERCISE

7.1

For a series circuit with two resistors:

1. If $V_1 = 3.5$ V and $V_2 = 7.5$ V, find E.
2. If $V_1 = 100$ mV and $V_2 = 225$ mV, find E.
3. If $E = 12.2$ V and $V_1 = 4.8$ V, find V_2.
4. If $E = 1$ V and $V_2 = 850$ mV, find V_1.

For a series circuit with three resistors:

5. If $V_1 = 15$ V, $V_2 = 5$ V, and $V_3 = 30$ V, find E.
6. If $V_1 = 9$ V, $V_2 = 12.8$ V, and $V_3 = 22.2$ V, find E.
7. If $E = 3.3$ V, $V_1 = 600$ mV, and $V_2 = 1.2$ V, find V_3.
8. If $E = 2.25$ V, $V_2 = 720$ mV, and $V_3 = 550$ mV, find V_1.
9. For the series-parallel circuit shown, if $E = 3.6$ V and $V_2 = 2.1$ V, find V_1 and V_3.
10. For the circuit in Exercise 9, if $V_1 = 45$ V and $V_3 = 15$ V, find E and V_2.

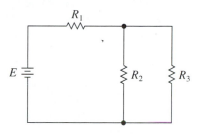

Circuit for Exercises 9 and 10.

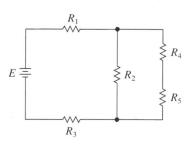

Circuit for Exercises 11 and 12.

11. For the series-parallel circuit shown above, if $E = 20$ V, $V_1 = 5$ V, $V_3 = 6$ V, and $V_4 = 8$ V, find V_2 and V_5.
12. For the circuit in Exercise 11, if $V_1 = 3$ V, $V_2 = 5.5$ V, $V_3 = 4.5$ V, and $V_4 = 2$ V, find E and V_5.
13. For the series-parallel circuit shown below, if $E = 9.5$ V, $V_2 = 2.2$ V, and $V_3 = 7.2$ V, find V_1, V_4, and V_5.
14. For the circuit in Exercise 13, if $V_1 = 35.5$ V, $V_2 = 11.5$ V, and $V_5 = 13$ V, find E, V_3, and V_4.
15. For the series-parallel circuit shown, if $E = 19$ V, $V_1 = 3.9$ V, $V_3 = 9$ V, and $V_4 = 4.8$ V, find V_2 and V_5.
16. For the circuit in Exercise 15, if $E = 2.15$ V, $V_1 = 300$ mV, $V_2 = 800$ mV, and $V_5 = 250$ mV, find V_3 and V_4.

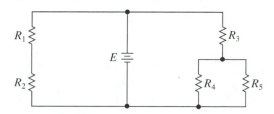

Circuit for Exercises 13 and 14.

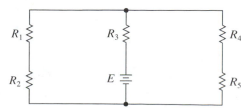

Circuit for Exercises 15 and 16.

17. The circuit shown is called a π-circuit. If $E = 10$ V and $V_2 = 6.6$ V, find V_1, V_3, and V_4:

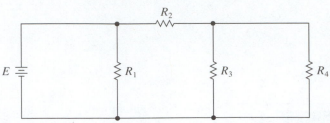

18. For the circuit in Exercise 17, if $V_2 = 15$ V and $V_3 = 25$ V, find E, V_1, and V_4.

19. For the voltage-divider circuit in Example 7.7 (page 173), if $E = 100$ V, $V_2 = 30$ V, and $V_3 = 40$ V, find V_1, V_4, V_5, and V_6.

20. For the voltage-divider circuit in Example 7.7, if $V_4 = 10$ V, $V_5 = 20$ V, and $V_6 = 30$ V, find E, V_1, V_2, and V_3.

<table>
<tr><td>**SECTION**
7.2</td></tr>
</table>

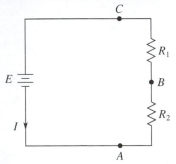

Current at some points in a series circuit.

Kirchhoff's Current Law

The current in electrical circuits consists of a flow of electrons from the negative terminal to the positive terminal of the voltage source. We can think of current as the flow of traffic on a limited-access highway, where the electrons travel like the cars (and just as erratically). Because the highway is limited-access, cars can enter the highway only at an entrance and leave only at an exit.

Series circuits have no entrances or exits. Therefore, the current I is constant throughout a series circuit; that is, the current is the same at every point of a series circuit. For example, if $I = 10$ mA at point A, then $I = 10$ mA at point B, and $I = 10$ mA at point C.

Parallel circuits have special points called **branch points** or **nodes**. In the parallel circuit shown below, points A and B are nodes. At nodes, current can go into or out of branches. Thus, the nodes are entrances or exits. Kirchhoff stated the relationship between currents entering a node and currents exiting a node.

Kirchhoff's Current Law: At any node, the sum of the currents into the node is equal to the sum of the currents out of the node.

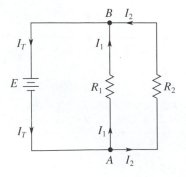

Nodes and currents in a parallel circuit.

For the parallel circuit with two branches, node A is an exit and node B is an entrance. At node A, some of the current exits into the branch containing R_2. According to Kirchhoff's current law, the current into node A is equal to the sum of the currents out of node A; that is,

$$I_T = I_1 + I_2.$$

At node B, current enters from the branch containing R_2. According to Kirchhoff's current law, the sum of the currents into node B is equal to the current out of node B; that is,

$$I_1 + I_2 = I_T.$$

Thus, at each node,

$$I_T = I_1 + I_2.$$

| **EXAMPLE 7.8** | |

In a parallel circuit with two resistors, the branch currents are $I_1 = 3$ mA and $I_2 = 5$ mA. Find the total current I_T.

SOLUTION   Using Kirchhoff's current law, we have

$$I_T = I_1 + I_2$$
$$= 3 \text{ mA} + 5 \text{ mA}$$
$$= 8 \text{ mA}.$$ ▲

| **EXAMPLE 7.9** | ▶ |

For a parallel circuit with two resistors, if $I_T = 1.02$ A and $I_1 = 660$ mA, find I_2.

SOLUTION   Using Kirchhoff's current law, we have

$$I_T = I_1 + I_2.$$

Thus, we solve for I_2 to obtain

$$I_2 = I_T - I_1$$
$$= 1.02 \text{ A} - 660 \text{ mA}.$$

Since 1.02 A = 1020 mA, we have

$$I_2 = 1020 \text{ mA} - 660 \text{ mA}$$
$$= 360 \text{ mA}.$$ ▲

| **EXAMPLE 7.10** | ▶ |

For the parallel circuit with three resistors shown below, if $I_1 = 3.6$ mA, $I_2 = 12.9$ mA, and $I_3 = 7.5$ mA, find I_T.

SOLUTION   We observe that node A is an exit, where the current splits into two parts. Node C is also an exit, where the current again splits into two parts:

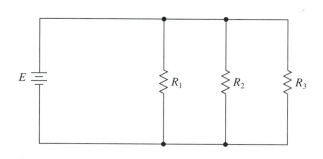

Circuit for Example 7.10.

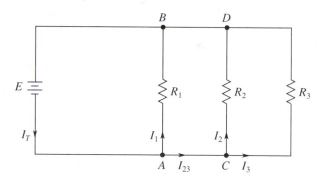

Nodes and currents for the circuit.

The currents out of node A are I_1 and another that continues on to resistors R_2 and R_3. We have labeled this second current I_{23}. Then, the current into node C is I_{23}, and the currents out are I_2 and I_3. Thus, using Kirchhoff's current law at node C,

$$I_{23} = I_2 + I_3$$
$$= 12.9 \text{ mA} + 7.5 \text{ mA}$$
$$= 20.4 \text{ mA}.$$

Then, using Kirchhoff's current law at node A,

$$I_T = I_1 + I_{23}$$

$$= 3.6 \text{ mA} + 20.4 \text{ mA}$$

$$= 24 \text{ mA}. \qquad \blacktriangle$$

For the parallel circuit with three resistors, we observe that

$$I_{23} = I_2 + I_3,$$

and

$$I_T = I_1 + I_{23}.$$

We may substitute for I_{23} to obtain

$$I_T = I_1 + I_2 + I_3.$$

Thus, the circuit in Example 7.10 is equivalent to a circuit where nodes A and C are the same point, and nodes B and D are the same point:

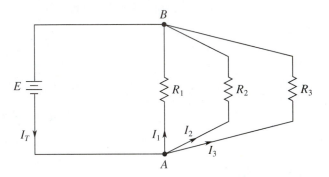

 EXAMPLE 7.11 ▶ For the parallel circuit in Example 7.10, if $I_T = 3.5$ A, $I_1 = 500$ mA, and $I_2 = 1$ A, find I_3.

SOLUTION We can use the preceding result to write

$$I_T = I_1 + I_2 + I_3.$$

Thus, we solve for I_3 to obtain

$$I_3 = I_T - I_1 - I_2$$

$$= 3.5 \text{ A} - 500 \text{ mA} - 1 \text{ A}.$$

Since 500 mA = 0.5 A, we have

$$I_3 = 3.5 \text{ A} - 0.5 \text{ A} - 1 \text{ A}$$

$$= 2 \text{ A}. \qquad \blacktriangle$$

SERIES-PARALLEL CIRCUITS

We can analyze currents in series-parallel circuits by finding currents in and out of each node of the circuit.

 EXAMPLE 7.12 ▶ For the series-parallel circuit shown at the top of page 179, if $I_T = 66$ mA and $I_2 = 23.4$ mA, find I_1 and I_3.

SOLUTION We label the nodes and currents:

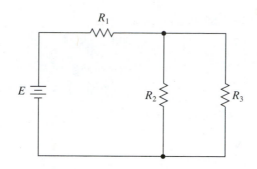

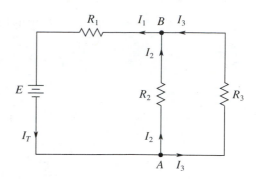

Circuit for Example 7.12.   Nodes and currents for the circuit.

Using Kirchhoff's current law at node A, we have

$$I_T = I_2 + I_3.$$

Solving for I_3, we obtain

$$I_3 = I_T - I_2$$

$$= 66 \text{ mA} - 23.4 \text{ mA}$$

$$= 42.6 \text{ mA}.$$

At node B, the currents in are I_2 and I_3, and the current out is I_1. Thus, at node B, we have

$$I_2 + I_3 = I_1.$$

Thus,

$$I_1 = I_2 + I_3$$

$$= 23.4 \text{ mA} + 42.6 \text{ mA}$$

$$= 66 \text{ mA}.$$

We observe that the current I_T through E and the current I_1 through R_1 are the same. When there are no nodes between two components, the components are in series and the currents through them are the same. ▲

EXAMPLE 7.13 ▶ For the series-parallel circuit shown below, if $I_T = 5.8$ mA, $I_1 = 2.75$ mA, and $I_3 = 1.05$ mA, find I_2 and I_4.

SOLUTION We label the nodes and currents:

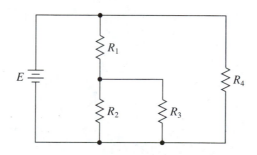

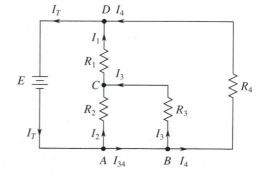

Circuit for Example 7.13.   Nodes and currents for the circuit.

We cannot start with node A or node B because we only know one current at each. Using Kirchhoff's current law at node C, we have

$$I_2 + I_3 = I_1.$$

Solving for I_2, we obtain

$$I_2 = I_1 - I_3$$
$$= 2.75 \text{ mA} - 1.05 \text{ mA}$$
$$= 1.7 \text{ mA}.$$

Using Kirchhoff's current law at node D, we have

$$I_1 + I_4 = I_T.$$

Solving for I_4, we obtain

$$I_4 = I_T - I_1$$
$$= 5.8 \text{ mA} - 2.75 \text{ mA}$$
$$= 3.05 \text{ mA}.$$

We may also observe that

$$I_{34} = I_3 + I_4$$

and

$$I_T = I_2 + I_{34},$$

so we have

$$I_T = I_2 + I_3 + I_4.$$

Therefore, we could find I_4 by writing

$$I_4 = I_T - I_2 - I_3$$
$$= 5.8 \text{ mA} - 1.7 \text{ mA} - 1.05 \text{ mA}.$$
$$= 3.05 \text{ mA}.$$ ▲

EXAMPLE 7.14 ▶ The series-parallel circuit shown below is called a Wheatstone bridge. If $I_T = 10$ A, $I_1 = 4$ A, and $I_3 = 4$ A, find $I_2, I_4,$ and I_5.

SOLUTION   When we label the nodes and currents, we find that we do not know the direction of the current I_5 through R_5. For now, we will assume that I_5 goes from left to right:

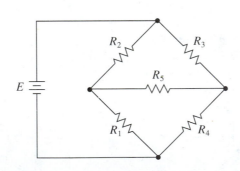

Circuit for Example 7.14.

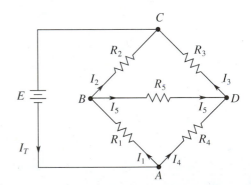

Nodes and currents for the circuit.

At node A, we have

$$I_T = I_1 + I_4.$$

Solving for I_4, we obtain

$$I_4 = I_T - I_1$$
$$= 10 \text{ A} - 4 \text{ A}$$
$$= 6 \text{ A}.$$

Then, assuming that I_5 goes from left to right, the currents into node D are I_4 and I_5, and the current out is I_3. Thus, at node D, we have

$$I_4 + I_5 = I_3,$$

and we can solve for I_5 to obtain

$$I_5 = I_3 - I_4$$
$$= 4\,\text{A} - 6\,\text{A}$$
$$= -2\,\text{A}.$$

When we have made an assumption concerning the direction of a current, and the current found by using Kirchhoff's law is negative, then the current actually goes in the opposite direction. Thus, this current is

$$I_5 = 2\,\text{A}$$

from right to left:

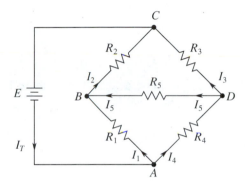

Then, using this direction, we find that the currents into node B are I_1 and I_5, and the current out is I_2. Thus, at node B, we have

$$I_1 + I_5 = I_2,$$

and so

$$I_2 = 4\,\text{A} + 2\,\text{A}$$
$$= 6\,\text{A}.$$

We observe that, at node C,

$$I_2 + I_3 = I_T,$$

and so

$$I_T = 6\,\text{A} + 4\,\text{A}$$
$$= 10\,\text{A}.$$ ▲

EXERCISE

7.2

For a parallel circuit with two resistors:

1. If $I_1 = 120\,\mu\text{A}$ and $I_2 = 90\,\mu\text{A}$, find I_T.
2. If $I_1 = 500\,\mu\text{A}$ and $I_2 = 1.2\,\text{mA}$, find I_T.
3. If $I_T = 1.02\,\text{A}$ and $I_1 = 220\,\text{mA}$, find I_2.
4. If $I_T = 0.115\,\text{mA}$ and $I_2 = 45\,\mu\text{A}$, find I_1.

For a parallel circuit with three resistors:

5. If $I_1 = 30.5\,\text{mA}$, $I_2 = 11.4\,\text{mA}$, and $I_3 = 6.2\,\text{mA}$, find I_T.
6. If $I_1 = 850\,\mu\text{A}$, $I_2 = 720\,\mu\text{A}$, and $I_3 = 930\,\mu\text{A}$, find I_T.

7. If $I_T = 3.91$ A, $I_1 = 440$ mA, and $I_2 = 660$ mA, find I_3.

8. If $I_T = 10$ mA, $I_1 = 7.5$ mA, and $I_3 = 910$ μA, find I_2.

9. For the series-parallel circuit shown, if $I_T = 3.45$ mA and $I_2 = 2.02$ mA, find I_1 and I_3.

10. For the circuit in Exercise 9, if $I_T = 120$ μA and $I_3 = 45$ μA, find I_1 and I_2.

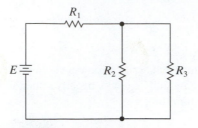

Circuit for Exercises 9 and 10.

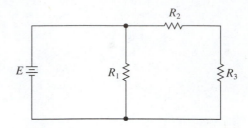

Circuit for Exercises 11 and 12.

11. For the series-parallel circuit shown, if $I_T = 16$ mA and $I_1 = 9.2$ mA, find I_2 and I_3.

12. For the circuit in Exercise 11, if $I_T = 1.2$ mA and $I_2 = 825$ μA, find I_1 and I_3.

13. For the series-parallel circuit shown, if $I_T = 25.5$ mA, $I_1 = 13.9$ mA, and $I_2 = 6.6$ mA, find I_3 and I_4.

14. For the circuit in Exercise 13, if $I_T = 111$ μA, $I_2 = 36.2$ μA, and $I_3 = 49.4$ μA, find I_1 and I_4.

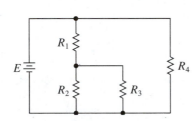

Circuit for Exercises 13 and 14.

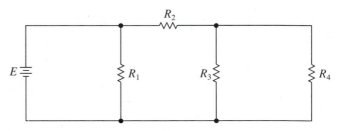

Circuit for Exercises 15 and 16.

15. For the π-circuit shown, if $I_T = 5.7$ A, $I_1 = 2.8$ A, and $I_4 = 1.6$ A, find I_2 and I_3.

16. For the circuit in Exercise 15, if $I_T = 84$ mA, $I_3 = 14$ mA, and $I_4 = 18$ mA, find I_1 and I_2.

17. For the voltage-divider circuit shown, if $I_T = 65$ mA, $I_1 = 5$ mA, $I_2 = 15$ mA, and $I_3 = 35$ mA, find I_4, I_5, and I_6:

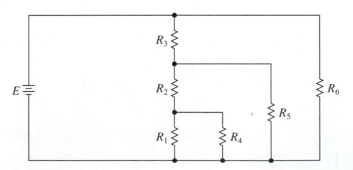

18. For the voltage-divider circuit in Exercise 17, if $I_1 = 5$ mA, $I_4 = 5$ mA, $I_5 = 5$ mA, and $I_6 = 5$ mA, find I_T, I_2, and I_3.

19. For the Wheatstone bridge in Example 7.14 (page 180), if $I_T = 15$ A, $I_1 = 6$ A, and $I_2 = 5$ A, find I_3, I_4, and I_5 including its direction.

20. For the Wheatstone bridge in Example 7.14, if $I_T = 6$ A, $I_2 = 5$ A, and $I_4 = 5$ A, find I_1, I_3, and I_5 including its direction.

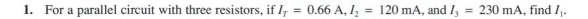

Self-Test

1. For a parallel circuit with three resistors, if $I_T = 0.66$ A, $I_2 = 120$ mA, and $I_3 = 230$ mA, find I_1.

    1. _____

2. For a parallel circuit with three resistors, if $E = 9.2$ V, find V_1, V_2, and V_3.

    2. _____

3. For a series circuit with three resistors, if $E = 1.75$ V, $V_1 = 370$ mV, and $V_2 = 0.69$ V, find V_3.

    3. _____

4. For this circuit, if $E = 10$ V and $V_2 = 4.3$ V, find V_1 and V_3:

    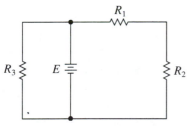

    4. _____

5. For this circuit, if $I_T = 39$ mA, $I_2 = 16$ mA, and $I_3 = 5.8$ mA, find I_1 and I_4:

    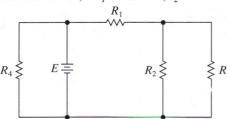

    5. _____

UNIT 8

DC Circuit Analysis

Introduction

In Unit 4, you learned about Ohm's law, which relates voltage, current, and resistance in electrical circuits. In Unit 7, you used Kirchhoff's voltage law to find voltages in series circuits, and Kirchhoff's current law to find currents in parallel circuits. In this unit, you will learn formulas to find resistances in series circuits and in parallel circuits. You will use all of these laws and formulas to find voltages, currents, and resistances in series circuits and in parallel circuits. Then you will use these laws and formulas to analyze series-parallel circuits, which contain resistors both in series and in parallel.

OBJECTIVES

When you have finished this unit you should be able to:

1. Use Ohm's law, Kirchhoff's voltage law, and the series resistance formula to find voltages, currents, and resistances in series circuits.

2. Use Ohm's law, Kirchhoff's current law, and the parallel resistance formula to find currents, voltages, and resistances in parallel circuits.

3. Find voltages, currents, and resistances in series-parallel circuits.

SECTION 8.1

Series Circuits

Suppose we have a circuit containing two resistors in series. Applying Ohm's law separately to each component of the circuit, we know that

$$V_1 = I_1 R_1, \ V_2 = I_2 R_2, \ \text{and} \ E = I_T R_T,$$

where R_T is the total resistance as seen by the source E. From Kirchhoff's voltage law we know that

$$E = V_1 + V_2,$$

and from Kirchhoff's current law we know that the current is the same throughout the circuit; that is,

$$I_T = I_1 = I_2.$$

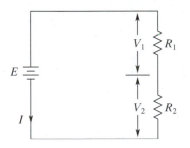

A series circuit with two resistors, with voltages and current.

We have called this constant current I in the circuit diagram.

From these laws we derive a formula for the resistance R_T in terms of the two resistances R_1 and R_2. We start with Kirchhoff's voltage law,

$$E = V_1 + V_2.$$

Replacing each term by its equivalent in Ohm's law, we have

$$I_T R_T = I_1 R_1 + I_2 R_2.$$

Then, from Kirchhoff's current law we can replace I_T, I_1, and I_2 by I to obtain

$$I R_T = I R_1 + I R_2.$$

We divide both sides by I to derive the **series resistance formula**,

$$R_T = R_1 + R_2.$$

Thus, for two resistors in series, the total resistance is the sum of the separate resistances.

The Series Resistance Formula:

$$R_T = R_1 + R_2$$

EXAMPLE 8.1 Find R_T for this series circuit with two resistors:

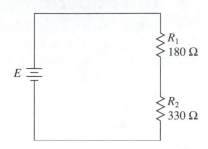

SOLUTION Using the series resistance formula, we have

$$R_T = R_1 + R_2$$

$$= 180\ \Omega + 330\ \Omega$$

$$= 510\ \Omega. \qquad \blacktriangle$$

EXAMPLE 8.2 For a series circuit with two resistors, find R_2 if $R_T = 6.6\ \text{k}\Omega$ and $R_1 = 3.9\ \text{k}\Omega$.

SOLUTION Using the series resistance formula,

$$R_T = R_1 + R_2,$$

and solving for R_2, we have

$$R_2 = R_T - R_1$$

$$= 6.6\ \text{k}\Omega - 3.9\ \text{k}\Omega$$

$$= 2.7\ \text{k}\Omega. \qquad \blacktriangle$$

The series resistance formula can be extended to three or more resistors. For any number of resistors in series, the total resistance is the sum of all the resistances.

EXAMPLE 8.3 Find R_T for this circuit:

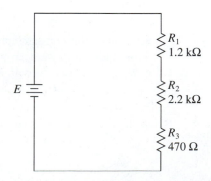

SOLUTION Writing R_3 in kilohms, we have

$$R_T = R_1 + R_2 + R_3$$

$$= 1.2\,k\Omega + 2.2\,k\Omega + 0.47\,k\Omega$$

$$= 3.87\,k\Omega.$$

▲

CURRENT AND VOLTAGES

If we are given sufficient information about a series circuit, we can find the current, all the voltages, and all the resistances for the circuit by using Ohm's law, Kirchhoff's voltage law, and the series resistance formula.

EXAMPLE 8.4   Find R_T, I, V_1, and V_2 for the circuit shown below.

SOLUTION We find R_T by using the series resistance formula:

$$R_T = R_1 + R_2$$

$$= 6.8\,k\Omega + 4.7\,k\Omega$$

$$= 11.5\,k\Omega.$$

Now, we can find I by applying Ohm's law to the source E:

$$I = \frac{E}{R_T}$$

$$= \frac{6\,V}{11.5\,k\Omega}$$

$$= 0.522\,mA,$$

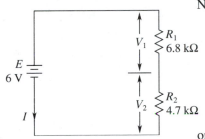

Circuit for Example 8.4.

or 522 μA. Then, applying Ohm's law to each resistor, we have

$$V_1 = IR_1$$

$$= (0.522\,mA)(6.8\,k\Omega)$$

$$= 3.55\,V,$$

and

$$V_2 = IR_2$$

$$= (0.522\,mA)(4.7\,k\Omega)$$

$$= 2.45\,V.$$

We can use Kirchhoff's voltage law to check:

$$E = V_1 + V_2$$

$$= 3.55\,V + 2.45\,V$$

$$= 6\,V.$$

▲

Often you will have several choices for ways to find the voltages, current, and resistances in a circuit, especially once you have found values for some of them. However, you might not always be able to find them in the order in which they are listed. With a little practice, you will see ways in which to proceed.

EXAMPLE 8.5   Find E, R_T, R_2, and V_1 for the circuit shown at the top of page 188.

SOLUTION We do not have sufficient information to find E or R_T. We can begin by applying Ohm's law to each resistor, to find R_2 and V_1:

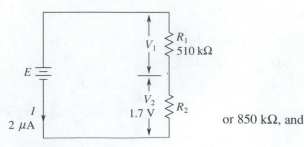

Circuit for Example 8.5.

$$R_2 = \frac{V_2}{I}$$

$$= \frac{1.7\text{ V}}{2\ \mu\text{A}}$$

$$= 0.85\text{ M}\Omega,$$

or 850 kΩ, and

$$V_1 = IR_1$$

$$= (2\ \mu\text{A})(510\text{ k}\Omega)$$

$$= 1020\text{ mV},$$

or 1.02 V. Now we can use Kirchhoff's voltage law to find E:

$$E = V_1 + V_2$$

$$= 1.02\text{ V} + 1.7\text{ V}$$

$$= 2.72\text{ V}.$$

Finally, we can use either Ohm's law or the series resistance formula to find R_T. Using Ohm's law, we have

$$R_T = \frac{E}{I}$$

$$= \frac{2.72\text{ V}}{2\ \mu\text{A}}$$

$$= 1.36\text{ M}\Omega.$$

Since we used Ohm's law to find R_T, we can use the series resistance formula to check:

$$R_T = R_1 + R_2$$

$$= 510\text{ k}\Omega + 850\text{ k}\Omega$$

$$= 1360\text{ k}\Omega$$

$$= 1.36\text{ M}\Omega.$$ ▲

EXAMPLE 8.6 ▶ Find I, R_T, R_3, V_1, and V_3 for the circuit shown.

SOLUTION   We start by applying Ohm's law to the second resistor to find I:

$$I = \frac{V_2}{R_2}$$

$$= \frac{3.7\text{ V}}{3.3\text{ k}\Omega}$$

$$= 1.12\text{ mA}.$$

We can now find R_T and V_1:

$$R_T = \frac{E}{I}$$

$$= \frac{12\text{ V}}{1.12\text{ mA}}$$

$$= 10.7\text{ k}\Omega,$$

Circuit for Example 8.6.

and

$$V_1 = IR_1$$

$$= (1.12 \text{ mA})(4.7 \text{ k}\Omega)$$

$$= 5.26 \text{ V}.$$

Finally, we can use the series resistance formula to find R_3, and Kirchhoff's voltage law to find V_3:

$$R_T = R_1 + R_2 + R_3,$$

and solving for R_3, we have

$$R_3 = R_T - R_1 - R_2$$

$$= 10.7 \text{ k}\Omega - 4.7 \text{ k}\Omega - 3.3 \text{ k}\Omega$$

$$= 2.7 \text{ k}\Omega.$$

Also,

$$E = V_1 + V_2 + V_3,$$

and solving for V_3, we have

$$V_3 = E - V_1 - V_2$$

$$= 12 \text{ V} - 5.26 \text{ V} - 3.7 \text{ V}$$

$$= 3.04 \text{ V}.$$

As we have not applied Ohm's law to the third resistor, we can use it to check:

$$I = \frac{V_3}{R_3}$$

$$= \frac{3.04 \text{ V}}{2.7 \text{ k}\Omega}$$

$$= 1.13 \text{ mA}.$$

Observe that our result is different by 0.01 mA from the value we found for I at the beginning of this example. Because we have rounded off many times between the first result and the last, this small difference is not surprising. We will not worry if results found by different methods differ by a small amount. ▲

EXERCISE

8.1

For Exercises 1– 4, refer to the circuit shown:

1. $R_1 = 120 \ \Omega$ and $R_2 = 390 \ \Omega$. Find R_T.
2. $R_1 = 68 \text{ k}\Omega$ and $R_2 = 22 \text{ k}\Omega$. Find R_T.
3. $R_T = 14.7 \text{ k}\Omega$ and $R_1 = 9.1 \text{ k}\Omega$. Find R_2.
4. $R_2 = 3.9 \text{ M}\Omega$ and $R_T = 9 \text{ M}\Omega$. Find R_1.

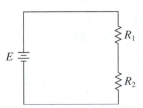

Circuit for Exercises 1–4.

For Exercises 5–8, refer to the circuit shown:

5. $R_1 = 1.2\ \text{k}\Omega$, $R_2 = 820\ \Omega$, and $R_3 = 750\ \Omega$. Find R_T.

6. $R_1 = 560\ \text{k}\Omega$, $R_2 = 1.2\ \text{M}\Omega$, and $R_3 = 1.5\ \text{M}\Omega$. Find R_T.

7. $R_T = 2.2\ \text{k}\Omega$, $R_1 = 910\ \Omega$, and $R_2 = 820\ \Omega$. Find R_3.

8. $R_2 = 510\ \text{k}\Omega$, $R_3 = 180\ \text{k}\Omega$, and $R_T = 1.2\ \text{M}\Omega$. Find R_1.

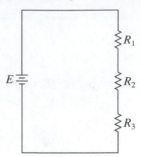

Circuit for Exercises 5–8.

For Exercises 9–14, refer to the circuit shown:

9. $E = 5\ \text{V}$, $R_1 = 18\ \text{k}\Omega$, and $R_2 = 22\ \text{k}\Omega$. Find I, R_T, V_1, and V_2.

10. $R_1 = 820\ \text{k}\Omega$, $R_2 = 470\ \text{k}\Omega$, and $E = 1.5\ \text{V}$. Find I, R_T, V_1, and V_2.

11. $I = 30\ \text{mA}$, $V_1 = 5.4\ \text{V}$, and $R_2 = 120\ \Omega$. Find E, R_T, R_1, and V_2.

12. $I = 1.9\ \text{mA}$, $E = 25\ \text{V}$, and $R_2 = 5.6\ \text{k}\Omega$. Find R_T, R_1, V_1, and V_2.

13. $E = 40\ \text{V}$, $R_2 = 47\ \text{k}\Omega$, and $V_1 = 16.5\ \text{V}$. Find I, R_T, R_1, and V_2.

14. $E = 12\ \text{V}$, $R_T = 1\ \text{M}\Omega$, and $V_2 = 4.8\ \text{V}$. Find I, R_1, R_2, and V_1.

For Exercises 15–20, refer to the circuit shown:

15. $E = 5\ \text{V}$, $R_1 = 120\ \Omega$, $R_2 = 150\ \Omega$, and $R_3 = 180\ \Omega$. Find I, R_T, V_1, V_2, and V_3.

16. $E = 14\ \text{V}$, $R_1 = 6.8\ \text{k}\Omega$, $R_2 = 5.1\ \text{k}\Omega$, and $R_3 = 9.1\ \text{k}\Omega$. Find I, R_T, V_1, V_2, and V_3.

17. $E = 40\ \text{V}$, $R_1 = 560\ \text{k}\Omega$, $R_2 = 470\ \text{k}\Omega$, and $V_2 = 13.3\ \text{V}$. Find I, R_T, R_3, V_1, and V_3.

18. $E = 25\ \text{V}$, $R_1 = 820\ \text{k}\Omega$, $R_3 = 750\ \text{k}\Omega$, and $V_1 = 8.3\ \text{V}$. Find I, R_T, R_2, V_2, and V_3.

19. $I = 20\ \text{mA}$, $R_T = 1.2\ \text{k}\Omega$, $R_2 = 330\ \Omega$, and $V_3 = 8\ \text{V}$. Find E, R_1, R_3, V_1, and V_2.

20. $I = 15\ \mu\text{A}$, $R_1 = 820\ \text{k}\Omega$, $R_2 = 2.2\ \text{M}\Omega$, and $V_3 = 14.7\ \text{V}$. Find E, R_T, R_3, V_1, and V_2.

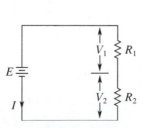

Circuit for Exercises 9–14.

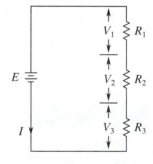

Circuit for Exercises 15–20.

SECTION 8.2

Parallel Circuits

Suppose we have a circuit containing two resistors in parallel. Applying Ohm's law separately to each component of the circuit, we know that

$$I_1 = \frac{V_1}{R_1}, \quad I_2 = \frac{V_2}{R_2}, \quad \text{and} \quad I_T = \frac{E}{R_T},$$

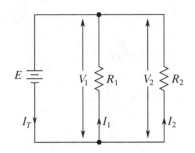

A parallel circuit with two resistors, with currents and voltage.

where R_T is the total resistance as seen by the source E. From Kirchhoff's current law we know that

$$I_T = I_1 + I_2,$$

and from Kirchhoff's voltage law we know that the voltage is the same in each branch; that is,

$$E = V_1 = V_2.$$

From these laws we derive a formula for the resistance R_T in terms of the two resistances R_1 and R_2. We start with Kirchhoff's current law,

$$I_T = I_1 + I_2.$$

Replacing each term by its equivalent in Ohm's law, we have

$$\frac{E}{R_T} = \frac{V_1}{R_1} + \frac{V_2}{R_2}.$$

Then, from Kirchhoff's voltage law we can replace V_1 and V_2 by E to obtain

$$\frac{E}{R_T} = \frac{E}{R_1} + \frac{E}{R_2}.$$

We divide both sides by E to derive the **parallel resistance formula**,

$$\frac{1}{R_T} = \frac{1}{R_1} + \frac{1}{R_2}.$$

In a parallel circuit, the *reciprocal* of the total resistance is equal to the sum of the *reciprocals* of the separate resistances.

The Parallel Resistance Formula:

$$\frac{1}{R_T} = \frac{1}{R_1} + \frac{1}{R_2}$$

EXAMPLE 8.7 ▶ Find R_T for this parallel circuit with two resistors:

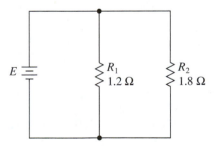

SOLUTION Using the parallel resistance formula, we have

$$\frac{1}{R_T} = \frac{1}{R_1} + \frac{1}{R_2}$$

$$= \frac{1}{1.2\ \Omega} + \frac{1}{1.8\ \Omega}.$$

The method for using a calculator to evaluate this formula was given in Section 5.1. We review the steps here:

		display:
Enter	1.2	1.2
Press	1/x	0.833333333
Press	+	0.833333333
Enter	1.8	1.8
Press	1/x	0.555555555
Press	=	1.388888889
Press	1/x	0.72

Therefore, $R_T = 0.72 \; \Omega$. You should recall that, when you press the equals key, you get $\frac{1}{R_T}$. You must press the reciprocal key one final time to get R_T.   ▲

We observe that R_T is less than either R_1 or R_2. Two resistors in *series* produce a resistance that is *larger* than either resistance alone. However, two resistors in *parallel* produce a resistance that is *smaller* than either resistance alone.

We can also use the parallel resistance formula to find R_1 or R_2, given R_T and one other resistance.

EXAMPLE 8.8 ▶ For a parallel circuit with two resistors, find R_2 if $R_T = 160 \; \Omega$ and $R_1 = 270 \; \Omega$.

SOLUTION Using the parallel resistance formula,

$$\frac{1}{R_T} = \frac{1}{R_1} + \frac{1}{R_2}$$

and solving for $\frac{1}{R_2}$, we have

$$\frac{1}{R_2} = \frac{1}{R_T} - \frac{1}{R_1}$$

$$= \frac{1}{160 \; \Omega} - \frac{1}{270 \; \Omega} \, .$$

To find R_2, you can use this variation of the calculator algorithm in Example 8.7:

		display:
Enter	160	160.
Press	1/x	0.00625
Press	–	0.00625
Enter	270	270.
Press	1/x	0.003703703
Press	=	0.002546296
Press	1/x	392.7272727

Therefore, $R_1 = 393 \; \Omega$ to three significant digits. We observe that R_T is less than either R_1 or R_2.   ▲

The parallel resistance formula can be extended to three or more resistors. For any number of resistors in parallel, the reciprocal of the total resistance is equal to the sum of the reciprocals of the separate resistances. We sometimes use the name **effective resistance** instead of total resistance to emphasize that the result is smaller than any of the separate resistances.

EXAMPLE 8.9 ▶ Find R_T for this circuit:

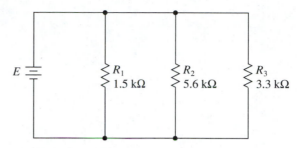

SOLUTION We extend the parallel resistance formula to three resistances:

$$\frac{1}{R_T} = \frac{1}{R_1} + \frac{1}{R_2} + \frac{1}{R_3}$$

$$= \frac{1}{1.5 \text{ k}\Omega} + \frac{1}{5.6 \text{ k}\Omega} + \frac{1}{3.3 \text{ k}\Omega}.$$

Now, you can add a few steps to the calculator algorithm given in Example 8.7 to find that

$$R_T = 0.871 \text{ k}\Omega$$

to three significant digits. Observe that this result is smaller than any of the separate resistances. ▲

VOLTAGE AND CURRENTS

If we are given sufficient information about a parallel circuit, we can find the voltage, all the currents, and all the resistances for the circuit by using Ohm's law, Kirchhoff's current law, and the parallel resistance formula.

EXAMPLE 8.10 ▶ Find R_T, I_T, I_1, and I_2 for the circuit shown below.

SOLUTION We find R_T by using the parallel resistance formula:

$$\frac{1}{R_T} = \frac{1}{R_1} + \frac{1}{R_2}$$

$$= \frac{1}{470 \ \Omega} + \frac{1}{680 \ \Omega}.$$

The calculation for R_T gives

$$R_T = 278 \ \Omega.$$

Now, we can find I_T by applying Ohm's law to the source E:

$$I_T = \frac{E}{R_T}$$

$$= \frac{10 \text{ V}}{278 \ \Omega}$$

$$= 0.036 \text{ A},$$

Circuit for Example 8.10.

or 36 mA. Then, we recall that $E = V_1 = V_2$. Therefore,

$$V_1 = V_2 = 10 \text{ V}.$$

We apply Ohm's law to each branch:

$$I_1 = \frac{V_1}{R_1}$$

$$= \frac{10\text{ V}}{470\ \Omega}$$

$$= 21.3\text{ mA,}$$

and

$$I_2 = \frac{V_2}{R_2}$$

$$= \frac{10\text{ V}}{680\ \Omega}$$

$$= 14.7\text{ mA.}$$

We can use Kirchhoff's current law to check:

$$I_T = I_1 + I_2$$

$$= 21.3\text{ mA} + 14.7\text{ mA}$$

$$= 36\text{ mA.} \qquad \blacktriangle$$

As with series circuits, you will often have several choices for ways to find the voltage, currents, and resistances in a parallel circuit, especially once you have found values for some of them.

EXAMPLE 8.11 ▶ Find E, R_T, R_2, and V_1 for the circuit shown.

SOLUTION We do not have sufficient information to find E or R_T. Since we know two currents, we can first use Kirchhoff's current law to find I_2:

$$I_2 = I_T - I_1$$

$$= 48\text{ mA} - 34.5\text{ mA}$$

$$= 13.5\text{ mA.}$$

Now, we can apply Ohm's law to R_2 to find V_2:

$$V_2 = I_2 R_2$$

$$= (13.5\text{ mA})(5.6\text{ k}\Omega)$$

$$= 75.6\text{ V.}$$

Since $E = V_1 = V_2$, we also have

$$V_1 = 75.6\text{ V}$$

and

$$E = 75.6\text{ V.}$$

Now, we can apply Ohm's law to R_1 and to E:

$$R_1 = \frac{V_1}{I_1}$$

$$= \frac{75.6\text{ V}}{34.5\text{ mA}}$$

$$= 2.19\text{ k}\Omega$$

Circuit for Example 8.11.

and

$$R_T = \frac{E}{I_T}$$

$$= \frac{75.6 \text{ V}}{48 \text{ mA}}$$

$$= 1.58 \text{ k}\Omega.$$

Since we did not use the parallel resistance formula to find R_T, we can use it to check:

$$\frac{1}{R_T} = \frac{1}{R_1} + \frac{1}{R_2}$$

$$\frac{1}{R_T} = \frac{1}{2.19 \text{ k}\Omega} + \frac{1}{5.6 \text{ k}\Omega}$$

$$R_T = 1.57 \text{ k}\Omega.$$

Recall that this result can differ slightly from our previous value for R_T because we have rounded off several times. ▲

EXAMPLE 8.12 ▶ Find R_T, R_1, R_3, I_2, and I_3 for this circuit:

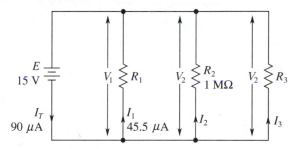

SOLUTION We start by applying Ohm's law to E:

$$R_T = \frac{E}{I_T}$$

$$= \frac{15 \text{ V}}{90 \text{ }\mu\text{A}}$$

$$= 0.167 \text{ M}\Omega,$$

or 167 kΩ. Since $E = V_1 = V_2$, we have

$$V_1 = V_2 = V_3 = 15 \text{ V}.$$

Therefore, we can also use Ohm's law to find R_1 and I_2:

$$R_1 = \frac{V_1}{I_1}$$

$$= \frac{15 \text{ V}}{45.5 \text{ }\mu\text{A}}$$

$$= 0.330 \text{ M}\Omega,$$

or 330 kΩ, and

$$I_2 = \frac{V_2}{R_2}$$

$$= \frac{15 \text{ V}}{1 \text{ M}\Omega}$$

$$= 15 \text{ }\mu\text{A}.$$

Now, we have several ways to find I_3 and R_3. We might use Kirchhoff's current law and then Ohm's law:

$$I_3 = I_T - I_1 - I_2$$

$$= 90\,\mu A - 45.5\,\mu A - 15\,\mu A$$

$$= 29.5\,\mu A$$

and

$$R_3 = \frac{V_3}{I_3}$$

$$= \frac{15\,V}{29.5\,\mu A}$$

$$= 508\,k\Omega.$$

Then, we can use the parallel resistance formula to check:

$$\frac{1}{R_T} = \frac{1}{R_1} + \frac{1}{R_2} + \frac{1}{R_3}$$

$$\frac{1}{R_T} = \frac{1}{330\,k\Omega} + \frac{1}{1\,M\Omega} + \frac{1}{508\,k\Omega}$$

$$R_T = 167\,k\Omega.$$

EXERCISE 8.2

For Exercises 1–4, refer to the circuit shown below:

1. $R_1 = 150\,\Omega$ and $R_2 = 470\,\Omega$. Find R_T.
2. $R_1 = 8.2\,k\Omega$ and $R_2 = 6.8\,k\Omega$. Find R_T.
3. $R_T = 170\,k\Omega$ and $R_1 = 750\,k\Omega$. Find R_2.
4. $R_2 = 33\,M\Omega$ and $R_T = 8.8\,M\Omega$. Find R_1.

For Exercises 5–8, refer to the circuit shown:

5. $R_1 = 390\,\Omega$, $R_2 = 150\,\Omega$, and $R_3 = 220\,\Omega$. Find R_T.
6. $R_1 = 820\,\Omega$, $R_2 = 1.5\,k\Omega$, and $R_3 = 1.8\,k\Omega$. Find R_T.
7. $R_T = 180\,\Omega$, $R_1 = 680\,\Omega$, and $R_2 = 470\,\Omega$. Find R_3.
8. $R_2 = 1\,M\Omega$, $R_3 = 910\,k\Omega$, and $R_T = 300\,k\Omega$. Find R_1.

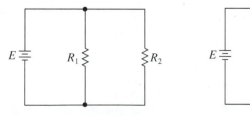 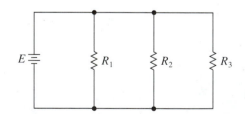

Circuit for Exercises 1–4.   Circuit for Exercises 5–8.

For Exercises 9–14, refer to the circuit shown on page 197:

9. $E = 5\,V$, $R_1 = 120\,\Omega$, and $R_2 = 220\,\Omega$. Find R_T, I_T, I_1, and I_2.
10. $R_1 = 5.6\,k\Omega$, $R_2 = 8.2\,k\Omega$, and $E = 6\,V$. Find R_T, I_T, I_1, and I_2.
11. $I_T = 75\,mA$, $R_1 = 2.7\,k\Omega$, and $R_2 = 6.8\,k\Omega$. Find E, R_T, I_1, and I_2.
12. $I_T = 2.42\,\mu A$, $I_1 = 1.91\,\mu A$, and $R_2 = 82\,M\Omega$. Find E, R_T, R_1, and I_2.
13. $E = 25\,V$, $I_T = 14\,\mu A$, and $R_1 = 3.9\,M\Omega$. Find R_T, R_2, I_1, and I_2.
14. $I_T = 26\,\mu A$, $R_1 = 470\,\Omega$, and $V_1 = 7.5\,mV$. Find R_T, R_2, I_1, and I_2.

For Exercises 15–20, refer to the circuit shown:

15. $E = 15$ V, $R_1 = 56$ kΩ, $R_2 = 39$ kΩ, and $R_3 = 33$ kΩ. Find R_T, I_T, I_1, I_2, and I_3.

16. $E = 30$ V, $R_1 = 2.2$ MΩ, $R_2 = 2.2$ MΩ, and $R_3 = 820$ kΩ. Find R_T. I_T, I_1, I_2, and I_3.

17. $E = 120$ V, $I_T = 1.8$ mA, $R_2 = 180$ kΩ, and $I_1 = 0.8$ mA. Find R_T, R_1, R_3, I_2, and I_3.

18. $E = 12$ V, $I_T = 28$ μA, $R_1 = 1.5$ MΩ, and $I_2 = 8$ μA. Find R_T, R_2, R_3, I_1, and I_3.

19. $I_T = 10.6$ mA, $R_1 = 47$ kΩ, $R_2 = 22$ kΩ, and $I_1 = 1.28$ mA. Find E, R_T, R_3, I_2, and I_3.

20. $I_T = 373$ μA, $R_T = 322$ Ω, $I_1 = 160$ μA, and $I_2 = 146$ μA. Find E, R_1, R_2, R_3, and I_3.

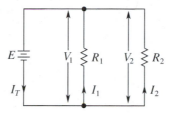

Circuit for Exercises 9–14.

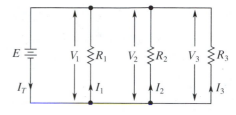

Circuit for Exercises 15–20.

Series-Parallel Circuits

A series-parallel circuit has some components in series, and some components in parallel branches. To find voltages, currents, and resistances in a series-parallel circuit, we may use any of the laws and formulas of the preceding sections: Ohm's law, Kirchhoff's laws, and the resistance formulas.

EXAMPLE 8.13 ▶

Find R_T, I_T, V_1, V_2, V_3, I_1, I_2, and I_3 for the circuit shown.

SOLUTION

Because we know all of the separate resistances, we can start by finding R_T. We observe that R_2 and R_3 are in series, but R_2 and R_3 are in a branch in parallel with R_1. Using the series resistance formula, we find a total resistance R_X for R_2 and R_3:

$$R_X = R_2 + R_3$$

$$= 4.7 \text{ k}\Omega + 3.3 \text{ k}\Omega$$

$$= 8 \text{ k}\Omega.$$

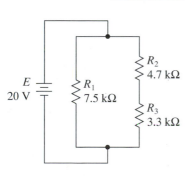

Circuit for Example 8.13.

We may replace R_2 and R_3 by R_X in the circuit diagram. This replacement reduces the circuit to a parallel circuit like those in Section 8.2. Then, using the parallel resistance formula, we find the effective resistance R_T for R_1 and R_X:

$$\frac{1}{R_T} = \frac{1}{R_1} + \frac{1}{R_X}$$

$$\frac{1}{R_T} = \frac{1}{7.5 \text{ k}\Omega} + \frac{1}{8 \text{ k}\Omega}$$

$$R_T = 3.87 \text{ k}\Omega.$$

Now, we use Ohm's law to find I_T:

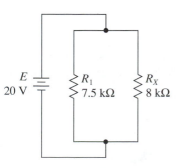

Reduced form of the circuit.

$$I_T = \frac{E}{R_T}$$

$$= \frac{20 \text{ V}}{3.87 \text{ k}\Omega}$$

$$= 5.17 \text{ mA}.$$

Because R_1 and R_X are in parallel with E,

$$V_1 = V_X = 20 \text{ V}.$$

We apply Ohm's law to each branch of the reduced circuit:

$$I_1 = \frac{V_1}{R_1}$$

$$= \frac{20 \text{ V}}{7.5 \text{ k}\Omega}$$

$$= 2.67 \text{ mA},$$

and

$$I_X = \frac{V_X}{R_X}$$

$$= \frac{20 \text{ V}}{8 \text{ k}\Omega}$$

$$= 2.5 \text{ mA}.$$

Observe that $I_T = I_1 + I_X$.

To complete the solution, we return to the original circuit. Since the current is constant within each branch,

$$I_2 = I_X$$

$$= 2.5 \text{ mA},$$

and

$$I_3 = I_X$$

$$= 2.5 \text{ mA}.$$

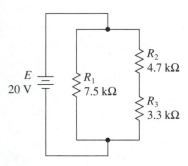

Circuit for Example 8.13.

Finally, we use Ohm's law to find the remaining voltages:

$$V_2 = I_2 R_2$$

$$= (2.5 \text{ mA})(4.7 \text{ k}\Omega)$$

$$= 11.75 \text{ V},$$

and

$$V_3 = I_3 R_3$$

$$= (2.5 \text{ mA})(3.3 \text{ k}\Omega)$$

$$= 8.25 \text{ V}.$$

As a final check, we observe that $V_X = V_2 + V_3 = 20 \text{ V}$. ▲

EXAMPLE 8.14 ▶ Find R_T, I_T, V_1, V_2, V_3, I_1, I_2, and I_3 for the circuit shown on page 199.

SOLUTION We start by finding R_T. We observe that R_2 and R_3 are in parallel branches, but the branches containing R_2 and R_3 are in series with R_1. Using the parallel resistance formula, we find the effective resistance R_X for R_2 and R_3:

$$\frac{1}{R_X} = \frac{1}{R_2} + \frac{1}{R_3}$$

$$\frac{1}{R_X} = \frac{1}{4.7 \text{ k}\Omega} + \frac{1}{3.3 \text{ k}\Omega}$$

$$R_X = 1.94 \text{ k}\Omega.$$

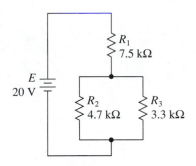

Circuit for Example 8.14.

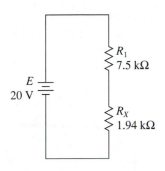

Reduced form of the circuit.

The part of the circuit containing R_2 and R_3 is called a **bank**. Since this bank is in series with R_1, when we replace it by R_X in the circuit diagram, the circuit is reduced to a series circuit like those in Section 8.1. Then, using the series resistance formula, we find the total resistance R_T for R_1 and R_X:

$$R_T = R_1 + R_X$$

$$= 7.5 \text{ k}\Omega + 1.94 \text{ k}\Omega$$

$$= 9.44 \text{ k}\Omega.$$

Now, we use Ohm's law to find I_T:

$$I_T = \frac{E}{R_T}$$

$$= \frac{20 \text{ V}}{9.44 \text{ k}\Omega}$$

$$= 2.12 \text{ mA}.$$

Because R_1 and R_X are in series with E,

$$I_1 = I_X = 2.12 \text{ mA}.$$

We apply Ohm's law to each resistor of the reduced circuit:

$$V_1 = I_1 R_1$$

$$= (2.12 \text{ mA})(7.5 \text{ k}\Omega)$$

$$= 15.9 \text{ V},$$

and

$$V_X = I_X R_X$$

$$= (2.12 \text{ mA})(1.94 \text{ k}\Omega)$$

$$= 4.11 \text{ V}.$$

Observe that $E = V_1 + V_X$, allowing for approximations.

To complete the solution, we return to the original circuit above. Since R_2 and R_3 are in parallel within the bank,

$$V_2 = V_X$$

$$= 4.11 \text{ V},$$

and

$$V_3 = V_X$$

$$= 4.11 \text{ V}.$$

Finally, we use Ohm's law to find the remaining currents:

$$I_2 = \frac{V_2}{R_2}$$

$$= \frac{4.11 \text{ V}}{4.7 \text{ k}\Omega}$$

$$= 0.874 \text{ mA},$$

and

$$I_3 = \frac{V_3}{R_3}$$

$$= \frac{4.11 \text{ V}}{3.3 \text{ k}\Omega}$$

$$= 1.25 \text{ mA}.$$

As a final check, we observe that $I_T = I_2 + I_3$, allowing for approximations. ▲

EXAMPLE 8.15 ▶ Find R_T, I_T, V_1, V_2, V_3, V_4, I_1, I_2, I_3, and I_4 for the circuit shown.

SOLUTION This circuit has two banks of resistors, where the resistors are in parallel within each bank. The banks themselves are in series. We can find the equivalent resistances R_A and R_B for each of the two banks. Thus,

Circuit for Example 8.15.

$$\frac{1}{R_A} = \frac{1}{R_1} + \frac{1}{R_2}$$

$$\frac{1}{R_A} = \frac{1}{1 \ \Omega} + \frac{1}{2 \ \Omega}$$

$$R_A = 0.667 \ \Omega,$$

and

$$\frac{1}{R_B} = \frac{1}{R_3} + \frac{1}{R_4}$$

$$\frac{1}{R_B} = \frac{1}{4 \ \Omega} + \frac{1}{5 \ \Omega}$$

$$R_B = 2.22 \ \Omega.$$

We may replace the two banks by R_A and R_B in the circuit diagram. This replacement reduces the circuit to a series circuit. Then, since banks A and B are in series,

$$R_T = R_A + R_B$$

$$= 0.667 \ \Omega + 2.22 \ \Omega$$

$$= 2.89 \ \Omega.$$

Now, we use Ohm's law to find I_T:

$$I_T = \frac{E}{R_T}$$

$$= \frac{10 \text{ V}}{2.89 \ \Omega}$$

$$= 3.46 \text{ A}.$$

Reduced form of the circuit.

Because R_A and R_B are in series with E,

$$I_A = I_B = 3.46 \text{ A}.$$

We apply Ohm's law to each resistor of the reduced circuit:

$$V_A = I_A R_A$$

$$= (3.46 \text{ A})(0.667 \ \Omega)$$

$$= 2.31 \text{ V},$$

and

$$V_B = I_B R_B$$

$$= (3.46 \text{ A})(2.22 \text{ Ω})$$

$$= 7.68 \text{ V}.$$

Observe that $E = V_A + V_B$, allowing for approximations.

To complete the solution, we return to the original circuit. Since R_1 and R_2 are in parallel within the bank represented by R_A,

$$V_1 = V_A = 2.31 \text{ V},$$

and

$$V_2 = V_A = 2.31 \text{ V}.$$

Also, since R_3 and R_4 are in parallel within the bank represented by R_B,

$$V_3 = V_B = 7.68 \text{ V},$$

and

$$V_4 = V_B = 7.68 \text{ V}.$$

Finally, we must find the individual currents. Using Ohm's law,

$$I_1 = \frac{V_1}{R_1}$$

$$= \frac{2.31 \text{ V}}{1 \text{ Ω}}$$

$$= 2.31 \text{ A},$$

and

$$I_2 = \frac{V_2}{R_2}$$

$$= \frac{2.31 \text{ V}}{2 \text{ Ω}}$$

$$= 1.16 \text{ A}.$$

Also,

$$I_3 = \frac{V_3}{R_3}$$

$$= \frac{7.68 \text{ V}}{4 \text{ Ω}}$$

$$= 1.92 \text{ A},$$

and

$$I_4 = \frac{V_4}{R_4}$$

$$= \frac{7.68 \text{ V}}{5 \text{ Ω}}$$

$$= 1.54 \text{ A}.$$

As a final check, we observe that $I_T = I_1 + I_2$, and also, $I_T = I_3 + I_4$, allowing for approximations. ▲

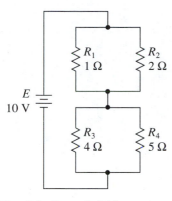

Circuit for Example 8.15.

EXAMPLE 8.16 ▶ This circuit is a π-circuit, where R_L is a "load resistance." Find R_T, I_T, V_1, V_2, V_3, V_L, I_1, I_2, I_3, and I_L:

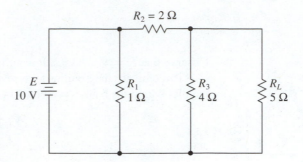

SOLUTION   We can derive a simpler circuit by combining R_3 and R_L to find a resistance we will call R_{3L}:

$$\frac{1}{R_{3L}} = \frac{1}{R_3} + \frac{1}{R_L}$$

$$\frac{1}{R_{3L}} = \frac{1}{4\ \Omega} + \frac{1}{5\ \Omega}$$

$$R_{3L} = 2.22\ \Omega.$$

Replacing R_3 and R_L by R_{3L}, we have this equivalent circuit:

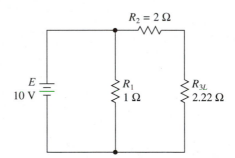

This circuit is the same as the circuit in Example 8.13: R_2 and R_{3L} are in series in one branch, and parallel to a branch containing R_1. Using the series resistance formula, we find that

$$R_X = R_2 + R_{3L}$$

$$= 2\ \Omega + 2.22\ \Omega$$

$$= 4.22\ \Omega.$$

Then, using the parallel resistance formula, we find the effective resistance R_T for R_1 and R_X:

$$\frac{1}{R_T} = \frac{1}{R_1} + \frac{1}{R_X}$$

$$\frac{1}{R_T} = \frac{1}{1\ \Omega} + \frac{1}{4.22\ \Omega}$$

$$R_T = 0.808\ \Omega.$$

Now, we use Ohm's law to find I_T:

$$I_T = \frac{E}{R_T}$$

$$= \frac{10 \text{ V}}{0.808 \text{ }\Omega}$$

$$= 12.4 \text{ A}.$$

Because R_1 and R_X are in parallel with E,

$$V_1 = V_X = 10 \text{ V}.$$

We apply Ohm's law to each branch of the reduced circuit:

$$I_1 = \frac{V_1}{R_1}$$

$$= \frac{10 \text{ V}}{1 \text{ }\Omega}$$

$$= 10 \text{ A},$$

and

$$I_X = \frac{V_X}{R_X}$$

$$= \frac{10 \text{ V}}{4.22 \text{ }\Omega}$$

$$= 2.37 \text{ A}.$$

Observe that $I_T = I_1 + I_X$. Also, in the branch represented by I_X,

$$I_2 = I_{3L} = 2.37 \text{ A}.$$

Then, using Ohm's law to find the corresponding voltages:

$$V_2 = I_2 R_2$$

$$= (2.37 \text{ A})(2 \text{ }\Omega)$$

$$= 4.74 \text{ V},$$

and

$$V_{3L} = I_{3L} R_{3L}$$

$$= (2.37 \text{ A})(2.22 \text{ }\Omega)$$

$$= 5.26 \text{ V}.$$

Observe that $V_X = V_2 + V_{3L}$.

Returning to the original circuit,

$$V_3 = V_L = 5.26 \text{ V}.$$

Finally, we find the remaining currents:

$$I_3 = \frac{V_3}{R_3}$$

$$= \frac{5.26 \text{ V}}{4 \text{ }\Omega}$$

$$= 1.32 \text{ A},$$

and

$$I_L = \frac{V_L}{R_L}$$

$$= \frac{5.26 \text{ V}}{5 \text{ }\Omega}$$

$$= 1.05 \text{ A},$$

and we observe that $I_2 = I_3 + I_L$.   ▲

EXERCISE 8.3

1. Find R_T, I_T, V_1, V_2, V_3, I_1, I_2, and I_3 for the circuit shown.
2. Repeat Exercise 1 for $E = 10$ V, $R_1 = 1$ MΩ, $R_2 = 1.5$ MΩ, and $R_3 = 7.5$ MΩ.
3. Find R_T, I_T, V_1, V_2, V_3, I_1, I_2, and I_3 for the circuit shown.
4. Repeat Exercise 3 for $E = 10$ V, $R_1 = 1$ MΩ, $R_2 = 1.5$ MΩ, and $R_3 = 7.5$ MΩ.
5. Find R_T, I_T, V_1, V_2, V_3, V_4, I_1, I_2, I_3, and I_4 for the circuit shown.
6. Repeat Exercise 5 for $E = 5$ V, $R_1 = 15$ kΩ, $R_2 = 22$ kΩ, $R_3 = 22$ kΩ, and $R_4 = 12$ kΩ.

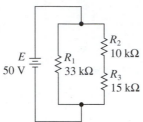
Circuit for Exercises 1 and 2.

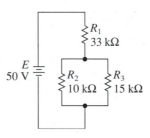
Circuit for Exercises 3 and 4.

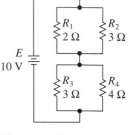
Circuit for Exercises 5 and 6.

7. Find R_T, I_T, V_1, V_2, V_3, V_4, I_1, I_2, I_3, and I_4 for the circuit shown.
8. Repeat Exercise 7 for $E = 60$ mV, $R_1 = 3.3$ kΩ, $R_2 = 1.2$ kΩ, $R_3 = 1.2$ kΩ, and $R_4 = 1.2$ kΩ.
9. Find R_T, I_T, V_1, V_2, V_3, V_4, I_1, I_2, I_3, and I_4 for the circuit shown.
10. Repeat Exercise 9 for $E = 60$ mV, $R_1 = 2.2$ kΩ, $R_2 = 1.5$ kΩ, $R_3 = 1.5$ kΩ, and $R_4 = 1.5$ kΩ.

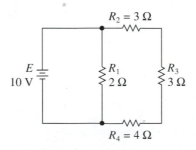
Circuit for Exercises 7 and 8.

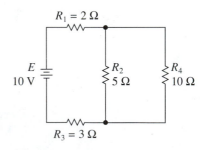
Circuit for Exercises 9 and 10.

11. Find R_T, I_T, V_1, V_2, V_3, V_L, I_1, I_2, I_3, and I_L for the π-circuit shown on page 205.
12. Repeat Exercise 11 for $E = 120$ V, $R_1 = 10$ kΩ, $R_2 = 10$ kΩ, $R_3 = 15$ kΩ, and $R_L = 20$ kΩ.
13. Find R_T, I_T, V_1, V_2, V_3, V_4, I_1, I_2, I_3, and I_4 for the circuit shown on page 205.
14. Repeat Exercise 13 for $E = 60$ V, $R_1 = 10$ kΩ, $R_2 = 10$ kΩ, $R_3 = 12$ kΩ, and $R_4 = 7.5$ kΩ.

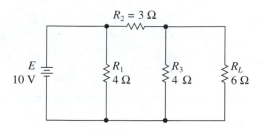

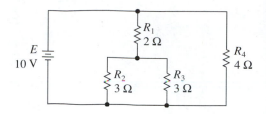

Circuit for Exercises 11 and 12.   Circuit for Exercises 13 and 14.

15. For the π-circuit, find the values needed for the resistors R_1, R_2, R_3, and R_L to produce the voltages and currents shown.

16. The circuit is a loaded voltage-divider circuit. Find the values needed for the resistors R_1, R_2, R_3, R_{L1}, R_{L2}, and R_{L3} to produce the voltages and currents shown.

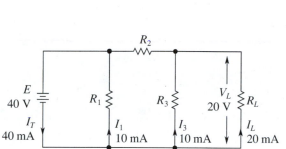

Circuit for Exercise 15.   Circuit for Exercise 16.

1. Find R_T for this circuit:

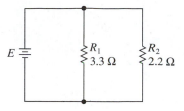

1. _____

2. Find R_T for this circuit:

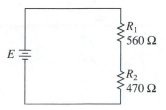

2. _____

3. Find R_T and I_T for this circuit:

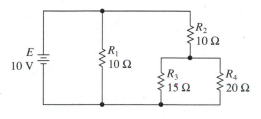

3. _____

4. Find V_2 for the circuit in Problem 3.

4. _____

5. Find I_3 for the circuit in Problem 3.

5. _____

UNIT 9

Rational Equations and Formulas

Introduction

In the preceding units, you have learned about several types of equations and formulas. You have solved some types of equations and formulas, and you have used certain formulas to solve applied problems. In this unit, you will learn how to solve other equations and formulas, those which involve rational expressions. To solve such equations and formulas, you first need to review the technique of factoring out a common factor. Then, you can use this technique to solve rational equations. You will learn how to derive and use some formulas that involve rational expressions, including an extension of the parallel resistance formula. Finally, you will learn how to derive and use voltage- and current-divider formulas.

OBJECTIVES

When you have finished this unit you should be able to:

1. Factor out common factors from expressions.

2. Solve equations and formulas involving rational expressions.

3. Use the alternate form of the parallel resistance formula to find equivalent resistance in parallel circuits.

4. Use the voltage-divider formulas to find voltages in series circuits, and use the current-divider formulas to find currents in parallel circuits.

SECTION 9.1

Common Factors

When two numbers are multiplied, the result is called the **product** of the numbers. The two numbers are called **factors** of the product. For example, since

$$12 = 3 \times 4,$$

we may say that the product of 3 and 4 is 12, and also that 3 and 4 are factors of 12. Since it is also true that

$$12 = 2 \times 6,$$

we may also say that the product of 2 and 6 is 12, and also that 2 and 6 are factors of 12.

The preceding factorizations of 12 are not complete. If we complete the factorizations we have

$$12 = 3 \times 4$$
$$= 3 \times 2 \times 2,$$

and

$$12 = 2 \times 6$$
$$= 2 \times 2 \times 3.$$

209

We see that, except for order, the factors are the same in each case. In general, when a factorization is complete, the factors will be the same regardless of the order in which they are derived.

Algebraic expressions can be factored in a similar way. We recall the **distributive property** given in Unit 3.

Distributive Property for Multiplication over Addition: For any three algebraic terms A, B, and C,

$$A(B + C) = AB + AC$$

We recall that we can use the distributive property from right to left:

$$AB + AC = A(B + C).$$

We say that A is a **common factor** of the expression $AB + AC$. We **factor out** the common factor A when we write $A(B + C)$.

EXAMPLE 9.1   Factor out the common factor from $2x + 2y$.

SOLUTION There is a common factor 2. Therefore, we write

$$2x + 2y = 2(x + y).$$ ▲

EXAMPLE 9.2   Factor out the common factor from $4x + 8y$.

SOLUTION There is a common factor 4. Therefore, we write

$$4x + 8y = 4(x + 2y).$$

Observe that factoring out a common factor 2 is not sufficient. For the factorization to be complete, we must factor out the largest possible common factor. ▲

EXAMPLE 9.3   Factor out the common factor from $4x^2 + 8x$.

SOLUTION There is a common factor 4. Therefore, we may write

$$4x^2 + 8x = 4(x^2 + 2x).$$

However, we observe that this factorization is not complete. We must also factor out any variables that are common factors. The variable x is a common factor, so we write

$$4(x^2 + 2x) = 4x(x + 2).$$ ▲

EXAMPLE 9.4   Factor out the common factor from $4x^2y + 8xy^2$.

SOLUTION There are common factors 4, x, and y. Therefore, the largest common factor is $4xy$. We can write each term as the product of $4xy$ and another factor:

$$4x^2y + 8xy^2 = 4xy(x) + 4xy(2y).$$

Then, factoring out the common factor $4xy$, we have

$$4xy(x) + 4xy(2y) = 4xy(x + 2y).$$ ▲

EXAMPLE 9.5   Factor out the common factor from $4x^3 - 4x^2 + 6x$.

SOLUTION When there are three terms, we look for factors common to all. There are factors 2 and x that are common to all of the terms. Therefore, the largest common factor is $2x$:

$$4x^3 - 4x^2 + 6x = 2x(2x^2) - 2x(2x) + 2x(3)$$
$$= 2x(2x^2 - 2x + 3).$$ ▲

EXAMPLE 9.6 ▶ Factor out the common factor from $4x^3 - 4x^2$.

SOLUTION There are common factors 4 and x^2. Therefore, the largest common factor is $4x^2$. Since $4x^2 = 4x^2(1)$, we write

$$4x^3 - 4x^2 = 4x^2(x) - 4x^2(1)$$
$$= 4x^2(x - 1).$$ ▲

EXAMPLE 9.7 ▶ Factor out the common factor from $2x + 2xy + 2xy^2$.

SOLUTION The largest common factor is $2x$. Since $2x = 2x(1)$, we write

$$2x + 2xy + 2xy^2 = 2x(1) + 2x(y) + 2x(y^2)$$
$$= 2x(1 + y + y^2).$$ ▲

EXERCISE 9.1

Factor out the common factor:

1. $6x + 6y$ 2. $12x - 12y$

3. $6x - 12y$ 4. $6x + 8y$

5. $3x^2 + 9x$ 6. $10x^2 - 15x$

7. $3x^2y - 12xy^2$ 8. $15x^2y - 5xy^2$

9. $2x^3 - 4x^2 + 4x$ 10. $12x^3 + 15x^2 - 9x$

11. $6x^3 + 6x^2$ 12. $8x^2 + 4x$

13. $8x - 16xy$ 14. $3y + 9xy$

15. $10x^3 + 10x^2 + 5x$ 16. $4xy - 8x^2y - 8xy^2$

SECTION 9.2

Solving Rational Equations

In Section 4.3, we solved some basic types of rational equations, which contain rational expressions. In these equations, the variable appears in one or more denominators. To solve such equations, we multiply both sides of the equation by a common denominator that will eliminate all of the denominators in the rational terms. Each denominator in a rational term must be a factor of the common denominator. For the examples in this book, such denominators can be found by inspection.

EXAMPLE 9.8 ▶ Solve the equation $\dfrac{2}{x} + \dfrac{1}{2} = \dfrac{3}{x}$ and check the solution.

SOLUTION The common denominator $2x$ will eliminate all of the denominators in the rational terms. Assuming x is not zero, we can multiply both sides of the equation by the common denominator $2x$:

$$\frac{2}{x} + \frac{1}{2} = \frac{3}{x}$$

$$2x\left(\frac{2}{x} + \frac{1}{2}\right) = 2x\left(\frac{3}{x}\right).$$

Then, we use the distributive property to multiply each term:

$$2x\left(\frac{2}{x}\right) + 2x\left(\frac{1}{2}\right) = 2x\left(\frac{3}{x}\right).$$

We can divide out the common factor from the numerator and the denominator in each term to obtain

$$2(2) + x(1) = 2(3)$$

$$4 + x = 6$$

$$x = 2.$$

To check this solution we substitute $x = 2$ in the original equation:

$$\frac{2}{x} + \frac{1}{2} = \frac{3}{x}$$

$$\frac{2}{2} + \frac{1}{2} \stackrel{?}{=} \frac{3}{2}$$

$$\frac{3}{2} = \frac{3}{2}.$$

▲

When the equation is a formula containing more than one variable, we can solve the formula for any of the variables. If the formula involves rational expressions, we multiply both sides by a common denominator.

EXAMPLE 9.9 Solve the formula $\dfrac{y}{x} = \dfrac{y + 3}{2}$ for x.

SOLUTION We multiply both sides by the common denominator $2x$:

$$\frac{y}{x} = \frac{y + 3}{2}$$

$$2x\left(\frac{y}{x}\right) = 2x\left(\frac{y + 3}{2}\right).$$

Observing that each side has just one term, we divide out the common factor from each side to obtain

$$2y = x(y + 3).$$

Then, to solve for x, we divide both sides by the factor $y + 3$:

$$\frac{2y}{y + 3} = \frac{x(y + 3)}{y + 3}$$

$$\frac{2y}{y + 3} = x$$

or

$$x = \frac{2y}{y + 3}.$$

▲

EXAMPLE 9.10 Solve the formula in Example 9.9 for y.

SOLUTION We begin as in Example 9.9, by multiplying both sides by $2x$:

$$\frac{y}{x} = \frac{y + 3}{2}$$

$$2x\left(\frac{y}{x}\right) = 2x\left(\frac{y + 3}{2}\right)$$

$$2y = x(y + 3).$$

In order to collect the terms involving y, we must remove the parentheses by using the distributive property:

$$2y = xy + 3x.$$

We collect the terms involving y by subtracting xy from both sides:

$$2y - xy = 3x.$$

Now, we can factor out the common factor y:

$$y(2 - x) = 3x.$$

Then, to solve for y, we divide both sides by the factor $2 - x$:

$$\frac{y(2 - x)}{2 - x} = \frac{3x}{2 - x}$$

$$y = \frac{3x}{2 - x}.$$

Observe that we must collect all terms involving the variable we are solving for, on one side of the equation. ▲

EXAMPLE 9.11 ▶ Solve the formula $\dfrac{1}{x} + \dfrac{1}{y} = 2$ for x.

SOLUTION We multiply both sides by the common denominator xy:

$$\frac{1}{x} + \frac{1}{y} = 2$$

$$xy\left(\frac{1}{x} + \frac{1}{y}\right) = xy(2).$$

Then, we use the distributive property to multiply each term by the common denominator:

$$xy\left(\frac{1}{x}\right) + xy\left(\frac{1}{y}\right) = xy(2).$$

Dividing out the common factor from the numerator and the denominator in each term, we have

$$y + x = 2xy.$$

We collect the terms involving x by subtracting x from both sides:

$$y = 2xy - x.$$

Now, we can factor out the common factor x:

$$y = x(2y - 1).$$

Then, we divide both sides by the factor $2y - 1$:

$$\frac{y}{2y - 1} = \frac{x(2y - 1)}{2y - 1}$$

$$\frac{y}{2y - 1} = x$$

or

$$x = \frac{y}{2y - 1}.$$

The solution of the original formula for y is essentially the same. You should solve the formula for y. ▲

FORMULAS WITH SEVERAL VARIABLES

EXAMPLE 9.12   In Section 9.4, we will derive the voltage-divider formula

$$V_1 = \frac{ER_1}{R_1 + R_2}.$$

Solve the formula for R_2.

SOLUTION We multiply both sides by the denominator $R_1 + R_2$:

$$V_1 = \frac{ER_1}{R_1 + R_2}$$

$$V_1(R_1 + R_2) = \left(\frac{ER_1}{R_1 + R_2}\right)(R_1 + R_2).$$

We divide out the common factor $R_1 + R_2$ from the numerator and the denominator on the right-hand side to obtain

$$V_1(R_1 + R_2) = ER_1.$$

Then, we use the distributive property to remove the parentheses:

$$V_1R_1 + V_1R_2 = ER_1.$$

We isolate the term involving R_2 by subtracting the term V_1R_1 from both sides:

$$V_1R_2 = ER_1 - V_1R_1.$$

Then, we divide both sides by V_1:

$$R_2 = \frac{ER_1 - V_1R_1}{V_1}.$$

▲

EXAMPLE 9.13   Solve the voltage-divider formula in Example 9.12 for R_1.

SOLUTION We can follow Example 9.12 for several steps:

$$V_1 = \frac{ER_1}{R_1 + R_2}$$

$$V_1(R_1 + R_2) = \left(\frac{ER_1}{R_1 + R_2}\right)(R_1 + R_2)$$

$$V_1(R_1 + R_2) = ER_1$$

$$V_1R_1 + V_1R_2 = ER_1$$

$$V_1R_2 = ER_1 - V_1R_1.$$

Now, we factor out the common factor R_1:

$$V_1R_2 = R_1(E - V_1).$$

Then, we divide both sides by the factor $E - V_1$:

$$\frac{V_1R_2}{E - V_1} = \frac{R_1(E - V_1)}{E - V_1}$$

$$\frac{V_1R_2}{E - V_1} = R_1$$

or

$$R_1 = \frac{V_1R_2}{E - V_1}.$$

▲

EXAMPLE 9.14   A formula for a current in a transistor circuit is

$$I_B = \frac{V_{BB} - V_{BE}}{R_B + (\beta + 1)R_E}.$$

Solve the formula for R_E.

SOLUTION We multiply both sides by the denominator $R_B + (\beta + 1)R_E$:

$$I_B = \frac{V_{BB} - V_{BE}}{R_B + (\beta + 1)R_E}$$

$$I_B[R_B + (\beta + 1)R_E] = \left[\frac{V_{BB} - V_{BE}}{R_B + (\beta + 1)R_E}\right][R_B + (\beta + 1)R_E].$$

We divide out the common factor $R_B + (\beta + 1)R_E$ from the numerator and the denominator on the right-hand side to obtain

$$I_B[R_B + (\beta + 1)R_E] = V_{BB} - V_{BE}.$$

Then, we use the distributive property to remove the brackets:

$$I_B R_B + (\beta + 1)I_B R_E = V_{BB} - V_{BE}.$$

We isolate the term involving $(\beta + 1)I_B R_E$ by subtracting the term $I_B R_B$ from both sides:

$$(\beta + 1)I_B R_E = V_{BB} - V_{BE} - I_B R_B.$$

Then, we divide both sides by $(\beta + 1)I_B$:

$$\frac{(\beta + 1)I_B R_E}{(\beta + 1)I_B} = \frac{V_{BB} - V_{BE} - I_B R_B}{(\beta + 1)I_B}$$

$$R_E = \frac{V_{BB} - V_{BE} - I_B R_B}{(\beta + 1)I_B}.$$

▲

**EXERCISE
9.2**

Solve the equation and check the solution:

1. $\dfrac{2}{x} + \dfrac{1}{3} = \dfrac{3}{x}$ **2.** $\dfrac{4}{x} - \dfrac{1}{2} = \dfrac{3}{x}$

3. $\dfrac{1}{2x} + \dfrac{2}{3} = \dfrac{1}{x}$ **4.** $\dfrac{5}{4} - \dfrac{4}{3x} = \dfrac{3}{4x}$

Solve the formula for x:

5. $\dfrac{y}{x} = \dfrac{y-4}{3}$ **6.** $\dfrac{y}{2x} = \dfrac{3-y}{6}$

Solve the formula for y:

7. $\dfrac{y}{x} = \dfrac{y-4}{3}$ **8.** $\dfrac{y}{2x} = \dfrac{3-y}{6}$

Solve the formula for x and for y:

9. $\dfrac{1}{x} + \dfrac{2}{y} = 3$ **10.** $\dfrac{2}{x} - \dfrac{4}{y} = 1$

11. $\dfrac{y}{x} = \dfrac{1}{x} + \dfrac{y}{2}$ **12.** $\dfrac{y}{2x} = \dfrac{y}{3} - \dfrac{1}{x}$

Solve the formula for the variable indicated:

13. $I_1 = \dfrac{I_T R_2}{R_1 + R_2}$, a current divider formula that will be derived in Section 9.4, for R_1

14. $A_i = \dfrac{\beta R_{BB}}{R_{BB} + \beta r_e}$, a current gain in a transistor circuit, for r_e

15. $I_1 = \dfrac{I_T R_2}{R_1 + R_2}$ for R_2

16. $A_i = \dfrac{\beta R_{BB}}{R_{BB} + \beta r_e}$ for R_{BB}

17. $R_T = \dfrac{R_1 R_2}{R_1 + R_2}$, an alternate form of the parallel resistor formula that will be derived in Section 9.3, for R_1

18. $A_i = \dfrac{R_B h_{fe}}{R_B + Z_b}$, a current gain in a transistor circuit, for R_B

19. $I_B = \dfrac{V_{BB} - V_{BE}}{R_B + (\beta + 1)R_E}$ for R_B

20. $A_i = \dfrac{(1 + h_{fe})R_B}{R_B + Z_b}$ for Z_b

21. $A_i = \dfrac{\beta R_B}{R_B + \beta(r_e + R_E)}$ for R_E

22. $A_i = \dfrac{\beta R_B}{R_B + \beta(r_e + R_E)}$ for R_B

SECTION 9.3

Alternate Parallel Resistance Formula

In Section 8.2, we derived the parallel resistance formula,

$$\frac{1}{R_T} = \frac{1}{R_1} + \frac{1}{R_2}.$$

The parallel resistance formula is often useful in this alternate form, where it is solved for R_T:

$$R_T = \frac{R_1 R_2}{R_1 + R_2}.$$

To derive the alternate form, we start with the reciprocal form and multiply both sides by the common denominator $R_T R_1 R_2$:

$$\frac{1}{R_T} = \frac{1}{R_1} + \frac{1}{R_2}$$

$$R_T R_1 R_2 \left(\frac{1}{R_T}\right) = R_T R_1 R_2 \left(\frac{1}{R_1} + \frac{1}{R_2}\right).$$

Then, we use the distributive property to multiply each term:

$$R_T R_1 R_2 \left(\frac{1}{R_T}\right) = R_T R_1 R_2 \left(\frac{1}{R_1}\right) + R_T R_1 R_2 \left(\frac{1}{R_2}\right).$$

We divide out the common factor from the numerator and the denominator in each term to obtain

$$R_1 R_2 = R_T R_2 + R_T R_1.$$

Now, we can factor out the common factor R_T:

$$R_1 R_2 = R_T (R_2 + R_1).$$

Then, to solve for R_T, we divide both sides by the factor $R_2 + R_1$:

$$\frac{R_1 R_2}{R_2 + R_1} = \frac{R_T (R_2 + R_1)}{R_2 + R_1}$$

$$\frac{R_1 R_2}{R_2 + R_1} = R_T$$

or

$$R_T = \frac{R_1 R_2}{R_1 + R_2}.$$

Alternate Form of the Parallel Resistance Formula:

$$R_T = \frac{R_1 R_2}{R_1 + R_2}$$

EXAMPLE 9.15 ▶ Use the alternate form of the parallel resistance formula to find R_T for the parallel circuit with two resistors shown.

SOLUTION To use the alternate form of the parallel resistance formula, we write

$$R_T = \frac{R_1 R_2}{R_1 + R_2}$$

$$= \frac{(1.8\ \Omega)(8.2\ \Omega)}{1.8\ \Omega + 8.2\ \Omega}$$

$$= \frac{(1.8\ \Omega)(8.2\ \Omega)}{10\ \Omega}$$

$$= \frac{(1.8)(8.2)}{10}\ \Omega$$

$$= 1.48\ \Omega. \qquad \blacktriangle$$

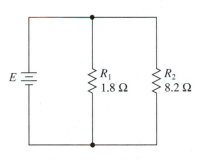

Circuit for Example 9.15.

EXAMPLE 9.16 ▶ For a parallel circuit with two resistors, find R_T if $R_1 = 39\ \text{k}\Omega$ and $R_2 = 68\ \text{k}\Omega$.

SOLUTION Using the alternate form of the parallel resistance formula, we write

$$R_T = \frac{R_1 R_2}{R_1 + R_2}$$

$$= \frac{(39\ \text{k}\Omega)(68\ \text{k}\Omega)}{39\ \text{k}\Omega + 68\ \text{k}\Omega}$$

$$= \frac{(39\ \text{k}\Omega)(68\ \text{k}\Omega)}{(39 + 68)\ \text{k}\Omega}$$

$$= \frac{(39)(68)}{39 + 68}\ \text{k}\Omega.$$

In Section 1.5, you used the parentheses keys on your calculator to do calculations of this type. Recall that the division line is a grouping symbol. The parentheses keys can be used to group the addition in the denominator by using these steps.

	display:
Enter 39	39.
Press $\times$	39.
Enter 68	68.
Press $\div$	2652.
Press $($	0.
Enter 39	39.
Press $+$	39.
Enter 68	68.
Press $)$	107.
Press $=$	24.78504673

The result is $R_T = 24.8$ kΩ. You can check this result by using the original reciprocal form of the parallel resistance formula. ▲

EXAMPLE 9.17 For a parallel circuit with two resistors, find R_2 if $R_1 = 47 \ \Omega$ and $R_T = 19.4 \ \Omega$.

SOLUTION We can solve the alternate form of the parallel resistance formula for R_2:

$$R_T = \frac{R_1 R_2}{R_1 + R_2}$$

$$R_T (R_1 + R_2) = \left(\frac{R_1 R_2}{R_1 + R_2} \right)(R_1 + R_2)$$

$$R_T (R_1 + R_2) = R_1 R_2$$

$$R_T R_1 + R_T R_2 = R_1 R_2$$

$$R_T R_1 = R_1 R_2 - R_T R_2$$

$$R_T R_1 = R_2 (R_1 - R_T)$$

$$\frac{R_T R_1}{R_1 - R_T} = \frac{R_2 (R_1 - R_T)}{R_1 - R_T}$$

$$\frac{R_T R_1}{R_1 - R_T} = R_2$$

or

$$R_2 = \frac{R_T R_1}{R_1 - R_T}.$$

Then, substituting the given values for R_T and R_1, we have

$$R_2 = \frac{(19.4 \ \Omega)(47 \ \Omega)}{47 \ \Omega - 19.4 \ \Omega}$$

$$= \frac{(19.4)(47)}{47 - 19.4} \ \Omega.$$

You can use a calculator algorithm similar to the one shown in Example 9.16 to find that $R_2 = 33 \ \Omega$. ▲

EXTENSION TO THREE RESISTORS

The alternate form of the parallel resistance formula can be extended to formulas for more than two resistors. For three resistors, we start with the reciprocal form

$$\frac{1}{R_T} = \frac{1}{R_1} + \frac{1}{R_2} + \frac{1}{R_3}.$$

We multiply both sides by the common denominator $R_T R_1 R_2 R_3$:

$$R_T R_1 R_2 R_3 \left(\frac{1}{R_T}\right) = R_T R_1 R_2 R_3 \left(\frac{1}{R_1} + \frac{1}{R_2} + \frac{1}{R_3}\right).$$

Then, we use the distributive property to multiply each term:

$$R_T R_1 R_2 R_3 \left(\frac{1}{R_T}\right) = R_T R_1 R_2 R_3 \left(\frac{1}{R_1}\right) + R_T R_1 R_2 R_3 \left(\frac{1}{R_2}\right) + R_T R_1 R_2 R_3 \left(\frac{1}{R_3}\right).$$

We divide out the common factor from the numerator and the denominator in each term:

$$R_1 R_2 R_3 = R_T R_2 R_3 + R_T R_1 R_3 + R_T R_1 R_2.$$

Now, we can factor out the common factor R_T:

$$R_1 R_2 R_3 = R_T (R_2 R_3 + R_1 R_3 + R_1 R_2).$$

Then, we divide both sides by the factor $R_2 R_3 + R_1 R_3 + R_1 R_2$:

$$\frac{R_1 R_2 R_3}{R_2 R_3 + R_1 R_3 + R_1 R_2} = \frac{R_T(R_2 R_3 + R_1 R_3 + R_1 R_2)}{R_2 R_3 + R_1 R_3 + R_1 R_2}$$

$$\frac{R_1 R_2 R_3}{R_2 R_3 + R_1 R_3 + R_1 R_2} = R_T$$

or

$$R_T = \frac{R_1 R_2 R_3}{R_2 R_3 + R_1 R_3 + R_1 R_2}.$$

EXAMPLE 9.18 Use the alternate form of the parallel resistance formula to find R_T if $R_1 = 4.7\ \Omega$, $R_2 = 3.3\ \Omega$, and $R_3 = 6.8\ \Omega$.

SOLUTION Using the alternate form of the parallel resistance formula for three resistors, we write

$$R_T = \frac{R_1 R_2 R_3}{R_2 R_3 + R_1 R_3 + R_1 R_2}$$

$$= \frac{(4.7\ \Omega)(3.3\ \Omega)(6.8\ \Omega)}{(3.3\ \Omega)(6.8\ \Omega) + (4.7\ \Omega)(6.8\ \Omega) + (4.7\ \Omega)(3.3\ \Omega)}$$

$$= \frac{(4.7)(3.3)(6.8)}{(3.3)(6.8) + (4.7)(6.8) + (4.7)(3.3)}\ \Omega.$$

You can use the parentheses keys on your calculator as in Example 9.16 to find R_T; however, it is easier just to calculate the numerator and the denominator separately:

$$R_T = \frac{105.468}{69.91}\ \Omega$$

$$= 1.51\ \Omega.$$

You can check this result by using the original reciprocal form of the parallel resistance formula for three resistors. We observe that, as the number of resistors increases, the alternate forms of the parallel resistance formulas become so long that they are not useful. ▲

Use the alternate form of the parallel resistance formula:

1. If $R_1 = 1.2\ \Omega$ and $R_2 = 1.8\ \Omega$, find R_T.
2. If $R_1 = 22\ k\Omega$ and $R_2 = 18\ k\Omega$, find R_T.
3. If $R_1 = 56\ k\Omega$ and $R_2 = 47\ k\Omega$, find R_T.
4. If $R_1 = 1.3\ M\Omega$ and $R_2 = 3.9\ M\Omega$, find R_T.
5. If $R_T = 23\ \Omega$ and $R_1 = 56\ \Omega$, find R_2.
6. If $R_T = 30.7\ \Omega$ and $R_2 = 68\ \Omega$, find R_1.
7. If $R_T = 1.6\ k\Omega$ and $R_2 = 2.7\ k\Omega$, find R_1.
8. If $R_T = 166\ \Omega$ and $R_1 = 220\ \Omega$, find R_2.
9. If $R_1 = 10\ \Omega$, $R_2 = 12\ \Omega$, and $R_3 = 15\ \Omega$, find R_T.
10. If $R_1 = 8.2\ \Omega$, $R_2 = 3.9\ \Omega$, and $R_3 = 5.6\ \Omega$, find R_T.

Voltage and Current Dividers

By using Ohm's law and the resistance formulas, we can derive the voltage- and current-divider formulas. The voltage-divider formulas provide an efficient way to find voltage drops across resistors in series circuits. We recall the series circuit containing two resistors R_1 and R_2, with their voltages V_1 and V_2, from Section 8.1. Applying Ohm's law to E and to R_1, we know that

$$I = \frac{E}{R_T}$$

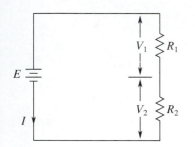

The series circuit with two resistors.

and

$$I = \frac{V_1}{R_1}.$$

Therefore,

$$\frac{E}{R_T} = \frac{V_1}{R_1}.$$

Multiplying both sides by R_1, we have

$$\frac{ER_1}{R_T} = V_1.$$

Now, we recall the series resistance formula derived in Section 8.1,

$$R_T = R_1 + R_2.$$

We substitute for R_T to derive the first voltage-divider formula:

$$\frac{ER_1}{R_1 + R_2} = V_1$$

or

$$V_1 = \frac{ER_1}{R_1 + R_2}.$$

We can apply Ohm's law to E and R_2, and follow a similar process, to derive the second voltage-divider formula. We know that

$$I = \frac{E}{R_T}$$

and

$$I = \frac{V_2}{R_2}.$$

Therefore,

$$\frac{E}{R_T} = \frac{V_2}{R_2}$$

$$\frac{ER_2}{R_T} = V_2$$

$$\frac{ER_2}{R_1 + R_2} = V_2$$

or

$$V_2 = \frac{ER_2}{R_1 + R_2}.$$

The Voltage-Divider Formulas:

$$V_1 = \frac{ER_1}{R_1 + R_2}$$

and

$$V_2 = \frac{ER_2}{R_1 + R_2}$$

EXAMPLE 9.19 Use the voltage-divider formulas to find V_1 and V_2 for the series circuit with two resistors shown.

SOLUTION   To use the voltage-divider formulas, we write

$$V_1 = \frac{ER_1}{R_1 + R_2}$$

$$= \frac{(9 \text{ V})(15 \text{ }\Omega)}{15 \text{ }\Omega + 75 \text{ }\Omega}$$

$$= \frac{(9 \text{ V})(15 \text{ }\Omega)}{90 \text{ }\Omega}$$

$$= 1.5 \text{ V},$$

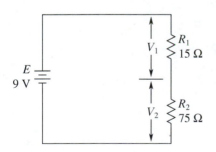

Circuit for Example 9.19.

and

$$V_2 = \frac{ER_2}{R_1 + R_2}$$

$$= \frac{(9 \text{ V})(75 \text{ }\Omega)}{15 \text{ }\Omega + 75 \text{ }\Omega}$$

$$= \frac{(9 \text{ V})(75 \text{ }\Omega)}{90 \text{ }\Omega}$$

$$= 7.5 \text{ V}.$$

We observe that

$$E = V_1 + V_2$$
$$= 1.5 \text{ V} + 7.5 \text{ V}$$
$$= 9 \text{ V}.$$ ▲

EXAMPLE 9.20 For a series circuit with two resistors, find V_1 and V_2 if $E = 9$ V, $R_1 = 15$ Ω, and $R_2 = 15$ Ω.

SOLUTION   Using the voltage-divider formulas, we write

$$V_1 = \frac{ER_1}{R_1 + R_2}$$
$$= \frac{(9 \text{ V})(15 \text{ Ω})}{15 \text{ Ω} + 15 \text{ Ω}}$$
$$= \frac{(9 \text{ V})(15 \text{ Ω})}{30 \text{ Ω}}$$
$$= 4.5 \text{ V},$$

and

$$V_2 = \frac{ER_2}{R_1 + R_2}$$
$$= \frac{(9 \text{ V})(15 \text{ Ω})}{15 \text{ Ω} + 15 \text{ Ω}}$$
$$= \frac{(9 \text{ V})(15 \text{ Ω})}{30 \text{ Ω}}$$
$$= 4.5 \text{ V}.$$

We observe that, if $R_1 = R_2$, then $V_1 = V_2 = \frac{1}{2} E$. ▲

EXTENSION TO THREE RESISTORS

We can extend the voltage-divider formulas to three or more resistors. For example, for a series circuit with three resistors, the series resistance formula is

$$R_T = R_1 + R_2 + R_3.$$

Therefore, the first voltage-divider formula becomes

$$V_1 = \frac{ER_1}{R_T}$$
$$= \frac{ER_1}{R_1 + R_2 + R_3}.$$

The second voltage-divider formula becomes

$$V_2 = \frac{ER_2}{R_1 + R_2 + R_3},$$

and we have a third voltage-divider formula,

$$V_3 = \frac{ER_3}{R_1 + R_2 + R_3}.$$

EXAMPLE 9.21 ▶ Use the voltage-divider formulas to find V_1, V_2, and V_3 for the circuit shown.

SOLUTION To use the voltage-divider formulas for three resistors, we write

$$V_1 = \frac{ER_1}{R_1 + R_2 + R_3}$$

$$= \frac{(20 \text{ V})(8.2 \text{ k}\Omega)}{8.2 \text{ k}\Omega + 3.9 \text{ k}\Omega + 6.8 \text{ }\Omega}.$$

You can use the parentheses keys on your calculator to find that

$$V_1 = 8.68 \text{ V}.$$

Similarly,

$$V_2 = \frac{ER_2}{R_1 + R_2 + R_3}$$

$$= \frac{(20 \text{ V})(3.9 \text{ k}\Omega)}{8.2 \text{ k}\Omega + 3.9 \text{ k}\Omega + 6.8 \text{ k}\Omega}$$

$$= 4.13 \text{ V},$$

and

$$V_3 = \frac{ER_3}{R_1 + R_2 + R_3}$$

$$= \frac{(20 \text{ V})(6.8 \text{ k}\Omega)}{8.2 \text{ k}\Omega + 3.9 \text{ k}\Omega + 6.8 \text{ k}\Omega}$$

$$= 7.2 \text{ V}.$$

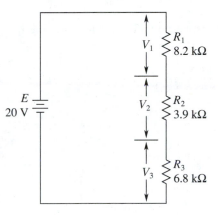

Circuit for Example 9.21.

We observe that

$$E = V_1 + V_2 + V_3$$

$$= 8.68 \text{ V} + 4.13 \text{ V} + 7.2 \text{ V}$$

$$= 20.01 \text{ V},$$

or approximately 20 V.   ▲

When you need to do several related calculations, it is often convenient to use the memory register provided by your calculator. First, find the keys marked

$$\boxed{\text{STO}} \quad \text{and} \quad \boxed{\text{RCL}}$$

or

$$\boxed{\text{MS}} \quad \text{and} \quad \boxed{\text{MR}}$$

or other codes meaning to store a number in a memory register and to recall it from memory. Now, follow the steps on page 224 to calculate

$$\frac{E}{R_1 + R_2 + R_3} = \frac{20}{8.2 + 3.9 + 6.8}$$

and store the result:

		display:
Enter 20		20.
Press	$\div$	20.
Press	(	0.
Enter 8.2		8.2
Press	+	8.2
Enter 3.9		3.9
Press	+	12.1
Enter 6.8		6.8
Press	)	18.9
Press	=	1.058201058
Press	STO	1.058201058

(On many calculators, a letter "M" appears on the display to indicate that something has been stored in the memory register.) To find

$$V_1 = \frac{(20 \text{ V})(8.2 \text{ k}\Omega)}{8.2 \text{ k}\Omega + 3.9 \text{ k}\Omega + 6.8 \text{ k}\Omega}$$

or

$$V_1 = 8.2 \left(\frac{20}{8.2 + 3.9 + 6.8} \right) \text{V},$$

continue with these steps:

		display:
Press	$\times$	1.058201058
Enter 8.2		8.2
Press	=	8.677248677

Therefore, $V_1 = 8.68$ V. To find

$$V_2 = \frac{(20 \text{ V})(3.9 \text{ k}\Omega)}{8.2 \text{ k}\Omega + 3.9 \text{ k}\Omega + 6.8 \text{ k}\Omega}$$

or

$$V_2 = 3.9 \left(\frac{20}{8.2 + 3.9 + 6.8} \right) \text{V},$$

follow these steps to recall the second factor from the memory register:

		display:
Press	RCL	1.058201058
Press	$\times$	1.058201058
Enter 3.9		3.9
Press	=	4.126984127

Therefore, $V_2 = 4.13$ V. To find

$$V_3 = \frac{(20 \text{ V})(6.8 \text{ k}\Omega)}{8.2 \text{ k}\Omega + 3.9 \text{ k}\Omega + 6.8 \text{ k}\Omega}$$

or

$$V_3 = 6.8 \left(\frac{20}{8.2 + 3.9 + 6.8} \right) \text{V},$$

follow these steps, again recalling the second factor from the memory register:

	display:
Press RCL	1.058201058
Press ×	1.058201058
Enter 6.8	6.8
Press =	7.195767196

Therefore, $V_3 = 7.2$ V.

CURRENT-DIVIDER FORMULAS

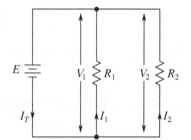

The parallel circuit with two resistors.

The current-divider formulas provide an efficient way to find branch currents in parallel circuits. We recall the parallel circuit containing two branches with resistors R_1 and R_2, with their currents I_1 and I_2, from Section 8.2. Applying Ohm's law to E and to R_1, we know that

$$E = I_T R_T$$

and

$$V_1 = I_1 R_1.$$

Therefore, since $E = V_1$,

$$I_T R_T = I_1 R_1.$$

Dividing both sides by R_1, we have

$$\frac{I_T R_T}{R_1} = I_1$$

or

$$\frac{I_T}{R_1} (R_T) = I_1.$$

Now, we use the alternate form of the parallel resistance formula,

$$R_T = \frac{R_1 R_2}{R_1 + R_2}.$$

We substitute for R_T to derive the first current-divider formula:

$$\frac{I_T}{R_1} \left(\frac{R_1 R_2}{R_1 + R_2} \right) = I_1$$

$$\frac{I_T R_2}{R_1 + R_2} = I_1$$

or

$$I_1 = \frac{I_T R_2}{R_1 + R_2}.$$

We can apply Ohm's law to E and R_2 and follow a similar process to derive the second current-divider formula. We know that

$$E = I_T R_T$$

and

$$V_2 = I_2 R_2.$$

Therefore, since $E = V_2$,

$$I_T R_T = I_2 R_2$$

$$\frac{I_T R_T}{R_2} = I_2$$

$$\frac{I_T}{R_2}\left(\frac{R_1 R_2}{R_1 + R_2}\right) = I_2$$

$$\frac{I_T R_1}{R_1 + R_2} = I_2$$

or

$$I_2 = \frac{I_T R_1}{R_1 + R_2}.$$

The Current-Divider Formulas:

$$I_1 = \frac{I_T R_2}{R_1 + R_2}$$

and

$$I_2 = \frac{I_T R_1}{R_1 + R_2}$$

Observe that to find I_1 we multiply by R_2, and to find I_2 we multiply by R_1.

EXAMPLE 9.22 ▶ Use the current-divider formulas to find I_1 and I_2 for the parallel circuit with two resistors shown.

SOLUTION   Using the current-divider formulas, and remembering that to find I_1 we multiply by R_2, we write

$$I_1 = \frac{I_T R_2}{R_1 + R_2}$$

$$= \frac{(2 \text{ mA})(5.6 \text{ k}\Omega)}{2.2 \text{ k}\Omega + 5.6 \text{ k}\Omega}$$

$$= 1.44 \text{ mA}.$$

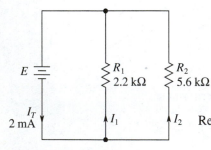

Circuit for Example 9.22.

Remembering that to find I_2 we multiply by R_1, we write

$$I_2 = \frac{I_T R_1}{R_1 + R_2}$$

$$= \frac{(2 \text{ mA})(2.2 \text{ k}\Omega)}{2.2 \text{ k}\Omega + 5.6 \text{ k}\Omega}$$

$$= 0.564 \text{ mA}.$$

We observe that

$$I_T = I_1 + I_2$$

$$= 1.44 \text{ mA} + 0.564 \text{ mA}$$

$$= 2.004 \text{ mA},$$

or approximately 2 mA.   ▲

EXAMPLE 9.23   For a parallel circuit with two resistors, find I_1 and I_2 if $I_T = 2$ mA, $R_1 = 2.2$ kΩ, and $R_2 = 2.2$ kΩ.

SOLUTION Using the current-divider formulas, we write

$$I_1 = \frac{I_T R_2}{R_1 + R_2}$$

$$= \frac{(2\text{ mA})(2.2\text{ k}\Omega)}{2.2\text{ k}\Omega + 2.2\text{ k}\Omega}$$

$$= 1\text{ mA},$$

and

$$I_2 = \frac{I_T R_1}{R_1 + R_2}$$

$$= \frac{(2\text{ mA})(2.2\text{ k}\Omega)}{2.2\text{ k}\Omega + 2.2\text{ k}\Omega}$$

$$= 1\text{ mA}.$$

We observe that, if $R_1 = R_2$, then $I_1 = I_2 = \frac{1}{2} I_T$. ▲

We recall that there is no simple extension of the alternate parallel resistance formula to three resistors. Therefore, we do not extend the current-divider formulas to three resistors.

EXERCISE 9.4

Use the voltage-divider formulas for a series with two resistors:

1. If $E = 40$ V, $R_1 = 33$ Ω, and $R_2 = 47$ Ω, find V_1 and V_2.
2. If $E = 5$ V, $R_1 = 2.2$ kΩ, and $R_2 = 5.1$ kΩ, find V_1 and V_2.
3. If $E = 40$ V, $R_1 = 33$ Ω, and $R_2 = 33$ Ω, find V_1 and V_2.
4. If $E = 5$ V, $R_1 = 5.1$ kΩ, and $R_2 = 5.1$ kΩ, find V_1 and V_2.

Use the voltage-divider formulas for a series circuit with three resistors:

5. If $E = 8$ V, $R_1 = 1.5$ MΩ, $R_2 = 2.2$ MΩ, and $R_3 = 1.8$ MΩ, find V_1, V_2, and V_3.
6. If $E = 65$ mV, $R_1 = 560$ Ω, $R_2 = 470$ Ω, and $R_3 = 270$ Ω, find V_1, V_2, and V_3.
7. If $E = 12$ V, $R_1 = 9.1$ kΩ, $R_2 = 9.1$ kΩ, and $R_3 = 9.1$ kΩ, find V_1, V_2, and V_3.
8. If $E = 12$ V, $R_1 = 1.5$ MΩ, $R_2 = 750$ kΩ, and $R_3 = 750$ kΩ, find V_1, V_2, and V_3.

Use the current-divider formulas for a parallel circuit with two resistors:

9. If $I_T = 5$ mA, $R_1 = 3.3$ kΩ, and $R_2 = 8.2$ kΩ, find I_1 and I_2.
10. If $I_T = 50$ μA, $R_1 = 560$ Ω, and $R_2 = 390$ Ω, find I_1 and I_2.
11. If $I_T = 5$ mA, $R_1 = 8.2$ kΩ, and $R_2 = 8.2$ kΩ, find I_1 and I_2.
12. If $I_T = 50$ μA, $R_1 = 560$ Ω, and $R_2 = 560$ Ω, find I_1 and I_2.

Self-Test

1. Factor out the common factor from $6x - 6x^2 + 9x^3$.

1. _____

2. Solve $\dfrac{3}{x} - \dfrac{y}{3} = \dfrac{y}{x}$ for y.

2. _____

3. Solve $A_v = \dfrac{g_m R_s}{1 + g_m R_s}$, a voltage gain in a transistor circuit, for R_s.

3. _____

4. For a parallel circuit with two resistors, use the alternate form of the parallel resistance formula to find R_T if $R_1 = 390 \ \Omega$ and $R_2 = 680 \ \Omega$.

4. _____

5. For a parallel circuit with two resistors, use the current-divider formulas to find I_1 and I_2 if $I_T = 25$ mA, $R_1 = 5.6$ kΩ, and $R_2 = 7.5$ kΩ.

5. _____

UNIT 10

Graphs

Introduction

In earlier units, you learned how to solve equations in one variable. Then, you learned how to use formulas, which are a type of equation in two or more variables. In this unit, you will learn about some specific types of equations in two variables. Such equations have many solutions, so many that it is not possible to write them all down. You can indicate the solutions of equations in two variables by means of a graph. In this unit, you will learn how to draw graphs of linear equations, and about two properties of these graphs, the intercepts and the slope. You will also draw the graphs of some nonlinear equations. Finally, you will learn a technique for estimating resistances by using the slope of a graph.

OBJECTIVES

When you have finished this unit you should be able to:

1. Find the x- and y-intercepts and draw graphs of linear equations in two variables.
2. Find slopes of lines joining two points.
3. Find the slope and y-intercept of lines given by linear equations in two variables.
4. Draw graphs of equations that are not linear by finding points on the graph.
5. Estimate resistances by approximating slopes of tangent lines.

SECTION 10.1

Graphs of Linear Equations

A method of describing equations by means of a graph was invented by a French mathematician and philosopher, Rene Descartes (1596–1650). This method is sometimes called the **Cartesian coordinate system** in honor of Descartes. It is more often called the **rectangular coordinate system** because lines to points in the system form a rectangle with the base lines of the system.

To construct the rectangular coordinate system, we recall the number line from Unit 1. We use two number lines as the base lines of the system. The two number lines are called the **axes**. To make a rectangular system, we place the axes **perpendicular** to one another, so they form right angles:

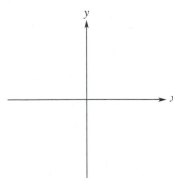

One line is horizontal, and the other line is vertical. We often call the horizontal number line the **x-axis** and the vertical number line the **y-axis**. Therefore, we have labelled the axes "*x*" and "*y*." In specific applications, we may use variables other than *x* and *y*.

We take the origin of each number line, or axis, to be the point where the axes cross. This point is called the **origin** of the rectangular coordinate system. We choose a unit, and rule off coordinates on each axis:

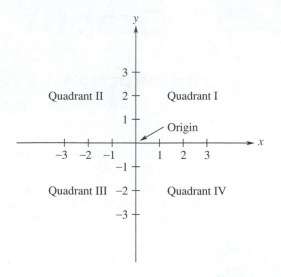

The axes divide the coordinate system into four sections called **quadrants**. The quadrants are numbered I, II, III, and IV, as shown. (The use of Roman numerals is not mandatory but is traditional.)

One number line forms a one-dimensional coordinate system. We can move only back and forth along the one line, and each point on the line is represented by one coordinate. Two number lines form a **two-dimensional** coordinate system. We can move along each of the two lines, and each point is represented by two coordinates.

We write the two coordinates as an **ordered pair** such as (1, 2). The first number is the **x-coordinate** and the second number is the **y-coordinate**. In the ordered pair (1, 2), the *x*-coordinate is 1 and the *y*-coordinate is 2:

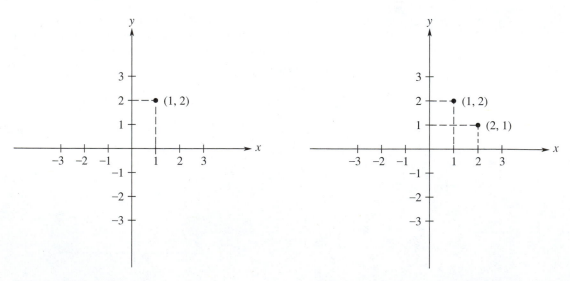

The ordered pair (1, 2).   Different ordered pairs (1, 2) and (2, 1).

The order of the numbers in an ordered pair is important. The ordered pair (1, 2) is not the same as the ordered pair (2, 1). In the ordered pair (2, 1), the *x*-coordinate is 2 and the *y*-coordinate is 1. Thus, (1, 2) and (2, 1) are different ordered pairs.

The **graph** of an ordered pair is a point. The graphs of two different ordered pairs such as (1, 2) and (2, 1) are different points. The dashed lines in the diagrams are only guidelines, and we will not generally show them. Observe, however, that the dashed lines, together with the axes, form the rectangle from which the rectangular coordinate system gets its name.

When we draw the graph of an ordered pair in the rectangular coordinate system, we **plot** the point.

EXAMPLE 10.1 ▶ Plot the points:

a. (3, 5)   b. (− 4, 2)   c. (5, − 4)

d. (−3, −2)   .   e. (4, 0)   f. (0, −3)

SOLUTIONS For each ordered pair, we locate the x-coordinate and the y-coordinate, and then plot the point.

a. For (3, 5), we locate 3 on the x-axis and 5 on the y-axis. Then, starting at 3, we go up until we are level with 5.

b. For (− 4, 2), we go left from the origin to − 4 on the x-axis, and then up until we are level with 2 on the y-axis.

c. For (5, − 4), we go right on the x-axis to 5, and then down until we are level with − 4 on the y-axis.

d. For (−3, −2), we go left on the x-axis to −3, and then down until we are level with −2 on the y-axis.

e. Any point with a zero coordinate is on an axis. For (4, 0), we go right on the x-axis to 4 and do not go up or down at all. The point is on the x-axis.

f. The point (0, −3) also has a zero coordinate. We do not go right or left at all, but we go down on the y-axis to −3. The point is on the y-axis.

The graphs of the points in parts **a–f** are

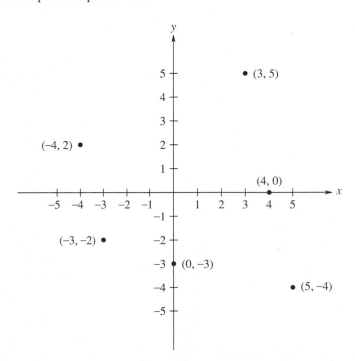

We see that, in quadrant I, the x-coordinate and y-coordinate both are positive. In quadrant II, the x-coordinate is negative, but the y-coordinate is positive. In quadrant III, the x-coordinate and y-coordinate both are negative. Finally, in quadrant IV, the x-coordinate is positive, but the y-coordinate is negative.   ▲

GRAPHS OF EQUATIONS

An ordered pair is a **solution** of an equation in two variables x and y if, when the x- and y-coordinates of the ordered pair are substituted for x and y in the equation, the two sides of the equation are equal. For example, consider the equation

$$y = \frac{1}{2}x + 2.$$

The ordered pair $(2, 3)$ is a solution of this equation because, when 2 is substituted for x and 3 is substituted for y, the result is

$$y = \frac{1}{2}x + 2$$

$$3 \overset{?}{=} \frac{1}{2}(2) + 2$$

$$3 = 3.$$

However, the ordered pair $(3, 2)$ is not a solution of the equation because

$$y = \frac{1}{2}x + 2$$

$$2 \overset{?}{=} \frac{1}{2}(3) + 2$$

$$2 \neq \frac{7}{2}.$$

Equations in two variables have many solutions. You should check that the equation

$$y = \frac{1}{2}x + 2$$

also has these solutions:

$$(0, 2), \quad \left(1, \frac{5}{2}\right), \quad \left(3, \frac{7}{2}\right), \quad \left(-1, \frac{3}{2}\right), \quad (-2, 1).$$

We could continue the list of solutions indefinitely. We say that the equation has infinitely many solutions. There is a solution for every real number x.

Clearly, we cannot list all the solutions of an equation in two variables. We indicate the solutions by drawing the **graph of the equation**. To draw the graph of an equation in two variables, we plot the points corresponding to some of the ordered pairs that are solutions of the equation.

EXAMPLE 10.2 Draw the graph of $y = \frac{1}{2}x + 2$.

SOLUTION We find some ordered pairs that are solutions of the equation. One way to do this is by choosing values of x and substituting to find values of y. By substituting $x = 0, 1, 2, 3, -1,$ and -2, we will find the ordered pairs listed previously. We may find some other solutions. For $x = -3$,

$$y = \frac{1}{2}x + 2$$

$$= \frac{1}{2}(-3) + 2$$

$$= \frac{1}{2},$$

which gives the ordered pair $(-3, \frac{1}{2})$. For $x = -4$,

$$y = \frac{1}{2}x + 2$$

$$= \frac{1}{2}(-4) + 2$$

$$= 0,$$

which gives the ordered pair $(-4, 0)$. When we plot the points corresponding to these ordered pairs, they form a straight line. The line is the graph of the equation:

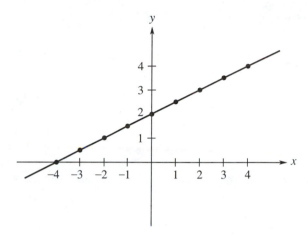

An equation in two variables is **linear** if both variables have the exponent 1. In general, a linear equation in two variables x and y has the form

$$Ax + By + C = 0.$$

We call Ax the **x-term**, By the **y-term**, and C the **constant term**. The graph of a linear equation in two variables is always a straight line.

INTERCEPTS

The points where a graph crosses the axes are called **intercepts**. The graph crosses the x-axis at the **x-intercept** and crosses the y-axis at the **y-intercept**. For the graph in Example 10.2, the x-intercept is $(-4, 0)$ and the y-intercept is $(0, 2)$. In general, at the x-intercept, the y-coordinate is 0, and at the y-intercept, the x-coordinate is 0. A line can be drawn by locating its x- and y-intercepts.

EXAMPLE 10.3 ▶ Find the x-intercept and y-intercept and draw the graph of $4x + 3y = 12$.

SOLUTION When $x = 0$,

$$4x + 3y = 12$$

$$3y = 12$$

$$y = 4.$$

The y-intercept is $(0, 4)$. When $y = 0$,

$$4x + 3y = 12$$

$$4x = 12$$

$$x = 3.$$

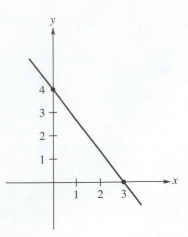

Graph for Example 10.3.

The x-intercept is $(3, 0)$. The graph is shown at the left. We find some other points on the line to check. For example, when $x = 1$,

$$4x + 3y = 12$$

$$4(1) + 3y = 12$$

$$3y = 8$$

$$y = \frac{8}{3}.$$

Thus, the point $(1, \frac{8}{3})$ should be on the graph. When $x = -1$,

$$4x + 3y = 12$$

$$4(-1) + 3y = 12$$

$$3y = 16$$

$$y = \frac{16}{3}.$$

Thus, the point $(-1, \frac{16}{3})$ should be on the graph. You can plot these points to see that they are on the graph. ▲

EXAMPLE 10.4 ▶ Find the x-intercept and y-intercept and draw the graph of $y = \frac{3}{2}x + \frac{9}{2}$.

SOLUTION When $x = 0$,

$$y = \frac{3}{2}x + \frac{9}{2}$$

$$= \frac{9}{2}.$$

The y-intercept is $(0, \frac{9}{2})$. When $y = 0$,

$$y = \frac{3}{2}x + \frac{9}{2}$$

$$0 = \frac{3}{2}x + \frac{9}{2}$$

$$-\frac{9}{2} = \frac{3}{2}x$$

$$-3 = x.$$

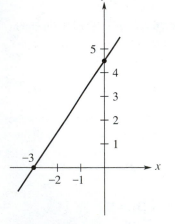

Graph for Example 10.4.

The x-intercept is $(-3, 0)$. The graph is shown at the left. You should check by finding some other points on this line. ▲

It is possible for the two intercepts to be identical. This situation occurs when both intercepts are at the origin.

EXAMPLE 10.5 ▶ Find the x-intercept and y-intercept and draw the graph of $3x + 2y = 0$.

SOLUTION When $x = 0$,

$$3x + 2y = 0$$

$$2y = 0$$

$$y = 0.$$

The y-intercept is $(0, 0)$. When $y = 0$,

$$3x + 2y = 0$$
$$3x = 0$$
$$x = 0.$$

The x-intercept is also $(0, 0)$. We have only the point $(0, 0)$, which is the origin. To draw the line we must find some other point. If $x = 2$,

$$3x + 2y = 0$$
$$3(2) + 2y = 0$$
$$y = -3,$$

so the point $(2, -3)$ is on the line. The graph is shown at the left. You should check by finding some other points on the line. ▲

We recall the general form of a linear equation in two variables x and y,

$$Ax + By + C = 0.$$

We observe that, when the constant term $C = 0$—that is, no constant term is present—the line goes through the origin.

When either $A = 0$ or $B = 0$—that is, either the x-term or the y-term is not present—the line has only one intercept. The other intercept does not exist.

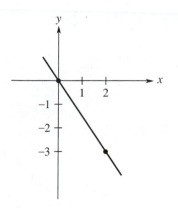

Graph for Example 10.5.

EXAMPLE 10.6   Find the x-intercept and y-intercept and draw the graph of $y + 3 = 0$.

SOLUTION Observe that the x-term does not appear in the equation. For any value of x, we have

$$y + 3 = 0$$
$$y = -3.$$

Therefore, y always has the value -3. In particular, when $x = 0$, $y = -3$. Thus the y-intercept is $(0, -3)$. But if y always has the value -3, then we cannot have $y = 0$. Thus, there is no x-intercept. The graph is

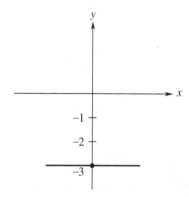

Observe that this line never crosses the x-axis. When the x-term is not present, the line is parallel to the x-axis.

EXAMPLE 10.7   Find the x-intercept and y-intercept and draw the graph of $2x - 3 = 0$.

SOLUTION Observe that the y-term does not appear in the equation. For any value of y, we have

$$2x - 3 = 0$$
$$x = \frac{3}{2}.$$

Therefore, x always has the value $\frac{3}{2}$. In particular, when $y = 0$, $x = \frac{3}{2}$. Thus the x-intercept is $(\frac{3}{2}, 0)$. But if x always has the value $\frac{3}{2}$, then we cannot have $x = 0$. Thus, there is no y-intercept. The graph is

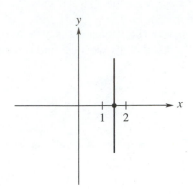

Observe that this line never crosses the y-axis. When the y-term is not present, the line is parallel to the y-axis. ▲

EXERCISE 10.1

Plot the points:

1. $(5, 3)$, $(-2, 5)$, $(4, -2)$, $\sim(-3, -4)$, $(0, 2)$, $(-5, 0)$

2. $\left(\frac{1}{2}, \frac{7}{2}\right)$, $\left(5, -\frac{1}{2}\right)$, $\left(-\frac{7}{2}, -2\right)$, $\left(-3, \frac{8}{3}\right)$, $\left(0, -\frac{4}{3}\right)$, $\left(\frac{5}{2}, 0\right)$

Draw the graph:

3. $y = \frac{1}{2}x - 1$ **4.** $y = 2 - \frac{1}{2}x$

Find the x-intercept and y-intercept and draw the graph:

5. $3x + 2y = 6$ **6.** $2x + y = 6$ **7.** $y = \frac{2}{3}x + 1$ **8.** $y = \frac{5}{2}x + \frac{5}{2}$

9. $2x - 4y = 9$ **10.** $y = \frac{3}{2}x - 4$ **11.** $4x + 3y = 0$ **12.** $5x - 2y = 0$

13. $y - 2 = 0$ **14.** $2y + 3 = 0$ **15.** $x + 3 = 0$ **16.** $2x - 5 = 0$

SECTION 10.2

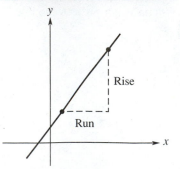

The "rise" and "run."

Slope

In the graphs in Section 10.1, some lines rise and some fall from left to right, some rise or fall more steeply than others, and some are parallel to one of the axes. The measure of how steeply a line rises or falls is called the **slope** m of the line. The slope of a line is the amount of change in the y-direction divided by the amount of change in the x-direction. We sometimes refer to this concept as "rise over run":

$$m = \frac{\text{rise}}{\text{run}}.$$

The rise, which is the change in y, is denoted by Δy (Δ is the Greek capital letter delta; Δy is a single variable name). The run, which is the change in x, is denoted by Δx.

Suppose (x_1, y_1) and (x_2, y_2) are two points on a line. The change in y is given by

$$\Delta y = y_2 - y_1,$$

and the change in x is given by

$$\Delta x = x_2 - x_1:$$

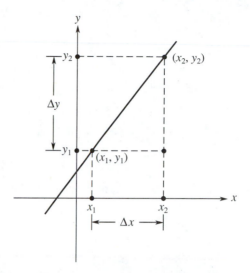

The slope m is Δy divided by Δx.

Slope formula: The slope m of a line is given by

$$m = \frac{\Delta y}{\Delta x},$$

where $\Delta y = y_2 - y_1$ and $\Delta x = x_2 - x_1$.

We use the slope formula to find the slope of a line when we know the coordinates of any two points on the line.

EXAMPLE 10.8 ▶   Find the slope of the line including the points $(1, 1)$ and $(3, 2)$.

SOLUTION   We can draw the graph of the line by plotting the two given points. The dashed lines in the figure show Δy and Δx. If we take $(x_1, y_1) = (1, 1)$ and $(x_2, y_2) = (3, 2)$, we have

$$\Delta y = y_2 - y_1$$
$$= 2 - 1$$
$$= 1,$$

and

$$\Delta x = x_2 - x_1$$
$$= 3 - 1$$
$$= 2.$$

Graph for Example 10.8.

Therefore, the slope of the line is

$$m = \frac{\Delta y}{\Delta x}$$

$$= \frac{1}{2}.$$

Observe that the line rises one unit vertically for each two units run horizontally. ▲

We will get the same result if we assign the points in Example 10.8 oppositely. If we take $(x_1, y_1) = (3, 2)$ and $(x_2, y_2) = (1, 1)$, we have

$$\Delta y = y_2 - y_1$$
$$= 1 - 2$$
$$= -1,$$

and

$$\Delta x = x_2 - x_1$$
$$= 1 - 3$$
$$= -2.$$

Therefore, the slope of the line again is

$$m = \frac{\Delta y}{\Delta x}$$
$$= \frac{-1}{-2}$$
$$= \frac{1}{2}.$$

EXAMPLE 10.9   Find the slope of the line including the points $(1, 3)$ and $(3, 2)$.

SOLUTION We draw the graph of the line, and again indicate Δy and Δx in the figure by the dashed lines. If we take $(x_1, y_1) = (1, 3)$ and $(x_2, y_2) = (3, 2)$, we have

$$\Delta y = y_2 - y_1$$
$$= 2 - 3$$
$$= -1,$$

and

$$\Delta x = x_2 - x_1$$
$$= 3 - 1$$
$$= 2.$$

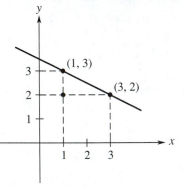

Graph for Example 10.9.

Therefore, the slope of the line is

$$m = \frac{\Delta y}{\Delta x}$$
$$= \frac{-1}{2}$$
$$= -\frac{1}{2}.$$

Observe that the line falls one unit vertically for each two units run horizontally. If the line rises from left to right, its slope is positive. If the line falls from left to right, its slope is negative. You should show that you get the same result if you assign the points oppositely. ▲

EXAMPLE 10.10 ▶ Find the slope of the line including the points $(2, 0)$ and $(0, 4)$.

SOLUTION The given points are the x- and y-intercepts. If we take $(x_1, y_1) = (2, 0)$ and $(x_2, y_2) = (0, 4)$, we have

$$\Delta y = y_2 - y_1$$
$$= 4 - 0$$
$$= 4,$$

and

$$\Delta x = x_2 - x_1$$
$$= 0 - 2$$
$$= -2.$$

Therefore, the slope of the line is

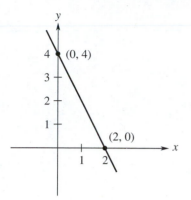

Graph for Example 10.10.

$$m = \frac{\Delta y}{\Delta x}$$

$$= \frac{4}{-2}$$

$$= -2.$$

The line falls from left to right, and the slope is negative.

You must be careful to take the coordinates of the points in a consistent order. If you start with $y_2 = 4$, you must then take $x_2 = 0$. You should try this example starting with $y_2 = 0$. You must then take $x_2 = 2$. ▲

When we stated the slope formula, we used two points in the first quadrant. However, the formula can be used for points in any quadrant.

EXAMPLE 10.11 ▶ Find the slope of the line including the points $(-1, 4)$ and $(3, -2)$.

SOLUTION If we take $(x_1, y_1) = (3, -2)$ and $(x_2, y_2) = (-1, 4)$, we have

$$\Delta y = y_2 - y_1$$
$$= 4 - (-2)$$
$$= 6,$$

and

$$\Delta x = x_2 - x_1$$
$$= -1 - 3$$
$$= -4.$$

Therefore, the slope of the line is

Graph for Example 10.11.

$$m = \frac{\Delta y}{\Delta x}$$

$$= \frac{6}{-4}$$

$$= -\frac{3}{2}.$$ ▲

EXAMPLE 10.12   Find the slope of the line including the points $(-2, -5)$ and $(4, -1)$.

SOLUTION   The graph is

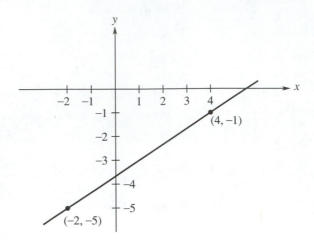

If we take $(x_1, y_1) = (-2, -5)$ and $(x_2, y_2) = (4, -1)$, we have

$$\Delta y = y_2 - y_1$$
$$= -1 - (-5)$$
$$= 4,$$

and

$$\Delta x = x_2 - x_1$$
$$= 4 - (-2)$$
$$= 6.$$

Therefore, the slope of the line is

$$m = \frac{\Delta y}{\Delta x}$$
$$= \frac{4}{6}$$
$$= \frac{2}{3}.$$

▲

EXAMPLE 10.13   Find the slope of the line including the points $(-2, 1)$ and $(3, 1)$.

SOLUTION   If we take $(x_1, y_1) = (-2, 1)$ and $(x_2, y_2) = (3, 1)$, we have

$$\Delta y = y_2 - y_1$$
$$= 1 - 1$$
$$= 0,$$

and

$$\Delta x = x_2 - x_1$$
$$= 3 - (-2)$$
$$= 5.$$

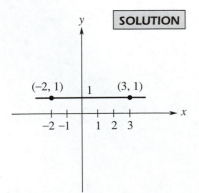

Graph for Example 10.13.

Therefore, the slope of the line is

$$m = \frac{\Delta y}{\Delta x}$$

$$= \frac{0}{5}$$

$$= 0.$$

Observe that the line is parallel to the x-axis. This line is a horizontal line that neither rises nor falls. The slope of a horizontal line is always zero. ▲

EXAMPLE 10.14 Find the slope of the line including the points $(3, 1)$ and $(3, -2)$.

SOLUTION If we take $(x_1, y_1) = (3, -2)$ and $(x_2, y_2) = (3, 1)$, we have

$$\Delta y = y_2 - y_1$$

$$= 1 - (-2)$$

$$= 3,$$

and

$$\Delta x = x_2 - x_1$$

$$= 3 - 3$$

$$= 0.$$

Therefore, the slope of the line is

$$m = \frac{\Delta y}{\Delta x}$$

$$= \frac{3}{0},$$

and so m is undefined. Observe that the line is parallel to the y-axis. This line is a vertical line that has no run. The slope of a vertical line is always undefined. ▲

Graph for Example 10.14.

EXERCISE

10.2

Find the slope of the line including the points:

1. $(2, 1)$ and $(5, 3)$ 2. $(1, 1)$ and $(3, 5)$

3. $(1, 6)$ and $(2, 3)$ 4. $(3, 5)$ and $(5, 2)$

5. $(1, 0)$ and $(0, 3)$ 6. $(4, 0)$ and $(0, 2)$

7. $(3, 2)$ and $(-3, -1)$ 8. $(-5, -2)$ and $(3, -6)$

9. $(-3, 4)$ and $(5, -2)$ 10. $(-2, 1)$ and $(2, 5)$

11. $(-3, 0)$ and $(0, 6)$ 12. $(-4, 0)$ and $(0, -2)$

13. $(-3, 2)$ and $(2, 2)$ 14. $(2, -4)$ and $(2, 3)$

15. $(-4, 1)$ and $(-4, 0)$ 16. $(0, -3)$ and $(4, -3)$

SECTION

10.3

Slope-Intercept Form

Suppose we have a linear equation in the general form

$$Ax + By + C = 0,$$

where neither A nor B is 0. Then, the equation has both an x-term and a y-term, so both intercepts exist. When $x = 0$,

$$Ax + By + C = 0$$
$$By + C = 0$$
$$y = -\frac{C}{B}.$$

The y-intercept is $(0, -\frac{C}{B})$. When $y = 0$,

$$Ax + By + C = 0$$
$$Ax + C = 0$$
$$x = -\frac{C}{A}.$$

The x-intercept is $(-\frac{C}{A}, 0)$. Using these two points, we can find the slope of the line. If we take $(x_1, y_1) = (-\frac{C}{A}, 0)$ and $(x_2, y_2) = (0, -\frac{C}{B})$, we have

$$\Delta y = y_2 - y_1$$
$$= -\frac{C}{B},$$

and

$$\Delta x = x_2 - x_1$$
$$= \frac{C}{A}.$$

Therefore, the slope of the line is

$$m = \frac{\Delta y}{\Delta x}$$
$$= \frac{-\frac{C}{B}}{\frac{C}{A}}$$
$$= -\frac{C}{B} \times \frac{A}{C}$$
$$= -\frac{A}{B}.$$

Now, suppose we solve the general form for y:

$$Ax + By + C = 0$$
$$By = -Ax - C$$
$$y = -\frac{A}{B}x - \frac{C}{B}.$$

Observe that, in this form, the x-term contains the slope of the line, and the constant term is the y-coordinate of the y-intercept. This form of a linear equation is called the **slope-intercept** form. In general, the slope-intercept form of a linear equation is

$$y = mx + b,$$

where m is the slope and b is the y-intercept.

EXAMPLE 10.15 Find the slope and y-intercept of the line given by $x - 2y + 4 = 0$.

SOLUTION We solve the equation for y:

$$x - 2y + 4 = 0$$

$$x + 4 = 2y.$$

Therefore,

$$2y = x + 4$$

$$y = \frac{1}{2}x + 2.$$

The equation is now in slope-intercept form, with $m = \frac{1}{2}$ and $b = 2$. Therefore, the slope is $\frac{1}{2}$ and the y-intercept is $(0, 2)$. The line includes the point $(0, 2)$, and rises one unit vertically for each two units run horizontally:

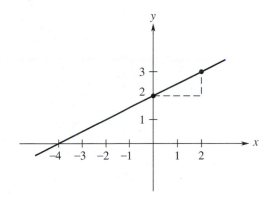

EXAMPLE 10.16 Find the slope and y-intercept of the line given by $3x + 4y - 6 = 0$.

SOLUTION We solve the equation for y:

$$3x + 4y - 6 = 0$$

$$4y = -3x + 6$$

$$y = -\frac{3}{4}x + \frac{6}{4}$$

$$y = -\frac{3}{4}x + \frac{3}{2}.$$

The equation is now in slope-intercept form, with $m = -\frac{3}{4}$ and $b = \frac{3}{2}$. Therefore, the slope is $-\frac{3}{4}$ and the y-intercept is $(0, \frac{3}{2})$. The line includes the point $(0, \frac{3}{2})$, and falls three units vertically for each four units run horizontally:

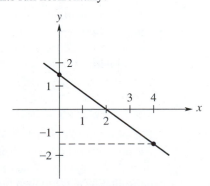

EXAMPLE 10.17 Find the slope and y-intercept of the line given by $3x + 4y = 0$.

SOLUTION   We solve the equation for y:

$$3x + 4y = 0$$

$$y = -\frac{3}{4}x.$$

When this equation is in slope-intercept form, $m = -\frac{3}{4}$ and $b = 0$. Therefore, the slope is $-\frac{3}{4}$ and the y-intercept is the origin $(0, 0)$. The line includes the origin, and falls three units vertically for each four units run horizontally:

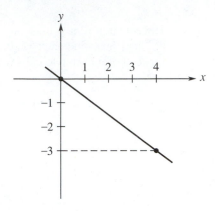

Recall that, if the equation of a line has no x-term, then the line has no x-intercept, is parallel to the x-axis, and has zero slope. We can write the equation of a line of this type in slope-intercept form.

EXAMPLE 10.18 Find the slope and y-intercept of the line given by $4y + 5 = 0$.

SOLUTION   We solve the equation for y:

$$4y + 5 = 0$$

$$y = -\frac{5}{4}.$$

Since no x-term appears in the slope-intercept form,

$$y = 0x - \frac{5}{4}.$$

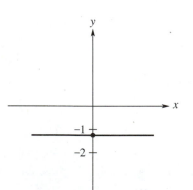

Graph for Example 10.18.

Therefore, $m = 0$, and the line is parallel to the x-axis. The constant term is $b = -\frac{5}{4}$, so the y-intercept is $(0, -\frac{5}{4})$. ▲

If the equation of a line has no y-term, then the slope of the line is undefined, the line has no y-intercept, and it is parallel to the y-axis. Also, if there is no y-term, the equation cannot be solved for y. Therefore, a linear equation that has no y-term cannot be written in slope-intercept form.

RELATIONS AND FUNCTIONS

Equations in two variables x and y are called **relations in x and y**. A linear relation that has no y-term is different from other relations in x and y. In mathematics, the distinction is made between those relations in x and y that have exactly one y-value for each x-value, and those that have more than one y-value for some or all x-values.

> **Function:** A relation in two variables x and y is a **function** if for each allowable value of x there is exactly one value of y.

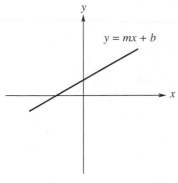

Graph of a function.

If a linear relation can be written in the slope-intercept form

$$y = mx + b,$$

then it has exactly one y-value for each x-value. Such a relation is a function.

If a linear relation in slope-intercept form has slope $m = 0$, then it can be written in the form

$$y = b.$$

Its graph is a line parallel to the x-axis. This relation has exactly one y-value for each x-value, and so it is a function.

Now, we consider a linear relation where the slope of the line is undefined. Its graph is a line parallel to the y-axis. Such a relation has many y-values for its one x-value. Therefore, this relation is *not a function*:

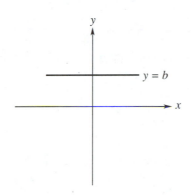

Graph of a function.

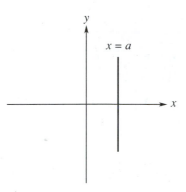

Graph of a relation that is not a function.

EXERCISE

10.3

Find the slope and y-intercept of the line given by the equation:

1. $2x - y + 4 = 0$ 2. $2x - 2y - 3 = 0$
3. $5x + 2y - 10 = 0$ 4. $3x + 4y + 8 = 0$
5. $x + 4y - 2 = 0$ 6. $2x - 4y + 3 = 0$
7. $3x - 6y - 4 = 0$ 8. $2x + 6y + 9 = 0$
9. $2x + 5y = 0$ 10. $4x - 3y = 0$
11. $3y + 2 = 0$ 12. $2y - 5 = 0$

SECTION

10.4

Other Graphs

Recall from Unit 4 that equations with an x^2 term are called quadratic equations, and equations containing $\sqrt{x}$ are related to quadratic equations. These are examples of equations in two variables that are quadratic, or are related to quadratic equations:

$$y = x^2, \quad y = x, \quad y = \sqrt{x}, \quad x^2 + y^2 = 1, \quad y = \frac{1}{x}.$$

The last example may not look quadratic. However, we can write it in the form $xy = 1$, and we consider xy to be quadratic just like $x^2 = xx$ and $y^2 = yy$.

Nonlinear equations in two variables have graphs that are not straight lines. Initially when we have a nonlinear equation, we do not know the shape of the graph. Therefore, we start the graph by plotting several points.

EXAMPLE 10.19   Draw the graph of $y = x^2$.

SOLUTION   We find several ordered pairs that are solutions of the equation. For $x = 0$,

$$y = x^2$$
$$= 0^2$$
$$= 0,$$

which gives the ordered pair $(0, 0)$. For $x = 1$,

$$y = x^2$$
$$= 1^2$$
$$= 1,$$

which gives the ordered pair $(1, 1)$. For $x = -1$,

$$y = x^2$$
$$= (-1)^2$$
$$= 1,$$

which gives the ordered pair $(-1, 1)$. You should check that these ordered pairs are also solutions:

$$(2, 4), \quad (-2, 4), \quad (3, 9), \quad (-3, 9).$$

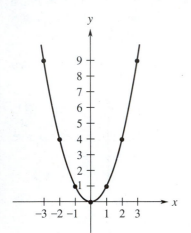

Graph for Example 10.19.

We plot the points corresponding to the ordered pairs we have found, and connect them by a smooth curve. The graph is shown at the left above. ▲

When we connect the points by a smooth curve, we make several assumptions. We assume that there are values of y defined for all values of x, and that they fall in a continuous curve. We also assume that the curve continues upward beyond $x = 3$ and $x = -3$. These assumptions can be proved by using methods from calculus.

The graph in Example 10.19 is called a **parabola**. Observe that the graph has a minimum value at the origin. This point is called the **vertex** of the parabola. We also observe that the relation $y = x^2$ is a function. For each x-value there is exactly one y-value.

There is a convenient test that can be used to determine whether a graph is the graph of a function. We draw a vertical line through the graph at any x for which the graph is defined. If every such vertical line crosses the graph exactly once, then there is exactly one y-value for each x-value and the graph is a graph of a function. When there is any point where such a line crosses the graph more than once, then the graph is not the graph of a function. We will refer to this method as the "vertical line test."

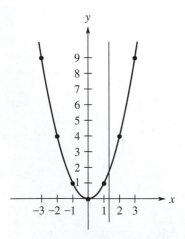

The "vertical line test" for the graph in Example 10.19.

For example, we can draw a line through the graph of $y = x^2$ at any x. Every such line crosses the graph exactly once, so the vertical line test shows that $y = x^2$ is a function.

EXAMPLE 10.20   Draw the graph of $y^2 = x$.

SOLUTION   We can find ordered pairs that are solutions of this equation by choosing values for y. For $y = 0$,

$$x = y^2$$
$$= 0^2$$
$$= 0,$$

which gives the ordered pair $(0, 0)$. For $y = 1$,

$$x = y^2$$
$$= 1^2$$
$$= 1,$$

which gives the ordered pair $(1, 1)$. For $y = -1$,

$$x = y^2$$
$$= (-1)^2$$
$$= 1,$$

which gives the ordered pair $(1, -1)$. You should check that these ordered pairs are also solutions:

$$(4, 2), \quad (4, -2), \quad (9, 3), \quad (9, -3).$$

We plot the points corresponding to the ordered pairs and connect them by a smooth curve. The graph is

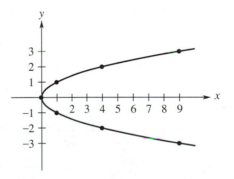

This graph is also a parabola with its vertex at the origin. However, the relation $y^2 = x$ is different from the function $y = x^2$ in one important aspect: the relation $y^2 = x$ is *not a function*. For each x-value other than 0, there are two y-values.

Using the vertical line test, we draw a line through the graph of $y^2 = x$ at any point where the graph is defined; that is, any $x \geq 0$:

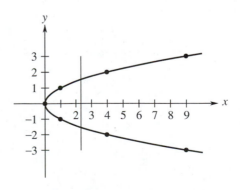

Except at $x = 0$, any such line crosses the graph twice, so the vertical line test shows that $y^2 = x$ is not a function.

INVERSE RELATIONS

We observe that the ordered pairs of the relation $y^2 = x$ are the same as the ordered pairs of the function $y = x^2$ with their coordinates reversed.

> **Inverse Relation:** The **inverse** of a function is the relation obtained by reversing the coordinates in the ordered pairs of the function.

The inverse of a function may be obtained by exchanging the variables of the function. When the variables of a function are exchanged, the coordinates of the ordered pairs reverse. The resulting relation might not be a function.

The graphs of a function and its inverse are also related. We observe that the graphs of $y = x^2$ and $y^2 = x$ are mirror images of one another, with the line $y = x$ as the mirror:

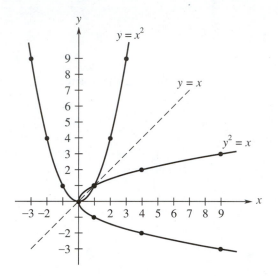

We say that the graphs are **symmetric** with respect to the line $y = x$. Graphs of inverse relations and functions are always symmetric with respect to the line $y = x$.

We can restrict the relation $y^2 = x$ so that it is a function and its graph is one side of a parabola.

EXAMPLE 10.21

Draw the graph of $y = \sqrt{x}$.

SOLUTION   We find ordered pairs for several values of x. For $x = 0$,

$$y = \sqrt{x}$$
$$= \sqrt{0}$$
$$= 0,$$

which gives the ordered pair $(0, 0)$. For $x = 1$,

$$y = \sqrt{x}$$
$$= \sqrt{1}$$
$$= 1,$$

which gives the ordered pair $(1, 1)$. For $x = 4$,

$$y = \sqrt{x}$$
$$= \sqrt{4}$$
$$= 2,$$

which gives the ordered pair $(4, 2)$. We cannot find ordered pairs for negative values of x because the square root of a negative number is not a real number. The graph is

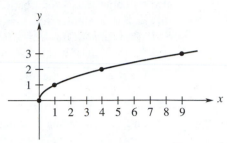

It might seem that the relations $y^2 = x$ and $y = \sqrt{x}$ should be the same. Mathematically, however, the relation
$$y^2 = x$$
is equivalent to
$$y = \pm\sqrt{x},$$
that is, the two functions
$$y = \sqrt{x} \text{ and } y = -\sqrt{x}.$$

The graph of the function $y = \sqrt{x}$ is the top half of the parabola in Example 10.20. You should show that the graph of the function $y = -\sqrt{x}$ is the bottom half of the parabola.

EXAMPLE 10.22 Draw the graph of $x^2 + y^2 = 1$.

SOLUTION In order to find some ordered pairs, we may solve the equation for y:
$$x^2 + y^2 = 1$$
$$y^2 = 1 - x^2$$
$$y = \pm\sqrt{1 - x^2}.$$

We must be careful to include both positive and negative solutions in order to obtain all parts of the graph. Now, we can find some ordered pairs. For $x = 0$,
$$y = \pm\sqrt{1 - x^2}$$
$$= \pm\sqrt{1 - 0^2}$$
$$= \pm\sqrt{1}$$
$$= \pm 1,$$
which gives two ordered pairs, $(0, 1)$ and $(0, -1)$. For $x = 1$,
$$y = \pm\sqrt{1 - x^2}$$
$$= \pm\sqrt{1 - 1^2}$$
$$= \pm\sqrt{1 - 1}$$
$$= \pm\sqrt{0}$$
$$= 0,$$
which gives one ordered pair $(1, 0)$. Also, for $x = -1$,
$$y = \pm\sqrt{1 - x^2}$$
$$= \pm\sqrt{1 - (-1)^2}$$
$$= \pm\sqrt{1 - 1}$$
$$= \pm\sqrt{0}$$
$$= 0,$$

Graph for Example 10.22.

which gives one ordered pair $(-1, 0)$. You should try some values for x greater than 1 or less than -1 to show that such values give a negative under the square root. You should also try some values for x between -1 and 1, using a calculator to get approximate values for y. For example, for $x = 0.5$, you should get the ordered pairs $(0.5, 0.87)$ and $(0.5, -0.87)$, and for $x = -0.5$, you should get the ordered pairs $(-0.5, 0.87)$ and $(-0.5, -0.87)$. The graph is a circle with its center at the origin and radius 1.   ▲

We observe that the relation

$$x^2 + y^2 = 1$$

is not a function, but each of the separate parts

$$y = \sqrt{1 - x^2} \text{ and } y = -\sqrt{1 - x^2}$$

is a function. The positive square root gives the top semicircle and the negative square root gives the bottom semicircle.

We also observe that the relation

$$x^2 + y^2 = 1,$$

is its own inverse because exchanging x and y gives

$$y^2 + x^2 = 1,$$

which does not change the relation.

EXAMPLE 10.23 ▶ Draw the graph of $y = \dfrac{1}{x}$.

SOLUTION   We cannot start with $x = 0$ because division by zero is undefined. There is no ordered pair of the relation that has $x = 0$. Beginning with $x = 1$, we find ordered pairs such as $(1, 1)$, $(2, 0.5)$, and $(4, 0.25)$. We can also use values for x between 0 and 1 to find ordered pairs such as $(0.5, 2)$ and $(0.25, 4)$. We can use negative values for x to find ordered pairs such as $(-1, -1)$, $(-2, -0.5)$, and $(-4, -0.25)$, and also $(-0.5, -2)$ and $(-0.25, -4)$. The graph is

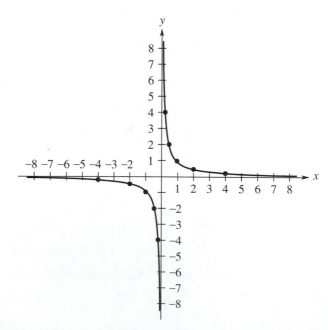

We observe that this relation is a function and also is its own inverse. The graph is called a **hyperbola**. A hyperbola consists of two identical branches. We further observe that the graph gets closer and closer to the x-axis from above as we go to the right and from below as we go to the left. We say that the graph approaches the x-axis **asymptotically**; or that the x-axis is an **asymptote** of the graph. The graph also gets closer and closer to the y-axis from the right as we go up and from the left as we go down. Thus, the graph also approaches the y-axis asymptotically; that is, the y-axis is also an asymptote of the graph. ▲

EXERCISE 10.4

Draw the graph:

1. $y = 2x^2$
2. $y = -x^2$
3. $y = x^2 + 1$
4. $y = (x + 1)^2$

5. $2y^2 = x$
6. $y^2 = -x$
7. $y = -\sqrt{x}$
8. $y = \sqrt{-x}$

9. $x^2 + y^2 = 4$
10. $x^2 + y^2 = 2$
11. $y^2 - x^2 = 4$
12. $x^2 - y^2 = 4$

13. $y = \dfrac{2}{x}$
14. $y = -\dfrac{1}{x}$
15. $y = \dfrac{1}{x} + 1$
16. $y = \dfrac{1}{x + 1}$

17. $y = x^3$
18. $y = -x^3$
19. $y = x^4$
20. $y = -x^4$

SECTION 10.5

Approximating Slopes

In Section 10.2, we found slopes of lines by using two points on the line. We can use the same method to approximate slopes of lines that are drawn from experimental data rather than from given equations.

Suppose that, in the laboratory, a simple circuit is constructed consisting of an energy source E and a resistance R, connected by wires carrying a current I. A **voltmeter** (for measuring voltage) is placed across the resistor, and an **ammeter** (for measuring current) is placed in the circuit:

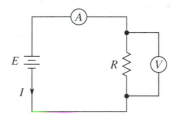

Then, the current I can be measured on the ammeter for different settings of the voltage V on the voltmeter.

EXAMPLE 10.24

In this chart are settings on the voltmeter from 0 V to 10 V in increments of 1 V, and the corresponding ammeter readings:

V	0	1	2	3	4	5	6	7	8	9	10
I	0	0.18	0.36	0.54	0.71	0.89	1.07	1.25	1.43	1.61	1.79

Draw the graph and find the slope of the line determined by the values in the chart.

SOLUTION

We make a graph of the values by using V as the horizontal axis and I as the vertical axis. Then, the coordinates of the points are the ordered pairs (V, I). We observe that the points lie along a straight line, and draw the line connecting them:

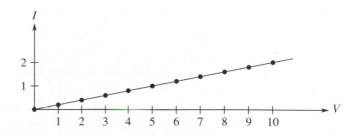

An analog multimeter called a VOM, for volt-ohm-milliammeter. Courtesy of *Simpson Electric Company*.

This graph is called the *I–V* **characteristic curve** for the resistor. (In mathematics, straight line graphs as well as nonlinear graphs are called "curves.")

We can calculate the slope of the line by using any two points (V, I) on the line. For example, for the points $(3, 0.54)$ and $(4, 0.71)$, we can write

$$\Delta I = I_2 - I_1$$
$$= 0.71 - 0.54$$
$$= 0.17,$$

and

$$\Delta V = V_2 - V_1$$
$$= 4 - 3$$
$$= 1.$$

Therefore, the slope of the line is

$$m = \frac{\Delta I}{\Delta V}$$
$$= \frac{0.17}{1}$$
$$= 0.17.$$

We know that the slope of a line should be the same regardless of which two points we use. However, small discrepancies might occur due to limitations in the number of places to which the meters can be read. Usually, we get a better approximation if we take two points further apart. Using the points $(3, 0.54)$ and $(6, 1.07)$, we have

$$\Delta I = I_2 - I_1$$
$$= 1.07 - 0.54$$
$$= 0.53,$$

and

$$\Delta V = V_2 - V_1$$
$$= 6 - 3$$
$$= 3.$$

Therefore, the slope of the line is

$$m = \frac{\Delta I}{\Delta V}$$
$$= \frac{0.53}{3}$$
$$= 0.177.$$

We will take the slope of this line to be approximately 0.18. You should try some more sets of points to substantiate this approximation.   ▲

We recall that the resistance R can be calculated by using a form of Ohm's law,

$$R = \frac{V}{I}.$$

A digital multimeter (DMM).
Courtesy of *Fluke Manufacturing Company*.

The slope of the I–V characteristic curve,

$$m = \frac{\Delta I}{\Delta V},$$

is the reciprocal of the resistance R. This reciprocal is called the **conductance**, G, of the resistor. Conductance is measured in **siemens**, S, named for Sir William Siemens (1823–1883), a German-born British engineer.

For the resistor in Example 10.24, we have found the conductance by finding the slope of the line. Since we found that

$$m = 0.18,$$

we have the conductance

$$G = 0.18 \text{ S}.$$

Then, the resistance is the reciprocal of the conductance. Therefore, we can find the approximate resistance of the resistor used in the circuit:

$$R = \frac{1}{G}$$

$$= \frac{1}{0.18 \text{ S}}$$

$$= 5.6 \ \Omega.$$

The numbers in the chart were found using a 5.6-Ω resistor.

STATIC RESISTANCE

The I–V characteristic curve for a semiconductor device called a **diode** is not a straight line. This might be the characteristic curve for a commercial silicon diode:

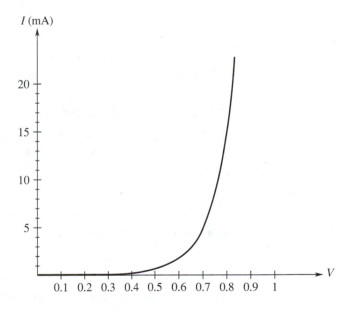

The **DC resistance** or **static resistance** of the diode at any point of the characteristic curve is

$$R_{DC} = \frac{V}{I}$$

where V and I are the coordinates of the point.

EXAMPLE 10.25

Find the DC resistance of the diode with the characteristic curve just described, at the point where $V = 0.7$ V.

SOLUTION

Referring to the diode I–V characteristic curve, we see that I is approximately 5 mA when $V = 0.7$ V:

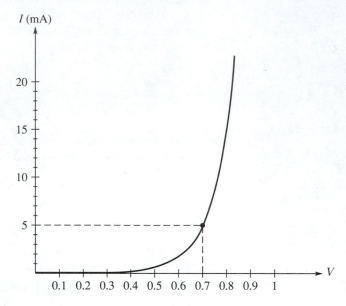

Therefore, we write

$$R_{DC} = \frac{V}{I}$$

$$= \frac{0.7 \text{ V}}{5 \text{ mA}}$$

$$= 0.14 \text{ k}\Omega$$

or approximately 140 Ω. Because we are reading values from a graph, even the second digit is approximate, and it does not make sense to try to write a third significant digit. ▲

DYNAMIC RESISTANCE

The DC resistance is determined by only a single point of the characteristic curve. It does not take into account the shape of the curve in the area of the point. The **AC resistance**, or **dynamic resistance**, of a diode at any point of the characteristic curve is

$$R_{AC} = \frac{\Delta V}{\Delta I},$$

where the slope of the **tangent line** to the curve at the point is

$$m = \frac{\Delta I}{\Delta V}.$$

The tangent line to a curve at a given point is a line that touches the curve at just that one point.

EXAMPLE 10.26

Find the AC resistance of the diode in Example 10.25, at the point where $V = 0.7$ V.

SOLUTION

We draw the tangent line to the curve at the point where $V = 0.7$ V and $I = 5$ mA. Then, we choose any two points on the tangent line. We have chosen the points where $I = 2$ mA and $I = 10$ mA:

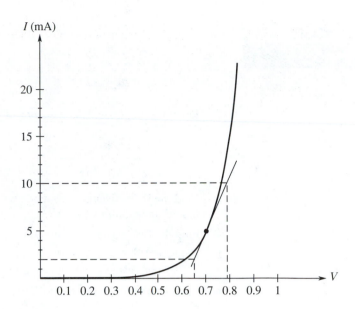

When $I = 2$ mA, we see that V is approximately 0.65 V, and when $I = 10$ mA, we see that V is approximately 0.79 V. Then, to find the slope of the tangent line, we write

$$\Delta I = I_2 - I_1$$

$$= 10 \text{ mA} - 2 \text{ mA}$$

$$= 8 \text{ mA},$$

and

$$\Delta V = V_2 - V_1$$

$$= 0.79 \text{ V} - 0.65 \text{ V}$$

$$= 0.14 \text{ V}.$$

Therefore, the slope is

$$m = \frac{\Delta I}{\Delta V}$$

$$= \frac{8 \text{ mA}}{0.14 \text{ V}}$$

$$= 57 \text{ mS}.$$

The AC resistance is

$$R_{AC} = \frac{\Delta V}{\Delta I}$$

$$= \frac{0.14 \text{ V}}{8 \text{ mA}}$$

$$= 0.0175 \text{ k}\Omega,$$

so the AC resistance is approximately 18 Ω. As before, our numbers are read from a graph, so this result is approximate. ▲

Diode I–V characteristic curves are related to **exponential functions**, which we will study in Unit 15. We observe that the curve in Examples 10.25 and 10.26 begins to increase very rapidly after approximately $V = 0.7$ V. When a curve is increasing rapidly, the slope of the tangent line is large, and its reciprocal is small. Thus, the AC resistance decreases rapidly after about $V = 0.7$ V.

1. Draw the graph and find the slope of the line determined by the values in the chart:

V	0	1	2	3	4	5	6	7	8	9	10
I	0	0.26	0.51	0.77	1.03	1.28	1.54	1.79	2.05	2.31	2.56

2. Draw the graph and find the slope of the line determined by the values in the chart:

V	0	1	2	3	4	5	6	7	8	9	10
I	0	0.12	0.24	0.37	0.49	0.61	0.73	0.85	0.98	1.10	1.21

3. Find the DC resistance of a diode with the I–V characteristic curve shown at the left below, at the point where $V = 0.7$ V.

4. Find the DC resistance of the diode in Exercise 3, at the point where $V = 0.8$ V.

5. Find the DC resistance of a commercial germanium diode, which might have the I–V characteristic curve shown at the right below, at the point where $V = 0.2$ V.

6. Find the DC resistance of the diode in Exercise 5, at the point where $V = 0.3$ V.

7. Find the AC resistance of the diode in Exercise 3, at the point where $V = 0.7$ V.

8. Find the AC resistance of the diode in Exercise 3, at the point where $V = 0.8$ V.

9. Find the AC resistance of the diode in Exercise 5, at the point where $V = 0.2$ V.

10. Find the AC resistance of the diode in Exercise 5, at the point where $V = 0.3$ V.

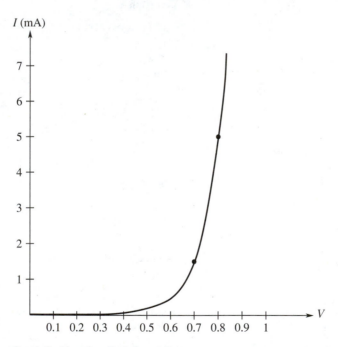

Graph for Exercises 3, 4, 7, and 8.

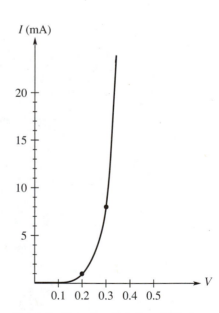

Graph for Exercises 5, 6, 9, and 10.

Self-Test

1. Find the x-intercept and y-intercept and draw the graph of $5x - 2y = 5$.

2. Draw the graph of $y = \frac{1}{2}x^2$.

3. Find the slope and y-intercept of the line given by $5x + 4y - 10 = 0$.

3. _____

4. Find the slope of the line including the points $(-3, 1)$ and $(3, -2)$.

4. _____

5. Find the AC resistance at the point where $V = 0.3$ V:

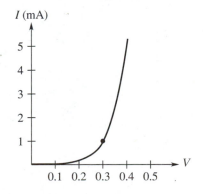

5. _____

UNIT 11

Cumulative Review

Introduction

In Units 1 through 5, you learned about real numbers, operations with signed numbers, and powers of ten, including scientific and engineering notations. You learned how to evaluate and simplify algebraic expressions. Then, you learned how to solve linear equations and some basic types of quadratic and related equations, and how to solve and evaluate formulas.

In Units 7 through 10, you applied these skills to basic applications involving formulas in electronics, specifically, Kirchhoff's laws and Ohm's law, and you applied these formulas to solving problems involving DC circuits. Then, you learned how to solve rational equations, and you used rational equations to derive more formulas of electronics, the alternate parallel resistance formula, and the voltage and current divider formulas. Finally, you learned about graphs, another of the main tools of mathematics. You drew graphs of linear equations in two variables, found the intercepts and slopes of lines, and drew graphs of some types of quadratic and related equations in two variables.

Now, you should review the material in all of the preceding units.

OBJECTIVE

When you have finished this unit you should be able to fulfill every objective of each of Units 1 through 10.

SECTION 11.1

Review Method

For this unit you should review the Self-Tests for Units 1 through 10. Do each problem in each Self-Test over again. If you cannot do a problem, or if you have the slightest doubt or difficulty, find the appropriate material to review:

1. You will find an objective number next to the answer to every Self-Test problem. Find the objective number of the problem you are working on (this is not necessarily the problem number). Go to the first page of the unit and reread the objective. For example, if you have difficulty with a Self-Test problem in Unit 2, and the objective number next to the answer is Objective 3, reread Objective 3 in the objectives list at the beginning of Unit 2 to find out what concept you need to review.

2. The material you should study to review the concept is in the section that has the same number as the objective. Reread the material in that section, rework the examples, and redo some of the exercises for the section. For example, if you are reviewing Unit 2, Objective 3, refer to Unit 2, Section 3; that is, Section 2.3. Reread Section 2.3, rework examples in that section, and redo problems in Exercise 2.3.

3. Try the Self-Test for the unit again to find out if there is any other objective you need to review in the unit. For example, if you have been reviewing Section 2.3 and Exercise 2.3, try the Self-Test for Unit 2 again.

Review each unit from Unit 1 through Unit 10 in this way. For some units, you may find that you can still do the Self-Test easily. For other units, you might need to review one or more of the objectives in more detail. As a final check, and only after you have reviewed every preceding unit, try the Self-Test for this Cumulative Review Unit 11.

1. Find the value of $10\sqrt{9 \times 10^2} - 10^3 \div 5$.

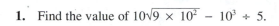

1. _____

2. Simplify $\dfrac{(10^{-5} \times 10^2)^{-1}}{10^{-6}}$.

2. _____

3. Simplify $2x^{-2}(3x^3 - x^2)$.

3. _____

4. Solve $3 - \dfrac{2\sqrt{x}}{3} = 1$ and check the solution.

4. _____

5. For the formula $F = \dfrac{km_1m_2}{d^2}$, find d (in meters) if $k = 6.67 \times 10^{-11}$, $m_1 = 7.36 \times 10^{22}$ kg, $m_2 = 100$ kg, and $F = 3.4 \times 10^{-3}$ N (newtons; d is the distance between earth and the moon).

5. _____

6. Solve the formula $\dfrac{y}{2x} = \dfrac{6-y}{3}$ for y.

6. _____

7. For this circuit, if $E = 25$ V, $V_1 = 10$ V, $V_3 = 3$ V, and $V_4 = 2$ V, find V_2 and V_5:

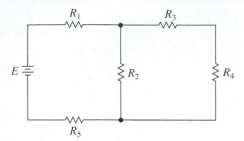

7. _____

8. Find R_T and I_T for this circuit:

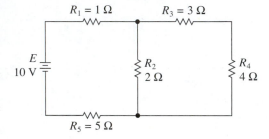

8. _____

9. Find the x-intercept and y-intercept and draw the graph of $y = \dfrac{1}{2}x - 2$.

10. Find the slope of the line including the points $(0.6, 2)$ and $(1, 10)$.

10. _____

UNIT 12

Systems of Equations

Introduction

In Unit 10, you learned about equations in two variables. These equations have solutions that are ordered pairs, and have infinitely many such solutions. However, two equations in two variables, taken together, have a limited number of solutions in common. In this unit, you will learn some ways to find the solutions of systems of two equations in two variables. You will learn how to identify solutions from the graphs of the equations, and how to find solutions by algebraic methods. Then, you will apply the algebraic methods to solving certain circuit problems by analyzing closed loops of the circuit.

OBJECTIVES

When you have finished this unit you should be able to:

1. Solve systems of two equations in two variables by graphing.

2. Solve systems of two equations in two variables algebraically by addition and by substitution.

3. Use the loop method to find currents in circuits.

SECTION 12.1

Graphical Solution

Suppose we have two lines in the Cartesian coordinate system. If the two lines cross—that is, if they are not parallel—then they cross at exactly one point. This point is called their **point of intersection**.

The equations of the two lines form a **system** of two linear equations in two variables. The ordered pair containing the coordinates of the point of intersection of the lines is the **solution of the system**. The solution of a system of two linear equations in two variables is also a solution of each of the individual equations. By drawing the graphs of each of the individual equations, and observing their point of intersection, we can find an approximate solution of the system.

EXAMPLE 12.1 ▶

Draw the graphs of the equations and find the solution of the system $x + 2y = 5$ and $x - y = 8$.

SOLUTION

First, we draw the graph of each equation by using the intercepts as in Unit 10. For the first equation, when $x = 0$,

$$x + 2y = 5$$

$$2y = 5$$

$$y = \frac{5}{2},$$

so the y-intercept is $(0, \frac{5}{2})$. When $y = 0$,

$$x + 2y = 5$$

$$x = 5,$$

so the x-intercept is $(5, 0)$. Similarly, for the second equation, when $x = 0$,

$$x - y = 8$$
$$-y = 8$$
$$y = -8,$$

so the y-intercept is $(0, -8)$. When $y = 0$,

$$x - y = 8$$
$$x = 8,$$

so the x-intercept is $(8, 0)$. We draw the graphs of both lines on one set of axes. This graph is the graph of the system:

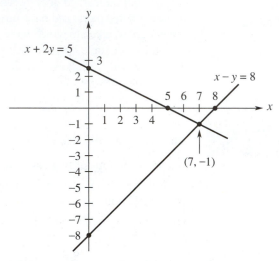

We determine the point of intersection of the two lines as closely as possible. The coordinates appear to be $x = 7$ and $y = -1$. Thus, the solution of the system appears to be the ordered pair $(7, -1)$.

We can verify this solution by substituting in each equation. For the first equation,

$$x + 2y = 5$$
$$7 + 2(-1) \overset{?}{=} 5$$
$$7 - 2 = 5.$$

For the second equation,

$$x - y = 8$$
$$7 - (-1) \overset{?}{=} 8$$
$$7 + 1 = 8.$$

Thus, the ordered pair $(7, -1)$ is a solution of each equation and is the solution of the system. ▲

You should be careful to distinguish between the words "intercept" and "intersect." Each separate line has two *intercepts*, an x-intercept and a y-intercept, independent of the other line. The two lines together *intersect*, that is, cross at a point of intersection.

EXAMPLE 12.2 ▶ Draw the graphs of the equations and find the solution of the system $x - y = 4$ and $3x + 5y = 0$.

SOLUTION For the first equation, $x - y = 4$, we find that the y-intercept is $(0, -4)$ and the x-intercept is $(4, 0)$. For the second equation, when $x = 0$, we also have $y = 0$. Both the x-intercept and the y-intercept of the equation $3x + 5y = 0$ are the origin $(0, 0)$. We may find any other solution of the equation, for example, $(5, -3)$. Then, we draw the graphs of both lines on one set of axes. The graph of the system is

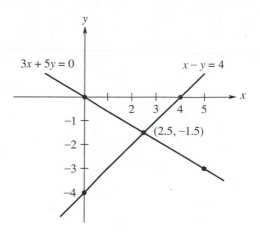

The coordinates of the point of intersection of the two lines appear to be $x = 2.5$ and $y = -1.5$. Thus, the solution of the system appears to be the ordered pair $(2.5, -1.5)$.

We cannot be sure that a solution we read from a graph is exact. The solution for this example could be slightly on either side of the point selected, for example, $(2.4, -1.6)$ or $(2.6, -1.4)$. We check the solution by substituting in each of the equations. For the first equation,

$$x - y = 4$$
$$2.5 - (-1.5) \overset{?}{=} 4$$
$$2.5 + 1.5 = 4.$$

For the second equation,

$$3x + 5y = 0$$
$$3(2.5) + 5(-1.5) \overset{?}{=} 0$$
$$7.5 - 7.5 = 0.$$

Therefore, the ordered pair $(2.5, -1.5)$ is the solution of the system.   ▲

SYSTEMS WITH NONLINEAR EQUATIONS

We can use the graphical method of solution to solve systems of equations where one or both of the equations are nonlinear. Such systems may have two or more solutions.

EXAMPLE 12.3 Draw the graphs of the equations and find the solutions of the system $y = x + 1$ and $y = \frac{1}{2}x^2$.

SOLUTION Since the first equation, $y = x + 1$, is a linear equation, we can draw its graph by using its intercepts $(0, 1)$ and $(-1, 0)$. The second equation, $y = \frac{1}{2}x^2$, is the equation of a parabola. We draw its graph by finding several points of the graph, for example, $(0, 0)$, $(2, 2)$, $(-2, 2)$, $(4, 8)$, and $(-4, 8)$. The graph of the system is shown at the top of page 268 on the left. We see that there are two points of intersection. We enlarge a portion of the graph to show the points of intersection more clearly. The enlarged portion is shown at the top of page 268 on the right:

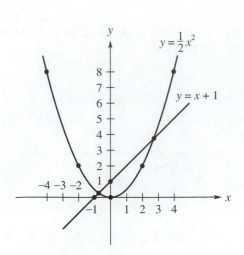

Graph for Example 12.3.

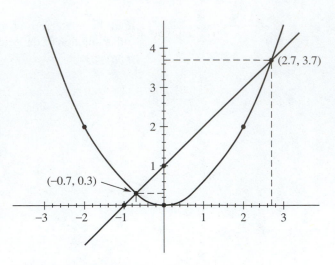

Enlarged portion showing the points of intersection.

The ordered pairs containing the coordinates of these two points appear to be (2.7, 3.7) and (− 0.7, 0.3). We check each of these solutions. For the first equation, we have

$$y = x + 1$$
$$3.7 = 2.7 + 1,$$

and

$$y = x + 1$$
$$0.3 = -0.7 + 1.$$

Thus, the ordered pairs are solutions for this equation. However, the ordered pairs must also satisfy the second equation:

$$y = \frac{1}{2}x^2$$
$$3.7 \stackrel{?}{=} \frac{1}{2}(2.7)^2$$
$$3.7 \approx 3.645,$$

and

$$y = \frac{1}{2}x^2$$
$$0.3 \stackrel{?}{=} \frac{1}{2}(-0.7)^2$$
$$0.3 \approx 0.245.$$

The ordered pairs are approximate solutions for this equation. We can estimate a decimal place between two whole numbers, but we cannot find exact solutions by this method. ▲

There are many special cases of systems of equations in two variables. If we move the line in Example 12.3 down a bit, it will touch the parabola exactly once. In this case we say the line is tangent to the parabola at the point where it touches the parabola.

EXAMPLE 12.4 Draw the graphs of the equations and find the solutions of the system $y = x - \frac{1}{2}$ and $y = \frac{1}{2}x^2$.

SOLUTION   The intercepts of the line are $(0, -0.5)$ and $(0.5, 0)$. The parabola is the same as the parabola in Example 12.3. Therefore, the graph of the system is

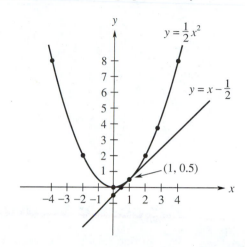

There appears to be only one solution, $(1, 0.5)$. You should check this solution in each equation to see that it is a solution of the system. The line is tangent to the parabola at the point $(1, 0.5)$.   ▲

It is possible for a system of equations in two variables to have no solution. If we move the line in Examples 12.3 and 12.4 down a bit more, it will not cross or touch the parabola at all. In this case, there are no points of intersection, and the system has no solution. If two equations both are linear equations and, when written in slope-intercept form have the same slope but different y-intercepts, then the lines are parallel. In this case also, the system has no solution.

EXERCISE 12.1   Draw the graphs of the equations and find the solutions of the system:

1. $x + 2y = 6$
   $x - 2y = 2$

2. $x - y = 5$
   $x + 2y = -4$

3. $x + 4y = 2$
   $x + y = 5$

4. $3x - 2y + 4 = 0$
   $x + y + 3 = 0$

5. $2y + x = 4$
   $2y - x = 6$

6. $x - 2y = 1$
   $2x - 2y = -1$

7. $3x + 2y = 6$
   $3x + 2y + 2 = 0$

8. $3y = 4x + 4$
   $4x = 3y + 4$

9. $y = 2x + 3$
   $y = x^2$

10. $x + 2y = 4$
    $y = \frac{1}{4}x^2$

11. $y = 2x + 2$
    $y = x^2$

12. $x + 2y = 2$
    $y = \frac{1}{4}x^2$

13. $y = 2x - 1$
  $y = x^2$

14. $2x + 4y + 1 = 0$
  $y = \dfrac{1}{4}x^2$

15. $y = 2x - 2$
  $y = x^2$

16. $2x + 4y + 3 = 0$
  $y = \dfrac{1}{4}x^2$

17. $x + y = 1$
  $x^2 + y^2 = 4$

18. $x + y = 2$
  $x^2 + y^2 = 2$

19. $x + y = 2$
  $y = \dfrac{1}{x}$

20. $y = x + 1$
  $y = \dfrac{1}{x}$

SECTION 12.2

Algebraic Solution

We can find solutions for some systems of two equations in two variables by using algebraic methods. We try to eliminate one variable, thus reducing the system to one equation in the remaining variable. Then we must be able to solve the resulting equation.

One way to eliminate a variable is by making its numerical coefficient zero. For example, if the y-term is $0y$, then the y-term will no longer appear in the equation. Suppose the coefficients of the y-terms in two equations have the *same absolute value* but *opposite signs*. If the like terms of the two equations are added, the y-term will become $0y$. This method is called the **addition method**.

EXAMPLE 12.5

Use the addition method to solve the system $x + y = 1$ and $x - y = 2$.

SOLUTION

The y-terms have the same absolute value 1, but opposite signs. We add the like terms so the y-term is $0y$:

$$
\begin{aligned}
x + \ y &= 1 \\
\underline{x - \ y} &= \underline{2} \\
2x + 0y &= 3 \\
2x &= 3.
\end{aligned}
$$

We have a resulting equation that we can solve for the remaining variable x:

$$2x = 3$$

$$x = \frac{3}{2}.$$

To find y, we may replace x by $\frac{3}{2}$ in either of the original equations. If we choose the first equation, we have

$$x + y = 1$$

$$\frac{3}{2} + y = 1$$

$$y = -\frac{1}{2}.$$

Therefore, the solution is $(\frac{3}{2}, -\frac{1}{2})$. We would get the same result by using the second equation. To check, we substitute in each equation as in Section 12.1. For the first equation,

$$x + y = 1$$

$$\frac{3}{2} + \left(-\frac{1}{2}\right) \stackrel{?}{=} 1$$

$$\frac{3}{2} - \frac{1}{2} = 1.$$

For the second equation,

$$x - y = 2$$

$$\frac{3}{2} - \left(-\frac{1}{2}\right) \stackrel{?}{=} 2$$

$$\frac{3}{2} + \frac{1}{2} = 2.$$

Of course, if you draw the graphs of the equations, you will have two lines that intersect at the point with coordinates $(\frac{3}{2}, -\frac{1}{2})$. ▲

Usually, we need to adjust one or both of the equations so that the coefficients of a variable have the same absolute value but opposite signs.

EXAMPLE 12.6 ▶

Use the addition method to solve the system $x + 2y = 5$ and $x - y = 8$.

SOLUTION

We will not eliminate either variable simply by adding the equations together. We may multiply both sides of the second equation by 2:

$$x + 2y = 5$$

$$2(x - y) = 2(8).$$

The resulting system is

$$x + 2y = 5$$

$$2x - 2y = 16.$$

Now, the y-terms have the same absolute value 2, but opposite signs. We add the like terms so the y-term is $0y$:

$$x + 2y = 5$$

$$\underline{2x - 2y = 16}$$

$$3x + 0y = 21$$

$$3x = 21$$

$$x = 7.$$

To find y, we may replace x by 7 in either equation. Using the first equation, we write

$$x + 2y = 5$$

$$7 + 2y = 5$$

$$2y = -2$$

$$y = -1.$$

Therefore, the solution is $(7, -1)$. You should check this solution by substituting in each of the original equations. Observe that this system is the same as the system in Example 12.1, and the solution $(7, -1)$ gives the coordinates of the point of intersection of the lines in that example. ▲

If the x-term is $0x$, then the x-term will no longer appear in the equation. If we choose to multiply both sides of the first equation in the preceding example by -1, we have

$$-1(x + 2y) = -1(5)$$
$$x - y = 8.$$

Then, we can solve for y:

$$-x - 2y = -5$$
$$\underline{\quad x - y = \quad 8\quad}$$
$$0x - 3y = \quad 3$$
$$-3y = \quad 3$$
$$y = -1.$$

To find x, we may replace y by -1 in either equation. Using the second equation, we write

$$x - y = 8$$
$$x - (-1) = 8$$
$$x = 7.$$

The solution is $(7, -1)$ as before.

EXAMPLE 12.7 Use the addition method to solve the system $x - y = 4$ and $3x + 5y = 0$.

SOLUTION   We can eliminate y by multiplying both sides of the first equation by 5, or we can eliminate x by multiplying both sides of the first equation by -3. If we choose to multiply the first equation by 5, we have

$$5(x - y) = 5(4)$$
$$3x + 5y = 0.$$

Then, we can solve for x:

$$5x - 5y = 20$$
$$\underline{\quad 3x + 5y = \quad 0\quad}$$
$$8x \quad = 20$$
$$x = \frac{20}{8}$$
$$x = \frac{5}{2}.$$

We replace x by $\frac{5}{2}$ in either equation. Using the first equation, we have

$$x - y = 4$$
$$\frac{5}{2} - y = 4$$
$$-y = \frac{3}{2}$$
$$y = -\frac{3}{2}.$$

Therefore, the solution is $(\frac{5}{2}, -\frac{3}{2})$. You should check this solution by substituting in each of the original equations. Also, you should solve the system by multiplying both sides of the second equation by -3. Observe that this system is the same as the system in Example 12.2. ▲

THE SUBSTITUTION METHOD

Another way to eliminate a variable is to solve one equation for one variable. Then, the resulting expression can be substituted for that variable in the other equation. This method is called the **substitution method**.

EXAMPLE 12.8

Use the substitution method to solve the system $x - y = 4$ and $3x + 5y = 0$.

SOLUTION

The first equation is easily solved for x:

$$x - y = 4$$
$$x = y + 4.$$

We substitute $y + 4$ for x in the second equation:

$$3x + 5y = 0$$
$$3(y + 4) + 5y = 0$$

Then, we remove the parentheses and solve for y:

$$3y + 12 + 5y = 0$$
$$8y = -12$$
$$y = \frac{-12}{8}$$
$$y = -\frac{3}{2}.$$

Replacing y by $-\frac{3}{2}$ in the first equation, we have

$$x = y + 4$$
$$= -\frac{3}{2} + 4$$
$$= \frac{5}{2}.$$

Therefore, the solution is $(\frac{5}{2}, -\frac{3}{2})$. This system is the same as the system in the preceding example, so the solution is the same. ▲

SYSTEMS WITH NONLINEAR EQUATIONS

Many systems of equations in two variables, where one or both of the equations are nonlinear, can also be solved by algebraic methods. Some cases of such systems result in equations that can be solved by methods in Unit 4.

EXAMPLE 12.9

Use the substitution method to solve the system $y = 2x$ and $y = \frac{1}{2}x^2$.

SOLUTION

Both equations are solved for y. Substituting $2x$ for y in the second equation, we have

$$y = \frac{1}{2}x^2$$

$$2x = \frac{1}{2}x^2,$$

and multiplying both sides by 2, we have

$$4x = x^2.$$

We consider two cases: $x = 0$ and $x \neq 0$.

1. If $x = 0$, then

$$4x = x^2$$
$$4(0) = 0^2$$
$$0 = 0.$$

Therefore, $x = 0$ is a solution. Replacing x by 0 in either of the original equations gives $y = 0$. Therefore, one solution of the system is $(0, 0)$.

2. If $x \neq 0$ then we can divide by x:

$$4x = x^2$$
$$4 = x.$$

Replacing x by 4 in either of the original equations gives $y = 8$. Therefore, a second solution of the system is $(4, 8)$.

You should check both solutions by substituting in each of the original equations, and also draw the graphs of the equations to show that the solutions are the coordinates of the points of intersection.   ▲

EXAMPLE 12.10   Use the addition method to solve the system $x^2 + y^2 = 3$ and $x^2 - y^2 = 1$.

SOLUTION Adding these equations eliminates the y^2-term:

$$x^2 + y^2 = 3$$
$$\underline{x^2 - y^2 = 1}$$
$$2x^2 \qquad = 4$$
$$x^2 = 2.$$

We recall from Unit 4 that there are two solutions for x,

$$x = \pm\sqrt{2}.$$

Replacing x by either $\sqrt{2}$ or $-\sqrt{2}$ in either of the original equations gives $y = \pm 1$. Therefore, the system has four solutions: $(\sqrt{2}, 1)$, $(\sqrt{2}, -1)$, $(-\sqrt{2}, 1)$, and $(-\sqrt{2}, -1)$.

You should check each solution in both of the original equations. The system consists of a circle and a hyperbola. The graph of the system is

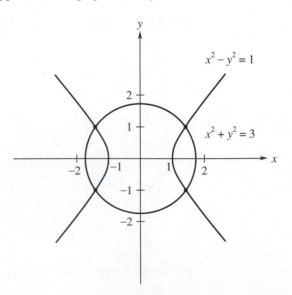

Recall from Section 12.1 that some systems have no solution. Examples are those where a line does not touch a curve at any point, and those where two lines are parallel. In these cases, the algebraic methods lead to some kind of impossible result.

EXAMPLE 12.11

Use addition or substitution to solve the system $4x - 2y = 1$ and $y = 2x - 3$.

SOLUTION

Since the second equation is solved for y, we choose the substitution method. Substituting $2x - 3$ for y in the first equation, we have

$$4x - 2y = 1$$
$$4x - 2(2x - 3) = 1$$
$$4x - 4x + 6 = 1$$
$$6 = 1.$$

Clearly, this result is not true for any value of x. Since this result is impossible, the system has no solution. We say that the system is **inconsistent**.   ▲

**EXERCISE
12.2**

Use addition or substitution to solve the system:

1. $x + y = 4$
   $x - y = 2$

2. $x + y = -1$
   $x - y = 2$

3. $x + 3y = 8$
   $x - y = -4$

4. $x + y = 3$
   $x - 2y = 9$

5. $x + y = 4$
   $4x - 2y = 1$

6. $x + y = 1$
   $3x - 6y = -3$

7. $2x + 3y = 5$
   $3x - 2y = 1$

8. $5x + 2y = 8$
   $4x - 3y = 11$

9. $3x + 2y = 7$
   $x + 4y = 4$

10. $8x + 3y = 17$
    $6x + 5y = 10$

11. $y = 4x - 2$
    $3x + y = 12$

12. $y = \frac{1}{2}x + \frac{3}{2}$
    $3x - 2y = 7$

13. $y = x - 3$
    $x - y = 5$

14. $y = \frac{2}{3}x + 3$
    $x = \frac{3}{2}y + 3$

15. $y = 2x$
    $y = x^2$

16. $y = -\frac{1}{2}x$
    $y = \frac{1}{4}x^2$

17. $y = x$
    $y = \frac{1}{x}$

18. $y = \frac{1}{4}x$
    $y = \frac{1}{x}$

19. $y = -x$

$y = \dfrac{1}{x}$

20. $y = 1$

$y = \dfrac{1}{x} + 1$

21. $x^2 + y^2 = 9$

$x^2 - y^2 = 1$

22. $x^2 + y^2 = 4$

$x^2 - y^2 = 4$

23. $x^2 + y^2 = 1$

$x^2 - y^2 = 3$

24. $x^2 - y^2 = 1$

$y^2 - x^2 = 1$

Loop Analysis

Many circuit problems can be solved by using methods that result in systems of linear equations. We will use these methods to solve problems involving circuits similar to some in Unit 8. We solved the problem in Example 8.13, involving a series-parallel circuit, by using the series and parallel resistance formulas to find R_T, and Ohm's law to find I_T. Then, we used Ohm's law again to find the voltage drops and the branch currents.

Now, we recall Kirchhoff's voltage and current laws from Unit 7. We apply Kirchhoff's laws to the circuit from Example 8.13.

EXAMPLE 12.12 ▶ Find the branch currents for this circuit:

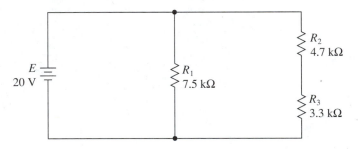

SOLUTION We draw two closed loops, each including the source E:

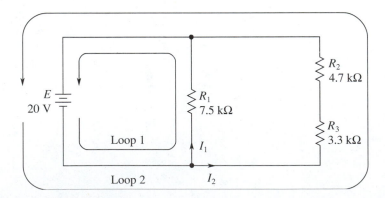

Loop 1 includes E and R_1. Loop 2, which goes around the outside of the circuit, includes E, R_2, and R_3. If we assume that the current flows from the negative terminal of E to the positive terminal of E, we can also assume that the branch currents I_1 and I_2 are directed as shown by the arrows. Now, we apply Kirchhoff's voltage law to each loop. In Loop 1, there is one voltage rise, E, and one voltage drop, V_1 across R_1. Therefore, for Loop 1,

$$E = V_1.$$

Applying Ohm's law, we have

$$E = I_1 R_1$$

$$20 \text{ V} = I_1(7.5 \text{ k}\Omega)$$

$$I_1 = 2.67 \text{ mA}.$$

In Loop 2, there is one voltage rise, E, but there are two voltage drops, V_2 across R_2 and V_3 across R_3. Therefore, for Loop 2,

$$E = V_2 + V_3$$

$$E = I_2 R_2 + I_2 R_3$$

$$20 \text{ V} = I_2(4.7 \text{ k}\Omega) + I_2(3.3 \text{ k}\Omega)$$

$$20 \text{ V} = I_2(8 \text{ k}\Omega)$$

$$I_2 = 2.5 \text{ mA}.$$

The branch currents are $I_1 = 2.67$ mA and $I_2 = 2.5$ mA. From here, we could go on to find I_T, R_T, and then V_1, V_2, and V_3 by using the methods shown in Unit 8.   ▲

In the preceding example, we did not actually solve a system of two equations in two variables because each of the two equations ended up involving only one variable. Now, we apply the same method to the circuit from Example 8.14.

EXAMPLE 12.13 ▶   Find the branch currents for this circuit:

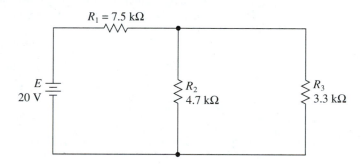

SOLUTION   We draw two closed loops, each including the source E, as in the preceding example:

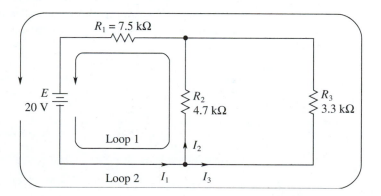

Loop 1 includes E, R_1, and R_2. Loop 2, which goes around the outside of the circuit, includes E, R_1, and R_3. If we assume that the current flows from the negative to the positive terminals of E, we can also assume that the currents I_1, I_2, and I_3 are directed as shown by the arrows.

In Loop 1, there is one voltage rise, E, and there are two drops, V_1 across R_1 and V_2 across R_2. The current through R_1 is I_1 and the current through R_2 is I_2. Therefore, for Loop 1,

$$E = V_1 + V_2$$

$$E = I_1R_1 + I_2R_2$$

$$20 \text{ V} = I_1(7.5 \text{ k}\Omega) + I_2(4.7 \text{ k}\Omega).$$

In Loop 2, there is one voltage rise, E, and there are two drops, V_1 across R_1 and V_3 across R_3. The current through R_1 is I_1 and the current through R_3 is I_3. Therefore, for Loop 2,

$$E = V_1 + V_3$$

$$E = I_1R_1 + I_3R_3$$

$$20 \text{ V} = I_1(7.5 \text{ k}\Omega) + I_3(3.3 \text{ k}\Omega).$$

If we note that the branch currents will be in milliamperes, we can write the system of equations

$$20 = 7.5I_1 + 4.7I_2$$

$$20 = 7.5I_1 + 3.3I_3.$$

Observe, however that this system is in *three* variables, $I_1, I_2,$ and I_3.

We need another relationship for the branch currents in order to solve the system. We may use Kirchhoff's current law from Unit 7. At either node of the circuit,

$$I_1 = I_2 + I_3.$$

We substitute for I_1 in each of the loop equations to obtain

$$20 = 7.5(I_2 + I_3) + 4.7I_2$$

$$20 = 7.5(I_2 + I_3) + 3.3I_3.$$

Then, simplifying each equation, we have

$$20 = 12.2I_2 + 7.5I_3$$

$$20 = 7.5I_2 + 10.8I_3.$$

This system has two equations in the two variables I_2 and I_3. To solve the system by the addition method, we can multiply both sides of the first equation by 10.8, multiply both sides of the second equation by -7.5, and add like terms:

$$216 = 131.76I_2 + 81I_3$$

$$\underline{-150 = -56.25I_2 - 81I_3}$$

$$66 = 75.51I_2$$

$$I_2 = 0.874 \text{ mA}.$$

We substitute this result in the second equation:

$$20 = 7.5I_2 + 10.8I_3$$

$$20 = 7.5(0.874) + 10.8I_3$$

$$13.445 = 10.8I_3$$

$$I_3 = 1.24 \text{ mA}.$$

Therefore, the branch currents are $I_2 = 0.874$ mA and $I_3 = 1.24$ mA. Since

$$I_1 = I_2 + I_3,$$

we also have $I_1 = 2.11$ mA.   ▲

The method we used to solve the circuit in Example 12.13 is sometimes called the **branch-current method**. In the branch-current method, we must use Kirchhoff's current law to reduce the number of variables from three to two.

THE LOOP-CURRENT METHOD

Another method is called the **loop-current** or **mesh-current method**. When we use the loop-current method to solve Example 12.13, we will get two equations in just two variables, and we will not need to use Kirchhoff's current law. We draw a closed loop, or mesh, in each "window" of the circuit:

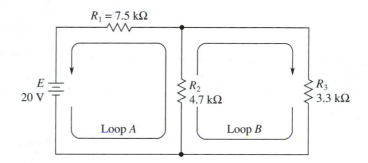

In the loop-current method, Loops A and B are not related to the branch currents. Instead, a current is associated with each loop. Thus, the current through R_1, in Loop A, is I_A. The current through R_3, in Loop B, is I_B. Since R_2 is in both Loops A and B, and both loops are in the same direction through R_2, the current through R_2 is $I_A + I_B$. Then, applying Kirchhoff's voltage law to Loop A, we have

$$E = V_1 + V_2$$

$$E = I_A R_1 + (I_A + I_B)R_2$$

$$20 = (I_A)\,7.5 + (I_A + I_B)\,4.7$$

$$20 = 7.5I_A + 4.7I_A + 4.7I_B$$

$$20 = 12.2I_A + 4.7I_B.$$

Applying Kirchhoff's voltage law to Loop B, we have

$$0 = V_2 + V_3$$

$$0 = (I_A + I_B)R_2 + I_B R_3$$

$$0 = (I_A + I_B)4.7 + (I_B)3.3$$

$$0 = 4.7I_A + 4.7I_B + 3.3I_B$$

$$0 = 4.7I_A + 8I_B.$$

Thus, we have the system of equations

$$20 = 12.2I_A + 4.7I_B$$

$$0 = 4.7I_A + 8I_B.$$

To solve the system by the substitution method, we can solve the second equation for I_B:

$$0 = 4.7I_A + 8I_B$$

$$-8I_B = 4.7I_A$$

$$I_B = -0.5875I_A.$$

Substituting this result in the first equation, we have

$$20 = 12.2I_A + 4.7I_B$$

$$20 = 12.2I_A + 4.7(-0.5875I_A)$$

$$20 = 9.43875I_A$$

$$I_A = 2.12 \text{ mA}.$$

Then, to find I_B, we can use

$$I_B = -0.5875I_A$$

$$= -0.5875(2.12 \text{ mA})$$

$$= -1.25 \text{ mA}.$$

In our diagram, Loop B is drawn so that I_B goes downward through R_3. This negative result means that the actual branch current through R_3 is

$$I_3 = 1.25 \text{ mA}$$

directed *upward*. Also, the actual branch current through R_2 is

$$I_2 = I_A + I_B$$

$$= 2.12 \text{ mA} + (-1.25 \text{ mA})$$

$$= 0.87 \text{ mA}$$

directed upward. Finally, I_1 is the same as I_A, so

$$I_1 = 2.12 \text{ mA}.$$

CIRCUITS WITH TWO SOURCES

We can solve circuits that have one voltage source by using methods from Unit 8 or by using systems of equations. Most circuits that have more than one voltage source cannot be solved by the methods in Unit 8. We can solve such circuits by using methods involving systems of equations. In the following examples, we will use the loop-current method.

EXAMPLE 12.14 ▶ Find the branch currents for this circuit:

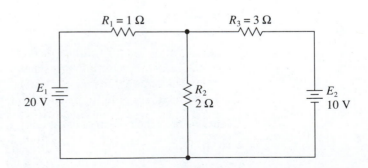

SOLUTION We draw a closed loop in each window of the circuit:

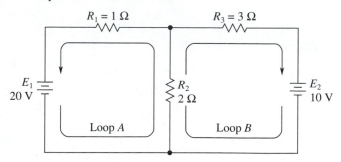

The current through R_1, in Loop A, is I_A. The current through R_3, in Loop B, is I_B. Since R_2 is in both Loops A and B, and both loops are in the same direction through R_2, the current through R_2 is $I_A + I_B$. Then, applying Kirchhoff's voltage law to Loop A, we have

$$E_1 = V_1 + V_2$$

$$E_1 = I_A R_1 + (I_A + I_B)R_2$$

$$20 = (I_A)1 + (I_A + I_B)2$$

$$20 = I_A + 2I_A + 2I_B$$

$$20 = 3I_A + 2I_B.$$

Applying Kirchhoff's voltage law to Loop B, we have

$$E_2 = V_2 + V_3$$

$$E_2 = (I_A + I_B)R_2 + I_B R_3$$

$$10 = (I_A + I_B)2 + (I_B)3$$

$$10 = 2I_A + 2I_B + 3I_B$$

$$10 = 2I_A + 5I_B.$$

Thus, we have the system of equations

$$20 = 3I_A + 2I_B$$

$$10 = 2I_A + 5I_B.$$

To solve the system by the addition method, we can multiply both sides of the first equation by 2, multiply both sides of the second equation by -3, and add like terms:

$$40 = 6I_A + 4I_B$$

$$\underline{-30 = -6I_A - 15I_B}$$

$$10 = \qquad -11I_B$$

$$I_B = -0.91 \text{ A}.$$

In our diagram, Loop B is directed **clockwise**, that is, in the same direction as the hands of a clock. The negative result means that Loop B actually goes **counterclockwise**, opposite to the direction of the hands of a clock. Thus, the current flows from the positive to the negative terminals of E_2. In practical terms, the battery E_2 is being charged.

To find I_A, we can substitute -0.91 A for I_B in the second equation:

$$10 = 2I_A + 5I_B$$

$$10 = 2I_A + 5(-0.91)$$

$$I_A = 7.28 \text{ A}.$$

To find the branch currents, we recall that $I_1 = I_A$; therefore,

$$I_1 = 7.28 \text{ A}.$$

Also, $I_2 = I_A + I_B$; therefore,

$$I_2 = 7.28 \text{ A} + (-0.91 \text{ A})$$

$$= 6.36 \text{ A}.$$

Because I_B is negative, $I_3 = I_B$ is taken oppositely; therefore,

$$I_3 = 0.91 \text{ A } \textit{to the left.}$$ ▲

EXAMPLE 12.15 ▶ Find the branch currents for this circuit:

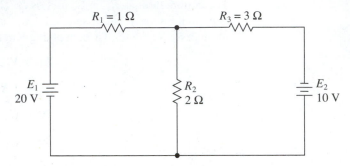

SOLUTION We draw a closed loop in each window of the circuit:

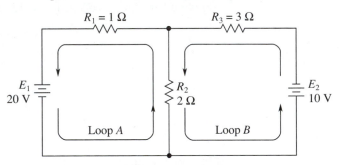

The orientation of E_2 is reversed from Example 12.14, so we have drawn Loop B counterclockwise. Loop A is directed upward through R_2 but Loop B is directed downward through R_2. Therefore, in Loop A, the current through R_2 is $I_A - I_B$. Thus, applying Kirchhoff's voltage law to Loop A, we have

$$E_1 = V_1 + V_2$$

$$E_1 = I_A R_1 + (I_A - I_B)R_2$$

$$20 = (I_A)1 + (I_A - I_B)2$$

$$20 = I_A + 2I_A - 2I_B$$

$$20 = 3I_A - 2I_B.$$

In Loop B, the current through R_2 is $I_B - I_A$. Thus, applying Kirchhoff's voltage law to Loop B, we have

$$E_2 = V_2 + V_3$$

$$E_2 = (I_B - I_A)R_2 + I_B R_3$$

$$10 = (I_B - I_A)2 + (I_B)3$$

$$10 = 2I_B - 2I_A + 3I_B$$

$$10 = -2I_A + 5I_B.$$

Thus, we have the system of equations

$$20 = 3I_A - 2I_B$$

$$10 = -2I_A + 5I_B.$$

To solve the system by the addition method, we can multiply both sides of the first equation by 2, multiply both sides of the second equation by 3, and add like terms:

$$40 = 6I_A - 4I_B$$

$$30 = -6I_A + 15I_B$$

$$\overline{70 = \qquad 11I_B}$$

$$I_B = 6.36 \text{ A}.$$

To find I_A, we can substitute 6.36 A for I_B in the first equation:

$$20 = 3I_A - 2I_B$$

$$20 = 3I_A - 2(6.36)$$

$$I_A = 10.9 \text{ A}.$$

To find the branch currents, we recall that $I_1 = I_A$; therefore

$$I_1 = 10.9 \text{ A}.$$

In Loop A, $I_2 = I_A - I_B$; therefore,

$$I_2 = 10.9 \text{ A} - 6.36 \text{ A}$$

$$= 4.54 \text{ A } upward.$$

Finally, $I_3 = I_B$; therefore,

$$I_3 = 6.36 \text{ A}.$$ ▲

EXERCISE
12.3

1. Find I_1 and I_2 for the circuit shown at the left below.
2. For the circuit in Exercise 1, $E = 5$ V, $R_1 = 1.8$ kΩ, $R_2 = 1$ kΩ, and $R_3 = 1.5$ kΩ. Find I_1 and I_2.
3. Find I_1, I_2, and I_3 for the circuit shown.
4. For the circuit in Exercise 3, $E = 5$ V, $R_1 = 1.8$ kΩ, $R_2 = 1$ kΩ, and $R_3 = 1.5$ kΩ. Find I_1, I_2, and I_3.

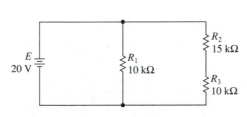

Circuit for Exercises 1 and 2.

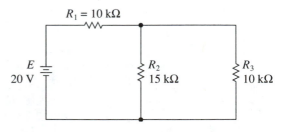

Circuit for Exercises 3 and 4.

5. Find I_1, I_2, and I_3 for the circuit shown at the left below.

6. For the circuit in Exercise 5, $E_1 = 20$ V, $E_2 = 10$ V, $R_1 = 1\ \Omega$, $R_2 = 2\ \Omega$, and $R_3 = 6\ \Omega$. Find I_1, I_2, and I_3.

7. Find I_1, I_2, and I_3 for the circuit shown.

8. For the circuit in Exercise 7, $E_1 = 10$ V, $E_2 = 6$ V, $R_1 = 5\ \Omega$, $R_2 = 2\ \Omega$, and $R_3 = 1\ \Omega$. Find I_1, I_2, and I_3.

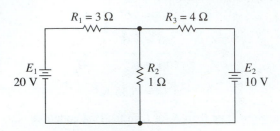

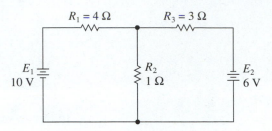

Circuit for Exercises 5 and 6.   Circuit for Exercises 7 and 8.

9. Find I_1, I_2, and I_3 for the circuit shown.
    (Hint: E_1 and E_2 each direct Loop A counterclockwise; therefore, in Loop A, $E = E_1 + E_2 = 30$ V.)

10. Find I_1, I_2, and I_3 for the circuit shown.
    (Hint: E_1 and E_2 direct Loop A oppositely; therefore, if Loop A is counterclockwise, $E = E_1 - E_2 = 10$ V.)

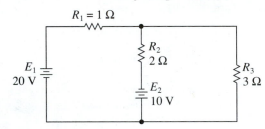

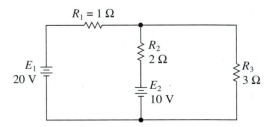

Circuit for Exercise 9.   Circuit for Exercise 10.

☐ **Self-Test** ☐

Draw the graphs of the equations and find the solutions of the system:

1. $3x - y + 3 = 0$

$x + y = 1$

2. $2x + y + 2 = 0$

$y = \dfrac{1}{2}x^2$

Use addition or substitution to solve the system:

3. $y = 2x - 1$

$4x - 2y = 1$

3. _____

4. $3x + 4y = 1$

$4x + 2y = 3$

4. _____

5. Find I_1, I_2, and I_3 for this circuit:

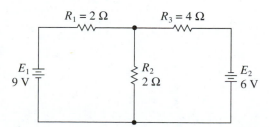

5. _____

285

Determinants

Introduction

In the preceding unit, you studied systems of linear equations in two variables. The generalization of such systems to those with several variables is the basis of an independent area of mathematics called linear algebra. From linear algebra, we get a tool called the determinant, which is used to solve systems of linear equations. In this unit, you will learn some basic techniques for using determinants. You will learn how to evaluate determinants, and how to use them to solve systems of linear equations in two variables and systems of linear equations in three variables. Then, you will apply these methods to analysis of circuits using three or more equations.

OBJECTIVES

When you have finished this unit you should be able to:

1. Evaluate second-order determinants and use second-order determinants to solve systems of linear equations in two variables.

2. Evaluate third-order determinants and use third-order determinants to solve systems of linear equations in three variables.

3. Evaluate third- and fourth-order determinants by using expansion by cofactors.

4. Use determinants to find currents in circuits by using three or more equations.

SECTION 13.1

Second-Order Determinants

A **second-order** or **two-by-two determinant** is a number that is represented by an arrangement of numbers in the form

$$\begin{vmatrix} a_1 & b_1 \\ a_2 & b_2 \end{vmatrix}.$$

The elements of the determinant are a_1, a_2, b_1 and b_2. The elements a_1 and b_1 form the first **row** of the determinant, and the elements a_2 and b_2 form the second row. The elements a_1 and a_2 form the first **column** of the determinant, and the elements b_1 and b_2 form the second column. The value of the second order determinant is given by

$$\begin{vmatrix} a_1 & b_1 \\ a_2 & b_2 \end{vmatrix} = a_1b_2 - a_2b_1.$$

The **diagonal** containing a_1 and b_2 is the **principal** or **major diagonal** and the diagonal containing a_2 and b_1 is the **secondary** or **minor diagonal**. Thus, to find the value of the determinant, we find the product of the elements in the principal diagonal and subtract the product of the elements in the secondary diagonal:

$$\begin{vmatrix} a_1 & b_1 \\ a_2 & b_2 \end{vmatrix} = a_1b_2 - a_2b_1.$$

EXAMPLE 13.1 Evaluate the determinant $\begin{vmatrix} 3 & 5 \\ 2 & 6 \end{vmatrix}$.

SOLUTION We find the product of the elements in the principal diagonal and subtract the product of the elements in the secondary diagonal:

$$\begin{vmatrix} 3 & 5 \\ 2 & 6 \end{vmatrix} = (3)(6) - (2)(5)$$

$$= 18 - 10$$

$$= 8. \qquad \blacktriangle$$

EXAMPLE 13.2 Evaluate the determinant $\begin{vmatrix} 4 & -5 \\ 2 & 1 \end{vmatrix}$.

SOLUTION The value of this determinant is

$$\begin{vmatrix} 4 & -5 \\ 2 & 1 \end{vmatrix} = (4)(1) - (2)(-5)$$

$$= 4 + 10$$

$$= 14. \qquad \blacktriangle$$

EXAMPLE 13.3 ▶ Evaluate the determinant $\begin{vmatrix} 1.2 & 3.9 \\ 3.3 & -1.5 \end{vmatrix}$.

SOLUTION The value of this determinant is

$$\begin{vmatrix} 1.2 & 3.9 \\ 3.3 & -1.5 \end{vmatrix} = (1.2)(-1.5) - (3.3)(3.9)$$

$$= -1.8 - 12.87$$

$$= -14.67. \qquad \blacktriangle$$

SYSTEMS OF EQUATIONS

Second-order determinants may be used to solve systems of two linear equations in two variables. We consider a system in the form

$$a_1x + b_1y = c_1$$

$$a_2x + b_2y = c_2.$$

To solve this system, we may multiply the first equation by b_2 and the second equation by $-b_1$:

$$b_2(a_1x + b_1y) = b_2c_1$$

$$-b_1(a_2x + b_2y) = -b_1c_2$$

or

$$a_1b_2x + b_1b_2y = b_2c_1$$

$$-a_2b_1x - b_1b_2y = -b_1c_2.$$

Adding these equations, we obtain

$$a_1b_2x - a_2b_1x = b_2c_1 - b_1c_2$$

$$(a_1b_2 - a_2b_1)x = b_2c_1 - b_1c_2$$

$$x = \frac{b_2c_1 - b_1c_2}{a_1b_2 - a_2b_1}$$

or

$$x = \frac{c_1 b_2 - c_2 b_1}{a_1 b_2 - a_2 b_1}.$$

The denominator of this solution for x is given by the determinant

$$\begin{vmatrix} a_1 & b_1 \\ a_2 & b_2 \end{vmatrix} = a_1 b_2 - a_2 b_1.$$

Similarly, the numerator of this solution for x is given by the determinant

$$\begin{vmatrix} c_1 & b_1 \\ c_2 & b_2 \end{vmatrix} = c_1 b_2 - c_2 b_1.$$

We observe that the numerator is found by replacing the elements a_1 and a_2 in the *first column* of the denominator determinant by the elements c_1 and c_2.

To solve for y, we may multiply the first equation by $-a_2$ and the second equation by a_1:

$$-a_2(a_1 x + b_1 y) = -a_2 c_1$$

$$a_1(a_2 x + b_2 y) = a_1 c_2$$

or

$$-a_1 a_2 x - a_2 b_1 y = -a_2 c_1$$

$$a_1 a_2 x + a_1 b_2 y = a_1 c_2.$$

Adding these equations we obtain

$$-a_2 b_1 y + a_1 b_2 y = -a_2 c_1 + a_1 c_2$$

$$a_1 b_2 y - a_2 b_1 y = a_1 c_2 - a_2 c_1$$

$$(a_1 b_2 - a_2 b_1) y = a_1 c_2 - a_2 c_1$$

$$y = \frac{a_1 c_2 - a_2 c_1}{a_1 b_2 - a_2 b_1}.$$

The denominator of this solution for y is also given by the determinant

$$\begin{vmatrix} a_1 & b_1 \\ a_2 & b_2 \end{vmatrix} = a_1 b_2 - a_2 b_1.$$

The numerator of this solution for y is given by the determinant

$$\begin{vmatrix} a_1 & c_1 \\ a_2 & c_2 \end{vmatrix} = a_1 c_2 - a_2 c_1.$$

We observe that the numerator is found by replacing the elements b_1 and b_2 in the *second column* of the denominator determinant by the elements c_1 and c_2.

In summary, when we have a system of two linear equations written in the form

$$a_1 x + b_1 y = c_1$$

$$a_2 x + b_2 y = c_2$$

the solutions are given by

$$x = \frac{\begin{vmatrix} c_1 & b_1 \\ c_2 & b_2 \end{vmatrix}}{D}$$

and

$$y = \frac{\begin{vmatrix} a_1 & c_1 \\ a_2 & c_2 \end{vmatrix}}{D},$$

where D is given by

$$D = \begin{vmatrix} a_1 & b_1 \\ a_2 & b_2 \end{vmatrix}.$$

The numerator of x is found by replacing the first column of D by c_1 and c_2, and the numerator of y is found by replacing the second column of D by c_1 and c_2.

EXAMPLE 13.4   Use determinants to solve the system $x + 2y = 5$ and $x - y = 8$.

SOLUTION   The system is in the form

$$x + 2y = 5$$
$$x - \ y = 8$$

so D is given by

$$D = \begin{vmatrix} a_1 & b_1 \\ a_2 & b_2 \end{vmatrix}$$

$$= \begin{vmatrix} 1 & 2 \\ 1 & -1 \end{vmatrix}$$

$$= (1)(-1) - (1)(2)$$

$$= -1 - 2$$

$$= -3.$$

Then, replacing the first column of D by c_1 and c_2, we have

$$x = \frac{\begin{vmatrix} c_1 & b_1 \\ c_2 & b_2 \end{vmatrix}}{D}$$

$$= \frac{\begin{vmatrix} 5 & 2 \\ 8 & -1 \end{vmatrix}}{-3}$$

$$= \frac{(5)(-1) - (8)(2)}{-3}$$

$$= \frac{-5 - 16}{-3}$$

$$= \frac{-21}{-3}$$

$$= 7.$$

Replacing the second column of D by c_1 and c_2, we have

$$y = \frac{\begin{vmatrix} a_1 & c_1 \\ a_2 & c_2 \end{vmatrix}}{D}$$

$$= \frac{\begin{vmatrix} 1 & 5 \\ 1 & 8 \end{vmatrix}}{-3}$$

$$= \frac{(1)(8) - (1)(5)}{-3}$$

$$= \frac{8 - 5}{-3}$$

$$= \frac{3}{-3}$$

$$= -1.$$

Thus, the solution of the system is $(7, -1)$. This system was solved by graphing in Example 12.1 and by the addition method in Example 12.6.   ▲

EXAMPLE 13.5   Use determinants to solve the system $x = y + 4$ and $3x + 5y = 0$.

SOLUTION   The system is in the form

$$x = y + 4$$
$$3x + 5y = 0.$$

We rewrite the system in the form

$$x - y = 4$$
$$3x + 5y = 0$$

so D is given by

$$D = \begin{vmatrix} a_1 & b_1 \\ a_2 & b_2 \end{vmatrix}$$

$$= \begin{vmatrix} 1 & -1 \\ 3 & 5 \end{vmatrix}$$

$$= (1)(5) - (3)(-1)$$

$$= 5 + 3$$

$$= 8.$$

Then, replacing the first column of D by c_1 and c_2, we have

$$x = \frac{\begin{vmatrix} c_1 & b_1 \\ c_2 & b_2 \end{vmatrix}}{D}$$

$$= \frac{\begin{vmatrix} 4 & -1 \\ 0 & 5 \end{vmatrix}}{8}$$

$$= \frac{(4)(5) - (0)(-1)}{8}$$

$$= \frac{20}{8}$$

$$= \frac{5}{2}.$$

Replacing the second column of D by c_1 and c_2, we have

$$y = \frac{\begin{vmatrix} a_1 & c_1 \\ a_2 & c_2 \end{vmatrix}}{D}$$

$$= \frac{\begin{vmatrix} 1 & 4 \\ 3 & 0 \end{vmatrix}}{-3}$$

$$= \frac{(1)(0) - (3)(4)}{8}$$

$$= \frac{-12}{8}$$

$$= -\frac{3}{2}.$$

Thus, the solution of the system is $(\frac{5}{2}, -\frac{3}{2})$. This system was solved by substitution in Example 12.8.   ▲

EXAMPLE 13.6   Use determinants to solve the system $3x = 2$ and $6x - 5y = -6$.

SOLUTION   The system is in the form

$$3x = 2$$
$$6x - 5y = -6.$$

Thus, we may write

$$3x + 0y = 2$$
$$6x - 5y = -6,$$

and D is given by

$$D = \begin{vmatrix} a_1 & b_1 \\ a_2 & b_2 \end{vmatrix}$$

$$= \begin{vmatrix} 3 & 0 \\ 6 & -5 \end{vmatrix}$$

$$= (3)(-5) - (6)(0)$$

$$= -15.$$

Then, replacing the first column of D by c_1 and c_2, we have

$$x = \frac{\begin{vmatrix} c_1 & b_1 \\ c_2 & b_2 \end{vmatrix}}{D}$$

$$= \frac{\begin{vmatrix} 2 & 0 \\ -6 & -5 \end{vmatrix}}{-15}$$

$$= \frac{(2)(-5) - (-6)(0)}{-15}$$

$$= \frac{-10}{-15}$$

$$= \frac{2}{3}.$$

Replacing the second column of D by c_1 and c_2, we have

$$y = \frac{\begin{vmatrix} a_1 & c_1 \\ a_2 & c_2 \end{vmatrix}}{D}$$

$$= \frac{\begin{vmatrix} 3 & 2 \\ 6 & -6 \end{vmatrix}}{-15}$$

$$= \frac{(3)(-6) - (6)(2)}{-15}$$

$$= \frac{-18 - 12}{-15}$$

$$= \frac{-30}{-15}$$

$$= 2.$$

Thus, the solution of the system is $(\frac{2}{3}, 2)$. To check this solution, we write

$$3x = 2$$

$$3\left(\frac{2}{3}\right) \stackrel{?}{=} 2$$

$$2 = 2,$$

and

$$6x - 5y = -6$$

$$6\left(\frac{2}{3}\right) - 5(2) \stackrel{?}{=} -6$$

$$4 - 10 = -6. \qquad \blacktriangle$$

EXAMPLE 13.7 Use determinants to solve the system $2.2x + 4.7y - 12 = 0$ and $6.3x - 8.2y - 2 = 0$.

SOLUTION The system is in the form

$$2.2x + 4.7y - 12 = 0$$

$$6.3x - 8.2y - 2 = 0.$$

We rewrite the system in the form

$$2.2x + 4.7y = 12$$

$$6.3x - 8.2y = 2$$

so D is given by

$$D = \begin{vmatrix} a_1 & b_1 \\ a_2 & b_2 \end{vmatrix}$$

$$= \begin{vmatrix} 2.2 & 4.7 \\ 6.3 & -8.2 \end{vmatrix}$$

$$= (2.2)(-8.2) - (6.3)(4.7)$$

$$= -18.04 - 29.61$$

$$= -47.65.$$

Then, replacing the first column of D by c_1 and c_2, we have

$$x = \frac{\begin{vmatrix} c_1 & b_1 \\ c_2 & b_2 \end{vmatrix}}{D}$$

$$= \frac{\begin{vmatrix} 12 & 4.7 \\ 2 & -8.2 \end{vmatrix}}{-47.65}$$

$$= \frac{(12)(-8.2) - (2)(4.7)}{-47.65}$$

$$= \frac{-98.4 - 9.4}{-47.65}$$

$$= \frac{-107.8}{-47.65}$$

$$= 2.26$$

to three significant digits. Replacing the second column of D by c_1 and c_2, we have

$$y = \frac{\begin{vmatrix} a_1 & c_1 \\ a_2 & c_2 \end{vmatrix}}{D}$$

$$= \frac{\begin{vmatrix} 2.2 & 12 \\ 6.3 & 2 \end{vmatrix}}{-47.65}$$

$$= \frac{(2.2)(2) - (6.3)(12)}{-47.65}$$

$$= \frac{4.4 - 75.6}{-47.65}$$

$$= \frac{-71.2}{-47.65}$$

$$= 1.49$$

to three significant digits. Thus, the solution of the system is $(2.26, 1.49)$. To check this solution, we use the original equations:

$$2.2x + 4.7y - 12 = 0$$

$$2.2(2.26) + 4.7(1.49) - 12 \stackrel{?}{=} 0$$

$$4.972 + 7.003 - 12 \stackrel{?}{=} 0$$

$$-0.025 \approx 0,$$

and

$$6.3x - 8.2y - 2 = 0$$

$$6.3(2.26) - 8.2(1.49) - 2 \stackrel{?}{=} 0$$

$$14.238 - 12.218 - 2 \stackrel{?}{=} 0$$

$$0.02 \approx 0.$$

We observe that the solution involves approximations, so we expect the checks to be approximate.   ▲

<table>
<tr><td>**EXERCISE**
13.1</td></tr>
</table>

Evaluate the determinant:

1. $\begin{vmatrix} 4 & 2 \\ 3 & 5 \end{vmatrix}$

2. $\begin{vmatrix} 10 & 15 \\ 8 & 14 \end{vmatrix}$

3. $\begin{vmatrix} 3 & -4 \\ 1 & 2 \end{vmatrix}$

4. $\begin{vmatrix} 6 & 8 \\ -9 & 12 \end{vmatrix}$

5. $\begin{vmatrix} 5 & 4 \\ -3 & 0 \end{vmatrix}$

6. $\begin{vmatrix} 8 & 12 \\ 0 & -2 \end{vmatrix}$

7. $\begin{vmatrix} 1 & -5.6 \\ 1.5 & 2.2 \end{vmatrix}$

8. $\begin{vmatrix} 1.1 & 0.63 \\ 3 & -0.75 \end{vmatrix}$

Use determinants to solve the system:

9. $2x + y = 5$
$x + 2y = 4$

10. $3x + 4y = 11$
$2x + 6y = 9$

11. $2x + 5y = 13$
$5x + 4y = 7$

12. $4x + 5y = 5$
$7x + 6y = 17$

13. $4x - 2y = 5$
$7x + 6y = 4$

14. $4x - 4y = 1$
$12x - 8y = 5$

15. $3x = y - 5$
$4x = 3y - 10$

16. $3x = 4y + 11$
$3y = 4x - 3$

17. $6x = 3$
$2x - 9y = -2$

18. $\dfrac{1}{2}y = 3$
$4x = y + 2$

19. $2x + 3y = 0$
$3x + 4y = \dfrac{1}{2}$

20. $4x - 9y = 0$
$6y + 4 = 0$

21. $x - 1.5y + 5 = 0$
$1.2x + 2.7y - 20 = 0$

22. $1.2x + 2.2y + 2 = 0$
$x + 1.5y + 2 = 0$

23. $-0.6x + 2.4y - 18 = 0$
$4.2x + 1.8y - 8 = 0$

24. $8x - 4.7y - 3 = 0$
$1.4x - 3.3y + 7 = 0$

<table>
<tr><td>**SECTION**
13.2</td></tr>
</table>

Third-Order Determinants

A **third-order** or **three-by-three determinant** is a number that is represented by an arrangement of numbers in the form

$$\begin{vmatrix} a_1 & b_1 & c_1 \\ a_2 & b_2 & c_2 \\ a_3 & b_3 & c_3 \end{vmatrix}.$$

The value of a third-order determinant is given by

$$\begin{vmatrix} a_1 & b_1 & c_1 \\ a_2 & b_2 & c_2 \\ a_3 & b_3 & c_3 \end{vmatrix} = a_1 b_2 c_3 + b_1 c_2 a_3 + c_1 a_2 b_3 - a_3 b_2 c_1 - b_3 c_2 a_1 - c_3 a_2 b_1.$$

To evaluate a third-order determinant, we may repeat the first and second columns:

$$\begin{vmatrix} a_1 & b_1 & c_1 \\ a_2 & b_2 & c_2 \\ a_3 & b_3 & c_3 \end{vmatrix} \begin{matrix} a_1 & b_1 \\ a_2 & b_2 \\ a_3 & b_3 \end{matrix}$$

Then, we draw three diagonals in the principal direction, and three diagonals in the secondary direction:

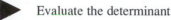

To find the value of the determinant, we add the products of the elements in the principal direction and subtract the products of the elements in the secondary direction.

EXAMPLE 13.8 ▶ Evaluate the determinant

$$\begin{vmatrix} 2 & 1 & 4 \\ 3 & 3 & 1 \\ 4 & 2 & 3 \end{vmatrix}.$$

SOLUTION   We repeat the first and second columns, and then draw three diagonals in the principal direction and three diagonals in the secondary direction:

Now, we add the products of the elements in the principal direction and subtract the products of the elements in the secondary direction:

$$\begin{vmatrix} 2 & 1 & 4 \\ 3 & 3 & 1 \\ 4 & 2 & 3 \end{vmatrix} = (2)(3)(3) + (1)(1)(4) + (4)(3)(2) - (4)(3)(4) - (2)(1)(2) - (3)(3)(1)$$

$$= 18 + 4 + 24 - 48 - 4 - 9$$

$$= -15.$$ ▲

EXAMPLE 13.9 ▶ Evaluate the determinant

$$\begin{vmatrix} 4 & -1 & 3 \\ 2 & -2 & 0 \\ 3 & 0 & 1 \end{vmatrix}.$$

SOLUTION   We redraw the determinant:

$$\begin{vmatrix} 4 & -1 & 3 \\ 2 & -2 & 0 \\ 3 & 0 & 1 \end{vmatrix} \begin{matrix} 4 & -1 \\ 2 & -2 \\ 3 & 0 \end{matrix}$$

Adding the products of the elements in the principal direction and subtracting the products of the elements in the secondary direction, we obtain the value

$$\begin{vmatrix} 4 & -1 & 3 \\ 2 & -2 & 0 \\ 3 & 0 & 1 \end{vmatrix} = (4)(-2)(1) + (-1)(0)(3) + (3)(2)(0) - (3)(-2)(3) - (0)(0)(4) - (1)(2)(-1)$$

$$= -8 + 0 + 0 + 18 - 0 + 2$$

$$= 12.$$ ▲

EXAMPLE 13.10   Evaluate the determinant

$$\begin{vmatrix} 1 & -2 & -2 \\ 0 & 2 & 3 \\ 0 & 0 & 1 \end{vmatrix}.$$

SOLUTION   We redraw the determinant:

$$\begin{array}{ccc|cc} \begin{vmatrix} 1 & -2 & -2 \\ 0 & 2 & 3 \\ 0 & 0 & 1 \end{vmatrix} & \begin{matrix} 1 & -2 \\ 0 & 2 \\ 0 & 0 \end{matrix} \end{array}$$

The value of this determinant is

$$\begin{vmatrix} 1 & -2 & -2 \\ 0 & 2 & 3 \\ 0 & 0 & 1 \end{vmatrix} = (1)(2)(1) + (-2)(3)(0) + (-2)(0)(0) - (0)(2)(-2) - (0)(3)(1) - (1)(0)(-2)$$

$$= 2. \qquad\blacktriangle$$

A determinant that has all zeros either below or above the principal diagonal is referred to as **triangular**. We observe that the value of a triangular determinant is simply the product of the elements in its principal diagonal. Furthermore, if a determinant has a row or column that contains all zeros, then its value is zero.

THREE EQUATIONS IN THREE VARIABLES

Third-order determinants may be used to solve systems of three linear equations in three variables. If we have a system of three linear equations written in the form

$$a_1x + b_1y + c_1z = d_1$$

$$a_2x + b_2y + c_2z = d_2$$

$$a_3x + b_3y + c_3z = d_3$$

the solutions are given by

$$x = \frac{\begin{vmatrix} d_1 & b_1 & c_1 \\ d_2 & b_2 & c_2 \\ d_3 & b_3 & c_3 \end{vmatrix}}{D}$$

$$y = \frac{\begin{vmatrix} a_1 & d_1 & c_1 \\ a_2 & d_2 & c_2 \\ a_3 & d_3 & c_3 \end{vmatrix}}{D}$$

$$z = \frac{\begin{vmatrix} a_1 & b_1 & d_1 \\ a_2 & b_2 & d_2 \\ a_3 & b_3 & d_3 \end{vmatrix}}{D}.$$

In these solutions, D is given by

$$D = \begin{vmatrix} a_1 & b_1 & c_1 \\ a_2 & b_2 & c_2 \\ a_3 & b_3 & c_3 \end{vmatrix}.$$

The numerator of x is found by replacing the first column of D by d_1, d_2, and d_3; the numerator of y is found by replacing the second column of D by d_1, d_2, and d_3; and the numerator of z is found by replacing the third column of D by d_1, d_2, and d_3.

EXAMPLE 13.11 Use determinants to solve the system $3x - 2y + 2z = 1$, $x + 3y - 2z = 2$, and $4x - y + z = 3$.

SOLUTION The system is in the form

$$3x - 2y + 2z = 1$$

$$x + 3y - 2z = 2$$

$$4x - y + z = 3$$

so D is given by

$$D = \begin{vmatrix} a_1 & b_1 & c_1 \\ a_2 & b_2 & c_2 \\ a_3 & b_3 & c_3 \end{vmatrix}$$

$$= \begin{vmatrix} 3 & -2 & 2 \\ 1 & 3 & -2 \\ 4 & -1 & 1 \end{vmatrix} \begin{matrix} 3 & -2 \\ 1 & 3 \\ 4 & -1 \end{matrix}$$

$$= (3)(3)(1) + (-2)(-2)(4) + (2)(1)(-1) - (4)(3)(2) - (-1)(-2)(3) - (1)(1)(-2)$$

$$= 9 + 16 - 2 - 24 - 6 + 2$$

$$= -5.$$

Then, replacing the first column of D by d_1, d_2, and d_3, we have

$$x = \frac{\begin{vmatrix} d_1 & b_1 & c_1 \\ d_2 & b_2 & c_2 \\ d_3 & b_3 & c_3 \end{vmatrix}}{D}$$

$$= \frac{\begin{vmatrix} 1 & -2 & 2 \\ 2 & 3 & -2 \\ 3 & -1 & 1 \end{vmatrix} \begin{matrix} 1 & -2 \\ 2 & 3 \\ 3 & -1 \end{matrix}}{-5}$$

$$= \frac{(1)(3)(1) + (-2)(-2)(3) + (2)(2)(-1) - (3)(3)(2) - (-1)(-2)(1) - (1)(2)(-2)}{-5}$$

$$= \frac{3 + 12 - 4 - 18 - 2 + 4}{-5}$$

$$= \frac{-5}{-5}$$

$$= 1.$$

Replacing the second column of D by d_1, d_2, and d_3 we have

$$y = \frac{\begin{vmatrix} a_1 & d_1 & c_1 \\ a_2 & d_2 & c_2 \\ a_3 & d_3 & c_3 \end{vmatrix}}{D}$$

$$= \frac{\begin{vmatrix} 3 & 1 & 2 \\ 1 & 2 & -2 \\ 4 & 3 & 1 \end{vmatrix} \begin{matrix} 3 & 1 \\ 1 & 2 \\ 4 & 3 \end{matrix}}{-5}$$

$$= \frac{(3)(2)(1) + (1)(-2)(4) + (2)(1)(3) - (4)(2)(2) - (3)(-2)(3) - (1)(1)(1)}{-5}$$

$$= \frac{6 - 8 + 6 - 16 + 18 - 1}{-5}$$

$$= \frac{5}{-5}$$

$$= -1.$$

Replacing the third column of D by d_1, d_2, and d_3 we have

$$z = \frac{\begin{vmatrix} a_1 & b_1 & d_1 \\ a_2 & b_2 & d_2 \\ a_3 & b_3 & d_3 \end{vmatrix}}{D}$$

$$= \frac{\begin{vmatrix} 3 & -2 & 1 \\ 1 & 3 & 2 \\ 4 & -1 & 3 \end{vmatrix} \begin{matrix} 3 & -2 \\ 1 & 3 \\ 4 & -1 \end{matrix}}{-5}$$

$$= \frac{(3)(3)(3) + (-2)(2)(4) + (1)(1)(-1) - (4)(3)(1) - (-1)(2)(3) - (3)(1)(-2)}{-5}$$

$$= \frac{27 - 16 - 1 - 12 + 6 + 6}{-5}$$

$$= \frac{10}{-5}$$

$$= -2.$$

Thus, the solution of the system is $(1, -1, -2)$. To check this solution, for the first equation we write

$$3x - 2y + 2z = 1$$

$$3(1) - 2(-1) + 2(-2) \overset{?}{=} 1$$

$$3 + 2 - 4 = 1,$$

for the second equation we write

$$x + 3y - 2z = 2$$

$$1 + 3(-1) - 2(-2) \overset{?}{=} 2$$

$$1 - 3 + 4 = 2,$$

and for the third equation we write

$$4x - y + z = 3$$

$$4(1) - (-1) + (-2) \overset{?}{=} 3$$

$$4 + 1 - 2 = 3.$$ ▲

EXAMPLE 13.12 Use determinants to solve the system $x - 2z = -1$, $2x - 3y = 3$, and $y - z = -1$.

SOLUTION The system is in the form

$$x \quad\quad - 2z = -1$$

$$2x - 3y \quad\quad = \quad 3$$

$$y - z = -1$$

so D is given by

$$D = \begin{vmatrix} a_1 & b_1 & c_1 \\ a_2 & b_2 & c_2 \\ a_3 & b_3 & c_3 \end{vmatrix}$$

$$= \begin{vmatrix} 1 & 0 & -2 \\ 2 & -3 & 0 \\ 0 & 1 & -1 \end{vmatrix} \begin{matrix} 1 & 0 \\ 2 & -3 \\ 0 & 1 \end{matrix}$$

Writing only the nonzero terms, we have

$$D = (1)(-3)(-1) + (-2)(2)(1)$$

$$= 3 - 4$$

$$= -1.$$

Then, replacing the first column of D by d_1, d_2, and d_3, we have

$$x = \frac{\begin{vmatrix} d_1 & b_1 & c_1 \\ d_2 & b_2 & c_2 \\ d_3 & b_3 & c_3 \end{vmatrix}}{D}$$

$$= \frac{\begin{vmatrix} -1 & 0 & -2 \\ 3 & -3 & 0 \\ -1 & 1 & -1 \end{vmatrix} \begin{matrix} -1 & 0 \\ 3 & -3 \\ -1 & 1 \end{matrix}}{-1}$$

$$= \frac{(-1)(-3)(-1) + (-2)(3)(1) - (-1)(-3)(-2)}{-1}$$

$$= \frac{-3 - 6 + 6}{-1}$$

$$= \frac{-3}{-1}$$

$$= 3.$$

Replacing the second column of D by d_1, d_2, and d_3, we have

$$y = \frac{\begin{vmatrix} a_1 & d_1 & c_1 \\ a_2 & d_2 & c_2 \\ a_3 & d_3 & c_3 \end{vmatrix}}{D}$$

$$= \frac{\begin{vmatrix} 1 & -1 & -2 \\ 2 & 3 & 0 \\ 0 & -1 & -1 \end{vmatrix} \begin{matrix} 1 & -1 \\ 2 & 3 \\ 0 & -1 \end{matrix}}{-1}$$

$$= \frac{(1)(3)(-1) + (-2)(2)(-1) - (-1)(2)(-1)}{-1}$$

$$= \frac{-3 + 4 - 2}{-1}$$

$$= \frac{-1}{-1}$$

$$= 1.$$

Replacing the third column of D by d_1, d_2, and d_3, we have

$$z = \frac{\begin{vmatrix} a_1 & b_1 & d_1 \\ a_2 & b_2 & d_2 \\ a_3 & b_3 & d_3 \end{vmatrix}}{D}$$

$$= \frac{\begin{vmatrix} 1 & 0 & -1 \\ 2 & -3 & 3 \\ 0 & 1 & -1 \end{vmatrix} \begin{matrix} 1 & 0 \\ 2 & -3 \\ 0 & 1 \end{matrix}}{-1}$$

$$= \frac{(1)(-3)(-1) + (-1)(2)(1) - (1)(3)(1)}{-1}$$

$$= \frac{3 - 2 - 3}{-1}$$

$$= \frac{-2}{-1}$$

$$= 2.$$

Thus, the solution of the system is $(3, 1, 2)$. You should check this solution by substituting in each of the original equations. ▲

EXERCISE 13.2

Evaluate the determinant:

1. $\begin{vmatrix} 1 & 3 & 2 \\ 2 & 1 & 1 \\ 3 & 4 & 2 \end{vmatrix}$

2. $\begin{vmatrix} 2 & 2 & 3 \\ 1 & 5 & 1 \\ 4 & 2 & 5 \end{vmatrix}$

3. $\begin{vmatrix} 1 & 2 & 3 \\ -1 & -3 & 1 \\ -1 & 2 & -4 \end{vmatrix}$

4. $\begin{vmatrix} -5 & 1 & -3 \\ 1 & -1 & 2 \\ -2 & 1 & -4 \end{vmatrix}$

5. $\begin{vmatrix} 2 & 0 & 1 \\ 0 & -1 & 2 \\ 3 & -4 & 4 \end{vmatrix}$

6. $\begin{vmatrix} 2 & -5 & 2 \\ -2 & 1 & 0 \\ 0 & 5 & -1 \end{vmatrix}$

7. $\begin{vmatrix} 3 & 2 & 4 \\ 0 & 1 & 2 \\ 0 & 5 & 6 \end{vmatrix}$

8. $\begin{vmatrix} 10 & 3 & 7 \\ -2 & 0 & 0 \\ 12 & 2 & 6 \end{vmatrix}$

9. $\begin{vmatrix} 3 & 5 & 1 \\ 0 & -2 & 5 \\ 0 & 0 & 3 \end{vmatrix}$

10. $\begin{vmatrix} 1 & 2 & -2 \\ 2 & -3 & 0 \\ 4 & 0 & 0 \end{vmatrix}$

11. $\begin{vmatrix} 2 & 1 & 3 \\ 0 & 0 & 0 \\ 1 & 3 & 5 \end{vmatrix}$

12. $\begin{vmatrix} 3 & -1 & 0 \\ -5 & 3 & 0 \\ 6 & 4 & 0 \end{vmatrix}$

Use determinants to solve the system:

13. $2x - y - z = 4$

$x - 2y + z = -1$

$3x - 3y + 2z = 1$

14. $2x - 2y + 3z = 2$

$x - 2y + 2z = 3$

$-x + 3y + 2z = 4$

15. $2x - 3z = 7$

$2x - 3y = 4$

$2y - z = -1$

16. $y - 3z = -3$

$x - z = 1$

$3x + y = 3$

17. $2x + 2y - z = 2$

$\quad\; 3x - y = 5$

$\quad\; 3x - z = 2$

18. $y - 2z = -2$

$\quad\; x - 2z = 1$

$\quad\; x + 4y + 4z = 0$

19. $2x - 3y = 0$

$\quad\; 2x + 3z = 4$

$\quad\; 6y - 3z = -1$

20. $5x - 4y + 2z = 0$

$\quad\; 3y - 5z = 3$

$\quad\; -5y + 4z = 8$

SECTION 13.3

Expansion by Cofactors

The pattern we have used to find values of third-order determinants does not work for higher-order determinants. There is a second method that, although sometimes lengthy, works for determinants of any order. This second method uses the **minors** of elements in any row or column of the determinant. The minor of an element is a determinant found by deleting the row and column in which the element is found. For example, to find the minor of a_1 in a third-order determinant, we delete the first row and the first column:

$$\begin{vmatrix} a_1 & b_1 & c_1 \\ a_2 & b_2 & c_2 \\ a_3 & b_3 & c_3 \end{vmatrix}.$$

Thus, the minor of a_1 is the determinant

$$\begin{vmatrix} b_2 & c_2 \\ b_3 & c_3 \end{vmatrix}.$$

To find the minor of c_2 in a third-order determinant, we delete the second row and the third column:

$$\begin{vmatrix} a_1 & b_1 & c_1 \\ a_2 & b_2 & c_2 \\ a_3 & b_3 & c_3 \end{vmatrix}.$$

Thus, the minor of c_2 is the determinant

$$\begin{vmatrix} a_1 & b_1 \\ a_3 & b_3 \end{vmatrix}.$$

We observe that the order of the minor is one less than the order of the original.

Each minor of a determinant has a sign factor. The sign factors for third-order determinants follow the pattern

$$\begin{vmatrix} + & - & + \\ - & + & - \\ + & - & + \end{vmatrix}.$$

The minor of an element, with its sign, is called the **cofactor** of the element.

To **expand** a third-order determinant by cofactors of a given row or column, we add the product of each element in the row or column with its cofactor. For example, to expand a third-order determinant by cofactors of its first row, we write

$$\begin{vmatrix} a_1 & b_1 & c_1 \\ a_2 & b_2 & c_2 \\ a_3 & b_3 & c_3 \end{vmatrix} = a_1 \begin{vmatrix} b_2 & c_2 \\ b_3 & c_3 \end{vmatrix} - b_1 \begin{vmatrix} a_2 & c_2 \\ a_3 & c_3 \end{vmatrix} + c_1 \begin{vmatrix} a_2 & b_2 \\ a_3 & b_3 \end{vmatrix}.$$

Now, evaluating each of the resulting second-order determinants, we obtain

$$a_1 \begin{vmatrix} b_2 & c_2 \\ b_3 & c_3 \end{vmatrix} - b_1 \begin{vmatrix} a_2 & c_2 \\ a_3 & c_3 \end{vmatrix} + c_1 \begin{vmatrix} a_2 & b_2 \\ a_3 & b_3 \end{vmatrix} = a_1(b_2c_3 - b_3c_2) - b_1(a_2c_3 - a_3c_2) + c_1(a_2b_3 - a_3b_2)$$

$$= a_1b_2c_3 - a_1b_3c_2 - b_1a_2c_3 + b_1a_3c_2 + c_1a_2b_3 - c_1a_3b_2$$

$$= a_1b_2c_3 + b_1a_3c_2 + c_1a_2b_3 - c_1a_3b_2 - a_1b_3c_2 - b_1a_2c_3$$

$$= a_1b_2c_3 + b_1c_2a_3 + c_1a_2b_3 - a_3b_2c_1 - b_3c_2a_1 - c_3a_2b_1.$$

This result is the same as our previous definition of the value of a third-order determinant. We may expand by cofactors of any row or column with the same result.

EXAMPLE 13.13 ▶ Use cofactors to evaluate the determinant

$$\begin{vmatrix} 2 & 1 & 4 \\ 3 & 3 & 1 \\ 4 & 2 & 3 \end{vmatrix}.$$

SOLUTION We may expand by cofactors of any row or column. To expand by cofactors of the first row, we write

$$\begin{vmatrix} 2 & 1 & 4 \\ 3 & 3 & 1 \\ 4 & 2 & 3 \end{vmatrix} = 2 \begin{vmatrix} 3 & 1 \\ 2 & 3 \end{vmatrix} - 1 \begin{vmatrix} 3 & 1 \\ 4 & 3 \end{vmatrix} + 4 \begin{vmatrix} 3 & 3 \\ 4 & 2 \end{vmatrix}$$

$$= 2[(3)(3) - (2)(1)] - 1[(3)(3) - (4)(1)] + 4[(3)(2) - (4)(3)]$$

$$= 2(9 - 2) - 1(9 - 4) + 4(6 - 12)$$

$$= 2(7) - 1(5) + 4(-6)$$

$$= 14 - 5 - 24$$

$$= -15.$$

We may expand by cofactors of any row or column. For example, suppose we expand by cofactors of the second row. Remembering to use the correct sign pattern, we obtain

$$\begin{vmatrix} 2 & 1 & 4 \\ 3 & 3 & 1 \\ 4 & 2 & 3 \end{vmatrix} = -3 \begin{vmatrix} 1 & 4 \\ 2 & 3 \end{vmatrix} + 3 \begin{vmatrix} 2 & 4 \\ 4 & 3 \end{vmatrix} - 1 \begin{vmatrix} 2 & 1 \\ 4 & 2 \end{vmatrix}$$

$$= -3[(1)(3) - (2)(4)] + 3[(2)(3) - (4)(4)] - 1[(2)(2) - (4)(1)]$$

$$= -3(3 - 8) + 3(6 - 16) - 1(4 - 4)$$

$$= -3(-5) + 3(-10) - 1(0)$$

$$= 15 - 30$$

$$= -15.$$

Of course, this result agrees with our first result in this example. It also agrees with the result we obtained by evaluating this determinant by diagonals in Example 13.8. ▲

EXAMPLE 13.14 ▶ Use cofactors to evaluate the determinant

$$\begin{vmatrix} 4 & -1 & 3 \\ 2 & -2 & 0 \\ 3 & 0 & 1 \end{vmatrix}.$$

SOLUTION When a determinant contains zeros, it is easiest to expand by the row or column containing the most zeros. To expand by cofactors of the second row, remembering to use the correct sign pattern, we write

$$\begin{vmatrix} 4 & -1 & 3 \\ 2 & -2 & 0 \\ 3 & 0 & 1 \end{vmatrix} = -2\begin{vmatrix} -1 & 3 \\ 0 & 1 \end{vmatrix} + (-2)\begin{vmatrix} 4 & 3 \\ 3 & 1 \end{vmatrix} - 0\begin{vmatrix} 4 & -1 \\ 3 & 0 \end{vmatrix}$$

$$= -2[(-1)(1) - (0)(3)] - 2[(4)(1) - (3)(3)] - 0[(4)(0) - (3)(-1)]$$

$$= -2(-1) - 2(4 - 9) - 0(3)$$

$$= 2 - 2(-5)$$

$$= 2 + 10$$

$$= 12.$$

This result agrees with the result we obtained by evaluating this determinant by diagonals in Example 13.9. ▲

EXAMPLE 13.15 ▶ Use cofactors to evaluate the determinant

$$\begin{vmatrix} 1 & -2 & -2 \\ 0 & 2 & 3 \\ 0 & 0 & 1 \end{vmatrix}.$$

SOLUTION This determinant has two zeros in the first column and in the third row. To expand by cofactors of the first column, we write

$$\begin{vmatrix} 1 & -2 & -2 \\ 0 & 2 & 3 \\ 0 & 0 & 1 \end{vmatrix} = 1\begin{vmatrix} 2 & 3 \\ 0 & 1 \end{vmatrix} - 0\begin{vmatrix} -2 & -2 \\ 0 & 1 \end{vmatrix} + 0\begin{vmatrix} -2 & -2 \\ 2 & 3 \end{vmatrix}.$$

Since the second two parts are zero, we need only write

$$\begin{vmatrix} 1 & -2 & -2 \\ 0 & 2 & 3 \\ 0 & 0 & 1 \end{vmatrix} = 1\begin{vmatrix} 2 & 3 \\ 0 & 1 \end{vmatrix}$$

$$= 1[(2)(1) - (0)(3)]$$

$$= 1(2)$$

$$= 2.$$

This result, which is the product of the elements in the principal diagonal, agrees with our result in Example 13.10. ▲

EXAMPLE 13.16 ▶ Use cofactors to evaluate the determinant

$$\begin{vmatrix} 2 & 0 & 3 \\ -4 & 0 & -1 \\ 1 & 0 & 2 \end{vmatrix}.$$

SOLUTION This determinant contains all zeros in the second column. To expand by cofactors of the second column, we write

$$\begin{vmatrix} 2 & 0 & 3 \\ -4 & 0 & -1 \\ 1 & 0 & 2 \end{vmatrix} = -0\begin{vmatrix} -4 & -1 \\ 1 & 2 \end{vmatrix} + 0\begin{vmatrix} 2 & 3 \\ 1 & 2 \end{vmatrix} - 0\begin{vmatrix} 2 & 3 \\ -4 & -1 \end{vmatrix}.$$

Since all parts are zero, we have

$$\begin{vmatrix} 2 & 0 & 3 \\ -4 & 0 & -1 \\ 1 & 0 & 2 \end{vmatrix} = 0.$$

If a determinant has a row or column that contains all zeros, then its value is zero. ▲

HIGHER-ORDER DETERMINANTS

Determinants of any order can be evaluated by the method of expansion by cofactors. The sign factors for fourth-order determinants follow the pattern

$$\begin{vmatrix} + & - & + & - \\ - & + & - & + \\ + & - & + & - \\ - & + & - & + \end{vmatrix}.$$

The pattern of sign factors can be extended similarly to any order:

$$\begin{vmatrix} + & - & + & - & + & \cdots \\ - & + & - & + & - & \cdots \\ + & - & + & - & + & \cdots \\ - & + & - & + & - & \cdots \\ + & - & + & - & + & \cdots \\ \vdots & \vdots & \vdots & \vdots & \vdots & \end{vmatrix}.$$

EXAMPLE 13.17 ▶

Use cofactors to evaluate the determinant

$$\begin{vmatrix} 1 & 0 & -1 & 2 \\ 0 & -1 & 2 & 0 \\ 3 & -3 & 1 & -1 \\ -1 & 0 & -2 & 1 \end{vmatrix}.$$

SOLUTION

We may expand by cofactors of any row or column. The second row and second column each contain two zeros. To expand by cofactors of the second row, remembering to use the correct sign pattern, we write

$$\begin{vmatrix} 1 & 0 & -1 & 2 \\ 0 & -1 & 2 & 0 \\ 3 & -3 & 1 & -1 \\ -1 & 0 & -2 & 1 \end{vmatrix} = -0\begin{vmatrix} 0 & -1 & 2 \\ -3 & 1 & -1 \\ 0 & -2 & 1 \end{vmatrix} + (-1)\begin{vmatrix} 1 & -1 & 2 \\ 3 & 1 & -1 \\ -1 & -2 & 1 \end{vmatrix} - 2\begin{vmatrix} 1 & 0 & 2 \\ 3 & -3 & -1 \\ -1 & 0 & 1 \end{vmatrix} + 0\begin{vmatrix} 1 & 0 & -1 \\ 3 & -3 & 1 \\ -1 & 0 & -2 \end{vmatrix}.$$

The first and last parts are zero, thus we need only write the terms

$$\begin{vmatrix} 1 & 0 & -1 & 2 \\ 0 & -1 & 2 & 0 \\ 3 & -3 & 1 & -1 \\ -1 & 0 & -2 & 1 \end{vmatrix} = -1\begin{vmatrix} 1 & -1 & 2 \\ 3 & 1 & -1 \\ -1 & -2 & 1 \end{vmatrix} - 2\begin{vmatrix} 1 & 0 & 2 \\ 3 & -3 & -1 \\ -1 & 0 & 1 \end{vmatrix}.$$

Expanding the third-order determinants either by diagonals or by cofactors, we obtain

$$\begin{vmatrix} 1 & 0 & -1 & 2 \\ 0 & -1 & 2 & 0 \\ 3 & -3 & 1 & -1 \\ -1 & 0 & -2 & 1 \end{vmatrix} = -1(-9) - 2(-9)$$

$$= 27.$$

You should try expanding this determinant by cofactors of the second column. ▲

EXAMPLE 13.18 ▶

Use determinants to solve the system $2x + y - z = 0, x + 2y + u = 1,$ $x - z + 2u = 1$, and $3y + 2u = -1$.

SOLUTION

The system is in the form

$$\begin{aligned} 2x + y - z & = 0 \\ x + 2y \quad + u & = 1 \\ x \quad - z + 2u & = 1 \\ 3y \quad + 2u & = -1 \end{aligned}$$

so D is given by

$$D = \begin{vmatrix} 2 & 1 & -1 & 0 \\ 1 & 2 & 0 & 1 \\ 1 & 0 & -1 & 2 \\ 0 & 3 & 0 & 2 \end{vmatrix}.$$

The fourth row and third column each contain two zeros. To expand by cofactors of the fourth row, we write

$$D = 3\begin{vmatrix} 2 & -1 & 0 \\ 1 & 0 & 1 \\ 1 & -1 & 2 \end{vmatrix} + 2\begin{vmatrix} 2 & 1 & -1 \\ 1 & 2 & 0 \\ 1 & 0 & -1 \end{vmatrix}.$$

Expanding the third-order determinants either by diagonals or by cofactors, we obtain

$$D = 3(3) + 2(-1)$$

$$= 7.$$

To find x, we replace the first column of D by the constants on the right-hand sides of the equations:

$$x = \frac{\begin{vmatrix} 0 & 1 & -1 & 0 \\ 1 & 2 & 0 & 1 \\ 1 & 0 & -1 & 2 \\ -1 & 3 & 0 & 2 \end{vmatrix}}{7}.$$

Now, the first row in the determinant has two zeros. Remembering to use the correct sign pattern, we write

$$\begin{vmatrix} 0 & 1 & -1 & 0 \\ 1 & 2 & 0 & 1 \\ 1 & 0 & -1 & 2 \\ -1 & 3 & 0 & 2 \end{vmatrix} = -1\begin{vmatrix} 1 & 0 & 1 \\ 1 & -1 & 2 \\ -1 & 0 & 2 \end{vmatrix} + (-1)\begin{vmatrix} 1 & 2 & 1 \\ 1 & 0 & 2 \\ -1 & 3 & 2 \end{vmatrix}$$

$$= -1(-3) + (-1)(-11)$$

$$= 14.$$

Therefore,

$$x = \frac{14}{7}$$

$$= 2.$$

To find y, we replace the second column of D by the constants on the right-hand sides of the equations:

$$y = \frac{\begin{vmatrix} 2 & 0 & -1 & 0 \\ 1 & 1 & 0 & 1 \\ 1 & 1 & -1 & 2 \\ 0 & -1 & 0 & 2 \end{vmatrix}}{7}.$$

Again using the first row, we write

$$\begin{vmatrix} 2 & 0 & -1 & 0 \\ 1 & 1 & 0 & 1 \\ 1 & 1 & -1 & 2 \\ 0 & -1 & 0 & 2 \end{vmatrix} = 2\begin{vmatrix} 1 & 0 & 1 \\ 1 & -1 & 2 \\ -1 & 0 & 2 \end{vmatrix} + (-1)\begin{vmatrix} 1 & 1 & 1 \\ 1 & 1 & 2 \\ 0 & -1 & 2 \end{vmatrix}$$

$$= 2(-3) + (-1)(1)$$

$$= -7.$$

Therefore,

$$y = \frac{-7}{7}$$

$$= -1.$$

At this point, we can find z and u by substitution. Using the first equation,

$$2x + y - z = 0$$

$$2(2) + (-1) - z = 0$$

$$3 - z = 0$$

$$z = 3.$$

Using the second equation,

$$x + 2y + u = 1$$

$$2 + 2(-1) + u = 1$$

$$2 - 2 + u = 1$$

$$u = 1.$$

Thus, the solution of the system is $x = 2$, $y = -1$, $z = 3$, and $u = 1$. You should find z and u by using determinants, and also check the solution by substituting in each of the original equations. ▲

Expansion by determinants is clearly easiest when there is a row or column that has many zeros. There are several properties of determinants that can be used to simplify determinants and to create zeros. These properties and their use are covered in some college algebra books, and in books on the field of mathematics called linear algebra.

EXERCISE 13.3

1.–12. Evaluate the determinants in Exercise 13.2, #1–#12, on page 301, by using cofactors.

Evaluate the determinant:

13.
$$\begin{vmatrix} 0 & 1 & 2 & 1 \\ -1 & 2 & 0 & 1 \\ 2 & 0 & 3 & -2 \\ -2 & 3 & -2 & 0 \end{vmatrix}$$

14.
$$\begin{vmatrix} 1 & 1 & 1 & 1 \\ 2 & 0 & -1 & 3 \\ -2 & 0 & 0 & 2 \\ 0 & 3 & 2 & -1 \end{vmatrix}$$

15.
$$\begin{vmatrix} 1 & -2 & 4 & 3 \\ 0 & 3 & -1 & 2 \\ 0 & 0 & 4 & -3 \\ 0 & 0 & 0 & -1 \end{vmatrix}$$

16.
$$\begin{vmatrix} 0 & 0 & 0 & 8 \\ 0 & 0 & -7 & 5 \\ 0 & 6 & 5 & -4 \\ 5 & -4 & 8 & -7 \end{vmatrix}$$

17.–24. Solve the systems in Exercise 13.2, #13–#20, on pages 301–302, by using cofactors to evaluate the determinants.

Use determinants to solve the system:

25. $x + y + z = 2$

$\quad\ x - z + u = 0$

$\quad\ y - z + u = 1$

$\quad\ y + z = 1$

26. $x + y + z = 3$

$\quad\ x + 2y + u = 4$

$\quad\ x + 2z + u = 4$

$\quad\ y + z + u = 2$

27. $x + 2y = -2$

$\quad\ y + 2z = 1$

$\quad\ z + 2u = 0$

$\quad\ x + 2u = 2$

28. $x - y + z - u = 3$

$\quad\ x + 2z = 3$

$\quad\ y - 2u = -1$

$\quad\ x + 2y = -1$

Loop Analysis Revisited

In Section 12.3, we used the branch-current and the loop-current methods to find currents in circuits. These methods lead to systems of two or more equations, where the variables are the currents. We can solve systems arising from such methods by using determinants.

EXAMPLE 13.19 ▶ Find the branch currents for this circuit:

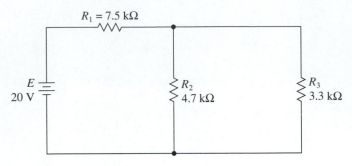

SOLUTION To use the branch-current method given in Section 12.3, we draw two closed loops, each including the source E. We also indicate directions for the branch currents at a node of the circuit:

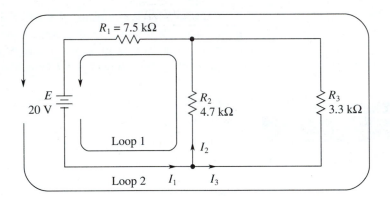

Then, applying Kirchhoff's voltage law to Loop 1, we have

$$E = V_1 + V_2$$

$$E = I_1R_1 + I_2R_2$$

$$20 = 7.5I_1 + 4.7I_2.$$

Similarly, in Loop 2, we have

$$E = V_1 + V_3$$

$$E = I_1R_1 + I_3R_3$$

$$20 = 7.5I_1 + 3.3I_3.$$

Thus, we have two equations in the three variables I_1, I_2, and I_3. To find a third equation, we use Kirchhoff's current law to write

$$I_1 = I_2 + I_3.$$

Now, we have a system of three equations in three variables:

$$I_1 - I_2 - I_3 = 0$$

$$7.5I_1 + 4.7I_2 \qquad = 20$$

$$7.5I_1 \qquad + 3.3I_3 = 20.$$

To solve the system by using determinants, D is given by

$$D = \begin{vmatrix} 1 & -1 & -1 \\ 7.5 & 4.7 & 0 \\ 7.5 & 0 & 3.3 \end{vmatrix}.$$

Expanding this determinant either by diagonals or by cofactors, we obtain

$$D = 75.51.$$

To find I_1, we replace the first column of D by the constants on the right-hand sides of the equations:

$$I_1 = \frac{\begin{vmatrix} 0 & -1 & -1 \\ 20 & 4.7 & 0 \\ 20 & 0 & 3.3 \end{vmatrix}}{75.51}$$

$$= \frac{160}{75.51}$$

$$= 2.12 \text{ mA}$$

to three significant digits. To find I_2, we replace the second column of D by the constants on the right-hand sides of the equations:

$$I_2 = \frac{\begin{vmatrix} 1 & 0 & -1 \\ 7.5 & 20 & 0 \\ 7.5 & 20 & 3.3 \end{vmatrix}}{75.51}$$

$$= \frac{66}{75.51}$$

$$= 0.874 \text{ mA}$$

to three significant digits. To find I_3, we replace the third column of D by the constants on the right-hand sides of the equations:

$$I_3 = \frac{\begin{vmatrix} 1 & -1 & 0 \\ 7.5 & 4.7 & 20 \\ 7.5 & 0 & 20 \end{vmatrix}}{75.51}$$

$$= \frac{94}{75.51}$$

$$= 1.25 \text{ mA}$$

to three significant digits. Allowing for approximations, these results agree with the currents we found for this circuit in Example 12.13. ▲

In Section 12.3, we also solved problems involving circuits by the loop-current method. When we use the loop-current method, we will not use Kirchhoff's current law and we will get fewer equations. To solve Example 13.19 by the loop-current method, we draw a closed loop in each "window" of the circuit:

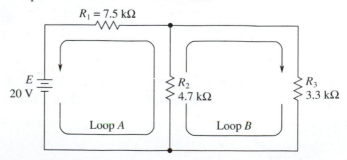

In Loop A, the current through R_1 is I_A. In Loop B, the current through R_3 is I_B. Since R_2 is in Loop A and Loop B, and both loops are in the same direction through R_2, the current through R_2 is $I_A + I_B$. Then, applying Kirchhoff's voltage law to Loop A, we have

$$E = V_1 + V_2$$

$$E = I_A R_1 + (I_A + I_B)R_2$$

$$20 = (I_A)7.5 + (I_A + I_B)4.7$$

$$20 = 12.2I_A + 4.7I_B.$$

Applying Kirchhoff's voltage law to Loop B, we have

$$0 = V_2 + V_3$$

$$0 = (I_A + I_B)R_2 + I_B R_3$$

$$0 = (I_A + I_B)4.7 + (I_B)3.3$$

$$0 = 4.7I_A + 8I_B.$$

Thus, we have a system of two equations in two variables:

$$12.2I_A + 4.7I_B = 20$$

$$4.7I_A + 8I_B = 0.$$

To solve this system by using determinants, D is given by

$$D = \begin{vmatrix} 12.2 & 4.7 \\ 4.7 & 8 \end{vmatrix}$$

$$= 75.51.$$

We observe that D has the same value as in the previous method. To find I_A, we replace the first column of D by the constants on the right-hand sides of the equations:

$$I_A = \frac{\begin{vmatrix} 20 & 4.7 \\ 0 & 8 \end{vmatrix}}{75.51}$$

$$= \frac{160}{75.51}$$

$$= 2.119 \text{ mA}.$$

We have included an extra digit to find I_2 at the end of the calculation. Since I_A is the same as I_1, we have

$$I_1 = 2.12 \text{ mA}$$

to three significant digits as before. To find I_B, we replace the second column of D by the constants on the right-hand sides of the equations:

$$I_B = \frac{\begin{vmatrix} 12.2 & 20 \\ 4.7 & 0 \end{vmatrix}}{75.51}$$

$$= \frac{-94}{75.51}$$

$$= -1.245 \text{ mA}.$$

The negative result means that Loop B actually goes *counter-clockwise*, opposite to the direction of the hands of a clock. Therefore, we have

$$I_3 = 1.25 \text{ mA } \textit{upward}$$

to three significant digits as before. Then, the branch current through R_2 is

$$I_2 = I_A + I_B$$

$$= 2.119 \text{ mA} + (-1.245 \text{ mA})$$

$$= 0.874 \text{ mA}.$$

EXAMPLE 13.20 ▶ Find the branch currents for this circuit:

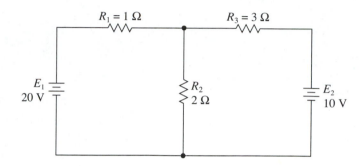

SOLUTION We can solve the circuit either by the branch-current or by the loop-current method. When we use the loop-current method, we will need fewer equations. We draw a closed loop in each window of the circuit:

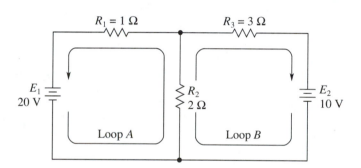

Applying Kirchhoff's voltage law to Loop A, we have

$$E_1 = V_1 + V_2$$

$$E_1 = I_A R_1 + (I_A + I_B)R_2$$

$$20 = (I_A)1 + (I_A + I_B)2$$

$$20 = 3I_A + 2I_B.$$

Applying Kirchhoff's voltage law to Loop B, we have

$$E_2 = V_2 + V_3$$

$$E_2 = (I_A + I_B)R_2 + I_B R_3$$

$$10 = (I_A + I_B)2 + (I_B)3$$

$$10 = 2I_A + 5I_B.$$

Thus, we have a system of two equations in two variables:

$$3I_A + 2I_B = 20$$

$$2I_A + 5I_B = 10.$$

To solve this system by using determinants, D is given by

$$D = \begin{vmatrix} 3 & 2 \\ 2 & 5 \end{vmatrix}$$

$$= 11.$$

To find I_A, we replace the first column of D by the constants on the right-hand sides of the equations:

$$I_A = \frac{\begin{vmatrix} 20 & 2 \\ 10 & 5 \end{vmatrix}}{11}$$

$$= \frac{80}{11}$$

$$= 7.27 \text{ A}.$$

Thus, $I_1 = 7.27$ A. To find I_B, we replace the second column of D by the constants on the right-hand sides of the equations:

$$I_B = \frac{\begin{vmatrix} 3 & 20 \\ 2 & 10 \end{vmatrix}}{11}$$

$$= \frac{-10}{11}$$

$$= -0.909 \text{ A}.$$

The negative result means that Loop B actually goes *counter-clockwise*. Therefore, we have

$$I_3 = 0.909 \text{ A } \textit{to the left.}$$

Finally, we find

$$I_2 = I_A + I_B$$

$$= 7.27 \text{ A} + (-0.91 \text{ A})$$

$$= 6.36 \text{ A}.$$

These results agree with the currents we found for this circuit in Example 12.14.   ▲

EXAMPLE 13.21 ▶ Find the branch currents for this circuit:

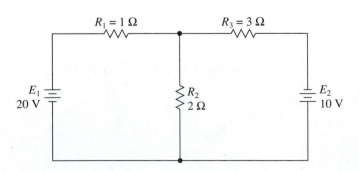

SOLUTION   To use the loop-current method, we draw a closed loop in each window of the circuit:

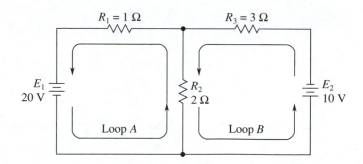

The loops are in opposite directions through R_2. Thus, in Loop A the current through R_2 is $I_A - I_B$, so we have

$$E_1 = V_1 + V_2$$

$$E_1 = I_A R_1 + (I_A - I_B) R_2$$

$$20 = (I_A)1 + (I_A - I_B)2$$

$$20 = 3I_A - 2I_B.$$

But, in Loop B the current through R_2 is $I_B - I_A$, so we have

$$E_2 = V_2 + V_3$$

$$E_2 = (I_B - I_A) R_2 + I_B R_3$$

$$10 = (I_B - I_A)2 + (I_B)3$$

$$10 = -2I_A + 5I_B.$$

Thus, we have a system of two equations in two variables:

$$3I_A - 2I_B = 20$$

$$-2I_A + 5I_B = 10.$$

To solve this system by using determinants, D is given by

$$D = \begin{vmatrix} 3 & -2 \\ -2 & 5 \end{vmatrix}$$

$$= 11.$$

To find I_A, we write

$$I_A = \frac{\begin{vmatrix} 20 & -2 \\ 10 & 5 \end{vmatrix}}{11}$$

$$= \frac{120}{11}$$

$$= 10.91 \text{ A.}$$

Thus, $I_1 = 10.9$ A. To find I_B, we write

$$I_B = \frac{\begin{vmatrix} 3 & 20 \\ -2 & 10 \end{vmatrix}}{11}$$

$$= \frac{70}{11}$$

$$= 6.36 \text{ A.}$$

Thus, $I_3 = 6.36$ A. Finally, from the point of view of Loop A, we find

$$I_2 = I_A - I_B$$

$$= 10.91 \text{ A} - 6.36 \text{ A}$$

$$= 4.55 \text{ A } upward.$$

These results agree with the currents we found for this circuit in Example 12.15.   ▲

EXAMPLE 13.22 ▶ Find I_5 for the Wheatstone bridge circuit shown at the left below.

SOLUTION We found currents in the Wheatstone bridge in Section 7.2. We observe that the circuit has three windows, and draw a closed loop in each:

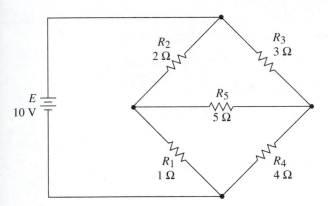

Circuit for Example 13.22.

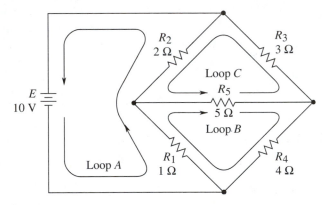

Closed loops in the three windows of the circuit.

In Loop A, the current through R_1 is $I_A + I_B$, but the current through R_2 is $I_A - I_C$. Therefore, for Loop A we have

$$E = V_1 + V_2$$

$$E = (I_A + I_B)R_1 + (I_A - I_C)R_2$$

$$10 = 1(I_A + I_B) + 2(I_A - I_C)$$

$$10 = 3I_A + I_B - 2I_C.$$

In Loop B, the current through R_1 is $I_A + I_B$, and the current through R_2 is $I_B + I_C$. Therefore, for Loop B we have

$$0 = V_1 + V_4 + V_5$$

$$0 = (I_A + I_B)R_1 + I_BR_4 + (I_B + I_C)R_5$$

$$0 = 1(I_A + I_B) + 4I_B + 5(I_B + I_C)$$

$$0 = I_A + 10I_B + 5I_C.$$

In Loop C, the current through R_2 is $I_C - I_A$. Therefore, for Loop C we have

$$0 = V_2 + V_3 + V_5$$

$$0 = (I_C - I_A)R_2 + I_C R_3 + (I_B + I_C)R_5$$

$$0 = 2(I_C - I_A) + 3I_C + 5(I_B + I_C)$$

$$0 = -2I_A + 5I_B + 10I_C.$$

Thus, we have a system of three equations in three variables:

$$3I_A + I_B - 2I_C = 10$$

$$I_A + 10I_B + 5I_C = 0$$

$$-2I_A + 5I_B + 10I_C = 0.$$

To solve this system by using determinants, D is given by

$$D = \begin{vmatrix} 3 & 1 & -2 \\ 1 & 10 & 5 \\ -2 & 5 & 10 \end{vmatrix}$$

$$= 155.$$

Since we are required to find only I_5, we do not need I_A. To find I_B, we replace the second column of D by the constants on the right-hand sides of the equations:

$$I_B = \frac{\begin{vmatrix} 3 & 10 & -2 \\ 1 & 0 & 5 \\ -2 & 0 & 10 \end{vmatrix}}{155}$$

$$= \frac{-200}{155}$$

$$= -1.290 \text{ A}.$$

The negative result means that Loop B actually goes counter-clockwise. To find I_C, we replace the third column of D by the constants on the right-hand sides of the equations:

$$I_C = \frac{\begin{vmatrix} 3 & 1 & 10 \\ 1 & 10 & 0 \\ -2 & 5 & 0 \end{vmatrix}}{155}$$

$$= \frac{250}{155}$$

$$= 1.613 \text{ A}.$$

Therefore, we have

$$I_5 = I_B + I_C$$

$$= -1.290 \text{ A} + 1.613 \text{ A}$$

$$= 0.323 \text{ A } \textit{to the right.}$$

▲

**EXERCISE
13.4**

1. Find I_1, I_2, and I_3 for the circuit shown.

2. For the circuit in Exercise 1, $E = 12$ V, $R_1 = 27$ kΩ, $R_2 = 56$ kΩ, and $R_3 = 82$ kΩ. Find I_1, I_2, and I_3.

3. Find I_1, I_2, and I_3 for the circuit shown.

4. For the circuit in Exercise 3, $E_1 = 2$ V, $E_2 = 5$ V, $R_1 = 10$ Ω, $R_2 = 22$ Ω, and $R_3 = 15$ Ω. Find I_1, I_2, and I_3.

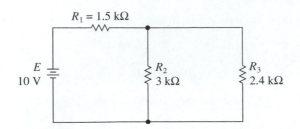

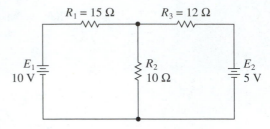

Circuit for Exercises 1 and 2.   Circuit for Exercises 3 and 4.

5. Find I_1, I_2, and I_3 for the circuit shown.

6. For the circuit in Exercise 5, $E_1 = 10$ V, $E_2 = 20$ V, $R_1 = 7.5$ Ω, $R_2 = 3$ Ω, and $R_3 = 1.5$ Ω. Find I_1, I_2, and I_3.

7. Find I_5 for the Wheatstone bridge circuit shown.

8. For the circuit in Exercise 7, $E = 10$ V, $R_1 = 1$ Ω, $R_2 = 1$ Ω, $R_3 = 2$ Ω, $R_4 = 2$ Ω, and $R_5 = 3$ Ω. Find I_5.

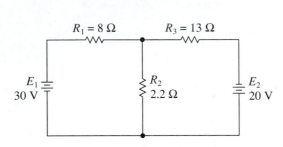

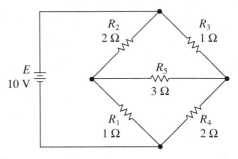

Circuit for Exercises 5 and 6.   Circuit for Exercises 7 and 8.

9. Find I_1, I_2, I_3, I_4, and I_5 for the circuit shown.

10. For the circuit in Exercise 9, $E_1 = 6$ V, $E_2 = 10$ V, and all the resistances are 1 Ω. Find I_1, I_2, I_3, I_4, and I_5.

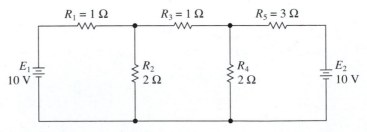

Circuit for Exercises 9 and 10.

Evaluate the determinant:

1. $\begin{vmatrix} 10 & 4 \\ -3 & 8 \end{vmatrix}$

1. _____

2. $\begin{vmatrix} -1 & 2 & 3 \\ 3 & 1 & 2 \\ 2 & -3 & 1 \end{vmatrix}$

2. _____

Use determinants to solve the system:

3. $3x - 4y = 11$

   $4x + 5y = -6$

3. _____

4. $x - y - 2z = 1$

   $3x - y + 4z = 2$

   $2x + y = 6$

4. _____

5. Find I_1, I_2, and I_3 for this circuit:

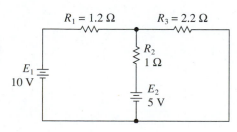

5. _____

UNIT 14

Quadratic Equations

Introduction

In this unit, you will learn more about a type of equation you encountered in Units 4, 10, and 12. In Unit 4, you solved basic quadratic equations. However, the technique you used in Unit 4 does not apply to more complicated quadratic equations. In this unit, you will learn two more ways to solve quadratic equations. First, you will learn how to factor quadratic expressions, and how to solve quadratic equations by factoring. Then, you will learn the most general method for solving quadratic equations, the quadratic formula. In Unit 10, you drew graphs of some equations in two variables, including graphs of parabolas. In this unit, you will learn some properties of parabolas and draw graphs of more general cases. In Unit 12, you solved systems involving a quadratic and a linear equation by graphing, and you could solve a few such systems algebraically. In this unit, you will apply the techniques for solving quadratic equations to algebraic solutions of systems of equations.

OBJECTIVES

When you have finished this unit you should be able to:

1. Factor quadratic expressions and solve quadratic equations by factoring.
2. Solve quadratic equations by using the quadratic formula.
3. Find the vertex of a parabola, determine whether the vertex is a minimum or a maximum, and draw the graph of the parabola.
4. Solve systems of two equations in two variables where one equation is a quadratic equation and one is linear.

SECTION 14.1

Solutions by Factoring

We have used the distributive property to multiply polynomials by monomials, and also to factor out monomials from polynomials. For example, we may use the distributive property to write

$$2(x + y) = 2x + 2y.$$

Also, we may factor out the common factor to write

$$2x + 2y = 2(x + y).$$

We may use factoring to solve quadratic equations.

First, we state a rule we will call the **zero product rule**.

Zero Product Rule: If $ab = 0$ then $a = 0$ or $b = 0$.

To use the zero product rule to solve quadratic equations, we collect all of the nonzero terms on one side of the equation, and then factor the resulting expression.

319

EXAMPLE 14.1 Solve the equation $2x^2 = x$ and check the solutions.

SOLUTION To use the zero product rule, we collect all of the nonzero terms on one side of the equation:

$$2x^2 = x$$

$$2x^2 - x = 0.$$

Then, we factor out the common factor x, to obtain

$$x(2x - 1) = 0.$$

Now, using the zero product rule, we derive two linear equations:

$$x = 0 \text{ and } 2x - 1 = 0.$$

One solution is $x = 0$. To find a second solution, we solve the second linear equation:

$$2x - 1 = 0$$

$$x = \frac{1}{2}.$$

Thus, the quadratic equation has two solutions,

$$x = 0 \text{ and } x = \frac{1}{2}.$$

To check, we substitute each solution in the original equation. For $x = 0$,

$$2x^2 = x$$

$$2(0)^2 \stackrel{?}{=} 0$$

$$2(0) = 0.$$

For $x = \frac{1}{2}$,

$$2x^2 = x$$

$$2\left(\frac{1}{2}\right)^2 \stackrel{?}{=} \frac{1}{2}$$

$$2\left(\frac{1}{4}\right) = \frac{1}{2}.$$

▲

MULTIPLYING BINOMIALS

The equation in Example 14.1 is a special case because it does not have a *constant term*; that is, a term with no variable. To solve more general quadratic equations, we must know how to multiply binomials and factor trinomials. To multiply binomials, we use the distributive property twice.

EXAMPLE 14.2 Multiply $(x - 3)(x + 2)$.

SOLUTION We may use the distributive property to write

$$(x - 3)(x + 2) = (x - 3)(x) + (x - 3)(2).$$

Then, we use the distributive property in each part to obtain

$$(x - 3)(x) + (x - 3)(2) = (x)(x) - 3x + (x)(2) - (3)(2)$$

$$= x^2 - 3x + 2x - 6$$

$$= x^2 - x - 6.$$

▲

We may shorten the process in Example 14.2 by using a diagram:

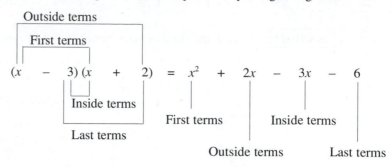

We observe that we have multiplied

the first terms: $(x)(x) = x^2$
the outside terms: $(x)(2) = 2x$
the inside terms: $(-3)(x) = -3x$
the last terms: $(-3)(2) = -6.$

Then, combining the results from the outside and inside terms, we have

$$(x - 3)(x + 2) = x^2 + 2x - 3x - 6$$
$$= x^2 - x - 6.$$

EXAMPLE 14.3 Multiply $(5x + 1)(2x - 3)$.

SOLUTION Using the diagram, we multiply

the first terms: $(5x)(2x) = 10x^2$
the outside terms: $(5x)(-3) = -15x$
the inside terms: $(1)(2x) = 2x$
the last terms: $(1)(-3) = -3.$

Then, combining the results from the outside and inside terms, we have

$$(5x + 1)(2x - 3) = 10x^2 - 15x + 2x - 3$$
$$= 10x^2 - 13x - 3. \qquad \blacktriangle$$

FACTORING TRINOMIALS

The results of the multiplications in Examples 14.2 and 14.3 are general trinomials of the form

$$ax^2 + bx + c.$$

Some trinomials of this form can be factored into two binomial factors. To **factor** a trinomial, we search for two binomials which, when multiplied, result in the trinomial.

EXAMPLE 14.4 Factor $x^2 + 5x + 6$.

SOLUTION The first terms must each be x. Therefore, we write

$$x^2 + 5x + 6 = (x \quad)(x \quad).$$

All of the constants are positive. Therefore, the possible last terms are the positive factors of 6, which are (1)(6) or (2)(3). We try these possible products:

$$(x + 1)(x + 6) = x^2 + 7x + 6$$

and

$$(x + 2)(x + 3) = x^2 + 5x + 6.$$

The second product gives the correct middle term. Thus, we write the factored form

$$x^2 + 5x + 6 = (x + 2)(x + 3). \qquad \blacktriangle$$

EXAMPLE 14.5 Factor $x^2 - 9x + 8$.

SOLUTION Again, the first terms must each be x, so we write

$$x^2 - 9x + 8 = (x \qquad)(x \qquad).$$

The middle term is negative but the constant term is positive. Therefore, the possible last terms are the negative factors of 8, which are $(-1)(-8)$ or $(-2)(-4)$. We try these possible products:

$$(x - 1)(x - 8) = x^2 - 9x + 8$$

and

$$(x - 2)(x - 4) = x^2 - 6x + 8.$$

The first product gives the correct middle term. Thus, we write the factored form

$$x^2 - 9x + 8 = (x - 1)(x - 8). \qquad \blacktriangle$$

EXAMPLE 14.6 Factor $x^2 + x - 6$.

SOLUTION Again, the first terms must each be x, so we write

$$x^2 + x - 6 = (x \qquad)(x \qquad).$$

The constant term is negative. Therefore, the possible last terms are the factors of 6 with opposite signs, which are $(1)(-6)$, $(-1)(6)$, $(2)(-3)$, and $(-2)(3)$. Trying the first set of possible products,

$$(x + 1)(x - 6) = x^2 - 5x - 6$$

and

$$(x - 1)(x + 6) = x^2 + 5x - 6.$$

We observe that changing the signs of the last terms changes the sign of the middle term, but not its absolute value. Trying the second set of possible products,

$$(x + 2)(x - 3) = x^2 - x - 6$$

and

$$(x - 2)(x + 3) = x^2 + x - 6.$$

The last product gives the correct middle term. Thus, we write the factored form

$$x^2 + x - 6 = (x - 2)(x + 3). \qquad \blacktriangle$$

EXAMPLE 14.7 Factor $2x^2 - 3x - 2$.

SOLUTION The first terms must be x and $2x$ so we write

$$2x^2 - 3x - 2 = (x \qquad)(2x \qquad).$$

The possible last terms are factors of 2 with opposite signs. We might try a possible product,

$$(x + 1)(2x - 2) = 2x^2 - 2.$$

When the first terms are not identical, we try the last terms in reverse order:

$$(x - 2)(2x + 1) = 2x^2 - 3x - 2.$$

This product gives the correct middle term. Thus, we write the factored form

$$2x^2 - 3x - 2 = (x - 2)(2x + 1). \qquad \blacktriangle$$

You should observe that the first try in Example 14.7,

$$(x + 1)(2x - 2) = 2x^2 - 2,$$

has a common factor in the second binomial. If you are careful always to factor out common factors first, you can then rule out any product that has a common factor in any binomial.

EXAMPLE 14.8   Factor $4x^2 - 16x + 12$.

SOLUTION   First, we factor out the common factor:

$$4x^2 - 16x + 12 = 4(x^2 - 4x + 3).$$

Now, we can factor the remaining trinomial to give

$$4(x^2 - 4x + 3) = 4(x - 1)(x - 3). \qquad \blacktriangle$$

SOLUTIONS BY FACTORING

To solve quadratic equations of the form

$$ax^2 + bx + c = 0,$$

if we can factor the quadratic expression, we may use the zero product rule.

EXAMPLE 14.9   Solve the equation $x^2 - 3x + 2 = 0$ and check the solutions.

SOLUTION   We factor the quadratic expression to obtain

$$x^2 - 3x + 2 = 0$$
$$(x - 1)(x - 2) = 0.$$

Now, using the zero product rule, we derive two linear equations:

$$x - 1 = 0 \text{ and } x - 2 = 0.$$

Solving the linear equations, we have the solutions

$$x = 1 \text{ and } x = 2.$$

You should check by substituting each solution in the original equation.   ▲

EXAMPLE 14.10   Solve the equation $2x^2 - x = 1$ and check the solutions.

SOLUTION   To use the zero product rule, we must collect all of the nonzero terms on one side of the equation:

$$2x^2 - x = 1$$
$$2x^2 - x - 1 = 0.$$

Now, we may factor the quadratic expression to obtain

$$(x - 1)(2x + 1) = 0.$$

Then, using the zero product rule, we derive two linear equations:

$$x - 1 = 0 \text{ and } 2x + 1 = 0$$

$$x = 1 \text{ and } x = -\frac{1}{2}.$$

You should check by substituting each solution in the original equation.   ▲

In Example 14.10, you might be tempted to factor out the common factor from the expression $2x^2 - x$, and then set each factor equal to 1. Remember that, to use the *zero product rule*, you must have a quadratic expression equal to *zero*.

EXAMPLE 14.11   Solve the equation $2(2 - x^2) = 3x + 2$ and check the solutions.

SOLUTION   We simplify the expression on the left-hand side, and collect all of the nonzero terms on one side of the equation:

$$2(2 - x^2) = 3x + 2$$
$$4 - 2x^2 = 3x + 2$$
$$0 = 2x^2 + 3x - 2.$$

Observe that we have collected the terms on the right-hand side of the equation. It is usually easiest to factor a quadratic expression if the x^2-term is positive. We factor the quadratic expression to obtain

$$0 = (x + 2)(2x - 1).$$

Then, using the zero product rule, we derive two linear equations:

$$x + 2 = 0 \text{ and } 2x - 1 = 0$$
$$x = -2 \text{ and } x = \frac{1}{2}.$$

To check, we substitute each solution in the original equation. For $x = -2$,

$$2(2 - x^2) = 3x + 2$$
$$2[2 - (-2)^2] \stackrel{?}{=} 3(-2) + 2$$
$$2(2 - 4) \stackrel{?}{=} -6 + 2$$
$$2(-2) = -4.$$

For $x = \frac{1}{2}$,

$$2(2 - x^2) = 3x + 2$$
$$2\left[2 - \left(\frac{1}{2}\right)^2\right] \stackrel{?}{=} 3\left(\frac{1}{2}\right) + 2$$
$$2\left(2 - \frac{1}{4}\right) \stackrel{?}{=} \frac{3}{2} + 2$$
$$2\left(\frac{7}{4}\right) = \frac{7}{2}.$$

▲

EXERCISE 14.1

Solve the equation and check the solutions:

1. $x^2 = 3x$ 2. $4x^2 = x$

3. $2x^2 = 4x$ 4. $3x^2 + 6x = 0$

Multiply:

5. $(x + 2)(x + 4)$ 6. $(x - 3)(x - 4)$

7. $(x - 4)(x + 3)$ 8. $(x + 6)(x - 3)$

9. $(2x - 5)(x + 2)$ 10. $(3x + 2)(x - 4)$

11. $(2x - 3)(2x - 1)$ 12. $(4x - 1)(2x + 5)$

Factor:

13. $x^2 + 4x + 3$ 14. $x^2 + 11x + 18$

15. $x^2 - 11x + 24$ 16. $x^2 - 13x + 12$

17. $x^2 - x - 12$
18. $x^2 + 5x - 36$

19. $2x^2 + 5x - 12$
20. $3x^2 - 7x - 6$

21. $4x^2 + 4x - 3$
22. $4x^2 - 5x - 6$

23. $6x^2 - 3x - 3$
24. $8x^2 - 10x + 2$

Solve the equation and check the solutions:

25. $x^2 - 7x + 12 = 0$
26. $x^2 - 2x - 15 = 0$

27. $3x^2 + 2x = 1$
28. $2x^2 - x = 6$

29. $2x^2 + 7x = 2 - 2x^2$
30. $4x - 3x^2 = x^2 - 3$

31. $3(x - 4) = x(x + 10)$
32. $2 - (x^2 + 3) = 5x(x - 1)$

SECTION 14.2

The Quadratic Formula

Quadratic equations in the form

$$ax^2 + bx + c = 0$$

are in **standard form**. Quadratic equations in standard form can be solved by using the **quadratic formula**.

The Quadratic Formula: For quadratic equations in the standard form

$$ax^2 + bx + c = 0,$$

the solutions are given by

$$x = \frac{-b \pm \sqrt{b^2 - 4ac}}{2a}.$$

The proof of the quadratic formula is given at the end of this unit.

EXAMPLE 14.12   Solve the equation $x^2 - 3x + 2 = 0$ and check the solutions.

SOLUTION   The solutions are given by the quadratic formula

$$x = \frac{-b \pm \sqrt{b^2 - 4ac}}{2a}.$$

Since the equation is in standard form, $a = 1$, $b = -3$, and $c = 2$. Substituting in the quadratic formula, we write

$$x = \frac{-(-3) \pm \sqrt{(-3)^2 - 4(1)(2)}}{2(1)}$$

$$= \frac{3 \pm \sqrt{9 - 8}}{2}$$

$$= \frac{3 \pm \sqrt{1}}{2}$$

$$= \frac{3 \pm 1}{2}.$$

This result means

$$x = \frac{3 + 1}{2} \text{ and } x = \frac{3 - 1}{2}.$$

Thus, the solutions are

$$x = 2 \text{ and } x = 1.$$

These solutions agree with those in Example 14.9.   ▲

EXAMPLE 14.13 ▶   Solve the equation $x^2 + 2x - 2 = 0$ and check the solutions.

SOLUTION   We use the quadratic formula with $a = 1$, $b = 2$, and $c = -2$:

$$x = \frac{-b \pm \sqrt{b^2 - 4ac}}{2a}$$

$$= \frac{-2 \pm \sqrt{(2)^2 - 4(1)(-2)}}{2(1)}$$

$$= \frac{-2 \pm \sqrt{4 + 8}}{2}$$

$$= \frac{-2 \pm \sqrt{12}}{2}.$$

Since $\sqrt{12}$ is not a rational number, you can use a calculator to approximate the solution

$$x = \frac{-2 + \sqrt{12}}{2}$$

by following these steps:

	display:
Enter 2	2.
Press ⬚+/−	−2.
Press ⬚+	−2.
Enter 12	12.
Press ⬚√x	3.464101615
Press ⬚=	1.464101615
Press ⬚÷	1.464101615
Enter 2	2.
Press ⬚=	0.732050807

Thus, one solution is

$$x = 0.732$$

to three significant digits. You can use similar steps to find that the second solution is

$$x = \frac{-2 - \sqrt{12}}{2}$$

$$= -2.73$$

to three significant digits. You can also use a calculator to check the solutions. When the variable appears several times in the original equation, it is useful to store the solution in the memory register. Recall from Section 9.4 the keys marked

$$\boxed{\text{STO}} \text{ and } \boxed{\text{RCL}}$$

or

$$\boxed{\text{MS}} \text{ and } \boxed{\text{MR}}$$

or other codes meaning to store a number in a memory register and to recall it from memory. Now, you can follow these steps to check $x = 0.732$:

display:

Enter .732	0.732
Press $\boxed{\text{STO}}$	0.732
Press $\boxed{x^2}$	0.535824
Press $\boxed{+}$	0.535824
Enter 2	2.
Press $\boxed{\times}$	2.
Press $\boxed{\text{RCL}}$	0.732
Press $\boxed{-}$	1.999824
Enter 2	2.
Press $\boxed{=}$	-0.000176

This result is sufficiently close to zero, allowing for approximations. ▲

EXAMPLE 14.14 ▶ Solve the equation $x^2 + 4x = 2$ and check the solutions.

SOLUTION To use the quadratic formula we must write the equation in standard form, with all of the nonzero terms on one side:

$$x^2 + 4x = 2$$
$$x^2 + 4x - 2 = 0.$$

Now, we use the quadratic formula with $a = 1$, $b = 4$, and $c = -2$:

$$x = \frac{-b \pm \sqrt{b^2 - 4ac}}{2a}$$

$$= \frac{-4 \pm \sqrt{(4)^2 - 4(1)(-2)}}{2(1)}$$

$$= \frac{-4 \pm \sqrt{16 + 8}}{2}$$

$$= \frac{-4 \pm \sqrt{24}}{2}.$$

You should use a calculator to find that the solutions are

$$x = 0.449 \text{ and } x = -4.45$$

to three significant digits, and to check these solutions. ▲

EXAMPLE 14.15 ▶ Solve the equation $x(1 - 3x) = 3x - 2$ and check the solutions.

SOLUTION We simplify the expression on the left-hand side, and collect all of the nonzero terms on one side of the equation:

$$x(1 - 3x) = 3x - 2$$
$$x - 3x^2 = 3x - 2$$
$$0 = 3x^2 + 2x - 2.$$

Now, we use the quadratic formula with $a = 3$, $b = 2$, and $c = -2$:

$$x = \frac{-b \pm \sqrt{b^2 - 4ac}}{2a}$$

$$= \frac{-2 \pm \sqrt{(2)^2 - 4(3)(-2)}}{2(3)}$$

$$= \frac{-2 \pm \sqrt{4 + 24}}{6}$$

$$= \frac{-2 \pm \sqrt{28}}{6}.$$

You should use a calculator to find that the solutions are

$$x = 0.549 \text{ and } x = -1.22$$

to three significant digits, and to check these solutions.   ▲

EXAMPLE 14.16 ▶ Solve the equation $1.2x^2 = 3.3x - 1$ and check the solutions.

SOLUTION   We collect all of the nonzero terms on one side of the equation:

$$1.2x^2 = 3.3x - 1$$

$$1.2x^2 - 3.3x + 1 = 0.$$

Now, we use the quadratic formula with $a = 1.2$, $b = -3.3$, and $c = 1$:

$$x = \frac{-b \pm \sqrt{b^2 - 4ac}}{2a}$$

$$= \frac{-(-3.3) \pm \sqrt{(-3.3)^2 - 4(1.2)(1)}}{2(1.2)}$$

$$= \frac{3.3 \pm \sqrt{6.09}}{2.4}.$$

You should use a calculator to find that the solutions are

$$x = 2.88 \text{ and } x = 0.347$$

to three significant digits, and to check these solutions.   ▲

When $b^2 - 4ac$ is negative, the quadratic formula will result in a square root of a negative. In this case, the equation has no real number solutions.

EXAMPLE 14.17 ▶ Solve the equation $2x^2 - 2x + 1 = 0$ and check the solutions.

SOLUTION   We use the quadratic formula with $a = 2$, $b = -2$, and $c = 1$:

$$x = \frac{-b \pm \sqrt{b^2 - 4ac}}{2a}$$

$$= \frac{-(-2) \pm \sqrt{(-2)^2 - 4(2)(1)}}{2(2)}$$

$$= \frac{2 \pm \sqrt{4 - 8}}{4}$$

$$= \frac{2 \pm \sqrt{-4}}{4}.$$

The quadratic formula results in a square root of a negative, so the equation has no real number solutions. Numbers in the form of this result are complex numbers, which we will study in Unit 18. We will see one interpretation for such results in the next section.   ▲

When a quadratic equation is written in standard form, the expression $b^2 - 4ac$ is called its **discriminant**. The equation can be solved by factoring, and its solutions are rational numbers, when the discriminant is positive and a perfect square. The equation is solved by using the quadratic formula, and its solutions are irrational numbers, when the discriminant is positive but not a perfect square. The equation has no real number solutions when the discriminant is negative.

<table>
<tr><td>

EXERCISE

14.2

</td></tr>
</table>

Solve the equation by using the quadratic formula, and check the solutions:

1. $x^2 - 5x + 6 = 0$ **2.** $2x^2 + 3x = 2$

3. $x^2 + 2x - 1 = 0$ **4.** $x^2 - 2x - 4 = 0$

5. $x^2 + x = 1$ **6.** $x^2 - 2x = 5$

7. $1 + 2x^2 = 4x$ **8.** $3x + 3 = 2x^2$

9. $3x^2 + 6x = 2$ **10.** $5x^2 = 5x + 1$

11. $2x(1 - x) = x - 2$ **12.** $x(x + 6) = 3x^2 + 3$

13. $1.5 - 3x^2 = 2x$ **14.** $x^2 + 3.9 = 3.3 + 2.2x$

15. $x^2 + 4 = 2(x + 1)$ **16.** $1.5x^2 - 7.5 = x(2.2x - 3)$

<table>
<tr><td>

SECTION

14.3

</td></tr>
</table>

Parabolas

In Section 10.4, we drew graphs of several basic quadratic functions and relations. In particular, for the quadratic function

$$y = x^2,$$

we drew the graph shown below. This graph is called a parabola, and contains a minimum point called the vertex.

In general, a **quadratic function** is an equation in two variables of the form

$$y = ax^2 + bx + c.$$

The graph of any quadratic function is a parabola, and contains a point called its vertex. The vertex of a parabola is either its minimum point or its maximum point. When the vertex is a *minimum*, the parabola opens *upward*. When the vertex is a *maximum*, the parabola opens *downward*.

EXAMPLE 14.18 ▶ Draw the graph of $y = -x^2$.

SOLUTION We find several ordered pairs that are solutions of the equation. For $x = 0$,

$$y = -x^2$$
$$= -0^2$$
$$= 0,$$

which gives the ordered pair $(0, 0)$. For $x = 1$,

$$y = -x^2$$
$$= -1^2$$
$$= -1,$$

which gives the ordered pair $(1, -1)$. For $x = -1$,

$$y = -x^2$$
$$= -(-1^2)$$
$$= -1,$$

which gives the ordered pair $(-1, -1)$. You should check that these ordered pairs are also solutions:

$$(2, -4), \quad (-2, -4), \quad (3, -9), \quad (-3, -9).$$

Graph of $y = x^2$.

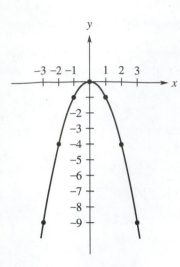

Graph for Example 14.18

We plot the points corresponding to the ordered pairs we have found, and connect them by a smooth curve. The graph is shown at the left. We observe that the vertex of this parabola is a maximum and the parabola opens downward. ▲

In general, when a parabola is given by the quadratic function

$$y = ax^2 + bx + c,$$

the x-coordinate of its vertex is given by the formula

$$x = \frac{-b}{2a}.$$

If $a > 0$, the vertex is a minimum and the parabola opens upward. If $a < 0$, the vertex is a maximum and the parabola opens downward.

For example, for the function

$$y = x^2,$$

$a = 1, b = 0,$ and $c = 0$. Therefore, we may find the vertex by writing

$$x = \frac{-b}{2a}$$

$$= \frac{-0}{2(1)}$$

$$= 0.$$

When $x = 0$, we also have $y = 0$. Thus, the vertex is $(0, 0)$. Moreover, $a > 0$ so the vertex is a minimum and the parabola opens upward.

Similarly, for the function

$$y = -x^2,$$

$a = -1, b = 0,$ and $c = 0$. Therefore, the vertex is also $(0, 0)$. But, $a < 0$ so the vertex is a maximum and the parabola opens downward.

EXAMPLE 14.19 ▶   Draw the graph of $y = x^2 - 4$.

SOLUTION   For this function, $a = 1, b = 0,$ and $c = -4$. Therefore, we can find the vertex by writing

$$x = \frac{-b}{2a}$$

$$= \frac{-0}{2(1)}$$

$$= 0.$$

When $x = 0$, we have

$$y = x^2 - 4$$

$$= 0^2 - 4$$

$$= -4.$$

Thus, the vertex is $(0, -4)$. Moreover, $a > 0$ so the vertex is a minimum and the parabola opens upward.

We may find other special points of the graph such as the x- and y-intercepts. At the y-intercept, $x = 0$. Thus, the y-intercept is the point $(0, -4)$, and the y-intercept is the same point as the vertex. At the x-intercept, $y = 0$. Therefore, we write

$$0 = x^2 - 4$$

$$x^2 = 4$$

$$x = \pm\sqrt{4}$$

$$x = \pm 2.$$

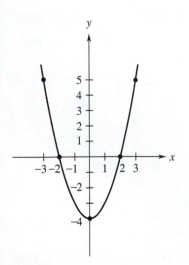

Graph for Example 14.19

Thus, there are two x-intercepts, $(2, 0)$ and $(-2, 0)$. (You should also try solving this equation by factoring.)

Three points are sufficient to draw a parabola. However, we will get a better graph by using two more points. You should check that $(3, 5)$ and $(-3, 5)$ are also solutions.   ▲

EXAMPLE 14.20 ▶ Draw the graph of $y = x^2 + 4x + 4$.

SOLUTION For this function, $a = 1, b = 4$, and $c = 4$. Therefore, we can find the vertex by writing

$$x = \frac{-b}{2a}$$

$$= \frac{-4}{2(1)}$$

$$= -2.$$

When $x = -2$, we have

$$y = x^2 + 4x + 4$$

$$= (-2)^2 + 4(-2) + 4$$

$$= 0.$$

Thus, the vertex is $(-2, 0)$. Since $a > 0$, the vertex is a minimum and the parabola opens upward.

At the y-intercept, $x = 0$. Thus, the y-intercept is the point $(0, 4)$. At the x-intercept, $y = 0$. Therefore, we write

$$0 = x^2 + 4x + 4$$

$$0 = (x + 2)(x + 2)$$

$$x + 2 = 0 \text{ and } x + 2 = 0$$

$$x = -2 \text{ and } x = -2.$$

Thus, we appear to have just one x-intercept, $(-2, 0)$. Actually, there are two identical x-intercepts, both $(-2, 0)$. We observe that this point is also the vertex. Our calculations have resulted in only two distinct points, so you should check that $(-4, 4)$, $(-1, 1)$ and $(-3, 1)$ are also solutions.   ▲

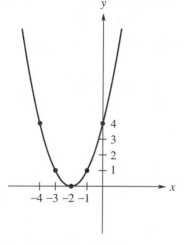

Graph for Example 14.20

EXAMPLE 14.21 ▶ Draw the graph of $y = x^2 + 2x - 4$.

SOLUTION For this function, $a = 1, b = 2$, and $c = -4$. Therefore, we can find the vertex by writing

$$x = \frac{-b}{2a}$$

$$= \frac{-2}{2(1)}$$

$$= -1.$$

When $x = -1$, we have

$$y = x^2 + 2x - 4$$

$$= (-1)^2 + 2(-1) - 4$$

$$= -5.$$

Thus, the vertex is $(-1, -5)$. Since $a > 0$, the vertex is a minimum and the parabola opens upward.

At the y-intercept, $x = 0$. Thus, the y-intercept is the point $(0, -4)$. At the x-intercept, $y = 0$. Therefore, we write

$$0 = x^2 + 2x - 4.$$

We use the quadratic formula with $a = 1$, $b = 2$, and $c = -4$:

$$x = \frac{-b \pm \sqrt{b^2 - 4ac}}{2a}$$

$$= \frac{-2 \pm \sqrt{(2)^2 - 4(1)(-4)}}{2(1)}$$

$$= \frac{-2 \pm \sqrt{20}}{2}.$$

You should use a calculator to find that the solutions are

$$x = 1.24 \text{ and } x = -3.24$$

to three significant digits. Thus, the x-intercepts are $(1.24, 0)$ and $(-3.24, 0)$. You should check that $(-2, -4)$ is also on the graph.   ▲

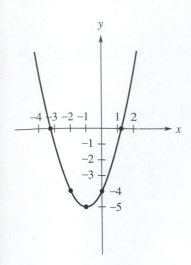

Graph for Example 14.21

EXAMPLE 14.22 ▶   Draw the graph of $y = x^2 - 2x + 3$.

SOLUTION   For this function, $a = 1$, $b = -2$, and $c = 3$. Therefore, we can find the vertex by writing

$$x = \frac{-b}{2a}$$

$$= \frac{-(-2)}{2(1)}$$

$$= 1.$$

When $x = 1$, we have

$$y = x^2 - 2x + 3$$

$$= 1^2 - 2(1) + 3$$

$$= 2.$$

Thus, the vertex is $(1, 2)$. Since $a > 0$, the vertex is a minimum and the parabola opens upward.

At the y-intercept, $x = 0$. Thus, the y-intercept is the point $(0, 3)$. At the x-intercept, $y = 0$. Therefore, we write

$$0 = x^2 - 2x + 3.$$

We use the quadratic formula with $a = 1$, $b = -2$, and $c = 3$:

$$x = \frac{-b \pm \sqrt{b^2 - 4ac}}{2a}$$

$$= \frac{-(-2) \pm \sqrt{(-2)^2 - 4(1)(3)}}{2(1)}$$

$$= \frac{2 \pm \sqrt{-8}}{2}.$$

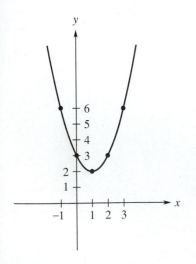

Graph for Example 14.22

The quadratic formula results in a square root of a negative, and so the equation has no real number solutions. This result means that there are no x-intercepts. We find some other points and draw the graph.   ▲

EXAMPLE 14.23 ▶ Draw the graph of $y = -x^2 - 2x + 4$.

SOLUTION For this function, $a = -1$, $b = -2$, and $c = 4$. Therefore, we can find the vertex by writing

$$x = \frac{-b}{2a}$$

$$= \frac{-(-2)}{2(-1)}$$

$$= -1.$$

When $x = -1$, we have

$$y = -x^2 - 2x + 4$$

$$= -(-1)^2 - 2(-1) + 4$$

$$= 5.$$

Thus, the vertex is $(-1, 5)$. Since $a < 0$, the vertex is a *maximum* and the parabola opens *downward*.

At the y-intercept, $x = 0$. Thus, the y-intercept is the point $(0, 4)$. At the x-intercept, $y = 0$. Therefore, we write

$$0 = -x^2 - 2x + 4$$

or

$$x^2 + 2x - 4 = 0.$$

This equation is the same as the one in Example 14.21. Thus,

$$x = 1.24 \quad \text{and} \quad x = -3.24,$$

and the x-intercepts are $(1.24, 0)$ and $(-3.24, 0)$. ▲

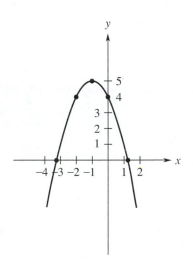

Graph for Example 14.23

We observe that, although the quadratic *equation* we used to find the x-intercepts in Example 4.23 is the same as that in Example 14.21, the quadratic *function* is different and so the graph is different. In Example 14.21, we observed that $a > 0$, the vertex is a minimum, and the parabola opens upward. In Example 14.23, however, $a < 0$, the vertex is a maximum, and the parabola opens downward.

**EXERCISE
14.3**

Draw the graph:

1. $y = \frac{1}{2}x^2$ 2. $y = -\frac{1}{2}x^2$

3. $y = x^2 - 1$ 4. $y = x^2 + 2$

5. $y = x^2 + 2x + 1$ 6. $y = x^2 - 4x + 4$

7. $y = x^2 - 4x + 3$ 8. $y = x^2 + 2x - 3$

9. $y = x^2 + 2x - 2$ 10. $y = x^2 - 4x + 2$

11. $y = x^2 + 2x + 2$ 12. $y = x^2 - x + 1$

13. $y = -x^2 - 2x + 3$ 14. $y = -x^2 - 4x - 3$

15. $y = 1 - 4x - 2x^2$ 16. $y = -2x^2 + 2x - 1$

SECTION

14.4

Systems with Quadratic Equations

In Section 12.1, we solved systems of two equations in two variables by graphing. The systems in that section included some that consist of a quadratic equation and a linear equation. In Section 12.2, we solved some such systems algebraically. The systems we could solve were limited, however, because our methods for solving the resulting quadratic equations were limited. By using the methods of this unit, we can solve many more systems consisting of a quadratic equation and a linear equation.

EXAMPLE 14.24 ▶

Solve the system $x + y = 2$ and $y = x^2$.

SOLUTION

We solve the system by substitution. Since the second equation is solved for y, we substitute x^2 for y in the first equation, to obtain

$$x + y = 2$$
$$x + x^2 = 2$$
$$x^2 + x - 2 = 0.$$

This quadratic equation can be solved by factoring:

$$(x - 1)(x + 2) = 0$$
$$x - 1 = 0 \text{ and } x + 2 = 0$$
$$x = 1 \text{ and } x = -2.$$

Replacing x by 1 in either of the original equations gives

$$y = 1.$$

Therefore, one solution is (1, 1). Replacing x by –2 in either of the original equations gives

$$y = 4.$$

Therefore, a second solution is (–2, 4). You should check both solutions by substituting in each of the original equations. We recall that the solutions are the coordinates of the points of intersection of the graphs of the equations. The graph of the system is

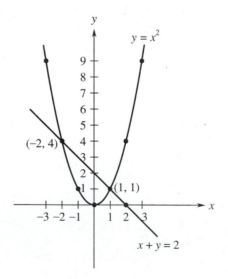

EXAMPLE 14.25 ▶

Solve the system $y = x + 1$ and $y = \frac{1}{2}x^2$.

SOLUTION

Both equations are solved for y. Substituting $\frac{1}{2}x^2$ for y in the first equation, we have

$$y = x + 1$$

$$\frac{1}{2}x^2 = x + 1.$$

We multiply both sides of the equation by 2, and collect the nonzero terms on one side:

$$x^2 = 2x + 2$$

$$x^2 - 2x - 2 = 0.$$

We use the quadratic formula with $a = 1$, $b = -2$, and $c = -2$:

$$x = \frac{-b \pm \sqrt{b^2 - 4ac}}{2a}$$

$$= \frac{-(-2) \pm \sqrt{(-2)^2 - 4(1)(-2)}}{2(1)}$$

$$= \frac{2 \pm \sqrt{12}}{2}.$$

You should use a calculator to find that the solutions are

$$x = 2.73 \quad \text{and} \quad x = -0.732$$

to three significant digits. Replacing x by 2.73 in either of the original equations gives

$$y = 3.73.$$

Therefore, one solution is $(2.73, 3.73)$. Replacing x by -0.732 in either of the original equations gives

$$y = 0.268.$$

Therefore, a second solution is $(-0.732, 0.268)$. You should check both solutions by substituting in each of the original equations. We solved this system by graphing in Example 12.3.

▲

EXAMPLE 14.26 Solve the system $y = x - \frac{1}{2}$ and $y = \frac{1}{2}x^2$.

SOLUTION Both equations are solved for y. Substituting $\frac{1}{2}x^2$ for y in the first equation, we have

$$y = x - \frac{1}{2}$$

$$\frac{1}{2}x^2 = x - \frac{1}{2}$$

$$x^2 = 2x - 1$$

$$x^2 - 2x + 1 = 0$$

$$(x - 1)(x - 1) = 0$$

$$x - 1 = 0 \quad \text{and} \quad x - 1 = 0$$

$$x = 1 \quad \text{and} \quad x = 1.$$

Replacing x by 1 in either of the original equations gives

$$y = \frac{1}{2}.$$

There is just one solution, $(1, \frac{1}{2})$. We solved this system by graphing in Example 12.4, and found that the line is tangent to the parabola at the point $(1, \frac{1}{2})$. ▲

EXAMPLE 14.27   Solve the system $y = x - 1$ and $y = \frac{1}{2}x^2$.

SOLUTION   Both equations are solved for y. Substituting $\frac{1}{2}x^2$ for y in the first equation, we have

$$y = x - 1$$

$$\frac{1}{2}x^2 = x - 1$$

$$x^2 = 2x - 2$$

$$x^2 - 2x + 2 = 0.$$

We use the quadratic formula with $a = 1$, $b = -2$, and $c = 2$:

$$x = \frac{-b \pm \sqrt{b^2 - 4ac}}{2a}$$

$$= \frac{-(-2) \pm \sqrt{(-2)^2 - 4(1)(2)}}{2(1)}$$

$$= \frac{2 \pm \sqrt{-4}}{2}.$$

The quadratic formula results in a square root of a negative, so the system has no real number solutions. This result means that the line and the parabola do not intersect at any point. You should check this conclusion by drawing the graph of the system.   ▲

EXAMPLE 14.28   Solve the system $x + y + 1 = 0$ and $x^2 + y^2 = 9$.

SOLUTION   Neither equation is solved for y. It is easier to solve the first equation:

$$x + y + 1 = 0$$

$$y = -x - 1$$

or

$$y = -1(x + 1).$$

Then, substituting $-1(x + 1)$ for y in the second equation, we have

$$x^2 + y^2 = 9$$

$$x^2 + [(-1)(x + 1)]^2 = 9.$$

To multiply the second term, we can write

$$x^2 + (-1)^2(x + 1)^2 = 9$$

$$x^2 + (1)(x + 1)(x + 1) = 9$$

$$x^2 + x^2 + 2x + 1 = 9.$$

Then, we collect the nonzero terms on one side and divide both sides by 2 to obtain

$$2x^2 + 2x - 8 = 0$$

$$x^2 + x - 4 = 0.$$

We use the quadratic formula with $a = 1$, $b = 1$, and $c = -4$:

$$x = \frac{-b \pm \sqrt{b^2 - 4ac}}{2a}$$

$$= \frac{-1 \pm \sqrt{(1)^2 - 4(1)(-4)}}{2(1)}$$

$$= \frac{-1 \pm \sqrt{17}}{2}.$$

You should use a calculator to find that the solutions are

$$x = 1.56 \text{ and } x = -2.56$$

to three significant digits. Replacing x by 1.56 in the equation $y = (-1)(x + 1)$ gives

$$y = -2.56.$$

Therefore, one solution is (1.56, –2.56). Replacing x by –2.56 in that equation gives

$$y = 1.56.$$

Therefore, a second solution is (–2.56, 1.56). You should check both solutions by substituting in each of the original equations. The graph of this system is a circle and a line:

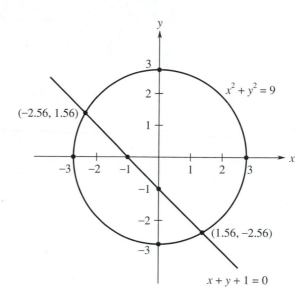

▲

<table>
<tr><td>

EXERCISE

14.4

</td></tr>
</table>

Solve the system:

1. $x + y = 2$

$\quad y = x^2$

2. $y = x + 1$

$\quad y = 2x^2$

3.–10. Solve the systems in Exercise 12.1, #9–#16, on pages 269–270, algebraically by using quadratic equations.

11. $y = x - 1$

$\quad x^2 + y^2 = 1$

12. $x + y = 1$

$\quad x^2 + y^2 = 5$

13. Solve the system in Exercise 12.1, #17, on page 270, algebraically by using a quadratic equation.

14. $x + y + 2 = 0$

$\quad x^2 + y^2 = 1$

15. Solve the system in Exercise 12.1, #18, on page 270, algebraically by using a quadratic equation.

16. $x + y + 4 = 0$

$\quad x^2 + y^2 = 8$

Proof of the Quadratic Formula

Starting with the quadratic equation in standard form,

$$ax^2 + bx + c = 0,$$

we divide both sides by a:

$$x^2 + \frac{b}{a}x + \frac{c}{a} = 0.$$

We subtract the constant term $\frac{c}{a}$ from both sides, and add the expression $\frac{b^2}{4a^2}$ to both sides:

$$x^2 + \frac{b}{a}x = -\frac{c}{a}$$

$$x^2 + \frac{b}{a}x + \frac{b^2}{4a^2} = -\frac{c}{a} + \frac{b^2}{4a^2}.$$

Now, the left-hand side can be factored:

$$\left(x + \frac{b}{2a}\right)\left(x + \frac{b}{2a}\right) = -\frac{c}{a} + \frac{b^2}{4a^2}$$

or

$$\left(x + \frac{b}{2a}\right)^2 = -\frac{c}{a} + \frac{b^2}{4a^2}.$$

Then, to add the fractions on the right-hand side, we write

$$\left(x + \frac{b}{2a}\right)^2 = -\frac{4ac}{4a^2} + \frac{b^2}{4a^2}$$

$$\left(x + \frac{b}{2a}\right)^2 = \frac{-4ac + b^2}{4a^2}$$

or

$$\left(x + \frac{b}{2a}\right)^2 = \frac{b^2 - 4ac}{4a^2}.$$

We complete the solution by taking the square root of each side:

$$x + \frac{b}{2a} = \pm\sqrt{\frac{b^2 - 4ac}{4a^2}}$$

$$x = -\frac{b}{2a} \pm\sqrt{\frac{b^2 - 4ac}{4a^2}}$$

$$= -\frac{b}{2a} \pm \frac{\sqrt{b^2 - 4ac}}{2a}$$

$$= \frac{-b \pm \sqrt{b^2 - 4ac}}{2a}.$$

☐ **Self-Test** ☐

Solve the equation and check the solutions:

1. $2x^2 + 4x = 3$

1. _____

2. $x^2 + 4x + 5 = 0$

2. _____

3. $4x^2 - 4x - 3 = 0$

3. _____

4. Draw the graph of $y = -x^2 + 2x + 3$.

4. _____

5. Solve the system $x + 2y = 1$ and $y = \frac{1}{2}x^2$.

5. _____

UNIT 15

Exponential and Logarithmic Functions

Introduction

In Unit 10, you learned about some types of equations in two variables. In this unit, you will learn about two more equations in two variables, the exponential function and the logarithmic function. First, you will learn how to graph exponential functions. Then, you will use exponential functions to define the logarithm and the logarithmic function. You will find values of logarithms, and graph logarithmic functions. There are properties of logarithms that you will use to solve exponential equations. Finally, you will learn about a circuit component called a capacitor and apply exponential equations to circuits containing capacitors.

OBJECTIVES

When you have finished this unit you should be able to:

1. Draw graphs of exponential functions by finding points on the graph.
2. Write exponential functions in logarithmic form, write logarithmic functions in exponential form, and find values of logarithmic functions when those values can be found by inspection.
3. Draw graphs of logarithmic functions by using exponential functions.
4. Use the properties of logarithms to solve exponential equations, and use natural logarithms to solve applied problems involving exponential growth and decay.
5. Use natural logarithms to solve problems involving charging and discharging capacitors in direct-current series *RC* circuits.

SECTION 15.1

Exponential Functions

Exponential equations have a variable in an exponent. A type of exponential equation in two variables has the form

$$y = b^x,$$

where $b > 0$ and $b \neq 1$. Equations of this form are not linear and have graphs that are not straight lines. We can draw their graphs by finding the coordinates of several points of the graph.

EXAMPLE 15.1 Draw the graph of $y = 2^x$.

SOLUTION We find ordered pairs containing the coordinates of several points on the graph. If $x = 0$, then

$$y = 2^0 = 1.$$

If $x = 1$, then

$$y = 2^1 = 2.$$

If $x = 2$, then

$$y = 2^2 = 4.$$

341

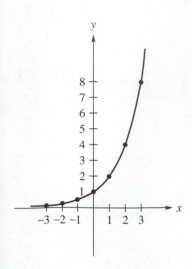

Graph for Example 15.1.

If $x = 3$, then

$$y = 2^3 = 8.$$

So far we have the ordered pairs $(0, 1)$, $(1, 2)$, $(2, 4)$, and $(3, 8)$. We should also find some ordered pairs where the x-value is negative. If $x = -1$, then

$$y = 2^{-1} = \frac{1}{2}.$$

If $x = -2$, then

$$y = 2^{-2} = \frac{1}{4}.$$

If $x = -3$, then

$$y = 2^{-3} = \frac{1}{8}.$$

Thus, we also have the ordered pairs $(-1, \frac{1}{2})$, $(-2, \frac{1}{4})$, and $(-3, \frac{1}{8})$. We plot the points we have found and draw a smooth curve. The graph is shown at the left. We observe that the y-values are always positive; that is, y is never zero or negative.   ▲

Recall from Section 10.3 that a relation in two variables x and y is a function if for each allowable value of x there is exactly one value of y. We can use the vertical line test shown in Section 10.4 to see that the relation $y = 2^x$ is a function.

Also recall from Section 10.4 that a graph approaches a line asymptotically, or has a line as an asymptote, if the graph gets closer and closer to the line without crossing or touching it. We observe that the graph of $y = 2^x$ gets closer and closer to the x-axis as we go to the left. Thus the graph approaches the x-axis asymptotically; that is, the x-axis is an asymptote of the graph.

EXAMPLE 15.2 ▶ Draw the graph of $y = \left(\frac{1}{2}\right)^x$.

SOLUTION We find ordered pairs containing the coordinates of several points on the graph. If $x = 0$, then

$$y = \left(\frac{1}{2}\right)^0 = 1.$$

If $x = 1$, then

$$y = \left(\frac{1}{2}\right)^1 = \frac{1}{2}.$$

If $x = 2$, then

$$y = \left(\frac{1}{2}\right)^2 = \frac{1}{4}.$$

If $x = 3$, then

$$y = \left(\frac{1}{2}\right)^3 = \frac{1}{8}.$$

So far, we have the ordered pairs $(0, 1)$, $(1, \frac{1}{2})$, $(2, \frac{1}{4})$, and $(3, \frac{1}{8})$. We also find some ordered pairs where the x-value is negative. If $x = -1$, then

$$y = \left(\frac{1}{2}\right)^{-1}$$

$$= \frac{1}{\frac{1}{2}}$$

$$= 2.$$

If $x = -2$, then

$$y = \left(\frac{1}{2}\right)^{-2}$$

$$= \frac{1}{\left(\frac{1}{2}\right)^2}$$

$$= \frac{1}{\frac{1}{4}}$$

$$= 4.$$

If $x = -3$, then

$$y = \left(\frac{1}{2}\right)^{-3}$$

$$= \frac{1}{\left(\frac{1}{2}\right)^3}$$

$$= \frac{1}{\frac{1}{8}}$$

$$= 8.$$

Graph for Example 15.2.

Thus, we also have the ordered pairs $(-1, 2)$, $(-2, 4)$, and $(-3, 8)$. We plot the points we have found and draw a smooth curve. The graph is shown at the left. Again, we observe that y is never zero or negative. ▲

The relation

$$y = \left(\frac{1}{2}\right)^x$$

is also a function. In general, every exponential equation of the form

$$y = b^x,$$

with $b > 0$ and $b \neq 1$, fits the definition of a function. Therefore, such equations are called **exponential functions**.

The graph of the exponential function $y = b^x$, where $b > 1$, approaches the x-axis asymptotically to the left, and rises very rapidly to the right. The graph of the exponential function $y = b^x$, where $0 < b < 1$, approaches the x-axis asymptotically to the right, and rises very rapidly to the left. All y-values of exponential functions are positive.

EXERCISE

15.1

Draw the graph:

1. $y = 3^x$ 2. $y = 4^x$

3. $y = 5^x$ 4. $y = 10^x$ (Use the scale 10, 20, 30, . . . on the y-axis.)

5. $y = \left(\frac{1}{3}\right)^x$ 6. $y = \left(\frac{1}{4}\right)^x$

7. $y = \left(\frac{1}{5}\right)^x$ 8. $y = \left(\frac{1}{10}\right)^x$ (Use the scale 10, 20, 30, . . . on the y-axis.)

<table>
<tr><td>SECTION
15.2</td></tr>
</table>

Definition of the Logarithm

In Section 15.1, exponential functions are defined by the equation

$$y = b^x,$$

where $b > 0$ and $b \neq 1$. Now, we recall the inverse of a function from Section 10.4. The inverse of a function is the relation obtained by reversing the coordinates of the ordered pairs of the function. We can form the inverse of a function by exchanging x and y. Thus the exponential function

$$y = b^x$$

has an inverse exponential relation

$$x = b^y.$$

We will see in Section 15.3 that this inverse exponential relation is also a function. There is no algebraic method by which the inverse relation

$$x = b^y$$

can be solved for y. Therefore, we define an equation that is the solution for y. This equation is called a **logarithmic function**.

Logarithmic Function: The **logarithmic function** $y = \log_b x$, where $b > 0$ and $b \neq 1$, is the solution of the exponential relation $x = b^y$.

The equation $y = \log_b x$ is read "y is the logarithm to the base b of x," or "y is log b of x." We say that the equation $x = b^y$ has the **logarithmic form**

$$y = \log_b x.$$

The equation $y = \log_b x$ has the **exponential form**

$$x = b^y.$$

EXAMPLE 15.3   Write in logarithmic form:

a. $2^4 = 16$ b. $16^{\frac{1}{4}} = 2$ c. $2^{-4} = \dfrac{1}{16}$

SOLUTIONS   We use the definition of $y = \log_b x$ as the logarithmic form of $x = b^y$, or $b^y = x$.

a.   For the exponential form

$$2^4 = 16,$$

we have

$$b^y = x,$$

where $b = 2$, $y = 4$, and $x = 16$. Therefore, the logarithmic form is

$$y = \log_b x$$

$$4 = \log_2 16$$

or

$$\log_2 16 = 4.$$

b. For the exponential form

$$16^{\frac{1}{4}} = 2,$$

we have

$$b^y = x,$$

where $b = 16$, $y = \frac{1}{4}$, and $x = 2$. Therefore, the logarithmic form is

$$y = \log_b x$$

$$\frac{1}{4} = \log_{16} 2$$

or

$$\log_{16} 2 = \frac{1}{4}.$$

c. For the exponential form

$$2^{-4} = \frac{1}{16},$$

we have

$$b^y = x,$$

where $b = 2$, $y = -4$, and $x = \frac{1}{16}$. Therefore, the logarithmic form is

$$y = \log_b x$$

$$-4 = \log_2 \frac{1}{16}$$

or

$$\log_2 \frac{1}{16} = -4.$$

▲

EXAMPLE 15.4 ▶ Write in exponential form:

a. $\log_5 25 = 2$ **b.** $\log_{25} 5 = \frac{1}{2}$ **c.** $\log_5 \frac{1}{25} = -2$

SOLUTIONS We use the definition of $x = b^y$ as the exponential form of $y = \log_b x$, or $\log_b x = y$.

a. For the logarithmic form

$$\log_5 25 = 2,$$

we have

$$\log_b x = y,$$

where $b = 5$, $x = 25$, and $y = 2$. Therefore, the exponential form is

$$x = b^y$$

$$25 = 5^2$$

or

$$5^2 = 25.$$

b. For the logarithmic form

$$\log_{25} 5 = \frac{1}{2},$$

we have

$$\log_b x = y,$$

where $b = 25$, $x = 5$, and $y = \frac{1}{2}$. Therefore, the exponential form is

$$x = b^y$$

$$5 = 25^{\frac{1}{2}}$$

or

$$25^{\frac{1}{2}} = 5.$$

c. For the logarithmic form

$$\log_5 \frac{1}{25} = -2,$$

we have

$$\log_b x = y,$$

where $b = 5$, $x = \frac{1}{25}$, and $y = -2$. Therefore, the exponential form is

$$x = b^y$$

$$\frac{1}{25} = 5^{-2}$$

or

$$5^{-2} = \frac{1}{25}.$$ ▲

VALUES BY INSPECTION

Values of certain logarithmic expressions can be found by using the definition of the logarithm. These expressions have exponential forms in which the exponent can be found by inspection.

| EXAMPLE 15.5 | Find the value:

a. $\log_2 32$ **b.** $\log_{125} 5$ **c.** $\log_5 \dfrac{1}{125}$

| SOLUTIONS | We write the expression $\log_b x$ as the logarithmic function $\log_b x = y$, and then rewrite this equation in the exponential form $b^y = x$.

a. We write

$$\log_2 32$$

as the logarithmic function

$$\log_2 32 = y.$$

Then, we rewrite this equation in the exponential form

$$2^y = 32.$$

We can find the exponent y if we have the same base on each side. Thus, we write 32 as a power of 2 to obtain

$$2^y = 2^5.$$

By inspection it is clear that $y = 5$, and therefore

$$\log_2 32 = 5.$$

b. We write

$$\log_{125} 5$$

as the logarithmic function

$$\log_{125} 5 = y.$$

Then we rewrite this equation in the exponential form

$$125^y = 5.$$

We can find the exponent y if we have base 125 on each side. Since 5 is the cube root of 125, we can write

$$125^y = \sqrt[3]{125}.$$

Then, we write the cube root as a fractional exponent to obtain

$$125^y = 125^{\frac{1}{3}}.$$

By inspection it is clear that $y = \frac{1}{3}$, and therefore

$$\log_{125} 5 = \frac{1}{3}.$$

c. We write

$$\log_5 \frac{1}{125}$$

as the logarithmic function

$$\log_5 \frac{1}{125} = y.$$

Then we rewrite this equation in the exponential form

$$5^y = \frac{1}{125}.$$

We write 125 as a power of 5 to obtain

$$5^y = 5^{-3}.$$

By inspection it is clear that $y = -3$, and therefore

$$\log_5 \frac{1}{125} = -3.$$

▲

EXAMPLE 15.6 ▶ Find the value of $\log_{10} 0.001$.

SOLUTION We write

$$\log_{10} 0.001 = y,$$

and then rewrite this equation in the exponential form

$$10^y = 0.001$$

$$10^y = 10^{-3}.$$

By inspection it is clear that $y = -3$, and therefore

$$\log_{10} 0.001 = -3.$$

▲

COMMON LOGARITHMS

We observe that logarithms to the base 10 are related to scientific and engineering notations. Logarithms to the base 10 are called **common logarithms**, and may be written without stating the base. Thus,

$$\log 0.001 = \log_{10} 0.001.$$

EXAMPLE 15.7 Find the value of log 1,000,000.

SOLUTION Since there is no stated base, the logarithm is a common logarithm with base 10. We write

$$\log_{10} 1,000,000 = y,$$

and then rewrite this equation in the exponential form

$$10^y = 1,000,000$$

$$10^y = 10^6.$$

By inspection it is clear that $y = 6$, and therefore

$$\log 1,000,000 = 6. \qquad \blacktriangle$$

EXERCISE 15.2

Write in logarithmic form:

1. $2^3 = 8$ 2. $3^2 = 9$ 3. $8^{\frac{1}{3}} = 2$ 4. $9^{\frac{1}{2}} = 3$

5. $2^{-3} = \dfrac{1}{8}$ 6. $3^{-2} = \dfrac{1}{9}$ 7. $10^{-2} = 0.01$ 8. $100^{\frac{1}{2}} = 10$

Write in exponential form:

9. $\log_3 81 = 4$ 10. $\log_4 64 = 3$ 11. $\log_{81} 3 = \dfrac{1}{4}$ 12. $\log_{64} 4 = \dfrac{1}{3}$

13. $\log_3 \dfrac{1}{81} = -4$ 14. $\log_4 \dfrac{1}{64} = -3$ 15. $\log_{10} 1000 = 3$ 16. $\log_{10} 0.0001 = -4$

Find the value:

17. $\log_6 36$ 18. $\log_5 125$ 19. $\log_2 64$ 20. $\log_7 343$

21. $\log_{36} 6$ 22. $\log_{27} 3$ 23. $\log_{64} 4$ 24. $\log_{64} 8$

25. $\log_2 \dfrac{1}{4}$ 26. $\log_4 \dfrac{1}{4}$ 27. $\log_5 \dfrac{1}{125}$ 28. $\log_6 \dfrac{1}{36}$

29. $\log 100,000$ 30. $\log 0.00001$ 31. $\log 10$ 32. $\log 1$

SECTION 15.3

Graphs of Logarithmic Functions

To draw the graph of the logarithmic function

$$y = \log_b x,$$

we can use its relationship to the exponential function

$$y = b^x.$$

Since the logarithmic function is the inverse of the exponential function, we can find ordered pairs of $y = \log_b x$ by reversing the coordinates of ordered pairs of $y = b^x$.

EXAMPLE 15.8   Draw the graph of $y = \log_2 x$.

SOLUTION   We must find ordered pairs containing the coordinates of several points on the graph. We can find such points by reversing the coordinates of ordered pairs of

$$y = 2^x.$$

We recall from Example 15.1 that some ordered pairs of this exponential function are

$$(0, 1), \ (1, 2), \ (2, 4), \ \text{and} \ (3, 8).$$

To draw the graph of

$$y = \log_2 x,$$

the coordinates of these ordered pairs may be reversed to become

$$(1, 0), \ (2, 1), \ (4, 2), \ \text{and} \ (8, 3).$$

Similarly, for negative values of x, the ordered pairs

$$\left(-1, \frac{1}{2}\right), \ \left(-2, \frac{1}{4}\right), \ \text{and} \ \left(-3, \frac{1}{8}\right)$$

of the exponential function may be reversed to become

$$\left(\frac{1}{2}, -1\right), \ \left(\frac{1}{4}, -2\right), \ \text{and} \ \left(\frac{1}{8}, -3\right).$$

We plot the points we have found and draw a smooth curve. The graph is

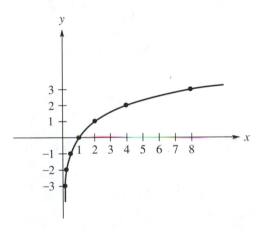

Observe that we can use the vertical line test to see that the logarithmic equation

$$y = \log_2 x$$

is a function. In general, every logarithmic equation with $b > 0$ and $b \neq 1$ fits the definition of a function.

EXAMPLE 15.9   Draw the graphs of $y = 2^x$ and $y = \log_2 x$ on one set of axes.

SOLUTION   We draw the graph of

$$y = 2^x$$

by finding several ordered pairs as in Example 15.1. Then, we can draw the graph of

$$y = \log_2 x$$

by reversing the coordinates of the ordered pairs. We show the graphs on one set of axes, with the line $y = x$ as a dashed line:

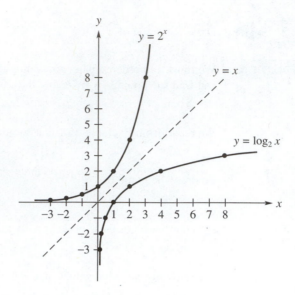

When we interchange x and y in the exponential function

$$y = b^x$$

to form its inverse, the logarithmic function

$$y = \log_b x,$$

we reverse all properties of x and y. For the exponential function, the y-values are always positive; that is, y is never zero or negative. For the logarithmic function, the x-values are always positive; that is, x is never zero or negative.

The graph of the exponential function approaches the x-axis asymptotically as we go to the left; thus, the x-axis is an asymptote of the graph of the exponential function. The graph of the logarithmic function approaches the y-axis asymptotically as we go down; thus, the y-axis is an asymptote of the graph of the logarithmic function.

Finally, we recall from Unit 10 that graphs of inverse relations and functions are symmetric with respect to the line $y = x$. When we draw the graphs of the exponential and logarithmic functions on one set of axes, we observe that the graphs are mirror images of one another, with the line $y = x$ as the mirror.

EXERCISE

15.3

Draw the graph:

1. $y = \log_3 x$ **2.** $y = \log_4 x$

3. $y = \log_5 x$ **4.** $y = \log_{10} x$ (Use the scale 10, 20, 30, ... on the x-axis.)

Draw each pair of graphs on one set of axes with the line $y = x$:

5. $y = 3^x$ and $y = \log_3 x$ **6.** $y = 4^x$ and $y = \log_4 x$

SECTION

15.4

Exponential Equations

Logarithms were invented in the early 1600s by John Napier (1550–1617), a Scottish mathematician. The name "logarithm" comes from the Greek words for "ratio" and "number." Napier based his logarithms on ratios of terms of certain mathematical sequences. The relationship to the exponential function was discovered later.

Originally, logarithms were used as an aid to calculation. A calculating device invented by Napier, and famous in his time, was called "Napier's bones." Another calculating device that works on the principles of logarithms is the slide rule. Slide rules were used by scientists and engineers for centuries until recent decades in which the hand-held calculator has become widely available. Although logarithms are no longer used for calculations, logarithmic and exponential equations are still important for their many applications in science, economics, and other fields.

We recall that an exponential equation is an equation with a variable in an exponent. Exponential equations cannot be solved by algebraic methods. The problem is to get the variable out of the exponent. This can be accomplished by using some special properties of logarithms. The definition of the logarithm can be used to prove the three following properties.

Properties of Logarithms:

    1. $\log_b PQ = \log_b P + \log_b Q$, $P > 0$ and $Q > 0$.

    2. $\log_b \dfrac{P}{Q} = \log_b P - \log_b Q$, $P > 0$ and $Q > 0$.

    3. $\log_b P^x = x \log_b P$, $P > 0$.

The proofs of these properties are given at the end of this unit.

Property 3, in particular, is useful for solving exponential equations. This property is often used in connection with a property that can be shown by the methods in Section 15.2: For any base b, $\log_b b = 1$. We write

$$\log_b b = y,$$

and then rewrite this equation in the exponential form

$$b^y = b$$

$$b^y = b^1.$$

By inspection it is clear that $y = 1$, and therefore

$$\log_b b = 1.$$

For common logarithms, we note that

$$\log 10 = \log_{10} 10,$$

and so

$$\log 10 = 1.$$

EXAMPLE 15.10 Solve the equation $10^x = 30$ and check the solution.

SOLUTION We take the common logarithm of each side of the equation:

$$10^x = 30$$

$$\log 10^x = \log 30.$$

Using Property 3 of logarithms we can replace $\log 10^x$ by $x \log 10$. Thus, by using Property 3 we remove the variable from the exponent to obtain

$$\log 10^x = \log 30$$

$$x \log 10 = \log 30.$$

Furthermore, we know that $\log 10 = 1$, so we have

$$x \log 10 = \log 30$$

$$x(1) = \log 30$$

$$x = \log 30.$$

You can use a scientific calculator to find a numerical approximation for x. The common log key marked

$$\boxed{\log}$$

gives the common logarithm of any positive number:

	display:
Enter 30	30.
Press $\boxed{\log}$	1.477121255

Thus, $x = 1.48$ to three significant digits.

For some calculators, such as those identified as direct algebraic logic (DAL) calculators, you must first press the common log key and then enter 30. Finally, you must press the equals key to display the value of the logarithm.

You can check this solution by writing

$$10^x = 30$$
$$10^{1.48} \overset{?}{\approx} 30.$$

Since the exponential function is the inverse of the logarithmic function, most calculators find 10^x by using the inverse, or second function, of the log key. Thus, on most calculators, you should press

$$\boxed{\text{INV}} \quad \text{or} \quad \boxed{2^{\text{nd}}}$$

and

$$\boxed{\log}$$

when we say to press

$$\boxed{10^x}.$$

You can find the value of $10^{1.48}$ by this method:

	display:
Enter 1.48	1.48
Press $\boxed{10^x}$	30.1995172

Therefore, $10^{1.48} \approx 30$. (For a DAL calculator, you press the inverse or second function and log keys, enter 1.48, and press the equals key.) ▲

Recall that x must always be positive in a logarithmic function $y = \log_b x$. You can enter a number on your calculator and press the change-sign key, and then press the log key. You calculator will give some type of error indication.

It is possible, however, to find values for the exponential function $y = 10^x$ when x is negative. For example, you can follow these steps:

	display:
Enter 1	1.
Press $\boxed{+/-}$	$-1.$
Press $\boxed{10^x}$	0.1

You have shown that $10^{-1} = 0.1$.

We can use common logarithms to solve exponential equations in which the base b is some number other than 10.

EXAMPLE 15.11 Solve the equation $15^x = 25$ and check the solution.

SOLUTION We take the common logarithm of each side of the equation:

$$15^x = 25$$

$$\log 15^x = \log 25.$$

We apply Property 3 of logarithms to obtain

$$x \log 15 = \log 25.$$

Dividing both sides by log 15, we have

$$x = \frac{\log 25}{\log 15}.$$

You can use a calculator to find an approximation for x:

		display:
Enter	25	25.
Press	log	1.397940009
Press	÷	1.397940009
Enter	15	15.
Press	log	1.176091259
Press	=	1.188632258

Thus, $x = 1.19$ to three significant digits.

You can check this solution, by writing

$$15^x = 25$$

$$15^{1.19} \stackrel{?}{\approx} 25.$$

But, because the base is not 10, you cannot use the inverse of the common logarithm to find the value of $15^{1.19}$. Your scientific calculator also has a power key

$$\boxed{y^x}.$$

On some calculators, this key might require the use of

$$\boxed{INV} \quad \text{or} \quad \boxed{2^{nd}}.$$

Also, on some calculators, it might be marked

$$\boxed{x^y}.$$

Now, using the power key on the calculator, follow these steps:

		display:
Enter	15	15.
Press	y^x	15.
Enter	1.19	1.19
Press	=	25.09276956

Therefore, $15^{1.19} \approx 25$. ▲

Common logarithms are also useful in solving some types of equations that are not actually exponential equations, because the variable is not in an exponent. These equations have numerical exponents that cannot be handled by ordinary algebraic methods.

EXAMPLE 15.12 Solve the equation $x^{2.5} = 1.43$ and check the solution.

SOLUTION We take the common logarithm of each side of the equation:

$$x^{2.5} = 1.43$$

$$\log x^{2.5} = \log 1.43.$$

Then, we apply Property 3 of logarithms and divide both sides by 2.5 to solve for log x:

$$2.5 \log x = \log 1.43$$

$$\log x = \frac{\log 1.43}{2.5}.$$

You can use a calculator to find log x:

		display:
Enter	1.43	1.43
Press	log	0.155336037
Press	÷	0.155336037
Enter	2.5	2.5
Press	=	0.062134414

So far, you have found

$$\log x = 0.062134414.$$

To find x, you need the exponential form

$$x = 10^{0.062134414}.$$

Thus, continuing from the preceding steps, you can find

	display:
(currently displayed)	0.062134414
Press 10^x	1.153810309

Therefore, $x = 1.15$ to three significant digits. (For a DAL calculator, you press the log key before entering 1.43. Then, you find 10^x last, by pressing the inverse or second function, log, and equals keys.)

You can now use the power key to check this result. The process of finding x from log x by using the inverse and log keys is often referred to as finding the **antilog** of x. ▲

NATURAL LOGARITHMS

Another kind of logarithm often encountered in applications is the **natural logarithm**. The base of natural logarithms is a constant called e. Like π, the constant e is a number that is found in many places in mathematics and science. Also like π, the constant e is an irrational number, which may be approximated to as many places as we wish by a decimal. A common approximation for e is

$$e \approx 2.718.$$

We use ln x to indicate the natural logarithm of x. When we write ln x, we mean $\log_e x$. We note that

$$\ln e = \log_e e,$$

and so

$$\ln e = 1.$$

On your calculator, the natural logarithm of a number is found by using a key marked

$$\boxed{\text{lnx}} \quad \text{or} \quad \boxed{\text{ln}}.$$

Also, values of the corresponding exponential function e^x are found by using

$$\boxed{\text{INV}} \quad \text{or} \quad \boxed{2^{\text{nd}}}$$

and

$$\boxed{\text{lnx}}.$$

We mean this combination when we say to press

$$\boxed{e^x}$$

on your calculator.

The base e occurs in formulas describing growth and decay. One type of example describes the growth of money in a savings account on which the interest is compounded continuously. The formula for such examples is

$$A = Pe^{rt},$$

where A is the total amount in the account, P is the principal, r is the interest rate written in decimal form, and t is the time in years.

EXAMPLE 15.13

You put \$500 in a savings account. The interest rate is 6% compounded continuously. How much will there be in the account after 2 years?

SOLUTION We have $P = 500$, $r = 0.06$, and $t = 2$. We must find A:

$$A = Pe^{rt}$$
$$= 500e^{(0.06)(2)}$$
$$= 500e^{0.12}.$$

This equation is not an exponential equation because A is not in an exponent. You can use a calculator to find the value of $500e^{0.12}$:

		display:
Enter	500	500.
Press	$\times$	500.
Enter	0.12	0.12
Press	e^x	1.127496852
Press	$=$	563.7484258

There will be approximately \$564 in the account. (As different banks use slightly different conventions, do not hold your bank to exactly this amount. For a DAL calculator, you press the inverse or second function and natural log keys for e^x, enter 0.12, and press the equals key.) ▲

EXAMPLE 15.14

A principal of \$1000 is invested at 8% compounded continuously. How long will it take to double?

SOLUTION Doubling the principal gives \$2000, and so $A = \$2000$, $P = \$1000$, and $r = 0.08$. We must find t:

$$A = Pe^{rt}$$
$$2000 = 1000e^{0.08t}$$
$$2 = e^{0.08t}.$$

This equation is an exponential equation because t is in an exponent. We take the natural logarithm of each side of the equation and apply Property 3 of logarithms to obtain

$$\ln 2 = \ln e^{0.08t}$$
$$\ln 2 = 0.08t(\ln e).$$

Since $\ln e = 1$, we have

$$\ln 2 = 0.08t(1)$$
$$\ln 2 = 0.08t.$$

Then, dividing both sides by 0.08, we have

$$t = \frac{\ln 2}{0.08}.$$

You can use a calculator to find
$$t = 8.66 \text{ yr},$$

or approximately $8\frac{2}{3}$ years (8 years, 8 months). Also, you should use the calculator to check by showing that
$$1000e^{(0.08)(8.66)} \approx 2000. \qquad \blacktriangle$$

PREVIEW: RC CIRCUITS

Examples of growth and decay formulas occur in electronics. An example is the RC circuit, in which a component called a capacitor is connected in series with a DC source and a resistor. Such circuits are covered in more detail in Section 15.5.

EXAMPLE 15.15

In a certain RC circuit, the voltage v_R across the resistor decreases according to the formula

$$v_R = Ee^{-\frac{t}{200 \text{ ms}}}.$$

If $E = 10$ V, find v_R when $t = 100$ ms.

SOLUTION

Since $E = 10$ V and $t = 100$ ms, we have
$$v_R = Ee^{-\frac{t}{200 \text{ ms}}}$$
$$= 10e^{-\frac{100 \text{ ms}}{200 \text{ ms}}} \text{ V}$$
$$= 10e^{-0.5} \text{ V}.$$

You can follow steps similar to those in Example 15.13 to find the value of $10e^{-0.5}$:

		display:
Enter 10		10.
Press	$\times$	10.
Enter .5		0.5
Press	+/−	−0.5
Press	e^x	0.606530659
Press	=	6.065306597

Thus, $v_R = 6.07$ V. $\blacktriangle$

EXAMPLE 15.16

For the *RC* circuit in Example 15.15, how long will it take for v_R to decrease to 4 V?

SOLUTION

From Example 15.15, we have the formula
$$v_R = Ee^{-\frac{t}{200 \text{ ms}}},$$

with $E = 10$ V. We find t when $v_R = 4$ V by using
$$4 \text{ V} = 10e^{-\frac{t}{200 \text{ ms}}} \text{ V}$$
$$0.4 = e^{-\frac{t}{200 \text{ ms}}}.$$

We take the natural logarithm of each side of the equation and apply Property 3 of logarithms to obtain
$$\ln 0.4 = \ln e^{-\frac{t}{200 \text{ ms}}}$$
$$\ln 0.4 = -\frac{t}{200 \text{ ms}}(\ln e)$$
$$\ln 0.4 = -\frac{t}{200 \text{ ms}}.$$

Then, multiplying both sides by -200 ms, we have

$$t = -200 \ln 0.4 \text{ ms.}$$

You can use a calculator to find

$$t = 183 \text{ ms.}$$

Observe that $\ln 0.4$ is negative, so the result is positive. You should use the calculator to check by showing that

$$10e^{-\frac{183}{200}} \approx 4. \qquad\qquad \blacktriangle$$

EXERCISE

15.4

Solve the equation and check the solution:

1. $10^x = 15$ **2.** $10^x = 3.5$ **3.** $12^x = 50$ **4.** $7.4^x = 5.5$

5. $6^x = 0.488$ **6.** $3.2^x = 0.175$ **7.** $5^{x-2} = 15$ **8.** $3^{3x} = 8.4$

9. $x^{1.2} = 6.4$ **10.** $x^{3.3} = 13.6$ **11.** $x^{2.2} = 0.2$ **12.** $x^{-1.5} = 0.36$

13. If you invest $2000 at 7% compounded continuously, how much will you have at the end of 2 years?

14. If you invest $150 at 5% compounded continuously, how much will you have at the end of 6 months ($\frac{1}{2}$ year)?

15. A principal of $4000 is invested at $5\frac{1}{2}$% (0.055) compounded continuously. How long will it take to double?

16. A principal of $1000 is invested at $7\frac{1}{2}$% compounded continuously. How long will it take to triple?

17. If $500 is invested at 8% compounded continuously, how long will it take to increase to $600?

18. If $200 is invested at $6\frac{1}{2}$% compounded continuously, how long will it take to increase to $210?

19. In a certain RC circuit, the voltage v_R across the resistor decreases according to the formula

$$v_R = Ee^{-\frac{t}{100 \text{ ms}}}.$$

If $E = 50$ V, find v_R when $t = 10$ ms.

20. When the capacitor is discharging in a certain RC circuit, the voltage v_C across the capacitor decreases according to the formula

$$v_C = Ee^{-\frac{t}{2 \text{ s}}}.$$

If $E = 50$ V, find v_C when $t = 4$ s.

21. For the RC circuit in Exercise 19, if $E = 50$ V, how long will it take for v_R to decrease to 40 V?

22. For the RC circuit in Exercise 20, if $E = 100$ V, how long will it take for v_C to decrease to 25 V?

23. When the capacitor is charging in a certain RC circuit, the voltage v_C across the capacitor increases according to the formula

$$v_C = E(1 - e^{-\frac{t}{100 \text{ ms}}}).$$

If $E = 50$ V, find v_C when $t = 10$ ms.

24. For the RC circuit in Exercise 23, if $E = 50$ V, how long will it take for v_C to increase to 40 V?

25. Dating of ancient remains can be done by measuring the amount remaining of a radioactive isotope of carbon called carbon 14. This isotope decays according to the formula

$$N = N_0 e^{-0.0001245t},$$

where N is the amount remaining, N_0 is the initial amount, and t is the time in years. What is the half-life of carbon 14; that is, how long will it take the initial amount N_0 to reduce to $\frac{1}{2}N_0$?

26. If the initial amount of carbon 14 in a piece of wood is known to have been N_0, and now only $\frac{1}{3}N_0$ remains, how old is the piece of wood? (Use the formula in Exercise 25.)

27. A typical I–V characteristic curve for a *commercial* diode is shown in Example 10.25. The equation for the I–V characteristic curve of an *ideal* diode is

$$I = I_S(e^{19.46\,V} - 1)$$

(at room temperature 25° C, when I is positive). If $I_S = 1\ \mu A$, find I when $V = 0.5$ V.

28. For the ideal diode in Exercise 27, find I when

    a. $V = 0.6$ V   **b.** $V = 0.7$ V   **c.** $V = 0.8$ V

(Note: Like all exponential functions, the diode curves increase very quickly; thus, values of I become very large for even relatively small values of V.)

Series RC Circuits

A **capacitor** is a device that stores oppositely charged particles on plates separated by an insulating material. A circuit containing a resistor and a capacitor in series with a battery is a **series RC circuit**. When the switch S is closed, the current flow from the battery E deposits electrons on one side of the capacitor C, and removes them from the other. We say that the capacitor is charging. The ability of the capacitor to store charge is called its **capacitance**, and is measured in **farads** (F), named for the British chemist and physicist Michael Faraday (1791–1867).

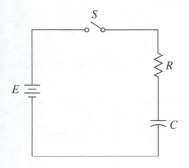

A series *RC* circuit

Because the components in the *RC* circuit are in series, Kirchhoff's voltage law is true for the circuit. If v_C and v_R are the voltages across the capacitor and the resistor, then

$$E = v_C + v_R.$$

The voltage E is constant. Lowercase letters are often used for quantities such as v_C and v_R, which vary with time.

When the switch is closed at time $t = 0$, the voltage across the capacitor is $v_C = 0$ and, therefore, the voltage across the resistor is $v_R = E$. As the capacitor charges, v_C increases from 0 while v_R decreases from E. The voltage v_C across the capacitor at any time t while the capacitor is charging is given by the formula

$$v_C = E(1 - e^{-\frac{t}{\tau}}).$$

The voltage v_R across the resistor is given by the formula

$$v_R = Ee^{-\frac{t}{\tau}}.$$

The letter τ (the Greek letter tau) in these formulas is the **time constant** for the *RC* circuit. The time constant is given by the formula

$$\tau = RC.$$

We observe that

$$v_C + v_R = E(1 - e^{-\frac{t}{\tau}}) + Ee^{-\frac{t}{\tau}}$$

$$= E - Ee^{-\frac{t}{\tau}} + Ee^{-\frac{t}{\tau}}$$

$$= E.$$

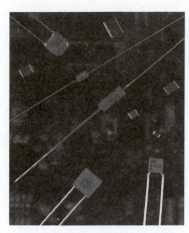

Ceramic capacitors. Courtesy of
Kemet Electronics.

THE UNIVERSAL TIME-CONSTANT CURVES

The **universal time-constant curves** express v_R and v_C as percents of E. To express v_R as a percent of E, we divide both sides of the formula for v_R by E:

$$v_R = Ee^{-\frac{t}{\tau}}$$

$$\frac{v_R}{E} = e^{-\frac{t}{\tau}}.$$

The graph of this exponential function decreases because the exponent is negative. To express v_C as a percent of E, we divide both sides of the formula for v_C by E:

$$v_C = E(1 - e^{-\frac{t}{\tau}})$$

$$\frac{v_C}{E} = 1 - e^{-\frac{t}{\tau}}.$$

The graph of this exponential function increases. These are the graphs of the universal time-constant curves through five time constants:

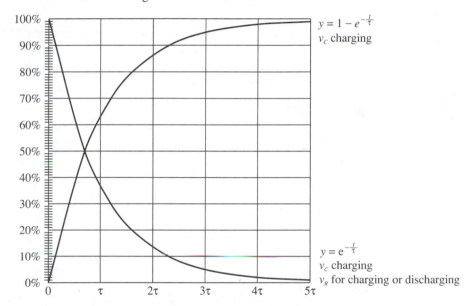

The numbers at the left are percents of the source voltage E. The curve that represents v_C starts from zero and increases, approaching 100% of E asymptotically. Thus, v_C gets closer and closer to E but never equals E. The curve that represents v_R starts from 100% of E and decreases, approaching zero asymptotically. Thus, v_R gets closer and closer to zero but never equals zero.

USING THE CHARGING FORMULAS

We can use the formulas for v_C and v_R to find their values at any time t while the capacitor is charging.

EXAMPLE 15.17

In a series RC circuit, $E = 10$ V, $R = 10$ kΩ, and $C = 10$ μF. Find v_C, v_R, and the current i when $t = 50$ ms.

SOLUTION

First, we find the time constant τ:

$$\tau = RC$$

$$= (10 \text{ k}\Omega)(10 \text{ μF})$$

$$= 100 \text{ ms.}$$

To find v_C, we use the formula

$$v_C = E(1 - e^{-\frac{t}{\tau}}).$$

Substituting for E, t, and τ, we have

$$v_C = 10\,(1 - e^{-\frac{50\,\text{ms}}{100\,\text{ms}}})\,\text{V}$$

$$= 10\,(1 - e^{-0.5})\,\text{V}.$$

At this point, you can use a calculator to find v_C. You can avoid using the parentheses keys by starting inside the parentheses and multiplying by 10 at the end of the calculation:

		display:
Enter	1	1.
Press	$-$	1.
Enter	.5	0.5
Press	+/−	− 0.5
Press	e^x	0.606530659
Press	=	0.39346934
Press	×	0.39346934
Enter	10	10.
Press	=	3.934693403

Thus, $v_C = 3.93$ V. Observe that at 50 ms, t is one-half of τ. Referring to the universal time-constant curve for v_C charging, you can see that v_C is approximately 40% of E.

To find v_R we use the formula

$$v_R = Ee^{-\frac{t}{\tau}}.$$

Substituting for E, t, and τ, we have

$$v_R = 10e^{-\frac{50\,\text{ms}}{100\,\text{ms}}}\,\text{V}$$

$$= 10e^{-0.5}\,\text{V}.$$

You should use a calculator to find that $v_R = 6.07$ V. Referring to the universal time-constant curve for v_R, you can see that v_R is approximately 60% of E.

To check, observe that

$$v_C + v_R = 3.93\,\text{V} + 6.07\,\text{V}$$

$$= 10\,\text{V}.$$

Thus, $v_C + v_R = E$.

Because the circuit is a series circuit, the current i is the same throughout the circuit at any time t. To find i, we use Ohm's law:

$$i = \frac{v_R}{R}$$

$$= \frac{6.07\,\text{V}}{10\,\text{k}\Omega}$$

$$= 0.607\,\text{mA},$$

or 607 μA.

▲

If we are given v_C, v_R, or i, for some time t while the capacitor is charging, we can use natural logarithms to find the value of t.

EXAMPLE 15.18

For the circuit described in Example 15.17, find t when $v_C = 7$ V.

SOLUTION

Because R and C are the same as in Example 15.17, the time constant is $\tau = 100$ ms as before. Using the formula for v_C, we write

$$v_C = E(1 - e^{-\frac{t}{\tau}})$$

$$7 \text{ V} = 10 \, (1 - e^{-\frac{t}{100 \text{ ms}}}) \text{ V.}$$

Dividing both sides by 10 V, we have

$$0.7 = 1 - e^{-\frac{t}{100 \text{ ms}}}.$$

Then, isolating the exponential expression, we have

$$e^{-\frac{t}{100 \text{ ms}}} = 0.3.$$

Now, we can take the natural logarithm of each side and apply Property 3 of logarithms from Section 15.4:

$$\ln e^{-\frac{t}{100 \text{ ms}}} = \ln 0.3$$

$$-\frac{t}{100 \text{ ms}} \ln e = \ln 0.3.$$

We recall that $\ln e = 1$ and solve for t:

$$-\frac{t}{100 \text{ ms}} = \ln 0.3$$

$$t = -100 \ln 0.3 \text{ ms.}$$

You should use a calculator to find that $t = 120$ ms. Observe that at 7 V, v_C is 70% of E. Referring to the universal time-constant curve for v_C charging, you can see that t is approximately 1.2τ.

You can solve this problem in slightly fewer steps if you subtract v_C from E to obtain v_R, and then find t by using the formula for v_R. You should check the preceding solution by this method. ▲

DISCHARGING THE CAPACITOR

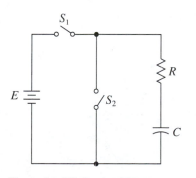

The series RC circuit with two switches.

Now, suppose we put a second switch S_2 in the RC circuit as shown. We close S_1 and leave S_2 open to charge the capacitor. If we then open S_1 to disconnect the battery, and close S_2, the capacitor is discharged. During discharge, the voltage v_C across the capacitor decreases from an initial value V_0 and approaches zero. This decrease is similar to the decrease of v_R when the capacitor is charged. Thus, the formula for v_C when the capacitor is discharged is the same as the formula for v_R when the capacitor is charged, with V_0 in place of E:

$$v_C = V_0 e^{-\frac{t}{\tau}}.$$

The formula for v_R when the capacitor is discharged is

$$v_R = -V_0 e^{-\frac{t}{\tau}}.$$

The current i is negative because the discharging current is opposite in direction to the charging current, and therefore v_R is negative.

EXAMPLE 15.19

In an RC circuit, $R = 20 \text{ k}\Omega$ and $C = 100 \text{ }\mu\text{F}$. The capacitor is discharging from an initial charge of $V_0 = 40 \text{ V}$. Find v_C after 10 s.

SOLUTION

First, we find the time constant τ:

$$\tau = RC$$
$$= (20 \text{ k}\Omega)(100 \text{ }\mu\text{F})$$
$$= 2 \text{ s}.$$

To find v_C, we use the formula

$$v_C = V_0 e^{-\frac{t}{\tau}}.$$

Substituting for V_0, t, and τ, we have

$$v_C = 40 \, e^{-\frac{10 \text{ s}}{2 \text{ s}}} \text{ V}$$
$$= 40 \, e^{-5} \text{ V}.$$

You should use a calculator to find that $v_C = 0.27 \text{ V}$, or 270 mV. Observe that at 10 s, $t = 5\tau$. Referring to the universal time-constant curve for v_C discharging, you can see that when the capacitor has discharged for five time constants, v_C is relatively small. ▲

EXAMPLE 15.20 ▶

For the circuit described in Example 15.19, with the capacitor discharging, find t when $v_C = 20 \text{ V}$.

SOLUTION

Since R and C are the same as in Example 15.19, the time constant is $\tau = 2 \text{ s}$ as before. Thus, we write

$$v_C = V_0 e^{-\frac{t}{\tau}}$$
$$20 \text{ V} = 40 e^{-\frac{t}{2 \text{ s}}} \text{ V}.$$

Dividing both sides by 40 V, we have

$$0.5 = e^{-\frac{t}{2 \text{ s}}}.$$

Now, we can take the natural logarithm of each side, apply Property 3 of logarithms, and solve for t:

$$\ln 0.5 = \ln e^{-\frac{t}{2 \text{ s}}}$$

$$\ln 0.5 = -\frac{t}{2 \text{ s}} \ln e$$

$$\ln 0.5 = -\frac{t}{2 \text{ s}}$$

$$t = -2 \ln 0.5 \text{ s}$$

$$t = 1.39 \text{ s}.$$

Observe that v_C is 50% of V_0; that is, v_C is at the point where the time-constant curves cross. At this point, t is approximately 0.7τ, or approximately $0.7(2 \text{ s}) = 1.4 \text{ s}$. ▲

EXERCISE

15.5

For Exercises 1–8, use the formulas for the series RC circuit when the capacitor is charging:

1. If $E = 100 \text{ V}$, $R = 50 \text{ k}\Omega$, and $C = 5 \text{ }\mu\text{F}$, find v_C, v_R, and i when $t = 25 \text{ ms}$.
2. If $E = 25 \text{ V}$, $R = 1.5 \text{ M}\Omega$, and $C = 10 \text{ }\mu\text{F}$, find v_C, v_R, and i when $t = 6 \text{ s}$.
3. If $E = 100 \text{ V}$, $R = 50 \text{ k}\Omega$, and $C = 5 \text{ }\mu\text{F}$, find t when $v_C = 20 \text{ V}$.
4. If $E = 25 \text{ V}$, $R = 1.5 \text{ M}\Omega$, and $C = 10 \text{ }\mu\text{F}$, find t when $v_C = 9 \text{ V}$.

5. If $E = 75$ V, $R = 2.2$ kΩ, and $C = 2.5$ μF, find t when $v_R = 24$ V.

6. If $E = 10$ V, $R = 1.2$ MΩ, and $C = 550$ pF, find t when $v_R = 8.2$ V.

7. If $E = 50$ V, $R = 3.3$ kΩ, and $C = 10$ μF, find t when $i = 8.5$ mA.

8. If $E = 60$ V, $R = 5$ MΩ, and $C = 10$ pF, find t when $i = 4.5$ μA.

For Exercises 9–12, use the formulas for the series RC circuit when the capacitor is discharging:

9. If $V_0 = 100$ V, $R = 50$ kΩ, and $C = 5$ μF, find v_C when $t = 0.5$ s.

10. If $V_0 = 30$ V, $R = 10$ MΩ, and $C = 500$ pF, find v_C when $t = 1.5$ ms.

11. If $V_0 = 15$ V, $R = 750$ Ω, and $C = 5$ μF, find t when $v_C = 1$ V.

12. If $V_0 = 80$ V, $R = 1.5$ kΩ, and $C = 25$ μF, find t when $v_C = 40$ V.

13. For a series RC circuit, $E = 100$ V, $R = 10$ kΩ, $C = 10$ μF and the capacitor is charging. Find the value of t where the universal time-constant curves cross; that is, $v_C = v_R$. (Hint: v_C and v_R are each 50% of E.)

14. Using the general formulas for the time-constant curves, find the value of t, in terms of τ, where the curves cross. (Hints: The curves cross where the y-values are each at 50%. Thus, the y-values are equal; that is,

$$1 - e^{-\frac{t}{\tau}} = e^{-\frac{t}{\tau}}.$$

Isolate the exponential expression $e^{-\frac{t}{\tau}}$ and then solve for t in terms of τ.)

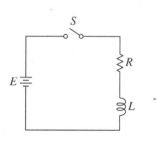

A series *RL* circuit.

15. An **inductor** or **coil** is a device in which changing current produces an opposing voltage. A **series *RL* circuit** contains a resistor and an inductor (labelled L) in series with a battery. When the switch S is closed, the current flow from the battery E causes a current increase in the coil, creating a magnetic field that induces voltage in the coil. In this phase, we say the current in the coil is built up. The ability of the inductor to induce voltage by current buildup is called its **inductance** (variable name L), and is measured in **henrys** (H), named for an American physicist Joseph Henry (1797–1878).

The time-constant τ for a series RL circuit is given by the formula

$$\tau = \frac{L}{R}.$$

The formula for current in an RL circuit during current buildup is

$$i = I_S(1 - e^{-\frac{t}{\tau}}),$$

where $I_S = \dfrac{E}{R}$. If $E = 10$ V, $R = 2.5$ kΩ, and $L = 5$ H, find i when $t = 1.5$ ms.

For Exercises 16–18, use the formulas for the series RL circuit given in Exercise 15:

16. If $E = 20$ V, $R = 100$ Ω, and $L = 1$ mH, find i when $t = 20$ μs.

17. If $E = 10$ V, $R = 10$ kΩ, and $L = 1$ H, find t when $i = 400$ μA.

18. If $E = 25$ V, $R = 50$ Ω, and $L = 500$ mH, find t when $i = 400$ mA.

19. If we put a second switch into the RL circuit in Exercise 15, open S_1 to disconnect the battery and close S_2, as we did with the RC circuit following Example 15.18, the current buildup in the inductor is said to collapse. The formula for current in an RL circuit during current collapse is

$$i = I_0 e^{-\frac{t}{\tau}}.$$

If $I_0 = 10$ mA, $R = 2$ Ω, and $L = 800$ mH, find i when $t = 100$ ms.

20. For an RL circuit during current collapse, if $I_0 = 18.2$ mA, $R = 330$ Ω, and $L = 2$ H, find t when $i = 3$ mA. (Use the formula given in Exercise 19.)

Proofs of the Properties of Logarithms

Property 1

Let $\log_b P = m$ and $\log_b Q = n$. Then, in exponential form,

$$P = b^m \text{ and } Q = b^n.$$

Thus, we can write

$$PQ = b^m b^n$$

$$PQ = b^{m+n}.$$

Returning to logarithmic form,

$$\log_b PQ = m + n$$

$$\log_b PQ = \log_b P + \log_b Q.$$

Property 2

Proceeding as in Property 1, we can write

$$\frac{P}{Q} = \frac{b^m}{b^n}$$

$$\frac{P}{Q} = b^{m-n}.$$

Returning to logarithmic form,

$$\log_b \frac{P}{Q} = m - n$$

$$\log_b \frac{P}{Q} = \log_b P - \log_b Q.$$

Property 3

Let $\log_b P = m$. Then, in exponential form,

$$P = b^m.$$

Thus, we can write

$$P^x = (b^m)^x$$

$$P^x = b^{xm}.$$

Returning to logarithmic form,

$$\log_b P^x = xm$$

$$\log_b P^x = x \log_b P.$$

Self-Test

1. Find the value of $\log_{64} 8$.

    1. _____

2. Solve $x^{5.5} = 0.5$ and check the solution.

    2. _____

3. For a series RC circuit, $E = 12$ V, $R = 2$ MΩ, and $C = 400$ pF. If the capacitor is charging, find t when $v_C = 2$ V.

    3. _____

4. Draw the graph of $y = 6^x$.

5. Draw the graph of $y = \log_6 x$.

UNIT 16

Cumulative Review

Introduction

In Units 1 through 11, you learned about real numbers, algebraic expressions, linear equations, and some basic types of quadratic and related equations. You used these skills to evaluate and solve formulas, and you applied Kirchhoff's and Ohm's laws to solving circuit problems. Then, you learned how to solve and apply rational equations. Finally, you learned to draw and interpret graphs of linear equations in two variables, and draw graphs of some types of quadratic and related equations in two variables.

In Units 12 through 15, you used both graphical and algebraic methods to solve systems of equations in two variables. You applied these methods to the loop method of solving circuit problems. You also learned how to solve systems of linear equations by using determinants, and how to solve systems of three equations in three variables. Then, you learned how to solve general quadratic equations, and specifically, how to use the quadratic formula. You learned how to draw parabolas, which are graphs of quadratic functions. Finally, you learned about one type of function that cannot be derived by algebraic methods, the logarithmic function, which is derived from the exponential function. You learned how to solve exponential equations, and you applied exponential equations to exponential growth and decay problems, and in particular, to charging and discharging capacitors.

Now, you should review the material in all of the preceding units.

OBJECTIVE

When you have finished this unit you should be able to fulfill every objective of each of Units 1 through 15.

SECTION 16.1

Review Method

For this unit, you should review the Self-Tests for Units 1 through 15. Do each problem in each Self-Test over again. If you cannot do a problem, or if you have the slightest doubt or difficulty, find the appropriate material to review:

1. You will find an objective number next to the answer to every Self-Test problem. Find the objective number of the problem you are working on (this is not necessarily the problem number). Go to the first page of the unit and reread the objective. For example, if you have difficulty with a Self-Test problem in Unit 2, and the objective number next to the answer is Objective 3, reread Objective 3 in the objectives list at the beginning of Unit 2 to find out what concept you need to review.

2. The material you should study to review the concept is in the section that has the same number as the objective. Reread the material in that section, rework the examples, and redo some of the exercises for the section. For example, if you are reviewing Unit 2, Objective 3, refer to Unit 2, Section 3; that is, Section 2.3. Reread Section 2.3, rework examples in that section, and redo problems in Exercise 2.3.

3. Try the Self-Test for the unit again to find out if there is any other objective you need to review in the unit. For example, if you have been reviewing Section 2.3 and Exercise 2.3, try the Self-Test for Unit 2 again.

Review each unit from Unit 1 through 15 in this way. For some units, you may find that you can still do the Self-Test easily. For other units, you might need to review one or more of the objectives in more detail. As a final check, and only after you have reviewed every preceding unit, try the Self-Test for this Cumulative Review Unit 16.

☐ Self-Test ☐

1. Solve $4(x - 3) - (x - 2) = 5$ and check the solution.

1. _____

2. Solve $\dfrac{4}{x} - \dfrac{x}{3} = 2$ and check the solution.

2. _____

3. Solve $Z_T = \dfrac{Z_1 Z_2}{Z_1 + Z_2}$ for Z_1.

3. _____

4. For the formula $I_B = \dfrac{V_{CC} - V_{BE}}{R_B}$, find V_{BE} if $I_B = 50\ \mu A$, $V_{CC} = 9$ V, and $R_B = 166$ kΩ.

4. _____

5. Find the slope and the y-intercept, and draw the graph of $2x - y = 2$.

6. Find the vertex, the y-intercept, and the x-intercepts, and draw the graph of $y = x^2 - 2x - 3$.

7. Solve the system:

$2x - 3y = 9$

$6x + 2y = 5$

7. _____

8. Find I_1, I_2, and I_3 for this circuit:

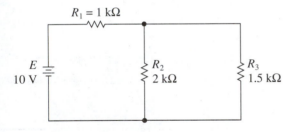

$R_1 = 1\ \text{k}\Omega$

E
10 V

R_2
2 kΩ

R_3
1.5 kΩ

8. _____

9. Solve $5^{\frac{1}{2}x} = 16$ and check the solution.

9. _____

10. For a series RC circuit, $E = 10$ V, $R = 50$ kΩ, and $C = 200$ μF. If the capacitor is charging, find v_C and v_R when $t = 4$ s.

10. _____

UNIT 17

Right Triangle Trigonometry

Introduction

Trigonometry is based on the study of triangles, in particular, right triangles. In this unit, you will review measurement of angles in degrees, and learn how the angles of triangles are related. You will learn special properties of right triangles, including a famous property called the Pythagorean theorem, which relates the sides of right triangles. You will learn the definitions of three trigonometric ratios, which relate the sides of right triangles to the angles. You will use all of these properties to solve right triangles; that is, to find the measures of their sides and angles. Then, you will learn the definition of vectors, quantities that have both magnitude and direction, and use the techniques of solving right triangles to study component and resultant vectors. Finally, you will learn how to add and subtract vectors.

OBJECTIVES

When you have finished this unit you should be able to:

1. Use the Pythagorean theorem to find the third side of right triangles, given any two sides.

2. Solve right triangles, given one acute angle and one side, or given two sides.

3. Find horizontal and vertical components of vectors, and find resultant vectors given their components.

4. Find sums and differences of vectors.

SECTION

17.1

Right Triangles

The use of right triangle trigonometry can be traced back to ancient civilizations. Isolated uses of the trigonometric ratios appear to have been developed by the Babylonians. A tablet from the height of the ancient Babylonian culture (about 1700 B.C.) shows they knew the rule that later came to be known as the Pythagorean theorem. The Pythagorean theorem is named for the Greek mathematician Pythagoras (about 540 B.C.), as he or his followers are thought to have proved it. Right triangle trigonometry as we know it was developed by Greek mathematicians and astronomers living in Alexandria about 200 B.C. to A.D. 300. The word *trigonometry* comes from Greek words for "triangle" and "measure."

We recall that angles are commonly measured in **degrees**. One degree is 1/360th of a circle; that is, there are 360° (360 degrees) in a circle. This unit of measure can be traced back through the Greeks of Alexandria to the ancient Babylonians. We also still use the Babylonians' units for time, with 60 seconds in a minute and 60 minutes in an hour, derived from their base 60 number system.

We can relate measures of angles to degrees in a circle. When a circle is cut in half by a diameter, two angles are formed. Each of these angles is a **straight angle**. Since half of 360° is 180°, each straight angle measures 180°.

If we cut each straight angle in half by another diameter, we have four angles. Each of these angles is a **right angle**. Since half of 180° is 90°, each right angle measures 90°:

360° in a circle.

371

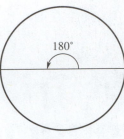
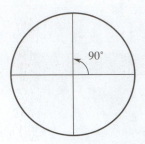

A 180° angle.   A 90° or right angle.

A **triangle** has three angles and three sides. The sum of the three angles of a triangle is 180°. If the three angles are A, B, and C, then

$$A + B + C = 180°.$$

A triangle is a **right triangle** if one of its angles is a right angle. If C is the right angle, then $C = 90°$. Therefore, the sum of the remaining angles is 90°; that is,

$$A + B = 90°:$$

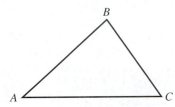
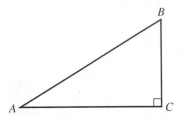

A triangle with angles A, B, and C.   A right triangle with right angle C.

We observe that a triangle cannot have more than one right angle. Two right angles would make 180°, leaving nothing for the third angle.

The side of a right triangle across from the right angle is called the **hypotenuse**. The word *hypotenuse* comes from Greek words meaning "to stretch under." When the angles of a right triangle are labeled A, B, and C, where C is the right angle, then the sides are labeled a, b, and c, where c, is the hypotenuse. Side a is across from angle A and side b is across from angle B. Sides a and b are sometimes called the **legs** of the right triangle to distinguish them from the hypotenuse.

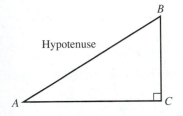

The hypotenuse of a right triangle.

THE PYTHAGOREAN THEOREM

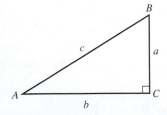

The hypotenuse c and legs a and b of a right triangle.

The **Pythagorean theorem** relates the hypotenuse c, and the legs a and b, of a right triangle. This rule states that the square of the hypotenuse is equal to the sum of the squares of the legs; that is,

$$c^2 = a^2 + b^2.$$

The sides of a triangle are positive, so we can take the square root of each side of the equation:

$$\sqrt{c^2} = \sqrt{a^2 + b^2}$$

$$c = \sqrt{a^2 + b^2}.$$

Pythagorean Theorem: If c is the hypotenuse of a right triangle, and a and b are the legs, then

$$c^2 = a^2 + b^2$$

or

$$c = \sqrt{a^2 + b^2}.$$

EXAMPLE 17.1 ▶   If the legs of a right triangle are 3 and 4, find the hypotenuse.

SOLUTION   It does not matter which leg we call a and which leg we call b. A diagram of the triangle is shown at the left. Using the Pythagorean theorem, we have

$$c = \sqrt{a^2 + b^2}$$
$$= \sqrt{3^2 + 4^2}$$
$$= \sqrt{25}$$
$$= 5. \qquad \blacktriangle$$

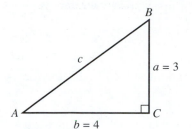

Triangle for Example 17.1

Observe that you must *not* take the separate square roots of a^2 and b^2; that is, $\sqrt{a^2 + b^2}$ is *not the same* as $\sqrt{a^2} + \sqrt{b^2}$. For example, we saw in Example 17.1 that

$$\sqrt{3^2 + 4^2} = 5,$$

but the separate square roots are

$$\sqrt{3^2} + \sqrt{4^2} = 3 + 4$$
$$= 7.$$

EXAMPLE 17.2 ▶   If one leg of a right triangle is 5.6 and the hypotenuse is 12.5, find the other leg.

SOLUTION   A diagram of the triangle is shown at the left. We may solve the Pythagorean theorem for b^2 by subtracting a^2 from both sides of the first form:

$$c^2 = a^2 + b^2$$
$$c^2 - a^2 = b^2$$

or

$$b^2 = c^2 - a^2.$$

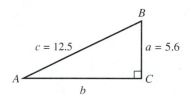

Then, taking the square root of each side, we have

$$b = \sqrt{c^2 - a^2}$$
$$= \sqrt{12.5^2 - 5.6^2}.$$

Triangle for Example 17.2

You can use a calculator to find that

$$b = 11.2$$

to three significant digits. The calculator algorithm for this type of calculation is given in Section 5.2.   ▲

A right triangle can be part of another figure. We find the right angle, and then find the hypotenuse across from the right angle.

EXAMPLE 17.3 ▶   The diagonal of a rectangle is 14 inches, and the width of the rectangle is 8.5 inches. Find the length of the rectangle.

SOLUTION   We draw the rectangle with its diagonal, and observe that the diagonal is the hypotenuse of a right triangle, with the width and length of the rectangle as the legs:

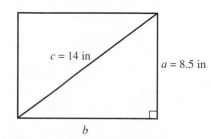

Using the Pythagorean theorem solved for b, as in Example 17.2, we have

$$b = \sqrt{c^2 - a^2}$$
$$= \sqrt{14^2 - 8.5^2}.$$

Then, using a calculator, you can find that

$$b = 11.1 \text{ in.}$$ ▲

| **EXAMPLE 17.4** ▶ | An equilateral triangle has three equal sides. If each side of an equilateral triangle is 2 meters, find the height of the triangle. |

SOLUTION   We draw the triangle with its height, and observe that the height is a leg of a right triangle. Moreover, the height cuts the base into two equal parts. Thus, the second leg of the right triangle is 1 meter, and the hypotenuse is 2 meters. Using the Pythagorean theorem solved for b, as in the preceding examples, we have

$$b = \sqrt{c^2 - a^2}$$
$$= \sqrt{2^2 - 1^2}$$
$$= \sqrt{3}.$$

Thus $b = \sqrt{3}$ meters, or 1.73 meters. ▲

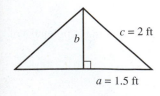

Triangle for Example 17.4

| **EXERCISE 17.1** | Find the missing side of the right triangle: |

1. The legs are 6 and 8.

2. The legs are 5 and 12.

3. One leg is 15 and the hypotenuse is 17.

4. One leg is 1.5 and the hypotenuse is 2.5.

5. One leg is 4.2 and the hypotenuse is 5.4.

6. One leg is 8.11 and the hypotenuse is 12.1.

7. The legs are 1.2 and 1.2.

8. The legs are 10 and 20.

9. The diagonal of a rectangle is 6 meters, and the length of the rectangle is 5 meters. Find the width of the rectangle.

10. The sides of a square are each 4 inches. Find the diagonal of the square.

11. If each side of an equilateral triangle is 4 meters, find the height of the triangle.

12. The sides of an equilateral triangle are each 20 inches. Find the height of the triangle.

13. An isosceles triangle has two equal sides. If the equal sides of an isosceles triangle are 2 feet, and the base is 3 feet, find the height of the triangle. (The height is a leg of a right triangle, and cuts the base into two equal parts.)

14. If the equal sides of an isosceles triangle are 12 feet, and the base is 8 feet, find the height of the triangle.

15. The height of an isosceles triangle is 2 meters and its base is 2 meters. Find the two equal sides.

16. The height of an isosceles triangle is 5.6 inches and each of the equal sides is 8.2 inches. Find the base.

Isosceles triangle for
Exercise 13.

The Trigonometric Ratios

In Section 17.1, we defined a right triangle as a triangle with one right angle. The other angles are called **acute angles**. Acute angles are angles that measure less than 90°.

Suppose θ (the Greek letter theta) is one of the acute angles. Then the leg across from θ is called the **opposite side** and the remaining leg, which with the hypotenuse forms θ, is called the **adjacent side**. For example, if θ is angle A, then the opposite side is a, and the adjacent side is b:

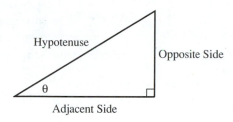

The opposite and adjacent sides.

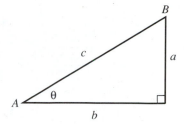

Sides a and b as opposite and adjacent sides.

Now, suppose we draw a line that doubles the hypotenuse and also doubles b. Another right triangle is formed that includes the angle θ and has opposite side a' (read "a-prime"):

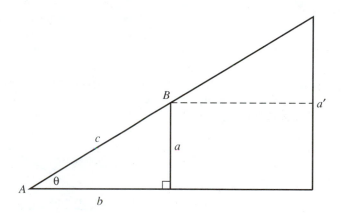

We can use the Pythagorean theorem to show that a' is also a doubled. By using the inside triangle, we can find that

$$a = \sqrt{c^2 - b^2}.$$

Then, by using the outside triangle, we have

$$a'^2 + (2b)^2 = (2c)^2$$

$$a'^2 = (2c)^2 - (2b)^2$$

$$a' = \sqrt{(2c)^2 + (2b)^2}$$

$$a' = 2\sqrt{c^2 - b^2}$$

$$a' = 2a.$$

Thus, the sides of the inside and outside triangles are proportional. Triangles that have the same acute angles are called **similar triangles**. It can be shown that the sides of any similar triangles are proportional.

The fact that the sides of similar triangles are proportional is used to define **trigono-metric ratios** for any acute angle θ.

The Trigonometric Ratios:

$$\text{sine of } \theta = \frac{\text{opposite side}}{\text{hypotenuse}}$$

$$\text{cosine of } \theta = \frac{\text{adjacent side}}{\text{hypotenuse}}$$

$$\text{tangent of } \theta = \frac{\text{opposite side}}{\text{adjacent side}}$$

(The ratio name "sine" is pronounced the same as "sign.") These definitions may be abbreviated

$$\sin \theta = \frac{\text{opp}}{\text{hyp}}$$

$$\cos \theta = \frac{\text{adj}}{\text{hyp}}$$

$$\tan \theta = \frac{\text{opp}}{\text{adj}}$$

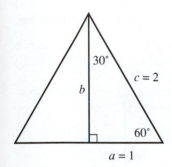

The equilateral triangle with sides 2.

(where "sin" is still pronounced like "sign," but "cos" may be pronounced like "coast" without the t).

Recall from Example 17.4 that an equilateral triangle has three equal sides. If the side of an equilateral triangle is $c = 2$, then one-half of the base is $a = 1$ and the height is $b = \sqrt{3}$. The three angles of an equilateral triangle are also equal. Each angle is 60°, and the height cuts the top angle into two 30° angles. Thus, we have a right triangle with acute angles 30° and 60°, and hypotenuse 2. If θ is the 30° angle, then the opposite side is 1, and the adjacent side is $\sqrt{3}$:

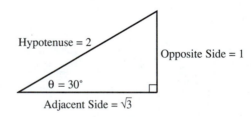

Then, we may write the trigonometric ratios for θ = 30°. For example,

$$\sin 30° = \frac{\text{opp}}{\text{hyp}}$$

$$= \frac{1}{2}$$

$$= 0.5.$$

Similarly,

$$\cos 30° = \frac{\text{adj}}{\text{hyp}}$$

$$= \frac{\sqrt{3}}{2}$$

$$= 0.866$$

to three significant digits, and

$$\tan 30° = \frac{\text{opp}}{\text{adj}}$$

$$= \frac{1}{\sqrt{3}}$$

$$= 0.577$$

to three significant digits.

You can find the value of the trigonometric ratio for any angle on your scientific calculator. First, make sure your calculator is in degree mode. Most calculators indicate that you are in degree mode by printing DEG (or just D) on the display. Other modes are RAD (radian mode, or R) and GRAD (or G). If one of these is on the display, there are two methods a calculator might use to change to degree mode.

One way to change to degree mode is to look for the degree-radian-grad key marked

$$\boxed{\text{DRG}} .$$

Press this key until DEG (or D) is printed on the display.

If your calculator does not have a degree-radian-grad key, look for the mode key marked

$$\boxed{\text{MODE}} .$$

Also, find a guide that matches modes such as DEG, RAD, and GRAD with number keys. After pressing the mode key, press the number key matched with degree mode.

To find sin 30°, find the sine key marked

$$\boxed{\text{sin}} .$$

Then, with your calculator in degree mode, follow these steps:

	display:
Enter 30	30.
Press $\boxed{\text{sin}}$	0.5

(For some calculators, such as those identified as DAL calculators, these steps are reversed; you must press the sine key first.)

To find cos 30°, find the cosine key marked

$$\boxed{\text{cos}} .$$

Then, with your calculator in degree mode, follow these steps:

	display:
Enter 30	30.
Press $\boxed{\text{cos}}$	0.866025403

(For a DAL calculator, you press the cosine key first.)

To find tan 30°, find the tangent key marked

$$\boxed{\text{tan}} .$$

Then, with your calculator in degree mode, follow these steps:

	display:
Enter 30	30.
Press $\boxed{\text{tan}}$	0.577350269

(For a DAL calculator, you press the tangent key first.)

SOLVING RIGHT TRIANGLES

We use the trigonometric ratios to **solve right triangles**. To solve a right triangle means to find all of its angles and sides. We must be given any side and an acute angle, or any two sides.

EXAMPLE 17.5 ▶

Solve the right triangle with an acute angle 30° and hypotenuse 10.

SOLUTION

First, we draw a diagram of the triangle, labelling the acute angle $A = 30°$, and the hypotenuse $c = 10$. We also label the opposite side a and the adjacent side b as shown. We note that the other acute angle is $B = 60°$.

From the definition of the sine, we may write

$$\sin \theta = \frac{\text{opp}}{\text{hyp}},$$

and therefore

$$\sin A = \frac{a}{c}.$$

Multiplying both sides by c, we have

$$c \sin A = a$$

or

$$a = c \sin A.$$

Substituting the values for A and c, we have

$$a = 10 \sin 30°.$$

Now, you can find side a by using a scientific calculator:

		display:
Enter	10	10.
Press	$\times$	10.
Enter	30	30.
Press	sin	0.5
Press	=	5.

Thus, $a = 5$. (Remember that, for a DAL calculator, you must reverse the third and fourth steps.)

To find side b, we use the definition,

$$\cos \theta = \frac{\text{adj}}{\text{hyp}},$$

and therefore

$$\cos A = \frac{b}{c}.$$

Multiplying both sides by c, we have

$$c \cos A = b$$

or

$$b = c \cos A.$$

Substituting the values for A and c, we have

$$b = 10 \cos 30°.$$

Triangle for Example 17.5

You can use a sequence of steps similar to those above to find that $b = 8.66$ to three significant digits.

We can check by using the Pythagorean theorem:

$$c = \sqrt{a^2 + b^2}$$

$$10 \overset{?}{=} \sqrt{5^2 + 8.66^2}$$

$$10 = 10.$$

▲

EXAMPLE 17.6 ▶ Solve the right triangle with an acute angle 56° and opposite side 32.

SOLUTION   If $A = 56°$, then the side opposite A is $a = 32$, and the adjacent side is b. We draw a diagram of the triangle as shown. We can find the other acute angle B by using

$$B = 90° - A$$

$$= 90° - 56°$$

$$= 34°.$$

From the definition of the trigonometric ratios, we have

$$\sin \theta = \frac{\text{opp}}{\text{hyp}},$$

and so

$$\sin A = \frac{a}{c}.$$

We must solve for c. First, we multiply both sides by c:

$$c \sin A = a.$$

Then, we observe that the entire expression $\sin A$ represents one quantity. Thus, we divide both sides by the quantity $\sin A$:

$$c = \frac{a}{\sin A}.$$

Substituting the values for A and a, we have

$$c = \frac{32}{\sin 56°}.$$

Now, you can find side c by using a scientific calculator:

		display:
Enter 32		32.
Press	÷	32.
Enter 56		56.
Press	sin	0.829037572
Press	=	38.59897435

Therefore, $c = 38.6$.

Since a is the opposite side for A, and b is the adjacent side, to find b we use the trigonometric ratio

$$\tan \theta = \frac{\text{opp}}{\text{adj}},$$

and so

$$\tan A = \frac{a}{b}.$$

Triangle for Example 17.6

B

c

$a = 32$

$56°$

A

b

We must solve for the unknown side b. First, we multiply both sides by b:

$$b \tan A = a.$$

Then, we divide both sides by the quantity $\tan A$:

$$b = \frac{a}{\tan A}.$$

Substituting the values for A and a, we have

$$b = \frac{32}{\tan 56°}.$$

You can use a sequence of steps similar to those above to find that $b = 21.6$. You should check by using the Pythagorean theorem. ▲

INVERSE TRIGONOMETRIC RATIOS

When we have the value of a trigonometric ratio for A, the solution for A is the **inverse trigonometric ratio**. For example, if we have

$$\sin A = 0.5,$$

then A is the inverse sine of 0.5. The inverse sine is written

$$A = \sin^{-1} 0.5.$$

We note that $\sin^{-1} x$ *does not mean* $\dfrac{1}{\sin x}$.

To find the inverse sine on most calculators, you press

$$\boxed{\text{INV}} \quad \text{or} \quad \boxed{2^{\text{nd}}}$$

and

$$\boxed{\text{sin}}.$$

We mean to press this combination whenever we say to press

$$\boxed{\sin^{-1}}.$$

(The symbol $\sin^{-1}$ might be above the sine key. Some calculators omit this symbol. Some older calculators use the obsolete term arcsin.) To find $\sin^{-1} 0.5$, follow these steps:

		display:
Enter 0.5		0.5
Press $\boxed{\sin^{-1}}$		30.

(For a DAL calculator, you must press the inverse sine key first.) You can find the inverse cosine and inverse tangent similarly.

EXAMPLE 17.7 ▶ Solve the right triangle with legs 3 and 4.

SOLUTION We identify the given parts as $a = 3$ and $b = 4$, and draw a diagram of the triangle. In Example 17.1, we used the Pythagorean theorem to find that

$$c = 5.$$

We must use trigonometric ratios to find the acute angles. To find angle A by using the given sides a and b, we use the trigonometric ratio

$$\tan \theta = \frac{\text{opp}}{\text{adj}},$$

or

$$\tan A = \frac{a}{b}.$$

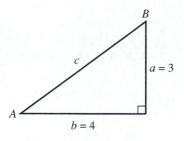

Triangle for Example 17.7.

We substitute the values for a and b:

$$\tan A = \frac{a}{b}$$

$$\tan A = \frac{3}{4}$$

$$\tan A = 0.75$$

$$A = \tan^{-1} 0.75.$$

To find A, be sure your calculator is in degree mode and follow these steps:

	display:
Enter 0.75	0.75
Press $\boxed{\tan^{-1}}$	36.86989765

Thus, $A = 36.9°$. We can find the other acute angle B by using

$$B = 90° - A$$

$$= 90° - 36.9°$$

$$= 53.1°.$$

We can check by using any combination we have not used previously. For example, we may write

$$\sin \theta = \frac{\text{opp}}{\text{hyp}}$$

$$\sin B = \frac{b}{c}$$

$$\sin B = \frac{4}{5}$$

$$B = \sin^{-1} 0.8.$$

You can use a calculator to find that $B = 53.1°$ as before. You can also try using

$$\cos \theta = \frac{\text{adj}}{\text{hyp}}$$

$$\cos B = \frac{a}{c}$$

$$\cos B = \frac{3}{5}$$

$$B = \cos^{-1} 0.6$$

and

$$\tan \theta = \frac{\text{opp}}{\text{adj}}$$

$$\tan B = \frac{b}{a}$$

$$\tan B = \frac{4}{3}$$

$$B = \tan^{-1} \left(\frac{4}{3}\right).$$

Observe that you should use the given parts of the triangle to solve the triangle, and then use the parts you have found to check. ▲

EXAMPLE 17.8   Solve the right triangle with one leg 560 and hypotenuse 820.

SOLUTION   We identify the given parts as $a = 560$ and $c = 820$, and draw a diagram of the triangle. We can use the Pythagorean theorem to find b:

$$b = \sqrt{c^2 - a^2}$$
$$= \sqrt{820^2 - 560^2}$$
$$= 599.$$

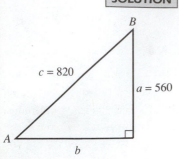

Triangle for Example 17.8.

Since a is the opposite side for A and c is the hypotenuse, we find angle A by using the trigonometric ratio

$$\sin \theta = \frac{\text{opp}}{\text{hyp}}$$

$$\sin A = \frac{a}{c}$$

$$\sin A = \frac{560}{820}$$

$$A = \sin^{-1}\left(\frac{560}{820}\right).$$

Be sure your calculator is in degree mode and follow these steps:

		display:
Enter 560		560.
Press $\div$		560.
Enter 820		820.
Press $=$		0.682926829
Press $\sin^{-1}$		43.07278135

Thus, $A = 43.1°$, and subtracting from 90°, $B = 46.9°$. You should check by using various combinations we did not use in the solution.   ▲

EXERCISE 17.2

Solve the right triangle:

1. An acute angle is 30° and the hypotenuse is 150.
2. An acute angle is 30° and the hypotenuse is 2.5.
3. An acute angle is 42° and the side opposite that angle is 75.
4. An acute angle is 69° and the side opposite that angle is 35.6.
5. An acute angle is 64° and the adjacent side for that angle is 7.8.
6. An acute angle is 18.4° and the adjacent side for that angle is 460.
7. An acute angle is 17° and the hypotenuse is 4.8.
8. An acute angle is 67° and the hypotenuse is 26.
9. The legs are 10 and 12.
10. The legs are 7.2 and 3.3.
11. One leg is 22 and the hypotenuse is 47.
12. One leg is 3.29 and the hypotenuse is 5.89.
13. Angle $B = 52°$ and side $b = 23$.
14. Angle $B = 25°$ and side $b = 3.74$.
15. Angle $B = 38°$ and side $a = 633$.
16. Angle $B = 75°$ and side $a = 35.5$.

17. Angle $B = 43.4°$ and the hypotenuse $c = 2.68$.

18. Angle $B = 51.5°$ and the hypotenuse $c = 85$.

19. Side $b = 11$ and the hypotenuse $c = 31$.

20. Side $b = 55.5$ and the hypotenuse $c = 70$.

SECTION

17.3

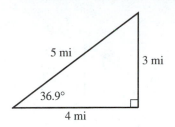

Five miles in the direction 36.9°.

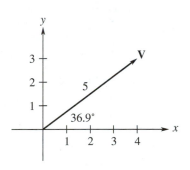

The vector with magnitude 5 and direction 36.9°.

Vectors

We have used real numbers to measure many types of quantities, for example, times such as 30 seconds, amounts of money such as 1000 dollars, and resistances such as 1.5 ohms. These measurements are examples of **scalars**. Scalar quantities have magnitude only.

Now, suppose you walk 4 miles east, and then walk 3 miles north. You have walked 7 miles in scalar terms. However, the measure from your starting point to your ending point "as the crow flies" is just 5 miles.

In Example 17.7, the angle between the side labelled 4 miles and the side labelled 5 miles was found to be 36.9°. Thus, the measure from the starting point to the ending point is 5 miles in a direction 36.9° north of east. The length 5 in the direction 36.9° is an example of a **vector**. Vector quantities have both *magnitude* and *direction*.

Vectors are often represented in the rectangular coordinate system. The magnitude of the vector is the length of a line segment, and the direction of the vector is the angle it makes with the positive *x*-axis. The vector with length 5 and in the direction 36.9° is represented as a line segment with length 5 and at an angle of 36.9° from the positive *x*-axis. This vector is written

$$\mathbf{V} = 5\underline{/36.9°},$$

and read "5 at an angle of 36.9°."

The magnitude of a vector is always positive. For this reason, the magnitude of a vector **V** is sometimes called its **absolute value**, $|\mathbf{V}|$. The absolute value of the vector $5\underline{/36.9°}$ is

$$|\mathbf{V}| = 5.$$

The direction of a vector is called the **argument**. The argument of the vector $5\underline{/36.9°}$ is

$$\theta = 36.9°.$$

Now, suppose we complete the rectangle formed by the axes, with the vector $5\underline{/36.9°}$ as a diagonal. We draw two vectors, one along the side of the rectangle lying on the *x*-axis, and the other along the side of the rectangle lying on the *y*-axis:

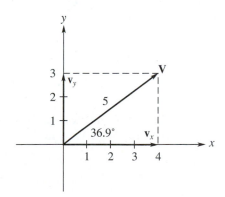

The vector along the *x*-axis is the **horizontal component** or **x-component** of the original vector. The vector along the *y*-axis is the **vertical component** or **y-component** of the original vector. The horizontal component is

$$\mathbf{v}_x = 4\underline{/0°},$$

that is, 4 at an angle of 0°. The vertical component is

$$\mathbf{v}_y = 3\underline{/90°},$$

that is, 3 at an angle of 90°.

 EXAMPLE 17.9 ▶ Find the horizontal and vertical components of the vector $10\underline{/35°}$.

SOLUTION We draw a diagram of the vector and its horizontal and vertical components. Then, we redraw $\mathbf{v}_y$ with its base at the tip of $\mathbf{v}_x$. Since its magnitude and direction are not changed, $\mathbf{v}_y$ is not changed:

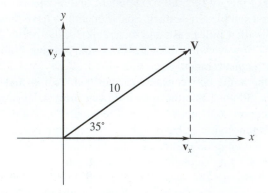

Vector diagram for Example 17.9.

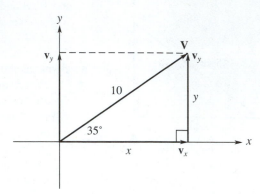

The diagram with $\mathbf{v}_x$ and $\mathbf{v}_y$ as legs of a right triangle.

Now, the vectors form a right triangle with an acute angle 35° and hypotenuse 10. To find $\mathbf{v}_x$ and $\mathbf{v}_y$, we find the adjacent side x and the opposite side y of the right triangle:

$$\cos 35° = \frac{x}{10}$$

$$x = 10 \cos 35°$$

$$x = 8.19,$$

and

$$\sin 35° = \frac{y}{10}$$

$$y = 10 \sin 35°$$

$$y = 5.74.$$

Thus, the horizontal and vertical components are

$$\mathbf{v}_x = 8.19\underline{/0°} \text{ and } \mathbf{v}_y = 5.74\underline{/90°}.$$

We often write the horizontal and vertical components of a vector as the coordinates of the point at its tip. Thus, the horizontal and vertical components can be written as

$$x = 8.19 \text{ and } y = 5.74. \qquad\qquad \blacktriangle$$

VECTORS IN OTHER QUADRANTS

We recall that the quadrants of the rectangular coordinate system are numbered I, II, III, and IV, counterclockwise from the upper right. We say that a vector is **in a quadrant** if its base is at the origin and the point at its tip lies in that quadrant. For example, the vector **V** in the diagram at the top of page 385 is in the second quadrant:

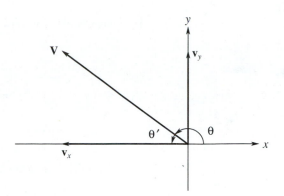

Directions of all vectors are in **standard position**, that is, taken from the positive x-axis. Therefore, this vector has a direction θ that is greater than 90°. However, the angle between the vector and its horizontal component is an acute angle. This acute angle is the **reference angle** θ' (theta-prime).

EXAMPLE 17.10 ▶

Find the horizontal and vertical components of the vector $\mathbf{V} = 10\underline{/120°}$.

SOLUTION

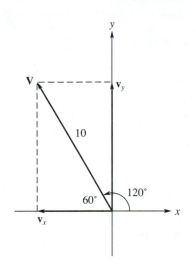

Vector diagram for Example 17.10

We draw a diagram of the vector, and its horizontal and vertical components. The vector is in the second quadrant, and its horizontal component is directed to the left, along the negative x-axis. The direction of the vector is 120°, so it cannot be an angle of a right triangle. The reference angle is

$$\theta' = 180° - 120°$$

$$= 60°.$$

The reference angle θ' is an acute angle of a right triangle. The legs of this right triangle are the horizontal and vertical components of the vector.

To find the horizontal component, we find its magnitude:

$$\cos 60° = \frac{x}{10}$$

$$x = 10 \cos 60°$$

$$x = 5;$$

therefore, the magnitude of $\mathbf{v}_x$ is 5. The direction of $\mathbf{v}_x$ is 180°. Thus,

$$\mathbf{v}_x = 5\underline{/180°}.$$

If we write the horizontal component as the x-coordinate of the point at the tip of the vector, however, we write

$$x = -5.$$

To find the vertical component, we may write

$$\sin 60° = \frac{y}{10}$$

$$y = 10 \sin 60°$$

$$y = 8.66.$$ ▲

We can use the 120° angle to find the horizontal component in Example 17.10 by writing

$$x = 10 \cos 120°.$$

You can use your calculator to find x by following these steps:

		display:
Enter	10	10.
Press	$\times$	10.
Enter	120	120.
Press	cos	-0.5
Press	$=$	$-5.$

Thus $x = -5$. The cosine is negative in the second quadrant. You can follow similar steps for

$$y = 10 \sin 120°$$

to find $y = 8.66$. The sine is positive in the second quadrant.

In general, we can define the horizontal and vertical components of a vector $|\mathbf{V}|\underline{/\theta}$ in terms of its absolute value $|\mathbf{V}|$ and its argument θ.

Horizontal and Vertical Components of $|\mathbf{V}|\underline{/\theta}$:

The horizontal component is given by $x = |\mathbf{V}| \cos \theta$.

The vertical component is given by $y = |\mathbf{V}| \sin \theta$.

EXAMPLE 17.11 ▶ Find the horizontal and vertical components of the vector $44\underline{/230°}$.

SOLUTION   We draw a diagram of the vector, and its horizontal and vertical components. The vector is in the third quadrant, its horizontal component is directed to the left along the negative x-axis, and its vertical component is directed downward along the negative y-axis. To find x, we may use

$$x = |\mathbf{V}| \cos \theta$$
$$= 44 \cos 230°$$
$$= -28.3.$$

The cosine is negative in the third quadrant. To find y, we may use

$$y = |\mathbf{V}| \sin \theta$$
$$= 44 \sin 230°$$
$$= -33.7.$$

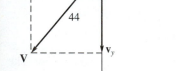

Vector diagram for Example 17.11

The sine is also negative in the third quadrant. The horizontal and vertical components may be written

$$\mathbf{v}_x = 28.3\underline{/180°} \text{ and } \mathbf{v}_y = 33.7\underline{/270°}$$

or

$$x = -28.3 \text{ and } y = -33.7. \qquad \blacktriangle$$

EXAMPLE 17.12 ▶ Find the horizontal and vertical components of the vector $15\underline{/315°}$.

SOLUTION   We draw a diagram of the vector, and its horizontal and vertical components. The vector is in the fourth quadrant, and its vertical component is directed downward along the negative y-axis. To find x, we may use

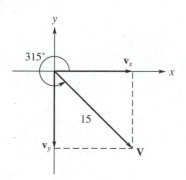

Vector diagram for Example 17.12.

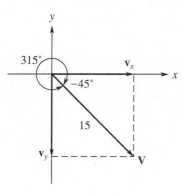

The diagram with a negative angle as the direction.

$$x = |\mathbf{V}| \cos \theta$$

$$= 15 \cos 315°$$

$$= 10.6.$$

The cosine is positive in the fourth quadrant. To find y, we may use

$$y = |\mathbf{V}| \sin \theta$$

$$= 15 \sin 315°$$

$$= -10.6.$$

The sine is negative in the fourth quadrant. The horizontal and vertical components may be written

$$\mathbf{v}_x = 10.6\underline{/0°} \quad \text{and} \quad \mathbf{v}_y = 10.6\underline{/270°}$$

or

$$x = 10.6 \quad \text{and} \quad y = -10.6. \qquad \blacktriangle$$

If an angle is taken *clockwise*, then the angle is *negative*. The negative angle is often used in the fourth quadrant. The vector $15\underline{/315°}$ has the reference angle

$$\theta' = 360° - 315°$$

$$= 45°.$$

The negative of this reference angle is $-45°$, which is the same direction as $315°$. Thus, the vector $15\underline{/-45°}$ is the same as the vector $15\underline{/315°}$.

RESULTANT VECTORS

If we are given a horizontal vector and a vertical vector, we can find a vector called their **resultant**. The given vectors are the horizontal and vertical components of the resultant.

EXAMPLE 17.13 ▶ Find the resultant of $x = 7.2$ and $y = 5.6$.

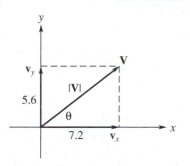

Vector diagram for Example 17.13.

SOLUTION We draw a diagram of the horizontal and vertical components and the resultant vector. The magnitude $|\mathbf{V}|$ of the resultant is the hypotenuse of a right triangle. Since we know the two legs, we can find $|\mathbf{V}|$ by using the Pythagorean theorem:

$$|\mathbf{V}| = \sqrt{x^2 + y^2}$$

$$= \sqrt{7.2^2 + 5.6^2}$$

$$= 9.12.$$

The direction θ of the resultant is the acute angle θ. We can find θ by using the tangent:

$$\tan \theta = \frac{\text{opp}}{\text{adj}}$$

$$= \frac{y}{x}$$

$$= \frac{5.6}{7.2}.$$

We use the inverse tangent to find

$$\theta = \tan^{-1}\left(\frac{5.6}{7.2}\right)$$

$$= 37.9°.$$

Therefore, the resultant vector is $9.12\underline{/37.9°}$.   ▲

In general, we can define the absolute value $|\mathbf{V}|$ and the reference angle θ' for the argument θ of a resultant vector in terms of the absolute values $|x|$ and $|y|$ of the horizontal and vertical components.

The Resultant of x and y:

The absolute value is given by $|\mathbf{V}| = \sqrt{|x|^2 + |y|^2}$.

The reference angle for the argument is given by $\theta' = \tan^{-1}\left(\frac{|y|}{|x|}\right)$.

EXAMPLE 17.14 ▶   Find the resultant of $x = -1.5$ and $y = 2.6$.

SOLUTION   Since x is negative, the horizontal component lies along the negative x-axis, and the vector is in the second quadrant. We draw a diagram of the horizontal and vertical components and the resultant vector. To find the magnitude $|\mathbf{V}|$, we write

$$|\mathbf{V}| = \sqrt{|x|^2 + |y|^2}$$

$$= \sqrt{1.5^2 + 2.6^2}$$

$$= 3.$$

To find the reference angle θ' for the direction θ, we write

$$\theta' = \tan^{-1}\left(\frac{|y|}{|x|}\right)$$

$$= \tan^{-1}\left(\frac{2.6}{1.5}\right)$$

$$= 60°.$$

Vector diagram for Example 17.14   Thus, the direction of the resultant vector is

$$\theta = 180° - 60°$$

$$= 120°.$$

Therefore, the resultant vector is $3\underline{/120°}$.   ▲

Because x and y are squared in the Pythagorean theorem, it does not matter whether we use x or $|x|$ to find $|\mathbf{V}|$ for Example 17.14. However, your calculator gives a different angle θ by using x in place of $|x|$:

$$\theta = \tan^{-1}\left(\frac{y}{x}\right)$$

$$= \tan^{-1}\left(\frac{2.6}{-1.5}\right)$$

$$= -60°.$$

Your calculator gives a fourth quadrant angle in the negative form. The tangent is negative in both the second and fourth quadrants, and the calculator cannot determine which quadrant is desired. You must determine the actual direction from the signs of the given components.

EXERCISE
17.3

Find the horizontal and vertical components of the vector:

1. $12/25°$ 2. $3.6/54°$ 3. $5/150°$ 4. $82/132°$

5. $2.3/240°$ 6. $40.5/209°$ 7. $3.39/320°$ 8. $660/298°$

Find the resultant of the horizontal and vertical components:

9. $x = 14$ and $y = 22$ 10. $x = 350$ and $y = 115$

11. $x = -22$ and $y = 15$ 12. $x = -6.8$ and $y = 6.8$

13. $x = -8.2$ and $y = -4.7$ 14. $x = -1.2$ and $y = -1.9$

15. $x = 430$ and $y = -770$ 16. $x = 11.5$ and $y = -9.3$

SECTION
17.4

Vector Addition

The operation of **vector addition** combines two vectors $\mathbf{V}_1$ and $\mathbf{V}_2$ by placing the base of one vector at the tip of the other vector. If the vectors are in the same direction, they can be added by adding their magnitudes.

| **EXAMPLE 17.15** | ▶ |

Find the sum of the vectors $\mathbf{V}_1 = 2/0°$ and $\mathbf{V}_2 = 3/0°$.

| SOLUTION |

The vectors are in the same direction, so we may add their magnitudes:

$$2/0° + 3/0° = 5/0°.$$

We observe that this result is the same as the result of placing the second vector at the tip of the first vector. ▲

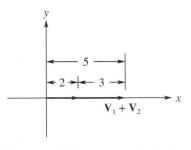

Vector diagram for Example 17.15.

ADDITION OF HORIZONTAL AND VERTICAL COMPONENTS

If $\mathbf{V}_1$ is directed along the x-axis, and $\mathbf{V}_2$ is directed along the y-axis, their sum is their resultant:

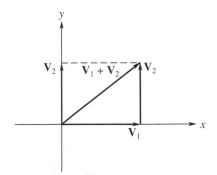

| **EXAMPLE 17.16** | ▶ |

Find the sum of the vectors $\mathbf{V}_1 = 2/0°$ and $\mathbf{V}_2 = 3/90°$.

| SOLUTION |

We draw a diagram of the vectors. Then, we place the base of $\mathbf{V}_2$ at the tip of $\mathbf{V}_1$ and draw the sum. Since $\mathbf{V}_1$ is directed along the x-axis, and $\mathbf{V}_2$ is directed along the y-axis, their sum is the resultant, as shown. We can find the magnitude $|\mathbf{V}|$ of the sum by using

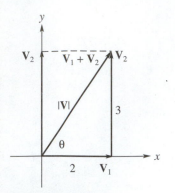

Vector diagram for Example 17.16

$$|\mathbf{V}| = \sqrt{x^2 + y^2}$$
$$= \sqrt{2^2 + 3^2}$$
$$= 3.61.$$

To find the direction θ of the sum, we write

$$\theta = \tan^{-1}\left(\frac{y}{x}\right)$$
$$= \tan^{-1}\left(\frac{3}{2}\right)$$
$$= 56.3°.$$

Therefore, the sum is the vector $3.61\underline{/56.3°}$. ▲

GENERAL ADDITION

If one or both vectors are not directed along the axes, we add them by placing the base of one vector at the tip of the other. For example, if the base of $\mathbf{V}_2$ is placed at the tip of $\mathbf{V}_1$, the sum $\mathbf{V}_1 + \mathbf{V}_2$ is the vector joining the base of $\mathbf{V}_1$ with the tip of $\mathbf{V}_2$. We observe that we have a parallelogram, with the sum as a diagonal:

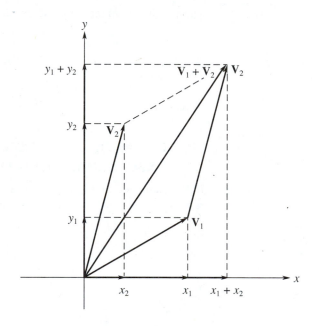

The horizontal component of the sum is the sum of the horizontal components of the two vectors. The vertical component of the sum is the sum of the vertical components of the two vectors. Therefore, to add two vectors, we find the horizontal and vertical components of each vector, and then we add the respective components.

EXAMPLE 17.17 ▶ Find the sum of the vectors $\mathbf{V}_1 = 6\underline{/30°}$ and $\mathbf{V}_2 = 8\underline{/75°}$.

SOLUTION   We draw a diagram of the vectors. Then, we place $\mathbf{V}_2$ at the tip of $\mathbf{V}_1$, and draw the sum from the base of $\mathbf{V}_1$ to the tip of $\mathbf{V}_2$:

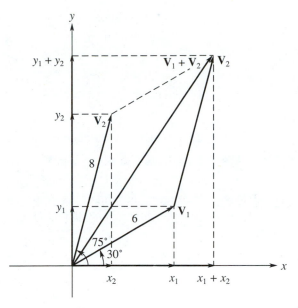

The horizontal component of V_1 is given by

$$x_1 = 6 \cos 30°$$

$$= 5.196.$$

(Because there are many calculations involved, we will reduce rounding errors in the final result by carrying four significant digits.) The vertical component of V_1 is given by

$$y_1 = 6 \sin 30°$$

$$= 3.$$

The horizontal component of V_2 is given by

$$x_2 = 8 \cos 75°$$

$$= 2.071.$$

The vertical component of V_2 is given by

$$y_2 = 8 \sin 75°$$

$$= 7.727.$$

We add the respective components to find the horizontal and vertical components of the sum. Therefore, the horizontal component of the sum is

$$x = x_1 + x_2$$

$$= 5.196 + 2.071$$

$$= 7.267.$$

The vertical component of the sum is

$$y = y_1 + y_2$$

$$= 3 + 7.727$$

$$= 10.73.$$

Now, we find the sum from its horizontal and vertical components. To find the magnitude, we write

$$|V| = \sqrt{x^2 + y^2}$$

$$= \sqrt{7.267^2 + 10.73^2}$$

$$= 13.$$

To find the direction, we write

$$\theta = \tan^{-1}\left(\frac{y}{x}\right)$$

$$= \tan^{-1}\left(\frac{10.73}{7.267}\right)$$

$$= 55.9°.$$

Therefore, the sum is the vector $13\underline{/55.9°}$.   ▲

EXAMPLE 17.18 Find the sum of the vectors $\mathbf{V}_1 = 8.5\underline{/60°}$ and $\mathbf{V}_2 = 11\underline{/140°}$.

SOLUTION   We draw a diagram of the vectors. Then, we place $\mathbf{V}_2$ at the tip of $\mathbf{V}_1$, and draw the sum from the base of $\mathbf{V}_1$ to the tip of $\mathbf{V}_2$:

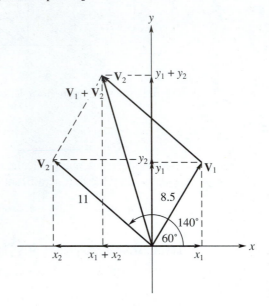

The horizontal and vertical components of $\mathbf{V}_1$ are given by

$$x_1 = 8.5 \cos 60°$$

$$= 4.25,$$

and

$$y_1 = 8.5 \sin 60°$$

$$= 7.361.$$

The horizontal and vertical components of $\mathbf{V}_2$ are given by

$$x_2 = 11 \cos 140°$$

$$= -8.426,$$

and

$$y_2 = 11 \sin 140°$$

$$= 7.071.$$

We observe that x_2 is negative. Then, the horizontal component of the sum is

$$x = x_1 + x_2$$

$$= 4.25 + (-8.426)$$

$$= -4.176,$$

and the vertical component of the sum is

$$y = y_1 + y_2$$

$$= 7.361 + 7.071$$

$$= 14.43.$$

Now, we find the sum from its horizontal and vertical components. To find the magnitude, we write

$$|\mathbf{V}| = \sqrt{|x|^2 + |y|^2}$$

$$= \sqrt{4.176^2 + 14.43^2}$$

$$= 15.$$

We find the reference angle θ' for the direction by writing

$$\theta' = \tan^{-1}\left(\frac{|y|}{|x|}\right)$$

$$= \tan^{-1}\left(\frac{14.43}{4.176}\right)$$

$$= 73.86°.$$

Since x is negative but y is positive, the sum is in the second quadrant. Therefore, the direction of the sum is

$$\theta = 180° - 73.86°$$

$$= 106°.$$

The sum is the vector $15\underline{/106°}$.   ▲

SUBTRACTION

To subtract one vector from another, we can subtract the respective components.

EXAMPLE 17.19 ▶ Find the difference $\mathbf{V}_1 - \mathbf{V}_2$ for the vectors $\mathbf{V}_1 = 4\underline{/45°}$ and $\mathbf{V}_2 = 6\underline{/110°}$.

SOLUTION   The horizontal and vertical components of $\mathbf{V}_1$ are given by

$$x_1 = 4 \cos 45°$$

$$= 2.828,$$

and

$$y_1 = 4 \sin 45°$$

$$= 2.828.$$

The horizontal and vertical components of $\mathbf{V}_2$ are given by

$$x_2 = 6 \cos 110°$$

$$= -2.052,$$

and

$$y_2 = 6 \sin 110°$$

$$= 5.638.$$

We subtract the respective components to find the horizontal and vertical components of the difference. Therefore, the horizontal component of $\mathbf{V}_1 - \mathbf{V}_2$ is

$$x = x_1 - x_2$$

$$= 2.828 - (-2.052)$$

$$= 4.88.$$

The vertical component of $\mathbf{V}_1 - \mathbf{V}_2$ is

$$y = y_1 - y_2$$
$$= 2.828 - 5.638$$
$$= -2.81.$$

Now, we find the difference from its horizontal and vertical components. To find the magnitude, we write

$$|\mathbf{V}| = \sqrt{|x|^2 + |y|^2}$$
$$= \sqrt{4.88^2 + 2.81^2}$$
$$= 5.63.$$

We find the reference angle θ' for the direction by writing

$$\theta' = \tan^{-1}\left(\frac{|y|}{|x|}\right)$$
$$= \tan^{-1}\left(\frac{2.81}{4.88}\right)$$
$$= 29.93°.$$

Since x is positive but y is negative, the difference is in the fourth quadrant. Therefore, we may use the negative of the reference angle as the direction. Thus, the difference $\mathbf{V}_1 - \mathbf{V}_2$ is the vector $5.63\underline{/-29.9°}$.   ▲

We can draw a diagram of the vectors in Example 17.19 by reversing the direction of the vector to be subtracted. The direction of a vector is reversed by either adding or subtracting 180°. The vector to be subtracted is

$$\mathbf{V}_2 = 6\underline{/110°}.$$

If we subtract 180° from the direction of $\mathbf{V}_2$, we have

$$-\mathbf{V}_2 = 6\underline{/-70°}.$$

Then, we may write

$$\mathbf{V}_1 - \mathbf{V}_2 = \mathbf{V}_1 + (-\mathbf{V}_2).$$

We draw $-\mathbf{V}_2$, then place $-\mathbf{V}_2$ at the tip of $\mathbf{V}_1$ and draw the difference from the base of $\mathbf{V}_1$ to the tip of $\mathbf{V}_2$.

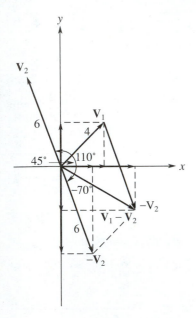

Vector diagram for Example 17.19.

EXERCISE

17.4

Find the sum $\mathbf{V}_1 + \mathbf{V}_2$ of the vectors:

1. $\mathbf{V}_1 = 5\underline{/90°}$ and $\mathbf{V}_2 = 8\underline{/90°}$ **2.** $\mathbf{V}_1 = 4.7\underline{/30°}$ and $\mathbf{V}_2 = 3.6\underline{/30°}$

3. $\mathbf{V}_1 = 4\underline{/0°}$ and $\mathbf{V}_2 = 6\underline{/180°}$ **4.** $\mathbf{V}_1 = 5\underline{/0°}$ and $\mathbf{V}_2 = 2\underline{/180°}$

5. $\mathbf{V}_1 = 2\underline{/0°}$ and $\mathbf{V}_2 = 2.5\underline{/90°}$ **6.** $\mathbf{V}_1 = 12.4\underline{/0°}$ and $\mathbf{V}_2 = 10.1\underline{/90°}$

7. $\mathbf{V}_1 = 3\underline{/180°}$ and $\mathbf{V}_2 = 1.5\underline{/90°}$ **8.** $\mathbf{V}_1 = 2.2\underline{/180°}$ and $\mathbf{V}_2 = 3.3\underline{/270°}$

9. $\mathbf{V}_1 = 5\underline{/10°}$ and $\mathbf{V}_2 = 4\underline{/70°}$ **10.** $\mathbf{V}_1 = 2\underline{/40°}$ and $\mathbf{V}_2 = 8\underline{/60°}$

11. $\mathbf{V}_1 = 10\underline{/45°}$ and $\mathbf{V}_2 = 7.5\underline{/120°}$ **12.** $\mathbf{V}_1 = 1.1\underline{/70°}$ and $\mathbf{V}_2 = 1.4\underline{/200°}$

13. $\mathbf{V}_1 = 3.9\underline{/60°}$ and $\mathbf{V}_2 = 2.3\underline{/-15°}$ **14.** $\mathbf{V}_1 = 8\underline{/30°}$ and $\mathbf{V}_2 = 12\underline{/-70°}$

Find the difference $\mathbf{V}_1 - \mathbf{V}_2$ of the vectors:

15. $\mathbf{V}_1 = 5\underline{/90°}$ and $\mathbf{V}_2 = 8\underline{/90°}$ **16.** $\mathbf{V}_1 = 4.7\underline{/30°}$ and $\mathbf{V}_2 = 3.6\underline{/30°}$

17. $\mathbf{V}_1 = 4\underline{/0°}$ and $\mathbf{V}_2 = 3\underline{/90°}$ **18.** $\mathbf{V}_1 = 1.2\underline{/180°}$ and $\mathbf{V}_2 = 2.2\underline{/270°}$

19. $\mathbf{V}_1 = 3\underline{/60°}$ and $\mathbf{V}_2 = 4\underline{/120°}$ **20.** $\mathbf{V}_1 = 10\underline{/70°}$ and $\mathbf{V}_2 = 6\underline{/100°}$

Self-Test

1. If each side of an equilateral triangle is 10 inches, find the height of the triangle.

1. _____

2. Solve the right triangle with legs 390 and 140.

2. _____

3. Solve the right triangle with an acute angle 40°, and the adjacent side for that angle 29.

3. _____

4. Find the horizontal and vertical components of the vector $\mathbf{V} = 45\underline{/-10.6°}$.

4. _____

5. Find the sum of the vectors $\mathbf{V}_1 = 2\underline{/45°}$ and $\mathbf{V}_2 = 4\underline{/120°}$.

5. _____

UNIT 18

Complex Numbers

Introduction

The real numbers, which you have used in all of the preceding units, are the coordinates of points on the real number line. These numbers are not sufficient, however, to describe all mathematical concepts or all types of scientific phenomena. Different numbers have been invented to describe different concepts and to study different phenomena. In this unit, you will learn about numbers called complex numbers that result from some types of equations. You will learn how complex numbers are defined and written, and how complex numbers can be related to vectors. Then, you will learn how to add and subtract complex numbers, and ways to multiply and divide complex numbers.

OBJECTIVES

When you have finished this unit you should be able to:

1. Use the j-operator to simplify square roots of negatives.

2. Represent complex numbers as vectors, and transfer complex numbers between rectangular form and polar form.

3. Add and subtract complex numbers in rectangular form, and by transferring polar form to rectangular form.

4. Multiply and divide complex numbers in rectangular and in polar form.

SECTION

18.1

The j-Operator

It is sometimes important to understand that numbers are invented, not discovered. Thus, we can invent any type of number we need. The counting numbers, or positive integers, were invented before the beginning of recorded history. The positive rational numbers or fractions, and even some positive irrational numbers such as simple square roots and π, were used by the earliest societies for which we have records, for example, the Babylonians. The properties of these numbers were fully understood by the ancient Greeks by about 200 B.C.

Zero, however, was not invented as a number until the Hindus used it in about the ninth century. Negative numbers did not come into use until as late as the sixteenth century, when they were invented to represent negative **roots**, or solutions of equations. For example, 1 is the root of the equation

$$x - 1 = 0.$$

The number –1 could be defined as the root of the equation

$$x + 1 = 0.$$

All of these types of numbers are real numbers, and can be represented by points on the real number line.

Another type of number was also defined in the sixteenth century as a result of efforts to solve equations. Consider, for example, a basic quadratic equation

$$x^2 - 1 = 0$$
$$x^2 = 1$$
$$x = \pm 1.$$

This equation has two roots, 1 and −1. But, consider a similar equation

$$x^2 + 1 = 0$$
$$x^2 = -1$$
$$x = \pm\sqrt{-1}.$$

This equation appears to have no roots, since square roots of negative numbers cannot be defined for real numbers. A type of number called an **imaginary number** is defined as the root of such an equation.

Imaginary Number: An **imaginary number** is a number of the form bi, where b is a real number and

$$i = \sqrt{-1},$$

or the form jb, where

$$j = \sqrt{-1}.$$

The name "imaginary number" is an unfortunate holdover from the sixteenth century, when even negative roots were considered to be "false" roots. René Descartes gave the name "imaginary" to roots that were square roots of negatives, which he considered to be even more "unreal" than negative roots. Interestingly, this discussion appeared in the same paper in which he described the foundations of the rectangular or Cartesian coordinate system named for him. We should keep in mind that all numbers are invented, and none are any less real than others.

While mathematicians usually use the form bi, the form jb is more likely to be used in electronics. The symbol j is called the **j-operator**. Since the j-operator is defined as

$$j = \sqrt{-1},$$

we can square each side of this equality to obtain

$$j^2 = (\sqrt{-1})^2$$
$$j^2 = -1.$$

The j-Operator:
$$j = \sqrt{-1}$$
and
$$j^2 = -1.$$

EXAMPLE 18.1   Write $\sqrt{-25}$ in terms of the j-operator.

SOLUTION   We replace $\sqrt{-1}$ with the j-operator by writing

$$\sqrt{-25} = \sqrt{(-1)(25)}$$
$$= \sqrt{-1}\sqrt{25}$$
$$= j5. \qquad \blacktriangle$$

EXAMPLE 18.2   Use the j-operator to simplify $\sqrt{-5}\sqrt{-5}$.

SOLUTION   Replacing $\sqrt{-1}$ with the j-operator, we have

$$\sqrt{-5} = \sqrt{(-1)(5)}$$
$$= \sqrt{-1}\sqrt{5}$$
$$= j\sqrt{5}.$$

Then we can write

$$\sqrt{-5}\sqrt{-5} = j\sqrt{5}j\sqrt{5}$$
$$= j^2 5$$
$$= (-1)5$$
$$= -5. \qquad \blacktriangle$$

Observe that you must *not* multiply the square roots in Example 18.2 without transferring to the *j*-operator; that is, $\sqrt{-5}\sqrt{-5}$ is *not the same as* $\sqrt{(-5)(-5)}$, which is $\sqrt{25}$ or 5. In general, if *a* and *b* are negative, $\sqrt{a}\sqrt{b}$ is *not the same as* $\sqrt{ab}$.

EXERCISE	Write in terms of the *j*-operator:
18.1	

1. $\sqrt{-4}$ **2.** $\sqrt{-36}$ **3.** $\sqrt{-20}$ **4.** $\sqrt{-72}$

5. $-\sqrt{-25}$ **6.** $-\sqrt{-10}$

Use the *j*-operator to simplify:

7. $\sqrt{-2}\sqrt{-2}$ **8.** $\sqrt{-4}\sqrt{-25}$ **9.** $\sqrt{-2}\sqrt{-8}$ **10.** $\sqrt{-2}\sqrt{-50}$

11. $\sqrt{4}\sqrt{-16}$ **12.** $\sqrt{-3}\sqrt{9}$

SECTION
18.2

Rectangular and Polar Forms

When real numbers and imaginary numbers are combined, they form **complex numbers**.

> **Complex number:** A **complex number** is a number of the form $a + jb$, where *a* is a real number and *jb* is an imaginary number.

Complex numbers written in the form $a + jb$ are in **rectangular form**. A complex number written in the rectangular form $a + jb$ has a **real part** *a*, and an **imaginary part** *jb*. For example, the complex number $5 + j5$ has a real part 5 and an imaginary part *j*5.

Real numbers and imaginary numbers are also complex numbers. For example, the number

$$5 + j0 = 5$$

is a real number, and also a complex number where the imaginary part is zero. The number

$$0 + j5 = j5$$

is an imaginary number, and also a complex number where the real part is zero.

Complex numbers written in rectangular form can be represented in the rectangular coordinate system, by using the *x*-axis for real numbers and the *y*-axis for imaginary numbers. A complex number

$$a + j0 = a,$$

which is also a real number, is represented by the point with coordinate *a* on the *x*-axis, called the **real axis**. A complex number

$$0 + jb = jb,$$

which is also an imaginary number, is represented by the point with coordinate *b* on the *y*-axis, called the **imaginary axis**.

Any complex number $a + jb$ can be represented as a vector with horizontal component $x = a$ on the real axis and vertical component $y = b$ on the imaginary axis.

EXAMPLE 18.3 ▶ Draw the vector that represents the complex number $3 + j4$.

SOLUTION The vector that represents $3 + j4$ has horizontal component $x = 3$ and vertical component $y = 4$. The vector is shown in the diagram at the left below. ▲

EXAMPLE 18.4 ▶ Draw the vector that represents the complex number $3 - j2$.

SOLUTION The vector that represents $3 - j2$ has horizontal component $x = 3$ and vertical component $y = -2$. The vector is shown at the right:

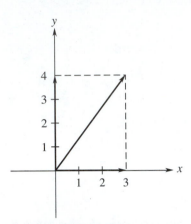
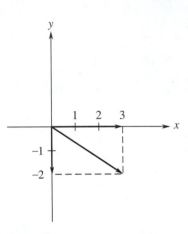

Vector diagram for Example 18.3.   Vector diagram for Example 18.4.   ▲

EXAMPLE 18.5 ▶ Draw the vector that represents the complex number 3.

SOLUTION The number 3 is the complex number $3 + j0$ in rectangular form. The vector that represents $3 + j0$ has horizontal component $x = 3$ and vertical component $y = 0$. The vector is shown at the left below. ▲

EXAMPLE 18.6 ▶ Draw the vector that represents the complex number $-j2$.

SOLUTION The number $-j2$ is the complex number $0 - j2$ in rectangular form. The vector that represents $0 - j2$ has horizontal component $x = 0$ and vertical component $y = -2$. The vector is shown at the right:

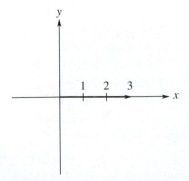

Vector diagram for Example 18.5.   Vector diagram for Example 18.6.   ▲

POLAR FORM

A complex number represented by a vector with magnitude r and direction θ.

Now, suppose $a + jb$ is a complex number represented by a vector with magnitude r and direction θ. We know from Section 17.3 that the horizontal component a and the vertical component b are given by

$$a = r \cos \theta$$

and

$$b = r \sin \theta.$$

Thus, the complex number $a + jb$ can be written in the form

$$a + jb = r \cos \theta + j(r \sin \theta).$$

Complex numbers written in terms of r and θ are in **polar form** or **trigonometric form**. The polar form of a complex number can be written in the vector notation

$$r\underline{/\theta}.$$

As in Section 17.3, r is called the absolute value of the complex number and θ is called the argument of the complex number.

We also know from Section 17.3 that the absolute value r and the reference angle θ' for the argument θ are given in terms of $|a|$ and $|b|$ by

$$r = \sqrt{|a|^2 + |b|^2}$$

and

$$\theta' = \tan^{-1}\left(\frac{|b|}{|a|}\right).$$

For the **rectangular form** $a + jb$:

$$a = r \cos \theta$$
$$b = r \sin \theta.$$

For the **polar form** $r\underline{/\theta}$:

$$r = \sqrt{|a|^2 + |b|^2}$$
$$\theta' = \tan^{-1}\left(\frac{|b|}{|a|}\right).$$

EXAMPLE 18.7 ▶ Write the complex number $-1 + j$ in polar form.

SOLUTION The vector that represents $-1 + j$ has horizontal component $x = -1$ and vertical component $y = 1$, as shown. The absolute value is given by

$$r = \sqrt{|a|^2 + |b|^2}$$
$$= \sqrt{1^2 + 1^2}$$
$$= \sqrt{2},$$

or $r = 1.41$. The reference angle of the argument is given by

$$\theta' = \tan^{-1}\left(\frac{|b|}{|a|}\right)$$
$$= \tan^{-1}\left(\frac{1}{1}\right)$$
$$= 45°.$$

Vector diagram for Example 18.7.

The vector is in the second quadrant; therefore,

$$\theta = 180° - 45°$$
$$= 135°.$$

Thus, the polar form is $1.41\underline{/135°}$. ▲

EXAMPLE 18.8 ▶ Write the complex number 4 in polar form.

SOLUTION The vector that represents 4 has horizontal component $x = 4$ and vertical component $y = 0$. The vector diagram is at the left below. The absolute value is $r = 4$, and the argument is $\theta = 0°$. Therefore, the polar form is $4\underline{/0°}$. ▲

EXAMPLE 18.9 ▶ Write the complex number $-j3$ in polar form.

SOLUTION The vector that represents $-j3$ has horizontal component $x = 0$ and vertical component $y = -3$. The vector diagram is at the right below. The absolute value is $r = 3$, and the argument is $\theta = 270°$, or $\theta = -90°$. Therefore, the polar form is $3\underline{/270°}$ or $3\underline{/-90°}$:

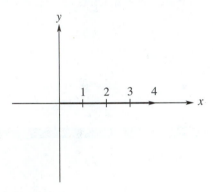

Vector diagram for Example 18.8.   Vector diagram for Example 18.9. ▲

POLAR TO RECTANGULAR FORM

Given the polar form of a complex number, we can also write the complex number in rectangular form.

EXAMPLE 18.10 ▶ Write the complex number given by $2\underline{/210°}$ in rectangular form.

SOLUTION The complex number is represented by a vector in the third quadrant. The horizontal and vertical components are given by

$$a = r \cos \theta$$
$$= 2 \cos 210°$$
$$= -1.73,$$

and

$$b = r \sin \theta$$
$$= 2 \sin 210°$$
$$= -1.$$

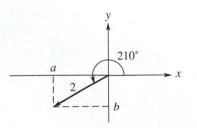

Vector diagram for Example 18.10.

Therefore, the rectangular form is $-1.73 - j$. ▲

EXAMPLE 18.11   Write the complex number given by $10\underline{/90°}$ in rectangular form.

SOLUTION   The complex number is represented by a vector lying on the y-axis, as shown at the left below. The horizontal and vertical components are $a = 0$ and $b = 10$. Therefore, the rectangular form is $0 + j10$, or $j10$.   ▲

EXAMPLE 18.12 ▶   Write the complex number given by $5.5\underline{/180°}$ in rectangular form.

SOLUTION   The complex number is represented by a vector lying on the negative x-axis, as shown at the right below. The horizontal and vertical components are $a = -5.5$ and $b = 0$. Therefore, the rectangular form is $-5.5 + j0$, or -5.5.

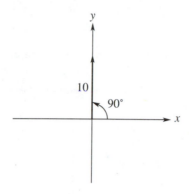

Vector diagram for Example 18.11.

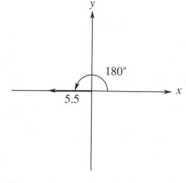

Vector diagram for Example 18.12.   ▲

EXERCISE 18.2

Draw the vector that represents the complex number:

1. $2 + j2$ 2. $1 + j3$ 3. $1 - j3$ 4. $-4 + j$

5. $-2 - j2$ 6. $-4 - j3$ 7. 5 8. -4

9. $j4$ 10. $-j5$

Write the complex number in polar form:

11. $1 + j$ 12. $3 + j4$ 13. $-3 + j2$ 14. $2 - j3$

15. $-4 - j4$ 16. $-2.5 - j1.5$ 17. 10 18. -1.5

19. j 20. $-j2.2$

Write the complex number in rectangular form:

21. $2\underline{/60°}$ 22. $1\underline{/45°}$ 23. $\sqrt{2}\underline{/135°}$ 24. $5\underline{/330°}$

25. $10\underline{/240°}$ 26. $2.2\underline{/-30°}$ 27. $3\underline{/180°}$ 28. $4\underline{/270°}$

29. $1.5\underline{/90°}$ 30. $1\underline{/0°}$

**SECTION
18.3**

Addition and Subtraction

Addition of complex numbers given in rectangular form is similar to addition of algebraic expressions. We combine the like terms of the complex numbers. The real parts are like terms and the imaginary parts are like terms.

EXAMPLE 18.13 ▶ Add $(1 + j3) + (4 + j2)$.

SOLUTION First, we remove the parentheses:

$$(1 + j3) + (4 + j2) = 1 + j3 + 4 + j2.$$

Then, we combine the real parts, which are like terms, and the imaginary parts, which are like terms. Thus, we have

$$1 + j3 + 4 + j2 = 1 + 4 + j(3 + 2)$$
$$= 5 + j5. \qquad ▲$$

Because complex numbers can be represented as vectors, addition of complex numbers is similar to the addition of vectors shown in Section 17.4. We add the real parts, which are the horizontal components, and the imaginary parts, which are the vertical components:

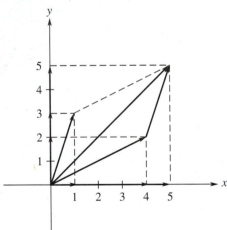

EXAMPLE 18.14 ▶ Add $(4 + j2) + (6 - j4)$.

SOLUTION First, we remove the parentheses:

$$(4 + j2) + (6 - j4) = 4 + j2 + 6 - j4.$$

Therefore, we have

$$4 + j2 + 6 - j4 = 4 + 6 + j(2 - 4)$$
$$= 10 - j2. \qquad ▲$$

EXAMPLE 18.15 ▶ Add $(5 - j2) + (1 + j2)$.

SOLUTION Removing parentheses and combining like terms, we have

$$(5 - j2) + (1 + j2) = 5 - j2 + 1 + j2$$
$$= 5 + 1 + j(-2 + 2)$$
$$= 6 + j0$$
$$= 6.$$

The sum is a real number. We recall that real numbers and imaginary numbers are also complex numbers. The sum of complex numbers is a complex number, but can also be a real number or an imaginary number.   ▲

To subtract two complex numbers, we follow the same procedure, adjusting the signs when we remove the parentheses.

EXAMPLE 18.16 Subtract $(3 - j) - (2 - j5)$.

SOLUTION First, we remove the parentheses, adjusting the signs of the second complex number:

$$(3 - j) - (2 - j5) = 3 - j - 2 + j5.$$

Then, we combine the real parts and the imaginary parts to obtain

$$3 - j - 2 + j5 = 1 + j4.$$ ▲

POLAR FORM

If the complex numbers to be added or subtracted are given in polar form, we must write them in rectangular form. Then, we can add or subtract, and return the result to polar form.

EXAMPLE 18.17 Add $2\underline{/30°} + 4\underline{/120°}$.

SOLUTION First, we write the complex numbers in rectangular form. The real and imaginary parts of the first complex number are given by

$$a_1 = 2 \cos 30°$$
$$= 1.732,$$

and

$$b_1 = 2 \sin 30°$$
$$= 1.$$

Therefore, the first complex number in rectangular form is

$$a_1 + jb_1 = 1.732 + j.$$

The real and imaginary parts of the second complex number are given by

$$a_2 = 4 \cos 120°$$
$$= -2,$$

and

$$b_2 = 4 \sin 120°$$
$$= 3.464.$$

Therefore, the second complex number in rectangular form is

$$a_2 + jb_2 = -2 + j3.464.$$

The sum of the complex numbers in rectangular form is

$$(a_1 + jb_1) + (a_2 + jb_2) = (1.732 + j) + (-2 + j3.464).$$

We remove the parentheses and combine like terms:

$$(1.732 + j) + (-2 + j3.464) = 1.732 + j - 2 + j3.464$$
$$= -0.268 + j4.464.$$

We observe that we have added the real parts to obtain

$$a = a_1 + a_2,$$

and the imaginary parts to obtain

$$b = b_1 + b_2.$$

The sum in rectangular form is

$$a + jb = -0.268 + j4.464.$$

Now, we return the sum to polar form. The absolute value is given by

$$r = \sqrt{|a|^2 + |b|^2}$$

$$= \sqrt{0.268^2 + 4.464^2}$$

$$= 4.47.$$

The reference angle of the argument is given by

$$\theta' = \tan^{-1}\left(\frac{|b|}{|a|}\right)$$

$$= \tan^{-1}\left(\frac{4.464}{0.268}\right)$$

$$= 86.6°.$$

Since a is negative but b is positive, the complex number is in the second quadrant. Therefore, the argument is

$$\theta = 180° - 86.6°$$

$$= 93.4°.$$

The sum in polar form is $4.47\underline{/93.4°}$.   ▲

We observe that the process we have used to solve Example 18.17 is similar to the process for addition of vectors given in Section 17.4. We can represent the complex numbers as vectors. Then, we can place one vector at the tip of the other and draw their sum. To write the complex numbers in rectangular form, we find the real parts, which are the horizontal components, and the imaginary parts, which are the vertical components. Then, we add the real parts and the imaginary parts, which is the same as adding the respective components. Finally, we return the sum to polar form.

Subtraction of complex numbers given in polar form is similar to subtraction of vectors.

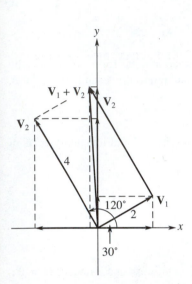

Vector diagram for Example 18.17.

EXAMPLE 18.18 ▶ Subtract $1.5\underline{/-20°} - 3\underline{/260°}$.

SOLUTION First, we write the complex numbers in rectangular form. The real and imaginary parts of the first complex number are given by

$$a_1 = 1.5 \cos(-20°)$$

$$= 1.41,$$

and

$$b_1 = 1.5 \sin(-20°)$$

$$= -0.513.$$

Therefore, the first complex number in rectangular form is

$$a_1 + jb_1 = 1.41 - j0.513.$$

The real and imaginary parts of the second complex number are given by

$$a_2 = 3 \cos 260°$$

$$= -0.5209,$$

and

$$b_2 = 3 \sin 260°$$

$$= -2.954.$$

Therefore, the second complex number in rectangular form is

$$a_2 + jb_2 = -0.5209 - j2.954.$$

The difference of the complex numbers in rectangular form is

$$(a_1 + jb_1) - (a_2 + jb_2) = (1.41 - j0.513) - (-0.5209 - j2.954).$$

We remove the parentheses, adjusting the signs of the second complex number, and combine the like terms:

$$(1.41 - j0.513) - (-0.5209 - j2.954) = 1.41 - j0.513 + 0.5209 + j2.954$$

$$= 1.931 + j2.441.$$

The difference in rectangular form is $1.931 + j2.441$.

Now, we return the difference to polar form. To find the absolute value, we write

$$r = \sqrt{|a|^2 + |b|^2}$$

$$= \sqrt{1.931^2 + 2.441^2}$$

$$= 3.11.$$

To find the argument, we write

$$\theta' = \tan^{-1}\left(\frac{|b|}{|a|}\right)$$

$$= \tan^{-1}\left(\frac{2.441}{1.931}\right)$$

$$= 51.7°.$$

Since a and b are both positive, the complex number is in the first quadrant and $\theta = \theta'$. Therefore, the difference in polar form is $3.11\underline{/51.7°}$. ▲

We can represent the complex numbers in Example 18.18 as vectors. Then, we may draw the negative of the second vector, recalling that the negative of a vector has the same magnitude but the opposite direction:

$$-3\underline{/260°} = 3\underline{/80°}.$$

We can place the negative of the second vector at the tip of the first vector and then draw the difference as shown in the diagram to the left.

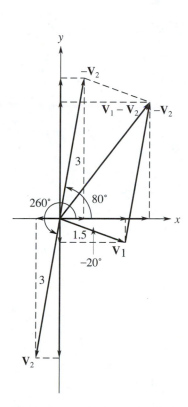

Vector diagram for Example 18.18.

EXERCISE 18.3

Add or subtract as indicated:

1. $(3 + j2) + (6 + j4)$

2. $(8 + j5) + (3 + j9)$

3. $(8 - j) + (8 + j4)$

4. $(3 - j2) + (5 - j7)$

5. $(7 - j8) + (5 - j2)$

6. $(1 + j8) + (4 - j9)$

7. $(3 + j3) + (7 - j3)$

8. $(4 - j8) + (j9 - 4)$

9. $(9 + j6) - (7 - j2)$

10. $(2.1 - j4.9) - (6.8 - j7.5)$

11. $(3.5 + j2.4) - (3.5 - j4.8)$

12. $\left(\dfrac{7}{2} - j\dfrac{1}{2}\right) - \left(\dfrac{3}{2} - j\dfrac{1}{2}\right)$

13. $2\underline{/45°} + 1\underline{/135°}$

14. $3\underline{/60°} + 4\underline{/120°}$

15. $6\underline{/90°} + 4.5\underline{/-30°}$

16. $\sqrt{2}\underline{/-45°} + \sqrt{2}\underline{/225°}$

17. $10\underline{/60°} - 4\underline{/150°}$

18. $2\underline{/45°} - 3\underline{/120°}$

19. $4\underline{/-30°} - 4\underline{/30°}$

20. $5\underline{/-45°} - 5\underline{/-135°}$

Multiplication and Division

One method for multiplication of complex numbers is similar to the method for multiplication of binomials given in Section 14.1. We multiply the first, outside, inside, and last terms of the complex numbers. Then, we combine like terms by using $j^2 = -1$.

EXAMPLE 18.19   Multiply $(3 + j)(2 + j4)$.

SOLUTION   We multiply each pair of terms:

$$(3 + j)(2 + j4) = (3)(2) + (3)(j4) + (j)(2) + (j)(j4)$$
$$= 6 + j12 + j2 + j^2 4.$$

Then, since $j^2 = -1$, we write

$$6 + j12 + j2 + j^2 4 = 6 + j12 + j2 + (-1)4$$
$$= 6 + j12 + j2 - 4.$$

We combine the real parts, observing that the term that included j^2 has become a real part, and the imaginary parts:

$$6 + j12 + j2 - 4 = 2 + j14. \qquad \blacktriangle$$

EXAMPLE 18.20   Multiply $(3 + j2)(5 - j4)$.

SOLUTION   We multiply each pair of terms, replace j^2 by -1, and combine like terms:

$$(3 + j2)(5 - j4) = (3)(5) - (3)(j4) + (j2)(5) - (j2)(j4)$$
$$= 15 - j12 + j10 - j^2 8$$
$$= 15 - j12 + j10 - (-1)(8)$$
$$= 15 - j12 + j10 + 8$$
$$= 23 - j2. \qquad \blacktriangle$$

EXAMPLE 18.21   Multiply $(3 + j2)(3 - j2)$.

SOLUTION   We multiply each pair of terms, replace j^2 by -1, and combine like terms:

$$(3 + j2)(3 - j2) = 9 - j6 + j6 - j^2 4$$
$$= 9 - j6 + j6 + 4$$
$$= 13.$$

The product is a real number. $\qquad \blacktriangle$

We observe that the absolute value of $3 + j2$ is

$$r = \sqrt{3^2 + 2^2}$$
$$= \sqrt{13},$$

and so

$$r^2 = 13.$$

The square of the absolute value of $3 - j2$ is also 13. The product in Example 18.21 is the square of the absolute value of each of the factors.

Two complex numbers of the form $a + jb$ and $a - jb$, which are the same except for the sign of their imaginary parts, are called **complex conjugates**. The product of complex conjugates is always a real number, and is the square of the absolute value of each of the conjugates.

To divide two complex numbers, we may multiply each complex number by the conjugate of the divisor. When the division is written as a fraction, the divisor is the complex number in the denominator. Thus, we obtain a real number in the denominator.

EXAMPLE 18.22 Divide $\dfrac{3 + j}{2 + j4}$.

SOLUTION We multiply the numerator and the denominator by the conjugate of $2 + j4$, which is $2 - j4$:

$$\frac{3 + j}{2 + j4} = \frac{(3 + j)(2 - j4)}{(2 + j4)(2 - j4)}.$$

Then, we multiply the factors in the numerator and the factors in the denominator:

$$\frac{(3 + j)(2 - j4)}{(2 + j4)(2 - j4)} = \frac{6 - j12 + j2 - j^2 4}{4 - j8 + j8 - j^2 16}$$

$$= \frac{6 - j12 + j2 + 4}{4 - j8 + j8 + 16}$$

$$= \frac{10 - j10}{20}.$$

We observe that the denominator is the square of the absolute value of $2 + j4$; that is, $r^2 = 2^2 + 4^2 = 20$. We can reduce the resulting fraction to obtain

$$\frac{10 - j10}{20} = \frac{1 - j}{2}.$$

Thus, the quotient is the complex number $\dfrac{1}{2} - j\dfrac{1}{2}$, or $0.5 - j0.5$.   ▲

EXAMPLE 18.23 Divide $\dfrac{3 + j2}{5 - j4}$.

SOLUTION We multiply the numerator and the denominator by the conjugate of $5 - j4$, which is $5 + j4$:

$$\frac{3 + j2}{5 - j4} = \frac{(3 + j2)(5 + j4)}{(5 - j4)(5 + j4)}$$

$$= \frac{15 + j12 + j10 + j^2 8}{25 + j20 - j20 - j^2 16}$$

$$= \frac{15 + j12 + j10 - 8}{25 + j20 - j20 + 16}$$

$$= \frac{7 + j22}{41}$$

$$= \frac{7}{41} + j\frac{22}{41}.$$

This quotient can be written $0.171 + j0.537$.   ▲

EXAMPLE 18.24 Divide $\dfrac{6 - j4}{j5}$.

SOLUTION The conjugate of $j5$ or $0 + j5$, is $0 - j5$ or $-j5$. However, to obtain a real number in the denominator, it is sufficient simply to multiply by j:

$$\frac{6 - j4}{j5} = \frac{(6 - j4)j}{(j5)j}$$

$$= \frac{j6 - j^2 4}{j^2 5}$$

$$= \frac{j6 + 4}{-5}$$

$$= -\frac{4}{5} - j\frac{6}{5}.$$

Thus, the quotient is $-0.8 - j1.2$.   ▲

POLAR FORM

It is easier to multiply or divide complex numbers in polar form than in rectangular form. To multiply complex numbers in polar form, we multiply the absolute values but *add* their arguments:

$$(r_1 \underline{/\theta_1})(r_2 \underline{/\theta_2}) = r_1 r_2 \underline{/\theta_1 + \theta_2}.$$

To divide complex numbers in polar form, we divide the absolute values but *subtract* their arguments:

$$\frac{r_1 \underline{/\theta_1}}{r_2 \underline{/\theta_2}} = \frac{r_1}{r_2} \underline{/\theta_1 - \theta_2}.$$

The proofs of these formulas require analytic trigonometry, and are given at the end of Unit 22. We can show by example that they appear to work.

EXAMPLE 18.25 Multiply $(3.16\underline{/18.4°})(4.47\underline{/63.4°})$.

SOLUTION Using the first formula described above, we write

$$(3.16\underline{/18.4°})(4.47\underline{/63.4°}) = (3.16)(4.47)\underline{/18.4° + 63.4°}$$

$$= 14.1\underline{/81.8°}.$$

Thus, the product in polar form is $14.1\underline{/81.8°}$.   ▲

To compare this result with the same multiplication in rectangular form, we transfer each of the complex numbers to rectangular form:

$$3.16\underline{/18.4°} = (3.16 \cos 18.4°) + j(3.16 \sin 18.4°)$$

$$= 3 + j$$

(we have rounded off 0.998 to 1), and

$$4.47\underline{/63.4°} = (4.47 \cos 63.4°) + j(4.47 \sin 63.4°)$$

$$= 2 + j4.$$

When we transfer the product to rectangular form, we have

$$14.1\underline{/81.8°} = (14.1 \cos 81.8°) + j(14.1 \sin 81.8°)$$

$$= 2 + j14.$$

(we have rounded off 2.01 to 2). In Example 18.19, we found that

$$(3 + j)(2 + j4) = 2 + j14.$$

Thus, our results in polar form and in rectangular form agree, allowing for approximations due to rounding off.

EXAMPLE 18.26 ▶ Divide $\dfrac{3.16\underline{/18.4°}}{4.47\underline{/63.4°}}$.

SOLUTION Using the formula for division in polar form, we write

$$\frac{3.16\underline{/18.4°}}{4.47\underline{/63.4°}} = \frac{3.16}{4.47}\underline{/18.4° - 63.4°}$$

$$= 0.707\underline{/-45°}.$$

Thus, the quotient in polar form is $0.707\underline{/-45°}$. ▲

To compare this result with the same division in rectangular form, we transfer each of the complex numbers to rectangular form. We know from the preceding example that

$$3.16\underline{/18.4°} = 3 + j,$$

and

$$4.47\underline{/63.4°} = 2 + j4.$$

When we transfer the quotient to rectangular form, we have

$$0.707\underline{/-45°} = [0.707 \cos(-45°)] + j[0.707 \sin(-45°)]$$

$$= 0.5 - j0.5.$$

In Example 18.22, we found that

$$\frac{3 + j}{2 + j4} = 0.5 - j0.5,$$

so our results in polar form and in rectangular form agree.

EXAMPLE 18.27 ▶ Divide $\dfrac{1\underline{/0°}}{5\underline{/-53.1°}}$.

SOLUTION We must be careful to subtract the negative angle:

$$\frac{1\underline{/0°}}{5\underline{/-53.1°}} = \frac{1}{5}\underline{/0° - (-53.1°)}$$

$$= 0.2\underline{/53.1°}.$$ ▲

Since the numerator of the division in Example 18.27 is

$$1\underline{/0°} = 1,$$

we observe that $0.2\underline{/53.1°}$ is the reciprocal of $5\underline{/-53.1°}$. In rectangular form, we have

$$5\underline{/-53.1°} = [5 \cos(-53.1°)] + j[5 \sin(-53.1°)]$$

$$= 3 - j4.$$

Therefore, in rectangular form this reciprocal is

$$\frac{1}{3 - j4} = \frac{1(3 + j4)}{(3 - j4)(3 + j4)}$$

$$= \frac{3 + j4}{25}$$

$$= 0.12 + j0.16.$$

You should check that this result agrees with the result of the division in polar form.

**EXERCISE
18.4**

Multiply or divide as indicated:

1. $(3 + j)(1 + j2)$

2. $(3 + j4)(2 + j5)$

3. $(2 - j)(5 + j4)$

4. $(5 - j)(4 - j)$

5. $(2 - j6)(4 - j5)$

6. $(1 + j5)(6 - j6)$

7. $(4 + j)(4 - j)$

8. $(2 - j6)(2 + j6)$

9. $(j2 + 7)(j2 - 7)$

10. $(j2 - 1)(2 + j4)$

11. $\dfrac{3 + j}{1 + j2}$

12. $\dfrac{1 + j3}{7 + j}$

13. $\dfrac{4 + j2}{5 - j3}$

14. $\dfrac{6 - j2}{2 - j3}$

15. $\dfrac{2 + j3}{j4}$

16. $\dfrac{j7 - 6}{j2}$

17. $(3.16\underline{/18.4°})(2.24\underline{/63.4°})$

18. $(6.32\underline{/-71.6°})(6.4\underline{/-51.3°})$

19. $\dfrac{3.16\underline{/18.4°}}{2.24\underline{/63.4°}}$

20. $\dfrac{4.47\underline{/26.6°}}{5.83\underline{/-31°}}$

21. $\dfrac{1\underline{/0°}}{5\underline{/36.9°}}$

22. $\dfrac{1\underline{/0°}}{0.707\underline{/-45°}}$

23. $\dfrac{5\underline{/53.1°}}{5\underline{/-90°}}$

24. $\dfrac{6\underline{/-90°}}{3\underline{/90°}}$

Self-Test

1. Use the *j*-operator to simplify $\sqrt{5}\sqrt{-20}$.

1. _____

2. Write $2 - j1.5$ in polar form.

2. _____

3. Add $4\underline{/45°} + 2\underline{/90°}$.

3. _____

4. Divide $\dfrac{2 + j3}{1 - j}$.

4. _____

5. Multiply $(1.5\underline{/270°})(2\underline{/225°})$.

5. _____

UNIT 19

Sine Waves

Introduction

In contemporary mathematics and science, the study of trigonometry goes far beyond triangle measurement. For example, the sine function, derived from the trigonometric ratio, describes a kind of motion called periodic motion. There are many applications of this type of function, ranging from the motion of springs and pendulums to the form of electricity called alternating current. In this unit, you will first learn how to measure angles in radians. You will use radian measure in generating the type of function called a sine function. You will study sine functions and their properties, including the amplitude, period, and phase angle. Then, you will learn some formulas involving angles in terms of time. Finally, you will apply the sine function defined in terms of time to alternating current and voltage.

OBJECTIVES

When you have finished this unit you should be able to:

1. Convert measures of angles from degree measure to radian measure, and from radian measure to degree measure.

2. Draw graphs of sine waves with given amplitudes and periods.

3. Draw graphs of sine waves with given phase angles.

4. Use angular motion formulas to find frequencies and times, periods, angular velocities, and angles.

5. Find instantaneous values of AC voltages or currents represented by sine waves, and AC voltages and currents related by leading or lagging phase angles.

**SECTION
19.1**

Radian Measure

In the preceding units, we have measured angles in the familiar degree measure. Another measure for angles, called **radian** measure, is often used in mathematics and science. To define the radian, we start with a **circle**. A circle is defined by all points at a fixed distance from a point at the center. The fixed distance is the **radius** r of the circle.

Now, suppose we have a piece of wire, or other material that we can bend, which is the exact length of the radius. We bend the wire around the circle to measure an arc that is the same length as the radius. The arc defines an angle θ at the center of the circle. This angle θ, whose sides are the radii at the ends of the arc, has the measure $\theta = 1$ radian.

The **circumference** c of the circle is the distance around the circle. The circumference is given by the formula

$$c = 2\pi r.$$

Exactly 2π pieces of wire with length r fit around the circle. Therefore, one revolution around the circle, or 360°, measures 2π radians. Thus, we have

$$360° = 2\pi \text{ rad}.$$

Dividing both sides by 2, we have a straight angle, which is

$$180° = \pi \text{ rad}.$$

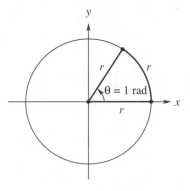

An angle with measure 1 radian.

We can find the remaining values for angles lying on the axes by dividing by 2 again to obtain a right angle, which is

$$90° = \frac{\pi}{2} \text{ rad.}$$

Then, we can multiply this result by 3 to obtain three right angles, which are

$$270° = \frac{3\pi}{2} \text{ rad.}$$

If the angle is 0°, there is no arc and so

$$0° = 0 \text{ rad.}$$

The angles on the axes, measuring 0°, 90°, 180°, 270°, and 360°, are the **quadrantal angles**. We summarize our results by a diagram of the quadrantal angles:

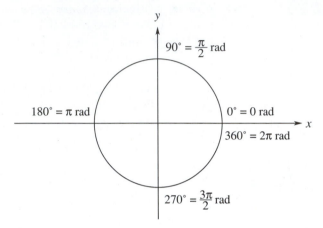

Radian and degree measure are related by the equation

$$180° = \pi \text{ rad.}$$

If we divide both sides by π radians, we have

$$\frac{180°}{\pi \text{ rad}} = \frac{\pi \text{ rad}}{\pi \text{ rad}},$$

and therefore,

$$\frac{180°}{\pi \text{ rad}} = 1.$$

To convert from radian measure to degree measure, we can multiply by $\frac{180°}{\pi \text{ rad}}$, which is the same as multiplying by 1.

EXAMPLE 19.1

Estimate the location of the angle with measure 1 radian, and then convert the radian measure to degree measure.

SOLUTION   We know that

$$180° = \pi \text{ rad.}$$

Approximating π by 3.14, we know that 180° is a little more than 3 radians. Therefore, we expect 60° to be a little more than 1 radian, or 1 radian to be a little less than 60° as shown in the diagram on p. 417. We multiply by $\frac{180°}{\pi \text{ rad}}$ to obtain the degree measure:

$$1 \text{ rad} = 1 \text{ rad} \cdot \frac{180°}{\pi \text{ rad}}$$

$$= \frac{1(180°)}{\pi}$$

$$= 57.3°.$$

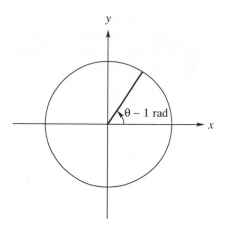

Angle in Example 19.1.   ▲

EXAMPLE 19.2 ▶ Estimate the location of the angle, and then convert the radian measure to degree measure:

a. 2 radians   **b.** 3 radians   **c.** 2.5 radians

SOLUTIONS We can use the approximation in Example 19.1 to locate the angle.

a. Since 1 radian is not quite 60°, we expect 2 radians to be in the second quadrant, at a little less than 120° as shown in the diagram at the left below. We multiply by $\frac{180°}{\pi \text{ rad}}$ to obtain the degree measure:

$$2 \text{ rad} = 2 \text{ rad} \cdot \frac{180°}{\pi \text{ rad}}$$

$$= \frac{2(180°)}{\pi}$$

$$= 115°.$$

b. We expect 3 radians to be near the end of the second quadrant, at not quite 180° as shown in the diagram at the right. We multiply by $\frac{180°}{\pi \text{ rad}}$ to obtain the degree measure:

$$3 \text{ rad} = 3 \text{ rad} \cdot \frac{180°}{\pi \text{ rad}}$$

$$= \frac{3(180°)}{\pi}$$

$$= 172°.$$

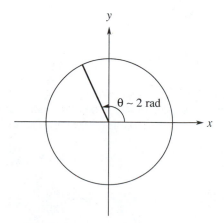

Angle in Example 19.2, part **a**.

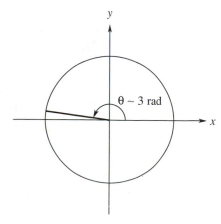

Angle in Example 19.2, part **b**.

c. We know that 2 radians are about one-third of the way through the second quadrant, and 3 radians are just at the end. Therefore, we expect 2.5 radians to be in the second quadrant, about two-thirds of the way through, or close to 150°:

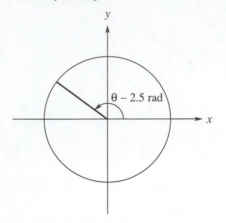

We multiply by $\frac{180°}{\pi\ \text{rad}}$ to obtain the degree measure:

$$2.5\ \text{rad} = 2.5\ \text{rad} \cdot \frac{180°}{\pi\ \text{rad}}$$

$$= \frac{2.5(180°)}{\pi}$$

$$= 143°. \qquad \blacktriangle$$

Now, we return to the basic relationship

$$180° = \pi\ \text{rad}.$$

If we divide both sides by 180°, we have

$$\frac{180°}{180°} = \frac{\pi\ \text{rad}}{180°},$$

and therefore,

$$1 = \frac{\pi\ \text{rad}}{180°}.$$

To convert from radian measure to degree measure, we can multiply by $\frac{\pi\ \text{rad}}{180°}$, which is the same as multiplying by 1.

 EXAMPLE 19.3 ▶ Approximate the radian measure of 280°, and then convert the degree measure to radian measure.

SOLUTION We know that

$$270° = \frac{3\pi}{2}\ \text{rad}.$$

Therefore, approximating $\frac{3\pi}{2}$ by 4.71, we expect 280° to be close to 5 radians as shown in the diagram on page 419. We multiply by $\frac{\pi\ \text{rad}}{180°}$ to obtain the radian measure:

$$280° = 280° \cdot \frac{\pi\ \text{rad}}{180°}$$

$$= \frac{280°(\pi)}{180°}$$

$$= 4.89\ \text{rad}.$$

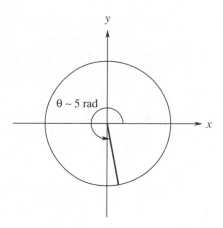

Angle in Example 19.3.

▲

Estimate the location of the angle, and then convert the radian measure to degree measure:

1. 0.5 radian **2.** 6 radians **3.** 4 radians **4.** 5 radians

5. 1.5 radians **6.** 4.5 radians **7.** 7 radians **8.** 10 radians

Approximate the radian measure, and then convert the degree measure to radian measure:

9. 60° **10.** 100° **11.** 225° **12.** 350°

13. 400° **14.** 500°

SECTION
19.2

Generating Sine Waves

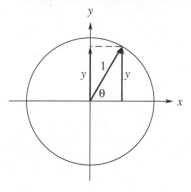

A radius vector in the unit circle.

The **unit circle** is a circle in the rectangular coordinate system with its center at the origin and with radius 1. Suppose we rotate a radius of the unit circle counterclockwise, starting at the positive x-axis. At any position, the radius is a vector called a **radius vector**. The magnitude of a radius vector in the unit circle is 1. The direction of a radius vector is the angle θ from the positive x-axis.

The horizontal and vertical components of the radius vector are the legs of a right triangle, with the radius vector as the hypotenuse. To define the **sine function**, we concentrate on the vertical component. By using the sine ratio, the vertical component is given by

$$\sin \theta = \frac{y}{1}$$

$$\sin \theta = y,$$

or

$$y = \sin \theta.$$

You can use your calculator to find values of the sine function for angles given in radians. Be sure your calculator is in *radian mode*, by using the degree-radian-grad key or the mode key, depending on the calculator as described on page 377. Then, for example, to find

$$y = \sin \frac{\pi}{6},$$

follow these steps:

		display:
Press	π	3.141592654
Press	÷	3.141592654
Enter	6	6.
Press	=	0.523598775
Press	sin	0.5

The values of y for angles θ in radians from 0 to $\frac{\pi}{2}$ are shown in this chart:

θ	0	$\frac{\pi}{6}$	$\frac{\pi}{4}$	$\frac{\pi}{3}$	$\frac{\pi}{2}$
y	0	0.5	0.707	0.866	1

You should use your calculator to check these values.

Now, we look at several positions of the radius vector, considering angles that are multiples of 30°, or $\frac{\pi}{6}$, and multiples of 45°, or $\frac{\pi}{4}$. We can use the equation

$$y = \sin\theta$$

to find values of the sine function for these angles:

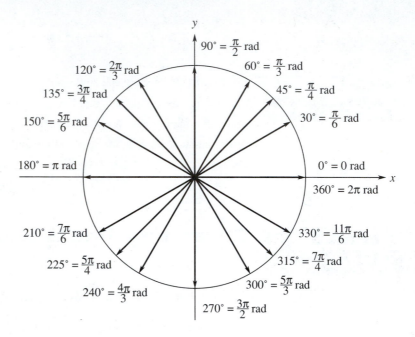

GRAPH OF THE SINE FUNCTION

To draw the graph of the sine function, we use the horizontal axis for angles θ in radians. An advantage of writing θ in radians is that radian measure relates easily to linear measure. Since π is approximately 3.14, we place π at approximately 3 units. Halving π and 3, we place $\frac{\pi}{2}$ at approximately 1.5 units. We take one-third of π and 3 to place $\frac{\pi}{3}$ at approximately 1 unit, one-quarter of π and 3 to place $\frac{\pi}{4}$ at approximately 0.75 unit, and one-sixth of π and 3 to place $\frac{\pi}{6}$ at approximately 0.5 unit:

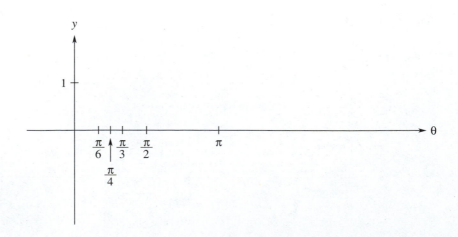

We continue this pattern from $\frac{\pi}{2}$ to π, then to $\frac{3\pi}{2}$, and to 2π:

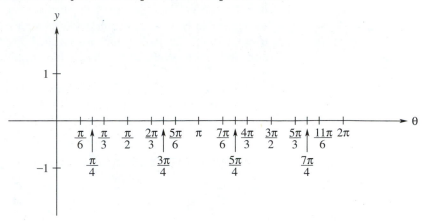

To draw the graph of the sine function,

$$y = \sin\theta,$$

we can follow the height, or *y*-component, of a radius vector in the unit circle. We observe that, for angles from $\theta = 0$ to $\theta = \frac{\pi}{2}$, this height increases from 0 to 1. The actual *y*-values are in the chart calculated previously:

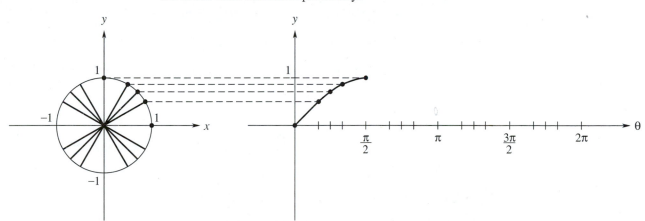

We continue the graph from $\frac{\pi}{2}$ to π, then to $\frac{3\pi}{2}$, and to 2π. We observe that, for $\theta = \frac{\pi}{2}$ to $\theta = \pi$, the height decreases through the same *y*-values from 1 to 0. Continuing around the circle, for $\theta = \pi$ to $\theta = \frac{3\pi}{2}$, the height decreases through negatives of the same *y*-values from 0 to -1. Finally, for $\theta = \frac{3\pi}{2}$ to $\theta = 2\pi$, the height increases through negatives of the same *y*-values from -1 to 0:

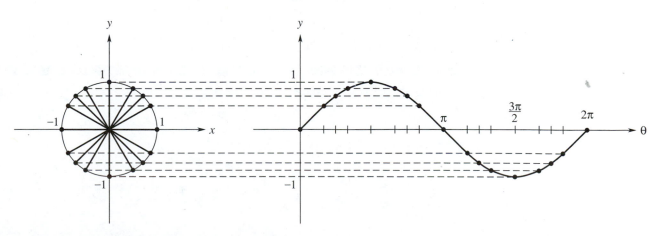

If we rotate the radius vector around the circle a second time, from $\theta = 2\pi$ to $\theta = 4\pi$, the y-values are repeated. They are repeated again for $\theta = 4\pi$ to $\theta = 6\pi$, and so on. Similarly, if we rotate the radius vector backwards, through negative angles from $\theta = 0$ to $\theta = -2\pi$, the y-values are repeated. Each rotation of the radius vector around the unit circle is one **cycle**, and each repetition of the graph is one cycle:

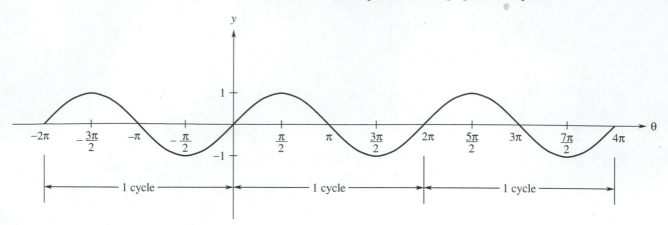

The graph generated is called a **sine wave**. This sine wave is the graph of the sine function, and has the equation

$$y = \sin \theta.$$

AMPLITUDE

We observe that the entire graph of the sine function falls between the y-values $y = -1$ and $y = 1$; that is,

$$-1 \le y \le 1.$$

We say that the sine function has **amplitude** 1.

We can easily change the amplitude of the sine function. For example, suppose we rotate a radius vector around a circle with radius 2. Then, the sine function is given by

$$\sin \theta = \frac{y}{2}$$

$$2 \sin \theta = y,$$

or

$$y = 2 \sin \theta.$$

This function has amplitude 2.

In general, the sine function

$$y = A \sin \theta,$$

where A can be positive or negative, has amplitude $|A|$.

Amplitude of a Sine Function: The sine function

$$y = A \sin \theta$$

has the amplitude $|A|$.

EXAMPLE 19.4   Draw the graph of $y = 2 \sin \theta$.

SOLUTION   Since $A = 2$, the function has amplitude 2. We draw one cycle of the graph:

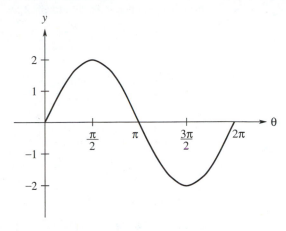

▲

If A is negative, then the first half of the cycle has negative values, decreasing to A and then increasing. The second half of the cycle has positive values, increasing to $|A|$ and then decreasing.

EXAMPLE 19.5   Draw the graph of $y = -2 \sin \theta$.

SOLUTION   Since $A = -2$, the function has amplitude $|-2| = 2$. The graph decreases to -2 and then increases to 2. We draw one cycle of the graph:

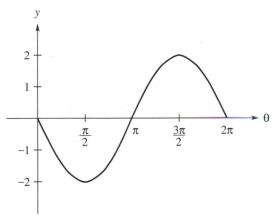

▲

PERIOD

We observe that each cycle of the graph of the sine function

$$y = \sin \theta$$

has length 2π. For example, one cycle falls between the θ-values $\theta = 0$ and $\theta = 2\pi$; that is,

$$0 \leq \theta \leq 2\pi.$$

We say that the sine function has **period** 2π.

To change the period of the sine function, suppose a radius vector completes a cycle from $\theta = 0$ to $\theta = \pi$; that is,

$$0 \leq \theta \leq \pi.$$

Then, if we multiply each part by 2, we have

$$2(0) \leq 2\theta \leq 2\pi$$

$$0 \leq 2\theta \leq 2\pi.$$

The sine function obtained is given by

$$y = \sin 2\theta.$$

This function has period $\frac{2\pi}{2} = \pi.$
  In general, the sine function

$$y = \sin B\theta$$

has period $\frac{2\pi}{B}.$

Period of a Sine Function: The sine function
$$y = \sin B\theta$$
has the period $\frac{2\pi}{B}.$

EXAMPLE 19.6   Draw the graph of $y = \sin 2\theta.$

SOLUTION   Since $B = 2$, the function has period

$$\frac{2\pi}{B} = \frac{2\pi}{2}$$

$$= \pi.$$

One cycle of the graph is shown. We note that the period is $\frac{2\pi}{2}$, and each quadrantal angle is divided by 2. Thus, the graph starts at

$$\theta = 0;$$

reaches 1 at

$$\theta = \frac{1}{2}\left(\frac{\pi}{2}\right) = \frac{\pi}{4};$$

returns to 0 at

$$\theta = \frac{1}{2}(\pi) = \frac{\pi}{2};$$

reaches -1 at

$$\theta = \frac{1}{2}\left(\frac{3\pi}{2}\right) = \frac{3\pi}{4};$$

and returns to 0 at

$$\theta = \frac{1}{2}(2\pi) = \pi.$$

Graph for Example 19.6.

EXAMPLE 19.7   Draw the graph of $y = \sin \frac{1}{2}\theta.$

SOLUTION   Since $B = \frac{1}{2}$, the function has period

$$\frac{2\pi}{B} = \frac{2\pi}{\frac{1}{2}}$$

$$= 2\pi\left(\frac{2}{1}\right)$$

$$= 4\pi.$$

We draw one cycle of the graph:

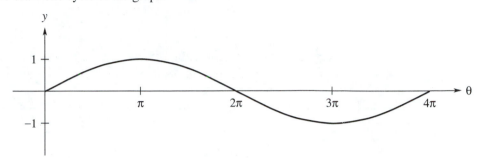

We note that the period is $2(2\pi)$, and each quadrantal angle is multiplied by 2. Thus, the graph starts at

$$\theta = 0;$$

reaches 1 at

$$\theta = 2\left(\frac{\pi}{2}\right) = \pi;$$

returns to 0 at

$$\theta = 2(\pi) = 2\pi;$$

reaches -1 at

$$\theta = 2\left(\frac{3\pi}{2}\right) = 3\pi;$$

and returns to 0 at

$$\theta = 2(2\pi) = 4\pi.$$ ▲

EXAMPLE 19.8 ▶ Draw the graph of $y = 2 \sin 3\theta$.

SOLUTION Since $A = 2$, the function has amplitude 2. Since $B = 3$, the function has period $\frac{2\pi}{3}$. One cycle of the graph is shown. We note that the period is $\frac{2\pi}{3}$, and each quadrantal angle is divided by 3. The graph starts at

$$\theta = 0.$$

Since the amplitude is 2, it reaches 2 at

$$\theta = \frac{1}{3}\left(\frac{\pi}{2}\right) = \frac{\pi}{6}.$$

It returns to 0 at

$$\theta = \frac{1}{3}(\pi) = \frac{\pi}{3};$$

reaches -2 at

$$\theta = \frac{1}{3}\left(\frac{3\pi}{2}\right) = \frac{\pi}{2};$$

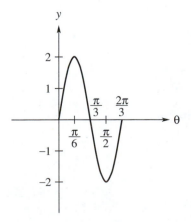

Graph for Example 19.8.

and returns to 0 at

$$\theta = \frac{1}{3}(2\pi) = \frac{2\pi}{3}.$$ ▲

Finally, we observe that, if the radius vector is rotated clockwise through the negative angles, then each angle θ is replaced by $-\theta$. Also, the first half of the cycle goes through the negative y-values:

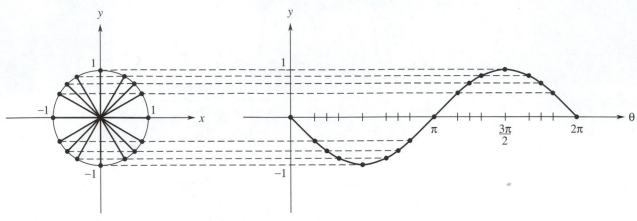

Thus, the function

$$y = \sin(-\theta)$$

is the same as the function

$$y = -\sin\theta,$$

and we may write

$$\sin(-\theta) = -\sin\theta.$$

Therefore, we can always write sine functions in the form

$$y = A \sin B\theta,$$

where B is positive, so the period is positive. However, A may be positive or negative, so the amplitude is $|A|$.

EXERCISE 19.2

Draw one cycle of the graph:

1. $y = 3\sin\theta$

2. $y = \frac{1}{2}\sin\theta$

3. $y = -\sin\theta$

4. $y = -1.5\sin\theta$

5. $y = \sin 3\theta$

6. $y = \sin 4\theta$

7. $y = \sin\frac{1}{3}\theta$

8. $y = \sin\frac{2}{3}\theta$

9. $y = \sin\pi\theta$

10. $y = \sin\frac{\pi}{2}\theta$

11. $y = 2\sin 2\theta$

12. $y = \frac{1}{2}\sin\frac{1}{2}\theta$

Phase Angles

In the preceding section, we generated sine waves by rotating a radius vector, starting on the positive x-axis. Now, suppose we rotate a radius vector starting at an angle $\theta = \phi$ (the Greek letter "phi," pronounced like "fi" in "hi-fi"). The angle ϕ is called the **phase angle**. If the phase angle ϕ is positive, the sine wave shifts to the left by the amount of ϕ. In this case, a cycle begins at $\theta = -\phi$. We say that ϕ is a **leading phase angle**. The shift is a **phase shift to the left** by the amount of ϕ:

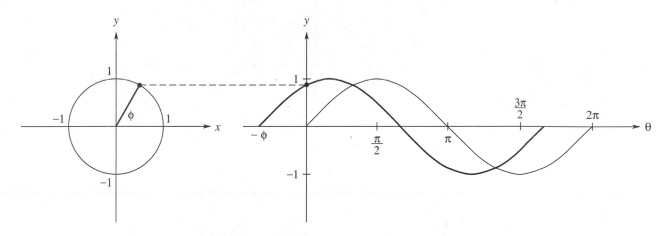

To change the phase angle of the sine function, suppose a radius vector starts on the positive y-axis; that is, at $\theta = \frac{\pi}{2}$. The phase angle is $\phi = \frac{\pi}{2}$, and a cycle begins at $\theta = -\phi$, or $\theta = -\frac{\pi}{2}$. The graph completes one cycle from $\theta = -\frac{\pi}{2}$ to $\theta = 2\pi - \frac{\pi}{2}$; that is,

$$-\frac{\pi}{2} \leq \theta \leq 2\pi - \frac{\pi}{2} .$$

Then, if we add $\frac{\pi}{2}$ to each part, we have

$$-\frac{\pi}{2} + \frac{\pi}{2} \leq \theta + \frac{\pi}{2} \leq 2\pi - \frac{\pi}{2} + \frac{\pi}{2}$$

$$0 \leq \theta + \frac{\pi}{2} \leq 2\pi.$$

The sine function obtained is given by

$$y = \sin\left(\theta + \frac{\pi}{2}\right).$$

This function has a leading phase angle $\phi = \frac{\pi}{2}$, and a phase shift to the left by $\frac{\pi}{2}$.

In general, the function

$$y = \sin(\theta + \phi)$$

has phase angle ϕ. If ϕ is positive, then ϕ is a leading phase angle and the phase shift is to the left by the amount of ϕ.

Leading Phase Angle: The sine function
$$y = \sin(\theta + \phi),$$
where ϕ is positive, has the leading phase angle ϕ, and shifts to the left by the amount of ϕ.

EXAMPLE 19.9 ▶ Draw the graph of $y = \sin\left(\theta + \frac{\pi}{2}\right)$.

SOLUTION   The phase angle is $\phi = \frac{\pi}{2}$, and the angle ϕ is a leading phase angle. The phase shift is to the left by $\frac{\pi}{2}$. We draw one cycle of the graph:

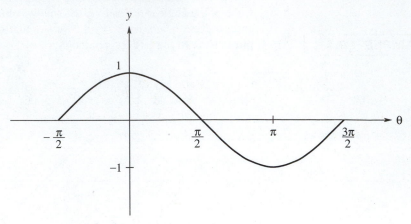

We note that the cycle starts at $\theta = -\frac{\pi}{2}$, and $\frac{\pi}{2}$ has been subtracted from each quadrantal angle. Thus, the graph starts at

$$\theta = 0 - \frac{\pi}{2} = -\frac{\pi}{2};$$

reaches 1 at

$$\theta = \frac{\pi}{2} - \frac{\pi}{2} = 0;$$

returns to 0 at

$$\theta = \pi - \frac{\pi}{2} = \frac{\pi}{2};$$

reaches −1 at

$$\theta = \frac{3\pi}{2} - \frac{\pi}{2} = \pi;$$

and returns to 0 at

$$\theta = 2\pi - \frac{\pi}{2} = \frac{3\pi}{2}.$$ ▲

Now, suppose we rotate a radius vector, starting at a negative angle $\theta = -\phi$. The sine wave shifts to the right by the amount of ϕ. In this case, a cycle begins at $\theta = \phi$. We say that ϕ is a **lagging phase angle**. The shift is a **phase shift to the right** by the amount of ϕ:

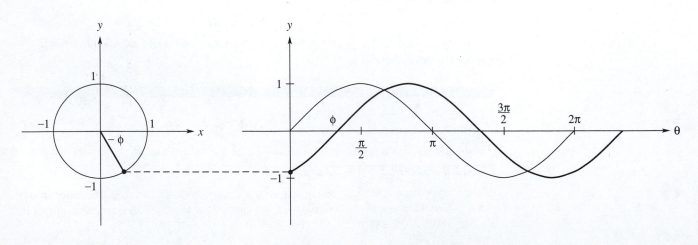

Lagging Phase Angle: The sine function
$$y = \sin(\theta - \phi),$$
where ϕ is positive, has the lagging phase angle ϕ, and shifts to the right by the amount of ϕ.

EXAMPLE 19.10   Draw the graph of $y = \sin\left(\theta - \dfrac{\pi}{4}\right)$.

SOLUTION   The phase angle is $\phi = \frac{\pi}{4}$, but the angle is subtracted so ϕ is a lagging phase angle. The phase shift is to the right by $\frac{\pi}{4}$. We draw one cycle of the graph and, for completeness, we draw a part of a cycle to its left so the graph crosses the y-axis:

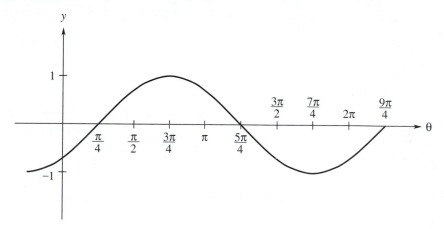

We note that the cycle starts at $\theta = \frac{\pi}{4}$, and $\frac{\pi}{4}$ has been added to each quadrantal angle. Thus, a cycle starts at

$$\theta = 0 + \frac{\pi}{4} = \frac{\pi}{4};$$

reaches 1 at

$$\theta = \frac{\pi}{2} + \frac{\pi}{4} = \frac{3\pi}{4};$$

returns to 0 at

$$\theta = \pi + \frac{\pi}{4} = \frac{5\pi}{4};$$

reaches -1 at

$$\theta = \frac{3\pi}{2} + \frac{\pi}{4} = \frac{7\pi}{4};$$

and returns to 0 at

$$\theta = 2\pi + \frac{\pi}{4} = \frac{9\pi}{4}.$$

PHASE ANGLES IN DEGREES

Phase angles are often measured in degrees, and the calculations are often easier in degrees. When the phase angle is in degrees, we change the labels of the quadrantal angles to $0°$, $90°$, $180°$, $270°$, and $360°$.

EXAMPLE 19.11 Draw the graph of $y = \sin(\theta + 30°)$.

SOLUTION The phase angle $\phi = 30°$ is a leading phase angle, and the phase shift is to the left by 30°. We draw one cycle of the graph:

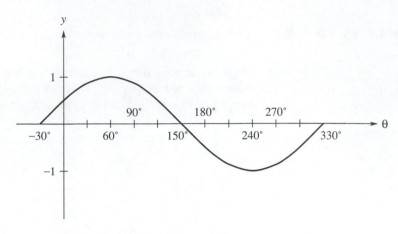

We note that the cycle starts at $\theta = 30°$, and 30° has been subtracted from each quadrantal angle. Thus, a cycle starts at

$$\theta = 0° - 30° = -30°;$$

reaches 1 at

$$\theta = 90° - 30° = 60°;$$

returns to 0 at

$$\theta = 180° - 30° = 150°;$$

reaches −1 at

$$\theta = 270° - 30° = 240°;$$

and returns to 0 at

$$\theta = 360° - 30° = 330°. \qquad \blacktriangle$$

EXAMPLE 19.12 Draw the graph of $y = \sin(\theta - 60°)$.

SOLUTION The phase angle $\phi = 60°$ is a lagging phase angle, and the phase shift is to the right by 60°. We draw one cycle of the graph, and another part of a cycle to its left so the graph crosses the y-axis:

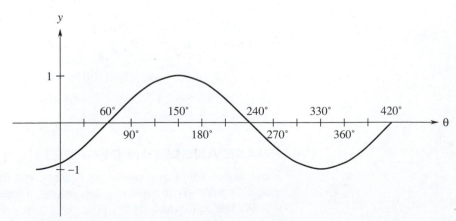

We note that the cycle starts at $\theta = 60°$, and $60°$ has been added to each quadrantal angle. You should check that $60°$ has been added to each of the angles $0°$, $90°$, $180°$, $270°$, and $360°$. ▲

EXAMPLE 19.13   Draw the graph of $y = 2 \sin (\theta + 30°)$.

SOLUTION   Since $A = 2$, the function has amplitude 2. Since $\phi = 30°$, the angle ϕ is a leading phase angle, and the phase shift is to the left by $30°$. We draw one cycle of the graph:

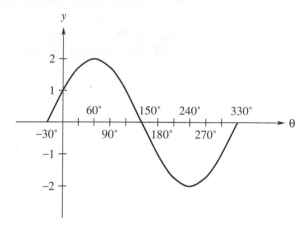

We note that the cycle starts $\theta = 30°$, and $30°$ has been subtracted from each quadrantal angle. Also, since the amplitude is 2, the graph reaches 2 at

$$\theta = 90° - 30° = 60°,$$

and reaches -2 at

$$\theta = 270° - 30° = 240°.$$ ▲

THE GENERAL SINE FUNCTION

The general form of the sine function is

$$y = A \sin (Bx \pm C).$$

The amplitude is $|A|$, and is not affected by changes in the period or phase angle. The period is $\frac{2\pi}{B}$, and the period *does affect* the phase angle. We must also divide the phase angle by B. Thus, the phase angle is

$$\phi = \frac{C}{B}.$$

The General Sine Function: The sine function
$$y = A \sin (Bx \pm C)$$
has amplitude $|A|$, period $\frac{2\pi}{B}$, and phase angle $\phi = \frac{C}{B}$, where B and C are positive. For the sine function
$$y = A \sin (Bx + C),$$
ϕ is a leading phase angle. For the sine function
$$y = A \sin (Bx - C),$$
ϕ is a lagging phase angle.

EXAMPLE 19.14   Draw the graph of $y = 2 \sin (2\theta + 30°)$.

SOLUTION   Since $A = 2$, the function has amplitude 2. Since $B = 2$, the function has period $\frac{360°}{2} = 180°$. Also, since $C = 30°$ and $B = 2$, the phase angle is

$$\phi = \frac{C}{B}$$

$$= \frac{30°}{2}$$

$$= 15°.$$

The angle $\phi = 15°$ is a leading phase angle, and the phase shift is to the left by 15°.

It is sometimes most efficient to draw the graph with the given amplitude and period first, and then to shift it. Using a light line, we draw one cycle of the graph with amplitude 2 and period 180°, as shown in the graph at the left. Then, we shift the cycle to start at $\theta = -15°$, as shown in the graph at the right:

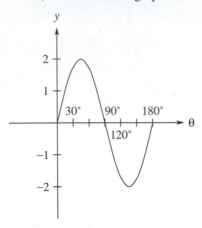

Amplitude and period for the graph for Example 19.14.

Completed graph for Example 19.14.

We can check the resulting graph by comparing the function

$$y = 2 \sin (2\theta + 30°)$$

with the basic sine function $y = \sin \theta$, which has its cycle from 0° to 360°. We write

$$0° \le 2\theta + 30° \le 360°.$$

Dividing each part by 2, we have

$$0° \le \theta + 15° \le 180°.$$

This result exhibits the period 180° and the phase angle 15°. Then, subtracting 15° from each part, we have

$$-15° \le \theta \le 165°.$$

This result shows that a cycle of the resulting graph goes from $\theta = -15°$ to $\theta = 165°$. ▲

EXAMPLE 19.15 ▶ Draw the graph of $y = 2 \sin (\pi\theta - \pi)$, where θ is in radians.

SOLUTION Since $A = 2$, the function has amplitude 2. Since $B = \pi$, the function has period $\frac{2\pi}{\pi} = 2$. Also, $C = \pi$ but C is subtracted. Therefore, the phase angle is

$$\phi = \frac{C}{B}$$

$$= \frac{\pi}{\pi}$$

$$= 1,$$

but $\phi = 1$ is a lagging phase angle, and the phase shift is to the right by 1 radian.

We can label the θ-axis with units 1, 2, and so on, like the y-axis. Using a light line, we draw one cycle of the graph with amplitude 2 and period 2, as shown in the graph at the left. Then, we shift the cycle to start at θ = 1, and draw another half of a cycle to the y-axis, as shown in the graph at the right:

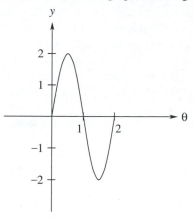

Amplitude and period for the graph
for Example 19.15.

Completed graph for Example 19.15.

We can check the resulting graph by comparing the function

$$y = 2 \sin (\pi\theta - \pi)$$

with the basic sine function $y = \sin \theta$, which has its cycle from 0 to 2π radians. We write

$$0 \text{ rad} \leq \pi\theta - \pi \text{ rad} \leq 2\pi \text{ rad}.$$

Dividing each part by π, we have

$$0 \text{ rad} \leq \theta - 1 \text{ rad} \leq 2 \text{ rad}.$$

This result exhibits the period 2 radians and the phase angle 1 radian. Then, adding 1 radian to each part, we have

$$1 \text{ rad} \leq \theta \leq 3 \text{ rad}.$$

This result shows that a cycle of the resulting graph goes from θ = 1 radian to θ = 3 radians. ▲

EXERCISE 19.3

Draw the graph:

1. $y = \sin \left(\theta + \dfrac{\pi}{4}\right)$ **2.** $y = \sin (\theta + \pi)$

3. $y = \sin \left(\theta - \dfrac{\pi}{2}\right)$ **4.** $y = \sin (\theta - \pi)$

5. $y = \sin (\theta + 60°)$ **6.** $y = \sin (\theta + 45°)$

7. $y = \sin (\theta - 45°)$ **8.** $y = \sin (\theta - 30°)$

9. $y = 3 \sin (\theta + 90°)$ **10.** $y = \dfrac{1}{2} \sin (\theta - 60°)$

11. $y = 2 \sin (3\theta + 60°)$ **12.** $y = \dfrac{1}{2} \sin (3\theta - 45°)$

13. $y = \dfrac{1}{2} \sin \left(\dfrac{1}{2} \theta - 45°\right)$ **14.** $y = 3 \sin \left(\dfrac{1}{3} \theta + 30°\right)$

15. $y = 2 \sin \left(\pi\theta - \dfrac{\pi}{2}\right)$ **16.** $y = \dfrac{1}{2} \sin \left(\dfrac{\pi}{2} \theta + \dfrac{\pi}{2}\right)$

SECTION	# Angular Motion
19.4	

An angle is often defined as an *amount of rotation*. Suppose a vector starts on the positive *x*-axis, and rotates counterclockwise. The vector generates a positive angle, which increases according to the *amount of time* that the vector is rotated. Thus, directions of vectors are often defined in terms of time. In electronics, vectors with directions defined in terms of time are sometimes called **phasors** (phase vectors).

Each time a phasor rotates through an entire circle, it completes one cycle. The number of cycles through which a phasor rotates in one second is its **frequency**, f, measured in cycles per second (cps), or hertz (Hz). If a phasor rotates through one cycle in one second, its frequency is $f = 1$ Hz.

The reciprocal of the frequency is the time the phasor takes to complete one cycle. This time is the **period**, T:

$$T = \frac{1}{f}.$$

The period is measured in seconds. Thus, if a phasor takes one second to complete one cycle, its period is $T = 1$ s. (We always use a *capital T* for the period to avoid confusion with time, t.)

EXAMPLE 19.16 ▶ If a phasor rotates with frequency $f = 4$ kHz, find its period.

SOLUTION Since the period is the reciprocal of the frequency, we write

$$T = \frac{1}{f}$$

$$= \frac{1}{4 \text{ kHz}}$$

$$= \frac{1}{4 \times 10^3} \text{ s}$$

$$= 0.25 \times 10^{-3} \text{ s}$$

$$= 250 \text{ μs.} \qquad \blacktriangle$$

EXAMPLE 19.17 ▶ If a phasor rotates with period $T = 200$ ms, find its frequency.

SOLUTION Since the frequency is the reciprocal of the period, we write

$$f = \frac{1}{T}$$

$$= \frac{1}{200 \text{ ms}}$$

$$= \frac{1}{200 \times 10^{-3}} \text{ Hz}$$

$$= 0.005 \times 10^3 \text{ Hz}$$

$$= 5 \text{ Hz.} \qquad \blacktriangle$$

ANGULAR VELOCITY

The rate at which a phasor rotates is the rate of change of the angle it generates. This rate is the **angular velocity**, ω (the lowercase Greek letter omega). Angular velocity may be measured in cycles per second, radians per second, or degrees per second. If a phasor rotates through one cycle in one second, its angular velocity is

$$\omega = 1 \text{ cps.}$$

Thus, when angular velocity is measured in cycles per second, it is the same as frequency. If a phasor rotates with an angular velocity of one cycle per second, its frequency is 1 Hz.

Since there are 2π radians or $360°$ in a circle, we can write

$$1 \text{ cps} = 2\pi \frac{\text{rad}}{\text{s}} \text{ (radians per second)}$$

or

$$1 \text{ cps} = 360 \frac{\text{deg}}{\text{s}} \text{ (degrees per second)}.$$

Thus, when angular velocity is measured in radians per second, it is equal to 2π times the frequency. When the angular velocity is measured in degrees per second, it is equal to $360°$ times the frequency. Therefore, we can find angular velocity in radians per second or in degrees per second by using

$$\omega = 2\pi f \text{ or } \omega = 360°f.$$

Angular Velocity Formulas: Angular velocity is given in radians per second by
$$\omega = 2\pi f,$$
or in degrees per second by
$$\omega = 360°f.$$

EXAMPLE 19.18 A phasor rotates with a frequency of 20 kHz. Find the angular velocity in

a. radians per second **b.** degrees per second

SOLUTIONS We use the formulas for angular velocities in terms of frequency.

a. To find the angular velocity in radians per second, we write

$$\omega = 2\pi f$$
$$= 2\pi(20 \text{ kHz})$$
$$= (2\pi)(20 \times 10^3) \frac{\text{rad}}{\text{s}}$$
$$= 126 \times 10^3 \frac{\text{rad}}{\text{s}}$$
$$= \frac{126}{10^{-3}} \frac{\text{rad}}{\text{s}}$$
$$= 126 \frac{\text{rad}}{\text{ms}}.$$

b. To find the angular velocity in degrees per second, we write

$$\omega = 360°f$$
$$= 360°(20 \text{ kHz})$$
$$= (360)(20 \times 10^3) \frac{\text{deg}}{\text{s}}$$
$$= 7.2 \times 10^6 \frac{\text{deg}}{\text{s}}$$
$$= \frac{7.2}{10^{-6}} \frac{\text{deg}}{\text{s}}$$
$$= 7.2 \frac{\text{deg}}{\mu\text{s}}.$$

We observe that angular velocities in radians per second and in degrees per second are related in the same way as angles in radians and in degrees. Thus, if we multiply the first result by $\frac{180°}{\pi\,\text{rad}}$, we have

$$\omega = 126\,\frac{\text{rad}}{\text{ms}} \cdot \frac{180°}{\pi\,\text{rad}}$$

$$= 7220\,\frac{\text{deg}}{\text{ms}},$$

or approximately $7.2\,\frac{\text{deg}}{\mu\text{s}}$.   ▲

ANGLES

In general terms, we know that distance is equal to rate multiplied by time. For example, if a car travels at 30 miles per hour for two hours, it will travel 60 miles. Similarly, if a phasor rotates with an angular velocity of π radians per second for two seconds, it will rotate through an angle of 2π radians. If it rotates at 180° per second for 2 seconds, it will rotate through an angle of 360°. For angular motion, the "distance" is an angle that is equal to angular velocity multiplied by time. Thus, angles are determined in terms of time by

$$\theta = \omega t.$$

Since we know that

$$\omega = 2\pi f \text{ or } \omega = 360° f,$$

we can write

$$\theta = 2\pi ft \text{ or } \theta = 360° ft.$$

Angle Formulas: Angles are given in radians by
$$\theta = 2\pi ft,$$
or in degrees by
$$\theta = 360° ft.$$

EXAMPLE 19.19 ▶ A phasor rotates with a frequency of 20 kHz for 100 μs. Find the angle through which it rotates in

a. radians   **b.** degrees

 We use the formulas for angles in terms of frequency and time.

a. To find the angle in radians, we write

$$\theta = 2\pi ft$$

$$= 2\pi(20\text{ kHz})(100\text{ μs})$$

$$= 12.6\text{ rad.}$$

We observe that this result is equivalent to 4π radians, which is a rotation of two cycles around the circle.

b. To find the angle in degrees, we write

$$\theta = 360° ft$$

$$= 360°(20\text{ kHz})(100\text{ μs})$$

$$= 720°.$$

We note that this result is also equivalent to a rotation of two cycles around the circle. Furthermore, if we multiply the first result by $\frac{180°}{\pi\,\text{rad}}$, we have

$$\theta = 12.6 \text{ rad} \cdot \frac{180°}{\pi \text{ rad}}$$

$$= 722°,$$

or approximately 720°.   ▲

EXAMPLE 19.20 ▶ If a phasor rotates through an angle of 225° in 500 ms, find the frequency.

SOLUTION Since the angle is given in degrees, we use

$$\theta = 360°ft.$$

Solving for the frequency, we have

$$f = \frac{\theta}{360°t}$$

$$= \frac{225°}{(360°)(500 \text{ ms})}$$

$$= 1.25 \text{ Hz.}$$ ▲

EXAMPLE 19.21 ▶ How long does it take a phasor to rotate through π radians at a frequency of $f = 50$ kHz?

SOLUTION Since the angle is given in radians, we use

$$\theta = 2\pi ft.$$

Solving for the time, we have

$$t = \frac{\theta}{2\pi f}$$

$$= \frac{\pi}{2\pi(50 \text{ kHz})}$$

$$= 10 \text{ μs.}$$ ▲

EXERCISE 19.4

Find the frequency:

1. $T = 50$ ms   **2.** $T = 100$ μs   **3.** $T = 25$ μs   **4.** $T = 2$ s

Find the period:

5. $f = 10$ kHz   **6.** $f = 0.8$ Hz   **7.** $f = 12$ Hz   **8.** $f = 16$ MHz

Find the angular velocity in

a. radians per second   **b.** degrees per second

9. $f = 4$ kHz   **10.** $f = 50$ kHz   **11.** $f = 10$ Hz   **12.** $f = 12$ MHz

Find the angle in

a. radians   **b.** degrees

13. $f = 50$ kHz and $t = 2$ μs   **14.** $f = 1$ MHz and $t = 10$ μs

15. $f = 250$ Hz and $t = 5$ ms   **16.** $f = 40$ Hz and $t = 650$ μs

Find the frequency:

17. $\theta = 90°$ and $t = 100$ μs   **18.** $\theta = 540°$ and $t = 2$ s

19. $\theta = 1$ radian and $t = 50$ μs   **20.** $\theta = 12.6$ radians and $t = 200$ ms

Find the time:

21. $\theta = 180°$ and $f = 5$ kHz   **22.** $\theta = 400°$ and $f = 1.5$ MHz

23. $\theta = 0.524$ radian and $f = 83.3$ kHz   **24.** $\theta = 10$ radians and $f = 60$ Hz

Alternating Current and Voltage

The electricity supplied by common household outlets in the United States is called **alternating current** (AC). Alternating current is produced by an alternating voltage generator. Simply put, this alternating voltage generator is a loop rotating in a magnetic field:

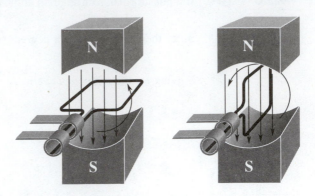

As the loop rotates through a quarter of a turn, the voltage generated varies from zero to a maximum value. In the next quarter of a turn, the voltage returns to zero. In the third quarter, the voltage varies through negatives to a minimum value. In the fourth quarter, the voltage returns again to zero.

The voltage produced by an alternating voltage generator is described by the sine function

$$e = E_M \sin \theta,$$

where E_M is the maximum value. The maximum voltage E_M is called the **peak voltage**. An AC voltage source E is indicated in circuits by a symbol containing a sine wave.

The alternating current produced by an AC voltage source is described by the sine function

$$i = I_M \sin \theta.$$

The maximum current I_M is called the **peak current**. The current reverses direction as it changes between positive and negative.

The AC voltage across the resistor is described by the sine function

$$v = V_M \sin \theta.$$

The maximum voltage V_M is also called the peak voltage. The resistance R remains constant; that is, R does not vary like the voltages and current.

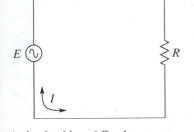

A circuit with an AC voltage source.

RMS VALUES

The magnitudes of AC voltages and currents can be measured in several different ways. The peak voltage and peak current are one type of measure of magnitude. Another common measure is the RMS voltage and the RMS current, where RMS stands for "root-mean-square." The RMS value is found by dividing the peak value by $\sqrt{2}$; that is,

$$E_{RMS} = \frac{E_M}{\sqrt{2}},$$

$$I_{RMS} = \frac{I_M}{\sqrt{2}},$$

and

$$V_{RMS} = \frac{V_M}{\sqrt{2}}.$$

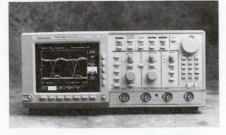

An oscilloscope.
Courtesy of *Tektronix, Inc.*

For example, if $E_M = 1$ V, then

$$E_{RMS} = \frac{E_M}{\sqrt{2}}$$

$$= \frac{1 \text{ V}}{\sqrt{2}}$$

$$= 0.707 \text{ V}.$$

The RMS value compares the electrical power produced by an AC voltage source with the power produced by a DC source. For example, an AC voltage source with $E_{RMS} = 120$ V produces power comparable to a DC source with $E = 120$ V.

We can multiply the E_{RMS} equation by $\sqrt{2}$ to solve for E_M:

$$E_{RMS} = \frac{E_M}{\sqrt{2}}$$

$$E_{RMS}\sqrt{2} = \left(\frac{E_M}{\sqrt{2}}\right)\sqrt{2}$$

$$E_{RMS}\sqrt{2} = E_M.$$

Thus, an AC voltage source with $E_{RMS} = 120$ V has peak voltage

$$E_M = E_{RMS}\sqrt{2}$$

$$= (120 \text{ V})\sqrt{2}$$

$$= 170 \text{ V}.$$

The RMS value $E_{RMS} = 120$ V is the AC voltage provided by common household outlets.

Other common measures of magnitudes of alternating voltage and current can be found in basic electronics books. In this section, we will use the peak voltage and current, which is the same as the amplitude of the sine wave.

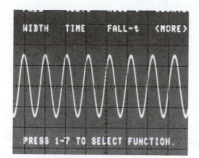

A sine wave shown on an oscilloscope.
Courtesy of *Tektronix, Inc.*

INSTANTANEOUS VALUES

An **instantaneous value** of an alternating voltage or current is its value at any time t. Instantaneous values can be found from the sine function that represents the alternating voltage or current. The angle θ is found at any time t by

$$\theta = 2\pi ft \quad \text{or} \quad \theta = 360°ft,$$

where f is the frequency of the AC source.

EXAMPLE 19.22 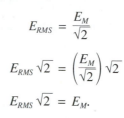 Find the instantaneous value of $i = 10 \sin \theta$ mA for frequency $f = 50$ kHz at $t = 2$ μs.

SOLUTION We can solve the problem in either radian measure or degree measure:

$$\theta = 2\pi ft \qquad\qquad \text{or} \qquad \theta = 360°ft$$

$$\theta = 2\pi(50 \text{ kHz})(2 \text{ μs}) \qquad \theta = 360°(50 \text{ kHz})(2 \text{ μs})$$

$$\theta = 0.628 \text{ rad} \qquad\qquad \theta = 36°.$$

Therefore, we find

$$i = 10 \sin 0.628 \text{ mA} \qquad \text{or} \qquad i = 10 \sin 36° \text{ mA}$$

$$i = 5.88 \text{ mA} \qquad\qquad\qquad i = 5.88 \text{ mA}.$$

(You must be sure your calculator is in radian mode to calculate $i = 10 \sin 0.628$ mA.) ▲

EXAMPLE 19.23 ▶ Find the instantaneous value of $v = 12.7 \sin \theta$ V for frequency $f = 35$ Hz at $t = 50$ ms.

SOLUTION   We can use radian or degree measure to write

$$\theta = 2\pi ft \qquad \text{or} \qquad \theta = 360° ft$$

$$\theta = 2\pi(35 \text{ Hz})(50 \text{ ms}) \qquad \theta = 360°(35 \text{ Hz})(50 \text{ ms})$$

$$\theta = 11 \text{ rad} \qquad \theta = 630°.$$

Therefore, we find

$$v = 12.7 \sin 11 \text{ V} \qquad \text{or} \qquad v = 12.7 \sin 630° \text{ V}$$

$$v = -12.7 \text{ V} \qquad v = -12.7 \text{ V}.$$

We note that this result is $-V_M$, the negative of the peak voltage or the minimum voltage, which occurs at 270° and again at 270° + 360° = 630°.   ▲

LEADING AND LAGGING SINE WAVES

Two sine waves are **in phase** if they have the same frequency and the same phase angle. When an AC circuit is **purely resistive**, that is, contains only the AC source and resistors, the voltages and currents are in phase:

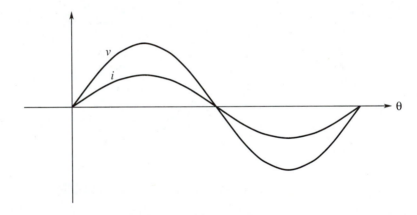

Two sine waves are **out of phase** if they have the same frequency but different phase angles. When an AC circuit contains components other than resistors, some voltages and currents are out of phase. If a voltage has a leading phase angle but the current starts at zero, then we say the voltage **leads** the current:

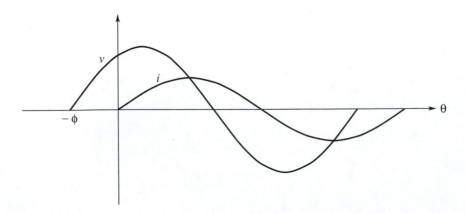

If a voltage has a lagging phase angle but the current starts at zero, then we say the voltage **lags** the current:

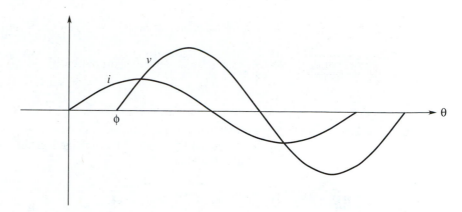

| EXAMPLE 19.24 | ▶ |

The current and voltage in an AC circuit have frequency $f = 50$ Hz. Their peak values are $I_M = 10$ mA and $V_M = 50$ V. The voltage leads the current by 30°. Find their instantaneous values at $t = 5$ ms.

| SOLUTION |

Since the phase angle is given in degrees, we find θ in degrees at $t = 5$ ms:

$$\theta = 360°ft$$

$$= 360°(50 \text{ Hz})(5 \text{ ms})$$

$$= 90°.$$

We draw the graphs of the two sine functions, with v leading i by 30°. Since there is no connection between the scales for milliamperes and volts, we may use any convenient scale for the peak values. We observe that, at $\theta = 90°$, the current i is at its peak value and v is somewhat less than its peak value:

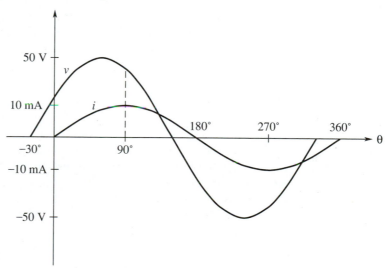

Now, since the voltage leads the current by 30°, the voltage has the equation

$$v = V_M \sin(\theta + \phi)$$

$$v = 50 \sin(\theta + 30°) \text{ V}.$$

The current has the equation

$$i = 10 \sin \theta \text{ mA}.$$

We find the instantaneous values by evaluating each sine function for $\theta = 90°$:

$$i = 10 \sin \theta \text{ mA}$$

$$= 10 \sin 90° \text{ mA}$$

$$= 10 \text{ mA},$$

and

$$v = 50 \sin (\theta + 30°) \text{ V}$$

$$= 50 \sin (90° + 30°) \text{ V}$$

$$= 50 \sin 120° \text{ V}$$

$$= 43.3 \text{ V.} \qquad \blacktriangle$$

EXAMPLE 19.25 ▶ The current and voltage in an AC circuit have frequency $f = 50$ Hz. Their peak values are $I_M = 10$ mA and $V_M = 50$ V. The voltage lags the current by 90°. Find their instantaneous values at $t = 12$ ms.

SOLUTION We find θ at $t = 12$ ms:

$$\theta = 360° ft$$

$$= 360° (50 \text{ Hz})(12 \text{ ms})$$

$$= 216°.$$

We draw the graphs of the two sine functions, with v lagging i by 90°. We observe that at $\theta = 216°$, the current i is negative but v is positive:

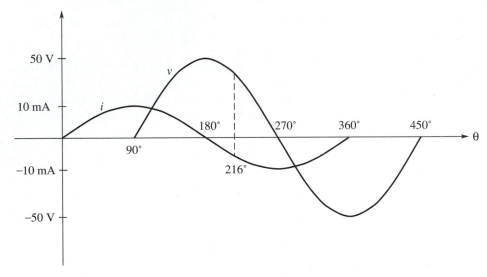

Now, since the voltage lags the current by 90°, the voltage has the equation

$$v = V_M \sin (\theta - \phi)$$

$$v = 50 \sin (\theta - 90°) \text{ V.}$$

The current has the equation

$$i = 10 \sin \theta \text{ mA.}$$

We find the instantaneous values by evaluating each sine function for $\theta = 216°$:

$$i = 10 \sin \theta \text{ mA}$$

$$= 10 \sin 216° \text{ mA}$$

$$= -5.88 \text{ mA,}$$

and

$$v = 50 \sin (\theta - 90°) \text{ V}$$

$$= 50 \sin (216° - 90°) \text{ V}$$

$$= 50 \sin 126° \text{ V}$$

$$= 40.5 \text{ V.} \qquad \blacktriangle$$

Of course, it does not have to be the voltage that is out of phase. The current can lead or lag the voltage.

EXAMPLE 19.26 ▶

The current and voltage in an AC circuit have frequency $f = 4.5$ kHz. Their peak values are $I_M = 20$ mA and $V_M = 15$ V. The current leads the voltage by 45°. Find their instantaneous values at $t = 200$ μs.

SOLUTION

We find θ at $t = 200$ μs:

$$\theta = 360°ft$$

$$= 360°(4.5 \text{ kHz})(200 \text{ μs})$$

$$= 324°.$$

We draw the graphs of the two sine functions, with i leading v by 45°. We observe that at $\theta = 324°$, the voltage v is negative but i is into its second cycle and is positive:

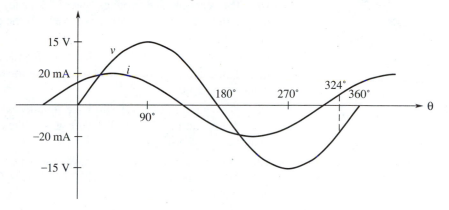

The current leads the voltage by 45°, so the current has the equation

$$i = I_M \sin (\theta + \phi)$$

$$i = 20 \sin (\theta + 45°) \text{ mA}.$$

The voltage has the equation

$$v = 15 \sin \theta \text{ V}.$$

We find the instantaneous values by evaluating each sine function for $\theta = 324°$:

$$v = 15 \sin \theta \text{ V}$$

$$= 15 \sin 324° \text{ V}$$

$$= -8.82 \text{ V},$$

and

$$i = 20 \sin (\theta + 45°) \text{ mA}$$

$$= 20 \sin (324° + 45°) \text{ mA}$$

$$= 20 \sin 369° \text{ mA}$$

$$= 3.13 \text{ mA}.$$

▲

EXERCISE
19.5

Find the instantaneous value:

1. $i = 20 \sin \theta$ mA, for $f = 15$ kHz and $t = 25\ \mu s$

2. $i = 21 \sin \theta$ A, for $f = 30$ Hz and $t = 10$ ms

3. $v = 100 \sin \theta$ V, for $f = 500$ kHz and $t = 2.5\ \mu s$

4. $v = 49 \sin \theta$ V, for $f = 250$ Hz and $t = 4$ ms

5. $e = 170 \sin \theta$ V, for $f = 60$ Hz and $t = 14$ ms

6. $e = 33 \sin \theta$ V, for $f = 5.5$ kHz and $t = 200\ \mu s$

7. $i = 15 \sin \theta\ \mu A$, for $f = 1.25$ MHz and $t = 0.5\ \mu s$

8. $v = 7 \sin \theta$ V, for $f = 8$ MHz and $t = 50$ ns

9. The current and voltage have frequency $f = 25$ Hz. Their peak values are $I_M = 5$ mA and $V_M = 25$ V. The voltage leads the current by 90°. Find their instantaneous values at $t = 4$ ms.

10. The current and voltage have frequency $f = 25$ Hz. Their peak values are $I_M = 5$ mA and $V_M = 25$ V. The voltage leads the current by 60°. Find their instantaneous values at $t = 10$ ms.

11. The current and voltage have frequency $f = 100$ Hz. Their peak values are $I_M = 5$ mA and $V_M = 25$ V. The voltage lags the current by 60°. Find their instantaneous values at $t = 4$ ms.

12. The current and voltage have frequency $f = 100$ Hz. Their peak values are $I_M = 5$ mA and $V_M = 25$ V. The voltage lags the current by 90°. Find their instantaneous values at $t = 5$ ms.

13. The current and voltage have frequency $f = 10$ kHz. Their peak values are $I_M = 10$ mA and $V_M = 100$ V. The voltage lags the current by 45°. Find their instantaneous values at $t = 2\ \mu s$.

14. The current and voltage have frequency $f = 10$ kHz. Their peak values are $I_M = 10$ mA and $V_M = 100$ V. The voltage leads the current by 180°. Find their instantaneous values at $t = 60\ \mu s$.

15. The current and voltage have frequency $f = 4$ kHz. Their peak values are $I_M = 10$ mA and $V_M = 20$ V. The current leads the voltage by 90°. Find their instantaneous values at $t = 25\ \mu s$.

16. The current and voltage have frequency $f = 4$ kHz. Their peak values are $I_M = 10$ mA and $V_M = 20$ V. The current lags the voltage by 45°. Find their instantaneous values at $t = 250\ \mu s$.

□ Self-Test □

1. Estimate the location of the angle with measure 3.5 radians, and then convert the radian measure to degree measure.

1. _____

2. A phasor rotates through an angle of 1.57 radians in 2 μs. What is its frequency?

2. _____

3. Draw the graph of $y = -\dfrac{1}{2} \sin \theta$.

4. Draw the graph of $y = \sin(2\theta - 90°)$.

5. The current and voltage have frequency $f = 15$ kHz. Their peak values are $I_M = 5$ mA and $V_M = 10$ V. The voltage lags the current by 30°. Find their instantaneous values at $t = 50$ μs.

5. _____

UNIT 20

AC Circuit Analysis

Introduction

In the preceding unit, you learned about leading and lagging phase angles in AC circuits. Inductors and capacitors are components of AC circuits that cause phase differences between voltages and currents. In this unit, you will learn how to do some types of calculations involving inductors and capacitors in AC circuits. First, you will learn basic analysis methods for AC series circuits. Then, you will learn how to use complex numbers in the analysis of AC series circuits that contain inductors and capacitors. You will also learn about a special frequency called the resonant frequency. Finally, you will learn some basic techniques for analysis of AC parallel circuits that contain inductors or capacitors or both.

OBJECTIVES

When you have finished this unit you should be able to:

1. Find voltages, inductive reactance, and impedance in AC series *RL* circuits.

2. Find voltages, capacitive reactance, and impedance in AC series *RC* circuits.

3. Use complex numbers to find voltages, reactances, and impedance in AC series *RLC* circuits.

4. Find resonant frequencies in AC series *RLC* circuits.

5. Use complex numbers to find currents, reactances, and impedance in AC parallel *RL*, *RC*, and *RLC* circuits, and total impedances in AC series-parallel circuits.

SECTION 20.1

AC Series RL Circuits

An inductor or coil is a device in which changing current produces an opposing voltage. The exercises for Section 15.4 include DC series circuits containing inductors. Now, suppose an inductor is placed with a resistor in an AC series circuit. The circuit formed is an **AC series *RL* circuit**. Since all circuits in Sections 20.1 through 20.4 are AC series circuits, we can refer to this circuit simply as an ***RL* circuit**.

The current is the same throughout the *RL* circuit at any time *t*. Thus, at any time *t*, the current is the same in each component. The voltage v_R across the resistor is in phase with the current. The voltage v_L across the inductor, however, *leads* the current by 90°.

Since v_L leads v_R by 90°, the equations of the voltage curves have the forms

$$v_R = V_R \sin \theta$$

and

$$v_L = V_L \sin (\theta + 90°).$$

The graphs of the voltage curves v_R and v_L are shown with the current curve *i* on page 448.

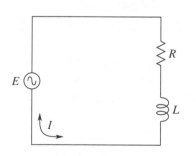

An AC series *RL* circuit.

447

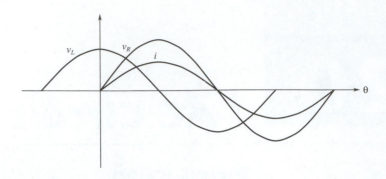

v_R in phase with i; v_L leading i by 90°.

The peak values of v_R and v_L are V_R and V_L. These values can be represented by a **phasor diagram**. We may draw V_R along the positive x-axis. Then, since V_L leads V_R by 90°, we draw V_L along the positive y-axis. Thus V_L is 90° more than V_R, as shown in the diagram at the left below. This phasor diagram shows that the total voltage E is given by

$$E = \sqrt{V_R^2 + V_L^2},$$

and the phase angle θ between E and V_R is given by

$$\theta = \tan^{-1}\left(\frac{V_L}{V_R}\right).$$

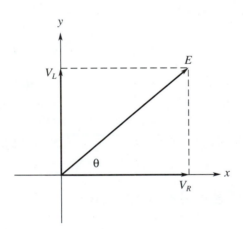

A phasor diagram showing voltages for the RL circuit.

The phasor diagram with E at an angle of 0°.

It is important to understand that the phasor diagram is a *mathematical model* of the circuit. This model simplifies the mathematics by showing the voltage E as a resultant, with V_R and V_L as its vertical and horizontal components. The actual circuit would have an applied voltage E at 0°, as shown in the diagram at the right. You should check that the mathematical formulas for E and θ are the same for both diagrams. In either case, the phase difference between E and V_R is θ.

EXAMPLE 20.1

In an RL circuit, $V_R = 10$ V and $V_L = 8.5$ V. Find the total voltage E and the phase angle θ.

SOLUTION

We draw a phasor diagram showing V_R, V_L, E, and θ. The total voltage E is given by

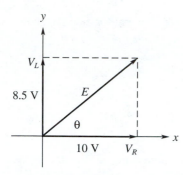

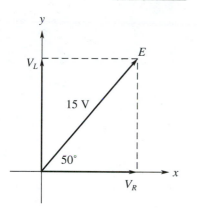

Phasor diagram for Example 20.1.

$$E = \sqrt{V_R^2 + V_L^2}$$
$$= \sqrt{10^2 + 8.5^2} \text{ V}$$
$$= 13.1 \text{ V}.$$

The phase angle θ is given by

$$\theta = \tan^{-1}\left(\frac{V_L}{V_R}\right)$$
$$= \tan^{-1}\left(\frac{8.5}{10}\right)$$
$$= 40.4°.$$ ▲

EXAMPLE 20.2

The total voltage in an *RL* circuit is $E = 15$ V and the phase angle is $\theta = 50°$. Find V_R and V_L.

SOLUTION

We draw a phasor diagram. Since V_R and V_L are the horizontal and vertical components of E, we write

$$V_R = E \cos \theta$$
$$= 15 \cos 50° \text{ V}$$
$$= 9.64 \text{ V},$$

and

$$V_L = E \sin \theta$$
$$= 15 \sin 50° \text{ V}$$
$$= 11.5 \text{ V}.$$ ▲

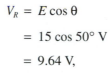

Phasor diagram for Example 20.2.

INDUCTIVE REACTANCE AND IMPEDANCE

Opposition to current provided by an inductor is called the **inductive reactance** X_L. Reactance, like resistance, is measured in ohms. We can draw a phasor diagram for the resistance R and the inductive reactance X_L that is similar to the diagram for the voltages V_R and V_L:

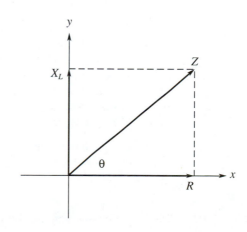

The total opposition to current is called the **impedance** Z. The phasor diagram shows that the impedance Z is given by

$$Z = \sqrt{R^2 + X_L^2},$$

and the phase angle θ between Z and R is given by

$$\theta = \tan^{-1}\left(\frac{X_L}{R}\right).$$

EXAMPLE 20.3

In an RL circuit, $R = 1.2\ \Omega$ and $X_L = 1.57\ \Omega$. Find the impedance Z and the phase angle θ.

SOLUTION

We draw a phasor diagram showing R, X_L, Z, and θ. The impedance Z is given by

$$Z = \sqrt{R^2 + X_L^2}$$
$$= \sqrt{1.2^2 + 1.57^2}\ \Omega$$
$$= 1.98\ \Omega.$$

The phase angle θ is given by

$$\theta = \tan^{-1}\left(\frac{X_L}{R}\right)$$
$$= \tan^{-1}\left(\frac{1.57}{1.2}\right)$$
$$= 52.6°.$$ ▲

Phasor diagram for Example 20.3.

In Unit 5, we solved algebra examples involving the formula

$$X_L = 2\pi f L.$$

This formula gives the value of the inductive reactance X_L, where the inductor has inductance L, and the voltage provided to the circuit has frequency f.

EXAMPLE 20.4

The inductor in an RL circuit has $L = 5$ mH, and the resistor has $R = 1\ \text{k}\Omega$. The frequency in the circuit is $f = 10$ kHz. Find the impedance Z and phase angle θ.

SOLUTION

First, we find the inductive reactance by using

$$X_L = 2\pi f L$$
$$= 2\pi(10\ \text{kHz})(5\ \text{mH})$$
$$= 314\ \Omega,$$

or $X_L = 0.314\ \text{k}\Omega$. Then, we find the impedance Z as in Example 20.3. If we use R and X_L both in kilohms, then Z is given by

$$Z = \sqrt{R^2 + X_L^2}$$
$$= \sqrt{1^2 + 0.314^2}\ \text{k}\Omega$$
$$= 1.05\ \text{k}\Omega.$$

The phase angle θ is given by

$$\theta = \tan^{-1}\left(\frac{X_L}{R}\right)$$
$$= \tan^{-1}\left(\frac{0.314}{1}\right)$$
$$= 17.4°.$$ ▲

CURRENT

Ohm's law is true for current, voltage, and resistance in AC circuits. Thus, for the resistor R in an RL circuit, we have

$$I = \frac{V_R}{R}.$$

We can also write versions of Ohm's law for the oppositions to current X_L and Z, and their corresponding voltages:

$$I = \frac{V_L}{X_L}$$

and

$$I = \frac{E}{Z}.$$

We can use any of these formulas to find I.

EXAMPLE 20.5

An RL circuit has source voltage $E = 15$ V at frequency $f = 100$ kHz. The inductor has $L = 20$ mH, and the resistor has $R = 33$ kΩ. Find X_L, Z, θ, and I.

SOLUTION

Since we know f and L, we can find X_L:

$$X_L = 2\pi f L$$

$$= 2\pi(100 \text{ kHz})(20 \text{ mH})$$

$$= 12.6 \text{ k}\Omega.$$

As before, we find Z by using

$$Z = \sqrt{R^2 + X_L^2}$$

$$= \sqrt{33^2 + 12.6^2} \text{ k}\Omega$$

$$= 35.3 \text{ k}\Omega,$$

and θ by using

$$\theta = \tan^{-1}\left(\frac{X_L}{R}\right)$$

$$= \tan^{-1}\left(\frac{12.6}{33}\right)$$

$$= 20.9°.$$

Now, we can find I by using

$$I = \frac{E}{Z}$$

$$= \frac{15 \text{ V}}{35.3 \text{ k}\Omega}$$

$$= 0.425 \text{ mA},$$

or $I = 425$ μA. At the beginning of this section we saw that I is in phase with V_R. The phase difference between I and E is given by the phase angle $\theta = 20.9°$.   ▲

EXERCISE
20.1

For an *RL* circuit:

1. If $V_R = 40$ V and $V_L = 30$ V, find E and θ.

2. If $V_R = 8$ V and $V_L = 10$ V, find E and θ.

3. If $E = 100$ V and $\theta = 30°$, find V_R and V_L.

4. If $E = 32$ V and $\theta = 65°$, find V_R and V_L.

5. If $R = 2.7\ \Omega$ and $X_L = 2.3\ \Omega$, find Z and θ.

6. If $R = 11.2\ \Omega$ and $X_L = 1.49\ \Omega$, find Z and θ.

7. If $Z = 550\ \Omega$ and $\theta = 53.1°$, find R and X_L.

8. If $Z = 8.5\ k\Omega$ and $\theta = 10°$, find R and X_L.

9. If the inductor has $L = 2$ H, the resistor has $R = 500\ \Omega$, and the frequency is $f = 50$ Hz, find X_L, Z, and θ.

10. If the inductor has $L = 10$ mH, the resistor has $R = 1.5\ k\Omega$, and the frequency is $f = 14$ kHz, find X_L, Z, and θ.

11. The source voltage is $E = 12$ V at frequency $f = 10$ Hz. The inductor has $L = 200$ mH, and the resistor has $R = 8\ \Omega$. Find X_L, Z, θ, and I.

12. The source voltage is $E = 5$ V at frequency $f = 20$ MHz. The inductor has $L = 250\ \mu H$, and the resistor has $R = 9.9\ k\Omega$. Find X_L, Z, θ, and I.

13. The source voltage is $E = 1.1$ V. The resistor has $R = 12\ \Omega$, and the voltage across the resistor is $V_R = 800$ mV. Find I, Z, and θ.

14. The component voltages are $V_R = 12$ V and $V_L = 12$ V. The resistor has $R = 1.5\ \Omega$. Find I, Z, and θ.

SECTION
20.2

AC Series RC Circuits

In Section 15.4, we showed the process of charging a capacitor in a DC series circuit. Now, suppose a capacitor is placed with a resistor in an AC series circuit. The circuit formed is an **AC series *RC* circuit**.

In the *RC* circuit, at any time *t*, the voltage v_R across the resistor is in phase with the current. The voltage v_C across the capacitor, however, *lags* the current by 90°:

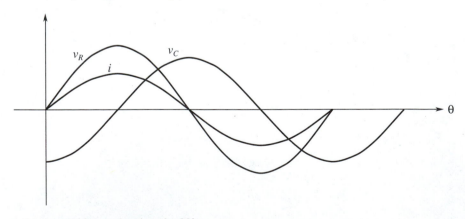

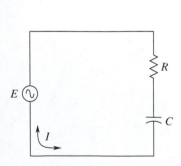

An AC series *RC* circuit.

v_R in phase with *i*; v_C lagging *i* by 90°.

Since v_C lags v_R by 90°, the equations of the voltage curves have the forms

$$v_R = V_R \sin \theta$$

and

$$v_C = V_C \sin (\theta - 90°).$$

The peak values of v_R and v_C are V_R and V_C. To represent these values by a phasor diagram, we draw V_R along the positive x-axis. Then, since V_C lags V_R by 90°, we draw V_C along the negative y-axis; that is, at 90° less than V_R:

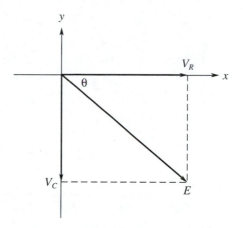

The phasor diagram shows that the total voltage E is given by

$$E = \sqrt{V_R^2 + V_C^2},$$

and the phase angle θ between E and R is given by

$$\theta = -\tan^{-1}\left(\frac{V_C}{V_R}\right).$$

We observe that θ is negative; therefore, we take the negative of the inverse tangent. As in Section 20.1, it is important to understand that the phasor diagram is a *mathematical model* of the circuit.

EXAMPLE 20.6

In an RC circuit, $V_R = 15$ V and $V_C = 12$ V. Find the total voltage E and the phase angle θ.

SOLUTION

We draw a phasor diagram showing V_R, V_C, E, and θ. The total voltage E is given by

$$E = \sqrt{V_R^2 + V_C^2}$$
$$= \sqrt{15^2 + 12^2} \text{ V}$$
$$= 19.2 \text{ V}.$$

The phase angle θ is given by

$$\theta = -\tan^{-1}\left(\frac{V_C}{V_R}\right)$$
$$= -\tan^{-1}\left(\frac{12}{15}\right)$$
$$= -38.7°.$$

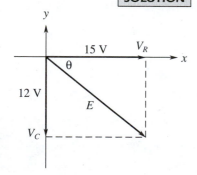

Phasor diagram for Example 20.6.

EXAMPLE 20.7

The total voltage in an *RC* circuit is $E = 6.8$ V and the phase angle is $\theta = -75°$. Find V_R and V_C.

SOLUTION

We draw a phasor diagram. Since V_R and V_C are the horizontal and vertical components of E, we write

$$V_R = E \cos \theta$$
$$= 6.8 \cos (-75°) \text{ V}$$
$$= 1.76 \text{ V},$$

and

$$V_C = E \sin \theta$$
$$= 6.8 \sin (-75°) \text{ V}$$
$$= -6.57 \text{ V}.$$

Since V_C is a peak value, we write $V_C = 6.57$ V. The negative result reflects the fact that the phasor representing V_C is directed downward, along the negative *y*-axis.   ▲

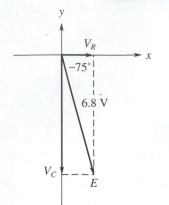

Phasor diagram for Example 20.7.

CAPACITIVE REACTANCE AND IMPEDANCE

Opposition to current provided by a capacitor is the **capacitive reactance** X_C. We can draw a phasor diagram for the resistance R and the capacitive reactance X_C that is similar to the diagram for the voltages V_R and V_C:

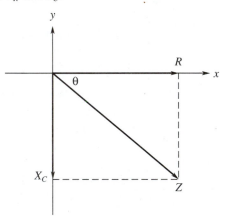

As in Section 20.1, the total opposition to current is the impedance *Z*. The phasor diagram shows that the impedance *Z* is given by

$$Z = \sqrt{R^2 + X_C^2},$$

and the phase angle θ between *Z* and *R* is given by

$$\theta = -\tan^{-1}\left(\frac{X_C}{R}\right).$$

EXAMPLE 20.8

In an *RC* circuit, $R = 5.6$ kΩ and $X_C = 3.78$ kΩ. Find the impedance *Z* and the phase angle θ.

SOLUTION

We draw a phasor diagram showing *R*, X_C, *Z*, and θ. The impedance *Z* is given by

$$Z = \sqrt{R^2 + X_C^2}$$
$$= \sqrt{5.6^2 + 3.78^2} \text{ Ω}$$
$$= 6.76 \text{ kΩ}.$$

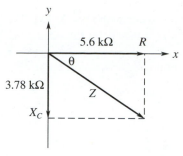

Phasor diagram for Example 20.8.

The phase angle θ is given by

$$\theta = -\tan^{-1}\left(\frac{X_C}{R}\right)$$

$$= -\tan^{-1}\left(\frac{3.78}{5.6}\right)$$

$$= -34°.$$
▲

In Unit 5, we solved algebra examples involving the formula

$$X_C = \frac{1}{2\pi f C}.$$

This formula gives the value of the capacitive reactance X_C, where the capacitor has capacitance C, and the voltage provided to the circuit has frequency f.

EXAMPLE 20.9

The capacitor in an RC circuit has $C = 0.006\ \mu\text{F}$, and the resistor has $R = 1\ \text{k}\Omega$. The frequency in the circuit is $f = 10\ \text{kHz}$. Find the impedance Z and phase angle θ.

SOLUTION

First, we find the capacitive reactance by using

$$X_C = \frac{1}{2\pi f C}$$

$$= \frac{1}{2\pi(10\ \text{kHz})(0.006\ \mu\text{F})}$$

$$= 2.65\ \text{k}\Omega.$$

Then, as in Example 20.8, the impedance Z is given by

$$Z = \sqrt{R^2 + X_C^2}$$

$$= \sqrt{1^2 + 2.65^2}\ \text{k}\Omega$$

$$= 2.83\ \text{k}\Omega.$$

The phase angle θ is given by

$$\theta = -\tan^{-1}\left(\frac{X_C}{R}\right)$$

$$= -\tan^{-1}\left(\frac{2.65}{1}\right)$$

$$= -69.3°.$$
▲

CURRENT

As in Section 20.1, the current is related to the voltages and the oppositions to current by the formulas

$$I = \frac{V_R}{R},$$

$$I = \frac{V_C}{X_C},$$

and

$$I = \frac{E}{Z}.$$

EXAMPLE 20.10

An *RC* circuit contains a capacitor with $C = 100$ pF and a resistor with $R = 33$ kΩ. The current in the circuit is $I = 410$ μA and the frequency is $f = 100$ kHz. Find X_C, Z, E, and θ.

SOLUTION

Since we know f and C, we can find X_C:

$$X_C = \frac{1}{2\pi f C}$$

$$= \frac{1}{2\pi(100 \text{ kHz})(100 \text{ pF})}$$

$$= 15.9 \text{ k}\Omega.$$

As before, we find Z by using

$$Z = \sqrt{R^2 + X_C^{\,2}}$$

$$= \sqrt{33^2 + 15.9^2} \text{ k}\Omega$$

$$= 36.6 \text{ k}\Omega,$$

and θ by using

$$\theta = -\tan^{-1}\left(\frac{X_L}{R}\right)$$

$$= -\tan^{-1}\left(\frac{15.9}{33}\right)$$

$$= -25.7°.$$

To find E, we can solve

$$I = \frac{E}{Z}$$

to obtain

$$E = IZ.$$

Then we have

$$E = (410 \text{ μA})(36.6 \text{ k}\Omega)$$

$$= 15 \text{ V}.$$

The current I is in phase with V_R, so the phase difference between I and E is given by the phase angle $\theta = -25.7°$. ▲

EXERCISE

20.2

For an *RC* circuit:

1. If $V_R = 15$ V and $V_C = 20$ V, find E and θ.
2. If $V_R = 160$ V and $V_C = 40$ V, find E and θ.
3. If $E = 10$ V and $\theta = -45°$, find V_R and V_C.
4. If $E = 120$ V and $\theta = -56°$, find V_R and V_C.
5. If $R = 23$ kΩ and $X_C = 16$ kΩ, find Z and θ.
6. If $R = 900$ Ω and $X_C = 1.2$ kΩ, find Z and θ.
7. If $Z = 1.22$ kΩ and $\theta = -82°$, find R and X_C.
8. If $Z = 1.1$ kΩ and $\theta = -41°$, find R and X_C.

9. If the capacitor has $C = 0.2 \, \mu\text{F}$, the resistor has $R = 27 \, \Omega$, and the frequency is $f = 50 \, \text{kHz}$, find X_C, Z, and θ.

10. If the capacitor has $C = 100 \, \text{pF}$, the resistor has $R = 10 \, \text{k}\Omega$, and the frequency is $f = 66 \, \text{kHz}$, find X_C, Z, and θ.

11. The voltage is $E = 30 \, \text{V}$ at frequency $f = 10 \, \text{kHz}$. The capacitor has $C = 0.2 \, \mu\text{F}$, and the resistor has $R = 100 \, \Omega$. Find X_C, Z, θ, and I.

12. The voltage is $E = 144 \, \text{V}$ at frequency $f = 60 \, \text{Hz}$. The capacitor has $C = 0.01 \, \mu\text{F}$ and the resistor has $R = 220 \, \text{k}\Omega$. Find X_C, Z, θ, and I.

13. The capacitor has $C = 250 \, \text{pF}$ and the resistor has $R = 1 \, \text{k}\Omega$. The current in the circuit is $I = 10 \, \text{mA}$ and the frequency is $f = 500 \, \text{kHz}$. Find X_C, Z, E, and θ.

14. The capacitor has $C = 0.075 \, \text{pF}$ and the resistor has $R = 500 \, \text{k}\Omega$. The current in the circuit is $I = 48 \, \mu\text{A}$ and the frequency is $f = 5 \, \text{MHz}$. Find X_C, Z, E, and θ.

15. The applied voltage is $E = 100 \, \text{V}$. The resistor has $R = 2 \, \text{k}\Omega$, and the voltage across the resistor is $V_R = 50 \, \text{V}$. Find I, Z, and θ.

16. The voltages across R and C are $V_R = 4 \, \text{V}$ and $V_C = 3 \, \text{V}$, and $R = 2.2 \, \text{k}\Omega$. Find I, Z, and θ.

SECTION
20.3

Complex Number Analysis

Complex numbers provide a convenient way to analyze AC circuits. We recall from Unit 18 that complex numbers can be written in rectangular form

$$a + jb,$$

where a is the real part and jb is the imaginary part. Each complex number in the rectangular form $a + jb$ is identified with a vector with horizontal component $x = a$ and vertical component $y = b$.

Since a phasor is a type of vector, complex numbers can be associated with phasors in the same way. In many areas of science and engineering, a letter with an arrow over it, such as $\vec{E}$, is used to represent such a vector or phasor.

In RL and RC circuits, the voltage V_R across the resistor is represented by a phasor on the positive x-axis. The x-axis is the real axis; therefore, to represent V_R we use a positive real number a, or a complex number of the form

$$V_R = a + j0.$$

In RL circuits, the voltage V_L across the inductor is represented by a phasor on the positive y-axis. The y-axis is the imaginary axis; therefore, to represent V_L we use an imaginary number jb, or a complex number of the form

$$V_L = 0 + jb.$$

The total voltage can then be represented by the complex number sum

$$\vec{E} = V_R + V_L$$

$$= (a + j0) + (0 + jb)$$

$$= a + jb.$$

Similarly, in RC circuits, the voltage V_R across the resistor is represented by a complex number of the form

$$V_R = a + j0.$$

The voltage V_C across the capacitor is represented by a phasor on the negative y-axis. Therefore, to represent V_C we use an imaginary number $-jb$, or a complex number of the form

$$V_C = 0 - jb.$$

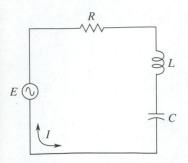

An AC series *RLC* circuit.

The total voltage can then be represented by the complex number sum

$$\vec{E} = V_R + V_C$$

$$= (a + j0) + (0 - jb)$$

$$= a - jb.$$

An **RLC circuit** contains a resistor, an inductor, and a capacitor. By using complex numbers, we can reduce *RLC* circuits to either *RL* or *RC* circuits. Then, we can represent the total voltage by a complex number

$$\vec{E} = a \pm jb.$$

EXAMPLE 20.11

In an *RLC* circuit, $V_R = 8$ V, $V_L = 10$ V, and $V_C = 6$ V. Find the total voltage as a complex number in rectangular form and in polar form.

SOLUTION

Using imaginary numbers, we represent V_L as $j10$ V and V_C as $-j6$ V. The phasor diagram is shown at the left below. Then, we can combine V_L and V_C by adding the imaginary numbers

$$V_L + V_C = j10 \text{ V} + (-j6 \text{ V})$$

$$= j4 \text{ V}.$$

Thus, V_L and V_C combined act like a voltage of 4 V across the inductor. The circuit is

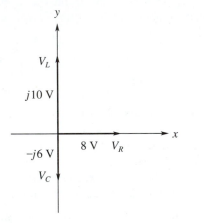

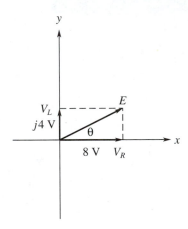

Phasor diagram for Example 20.11.

Phasor diagram for an equivalent *RL* circuit.

equivalent to an *RL* circuit with the phasor diagram shown at the right:
The total voltage $\vec{E}$ is then represented by the complex number

$$\vec{E} = 8 + j4 \text{ V}.$$

We can write this complex number in polar form. The absolute value is given by

$$E = \sqrt{V_R^2 + V_L^2}$$

$$= \sqrt{8^2 + 4^2} \text{ V}$$

$$= 8.94 \text{ V}.$$

The argument is given by

$$\theta = \tan^{-1}\left(\frac{V_L}{V_R}\right)$$

$$= \tan^{-1}\left(\frac{4}{8}\right)$$

$$= 26.6°.$$

Thus, in polar form the total voltage $\vec{E}$ is

$$\vec{E} = 8.94\underline{/26.6°}\text{ V.}$$

▲

EXAMPLE 20.12

In an *RLC* circuit, $V_R = 4.2$ V, $V_L = 2.8$ V, and $V_C = 4.5$ V. Find the total voltage as a complex number in rectangular form and in polar form.

SOLUTION

Using imaginary numbers, we represent V_L as $j2.8$ V and V_C as $-j4.5$ V, and draw the phasor diagram shown at the left below. Combining V_L and V_C, we have

$$V_L + V_C = j2.8\text{ V} + (-j4.5\text{ V})$$

$$= -j1.7\text{ V.}$$

Thus, V_L and V_C combined act like a voltage of 1.7 V across the capacitor. The circuit is equivalent to an *RC* circuit with the phasor diagram shown at the right:

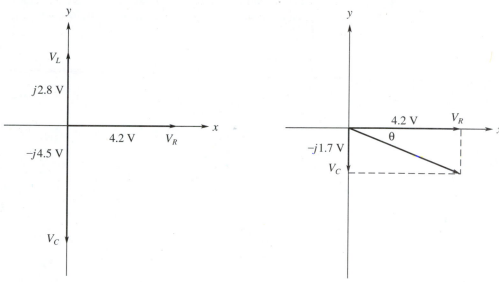

Phasor diagram for Example 20.12.   Phasor diagram for an equivalent *RC* circuit.

The total voltage $\vec{E}$ is then represented by the complex number

$$\vec{E} = 4.2 - j1.7\text{ V.}$$

We can write this complex number in polar form. The absolute value is given by

$$E = \sqrt{V_R^2 + V_C^2}$$

$$= \sqrt{4.2^2 + 1.7^2}\text{ V}$$

$$= 4.53\text{ V.}$$

The argument is given by

$$\theta = -\tan^{-1}\left(\frac{V_C}{V_R}\right)$$

$$= -\tan^{-1}\left(\frac{1.7}{4.2}\right)$$

$$= -22°.$$

Thus, in polar form the total voltage $\vec{E}$ is

$$\vec{E} = 4.53\underline{/-22°}\text{ V.}$$

▲

RESISTANCE, REACTANCES, AND IMPEDANCE

Resistance R, inductive reactance X_L, and capacitive reactance X_C are represented by phasors in the same directions as V_R, V_L, and V_C. Therefore, we can also represent R, X_L, and X_C by numbers of the form a, jb, and $-jb$.

EXAMPLE 20.13

In an RLC circuit, $R = 1.5\ \Omega$, $X_L = 5.5\ \Omega$, and $X_C = 8.5\ \Omega$. Find the impedance as a complex number in rectangular form and in polar form.

SOLUTION

Using imaginary numbers, we represent X_L as $j5.5\ \Omega$, and X_C as $-j8.5\ \Omega$, and draw the phasor diagram shown at the left below. Combining X_L and X_C, we have

$$X_L + X_C = j5.5\ \Omega + (-j8.5\ \Omega)$$

$$= -j3\ \Omega.$$

Thus, X_L and X_C combined act like a capacitive reactance of $3\ \Omega$. The circuit is equivalent to an RC circuit with the phasor diagram shown at the right:

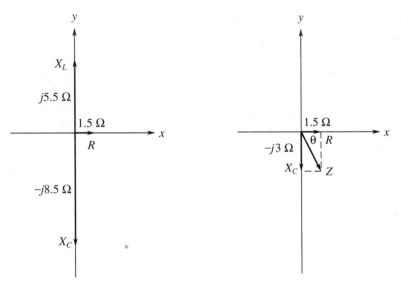

Phasor diagram for Example 20.13.

Phasor diagram for an equivalent RC circuit.

The impedance $\vec{Z}$ is then represented by the complex number

$$\vec{Z} = 1.5 - j3\ \Omega.$$

We can write this complex number in polar form. The absolute value is given by

$$Z = \sqrt{R^2 + X_C^2}$$

$$= \sqrt{1.5^2 + 3^2}\ \Omega$$

$$= 3.35\ \Omega.$$

The argument is given by

$$\theta = -\tan^{-1}\left(\frac{X_C}{R}\right)$$

$$= -\tan^{-1}\left(\frac{3}{1.5}\right)$$

$$= -63.4°.$$

Thus, in polar form the impedance $\vec{Z}$ is

$$\vec{Z} = 3.35\underline{/-63.4°}\ \Omega.$$ ▲

An *RL*, *RC*, or *RLC* circuit can have more than one resistor, inductor, or capacitor. As in DC series circuits, we can add resistances; thus, for AC *series* circuits

$$R_T = R_1 + R_2,$$

and similarly for three or more resistors.

We can add the imaginary numbers that represent two inductive reactances

$$jX_L = jX_{L_1} + jX_{L_2},$$

and divide by *j* to obtain

$$X_L = X_{L_1} + X_{L_2}.$$

Then, by using the formula $X_L = 2\pi fL$, we can write

$$2\pi fL = 2\pi fL_1 + 2\pi fL_2,$$

and dividing both sides by $2\pi f$, we have

$$L = L_1 + L_2.$$

This result can be extended to three or more inductors.

We can also add the imaginary numbers that represent two capacitive reactances

$$-jX_C = -jX_{C_1} + (-jX_{C_2}),$$

and divide by $-j$ to obtain

$$X_C = X_{C_1} + X_{C_2}.$$

When we use the formula $X_C = \dfrac{1}{2\pi fC}$, we find that

$$\frac{1}{2\pi fC} = \frac{1}{2\pi fC_1} + \frac{1}{2\pi fC_2};$$

and multiplying both sides by $2\pi f$, we have

$$\frac{1}{C} = \frac{1}{C_1} + \frac{1}{C_2}.$$

This result can be extended to three or more capacitors.

EXAMPLE 20.14

In an *RC* circuit, $R = 1\ \text{k}\Omega$, and two capacitors have $C_1 = 0.1\ \mu\text{F}$ and $C_2 = 0.15\ \mu\text{F}$:

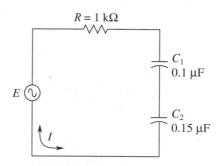

Find the impedance at frequency $f = 10\ \text{kHz}$. Write the impedance as a complex number in rectangular form and in polar form.

SOLUTION   We may find the total capacitance C by using

$$\frac{1}{C} = \frac{1}{C_1} + \frac{1}{C_2}$$

$$\frac{1}{C} = \frac{1}{0.1\ \mu F} + \frac{1}{0.15\ \mu F}$$

$$C = 0.06\ \mu F.$$

Then, we find X_C by using

$$X_C = \frac{1}{2\pi f C}$$

$$= \frac{1}{2\pi(10\ kHz)(0.06\ \mu F)}$$

$$= 265\ \Omega,$$

or 0.265 kΩ. The circuit is equivalent to an RC circuit with this phasor diagram:

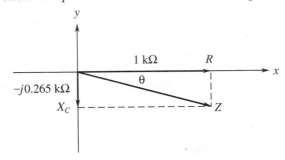

The impedance $\vec{Z}$ is then represented by the complex number

$$\vec{Z} = 1 - j0.265\ k\Omega.$$

We can find Z and θ as in the preceding example, and write the complex number in polar form:

$$\vec{Z} = 1.03\underline{/-14.8}\ k\Omega.$$

You should also solve this example by finding the separate capacitive reactances. Use C_1 to find $X_{C_1} = 159\ \Omega$ and C_2 to find $X_{C_2} = 106\ \Omega$. Then add to obtain $X_C = 265\ \Omega$.   ▲

EXERCISE
20.3

For the RLC circuit, find the total voltage $\vec{E}$ as a complex number in rectangular form and in polar form:

1. $V_R = 100$ V, $V_L = 120$ V, and $V_C = 60$ V
2. $V_R = 5$ V, $V_L = 57$ V, and $V_C = 33$ V
3. $V_R = 20$ V, $V_L = 69$ V, and $V_C = 90$ V
4. $V_R = 9.9$ V, $V_L = 4.4$ V, and $V_C = 7.4$ V

For the RLC circuit, find the impedance $\vec{Z}$ as a complex number in rectangular form and in polar form:

5. $R = 18.2\ \Omega$, $X_L = 193\ \Omega$, and $X_C = 141\ \Omega$
6. $R = 6.7\ k\Omega$, $X_L = 5.7\ k\Omega$, and $X_C = 1.2\ k\Omega$
7. $R = 9.1\ k\Omega$, $X_L = 5.14\ k\Omega$, and $X_C = 8.27\ k\Omega$
8. $R = 100\ \Omega$, $X_L = 7.25\ k\Omega$, and $X_C = 7.75\ k\Omega$
9. $R = 5\ \Omega$, $L = 40$ mH, $C = 500\ \mu F$, and $f = 100$ Hz
10. $R = 1.5\ k\Omega$, $L = 3\ \mu H$, $C = 12$ pF, and $f = 15$ MHz

For the *RL* or *RC* circuit, find the impedance $\vec{Z}$ as a complex number in rectangular form and in polar form:

11. $R = 15\ \Omega$, two inductors have $L_1 = 5$ mH and $L_2 = 10$ mH, and $f = 60$ Hz

12. $R = 1.2$ kΩ, two inductors have $L_1 = 540\ \mu$H and $L_2 = 310\ \mu$H, and $f = 150$ kHz

13. $R = 1\ \Omega$, two capacitors have $C_1 = 5\ \mu$F and $C_2 = 15\ \mu$F, and $f = 50$ kHz

14. $R = 1$ kΩ, two capacitors have $C_1 = 330$ pF and $C_2 = 660$ pF, and $f = 4$ MHz

---

SECTION
20.4

Series Resonance

In an AC series *RLC* circuit, suppose the inductive reactance X_L and the capacitive reactance X_C are equal. Then, the phasors on the y-axis are equal in magnitude and opposite in direction, as shown in the phasor diagram below. If we combine X_L and X_C by adding their imaginary number forms, we have

$$X_L + X_C = jb\ \Omega + (-jb\ \Omega)$$

$$= j0\ \Omega.$$

The impedance $\vec{Z}$ is represented by the complex number

$$\vec{Z} = R - j0\ \Omega.$$

Thus, the impedance has magnitude $Z = R$ and phase angle $\theta = 0°$, and the circuit is equivalent to a *purely resistive* circuit:

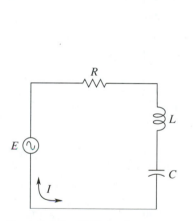

An AC series *RLC* circuit.

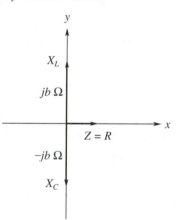

Phasor diagram for the *RLC* circuit with $X_L = X_C$.

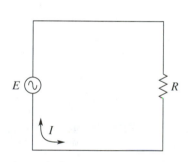

An equivalent purely resistive circuit.

EXAMPLE 20.15

In a series *RLC* circuit, the resistance $R = 1.5\ \Omega$, the inductance $L = 1.15$ mH, and the capacitance $C = 0.22\ \mu$F. Find X_L, X_C, and $\vec{Z}$, at the frequency $f = 10$ kHz.

SOLUTION We find X_L by using

$$X_L = 2\pi f L$$

$$= 2\pi(10\text{ kHz})(1.15\text{ mH})$$

$$= 72.3\ \Omega.$$

We find X_C by using

$$X_C = \frac{1}{2\pi f C}$$

$$= \frac{1}{2\pi(10\text{ kHz})(0.22\ \mu\text{F})}$$

$$= 72.3\ \Omega.$$

Therefore, $X_L = X_C$, but they are in opposite directions. Using imaginary numbers, we represent X_L as $j72.3$ V, and X_C as $-j72.3$ V. Thus, combining X_L and X_C, we have

$$X_L + X_C = j72.3\ \Omega + (-72.3\ \Omega)$$

$$= j0\ \Omega.$$

Then, the impedance $\vec{Z}$ is represented by the complex number

$$\vec{Z} = 1.5 + j0\ \Omega.$$

We see that the impedance has magnitude $Z = R$ and phase angle $\theta = 0°$. In polar form, we have

$$\vec{Z} = 1.5\underline{/0°}\ \Omega. \qquad \blacktriangle$$

RESONANT FREQUENCY

The frequency at which the inductive reactance and the capacitive reactance are equal is called the **resonant frequency**. The variable name for the resonant frequency is f_r, indicated by the subscript r. Thus, when $f = f_r$, we have $X_L = X_C$.

We can solve the system of equations

$$X_L = 2\pi f_r L$$

$$X_C = \frac{1}{2\pi f_r C},$$

when $X_L = X_C$, by setting the right-hand sides equal:

$$2\pi f_r L = \frac{1}{2\pi f_r C}.$$

To solve for f_r, we multiply both sides by f_r:

$$f_r(2\pi f_r L) = f_r\left(\frac{1}{2\pi f_r C}\right)$$

$$2\pi f_r^2 L = \frac{1}{2\pi C}.$$

Then, we divide both sides by $2\pi L$:

$$\frac{2\pi f_r^2 L}{2\pi L} = \frac{1}{(2\pi L)(2\pi C)}$$

$$f_r^2 = \frac{1}{(2\pi L)(2\pi C)}$$

$$f_r^2 = \frac{1}{(2\pi)^2 LC}.$$

We take the square root of each side to obtain

$$f_r = \sqrt{\frac{1}{(2\pi)^2 LC}}$$

$$= \frac{1}{2\pi\sqrt{LC}}.$$

We solved algebra examples involving this formula in Unit 5.

EXAMPLE 20.16

Find the resonant frequency for $L = 200$ mH and $C = 500$ pF.

SOLUTION

We use the formula

$$f_r = \frac{1}{2\pi \sqrt{LC}}$$

$$= \frac{1}{2\pi \sqrt{(200 \text{ mH})(500 \text{ pF})}} \, .$$

We replace the prefixes by powers of 10, to obtain

$$f_r = \frac{1}{2\pi \sqrt{(200 \times 10^{-3})(500 \times 10^{-12})}} \text{ Hz}$$

$$= \frac{1}{2\pi \sqrt{100 \times 10^{-12}}} \text{ Hz}$$

$$= \frac{1}{2\pi (10 \times 10^{-6})} \text{ Hz}$$

$$= \frac{10^6}{20\pi} \text{ Hz}.$$

Therefore, we find

$$f_r = 0.0159 \times 10^6 \text{ Hz}$$

$$= 15.9 \text{ kHz}. \qquad \blacktriangle$$

We can find XL for $L = 200$ mH, at the resonant frequency $f_r = 15.9$ kHz found in Example 20.16, by using

$$X_L = 2\pi f_r L$$

$$= 2\pi (15.9 \text{ kHz})(200 \text{ mH})$$

$$= 20 \text{ k}\Omega.$$

We can also find X_C for $C = 500$ pF, at the resonant frequency $f_r = 15.9$ kHz, by using

$$X_C = \frac{1}{2\pi f_r C}$$

$$= \frac{1}{2\pi (15.9 \text{ kHz})(500 \text{ pF})}$$

$$= 20 \text{ k}\Omega.$$

The reactances X_L and X_C are equal at the resonant frequency.

CURRENT

At the resonant frequency f_r, the magnitude of the impedance in a series RLC circuit is

$$Z = R.$$

If we know R and the applied voltage E, we can find the magnitude of the current by using

$$I = \frac{E}{Z}$$

or

$$I = \frac{E}{R} \, .$$

EXAMPLE 20.17

For the circuit in Example 20.16, with $L = 200$ mH and $C = 500$ pF, the resistance is $R = 100\ \Omega$ and the applied voltage is $E = 5$ V. Find the current in the circuit at the resonant frequency f_r.

SOLUTION

The magnitude of the current at the resonant frequency f_r is given by

$$I = \frac{E}{R}$$

$$= \frac{5\text{ V}}{100\ \Omega}$$

$$= 50\text{ mA.}$$

We observe that, at the resonant frequency f_r, the impedance $\vec{Z}$ has the phase angle $\theta = 0°$. The circuit is a purely resistive circuit, so the voltage and current are in phase, and both have the phase angle $\theta = 0°$. Therefore, we can write the current as the complex number in polar form

$$\vec{I} = 50\underline{/0°}\text{ mA.} \qquad \blacktriangle$$

Fractions decrease when their denominators increase. Since the magnitude of the current is given by the fraction

$$I = \frac{E}{Z},$$

the magnitude of the current I in the series RLC circuit decreases as the magnitude of the impedance Z increases; that is, the magnitude of the current I varies inversely as the magnitude of the impedance Z.

At the series resonant frequency, the magnitude of the impedance is $Z = R$, which is the minimum value of Z. When we change to a frequency smaller than or larger than the resonant frequency, Z increases. Therefore, since I varies inversely as Z, when we change to a frequency smaller than or larger than the resonant frequency, I decreases. Thus, I is at its maximum value at the series resonant frequency and decreases as we move away from the resonant frequency. (The effect of parallel resonant frequency on a parallel RLC circuit is different from the effect of series resonant frequency.)

EXAMPLE 20.18

For the circuit in Examples 20.16 and 20.17, with $L = 200$ mH, $C = 500$ pF, $R = 100\ \Omega$, and $E = 5$ V, find the current at $f = 16$ kHz.

SOLUTION

First, we find the reactances X_L and X_C. We find X_L at $f = 16$ kHz by using

$$X_L = 2\pi f L$$

$$= 2\pi(16\text{ kHz})(200\text{ mH})$$

$$= 20.1\text{ k}\Omega.$$

We find X_C at $f = 16$ kHz by using

$$X_C = \frac{1}{2\pi f C}$$

$$= \frac{1}{2\pi(16\text{ kHz})(500\text{ pF})}$$

$$= 19.9\text{ k}\Omega.$$

Then, combining X_L and X_C by adding their imaginary number forms, we have

$$X_L + X_C = j20.1\text{ k}\Omega + (-j19.9\text{ k}\Omega)$$

$$= j0.2\text{ k}\Omega.$$

By writing R in kilohms as

$$R = 0.1\text{ k}\Omega,$$

we can represent the impedance by the complex number

$$\vec{Z} = 0.1 + j0.2 \text{ k}\Omega,$$

or, in polar form,

$$\vec{Z} = 0.224\underline{/63.4°} \text{ k}\Omega.$$

Then, we find $\vec{I}$ by using

$$\vec{I} = \frac{\vec{E}}{\vec{Z}}$$

$$= \frac{5\underline{/0°} \text{ V}}{0.224\underline{/63.4°} \text{ k}\Omega}.$$

We divide the absolute values and subtract the arguments to find

$$\vec{I} = 22.3\underline{/-63.4°} \text{ mA}. \qquad \blacktriangle$$

The frequency $f = 16$ kHz in Example 20.18 is only 0.1 kHz away from the resonant frequency $f_r = 15.9$ kHz that we found in Example 20.16. However, the magnitude of the current $I = 22.3$ mA is less than half of the 50 mA at the resonant frequency, which we found in Example 20.17. In Exercises 9 and 10, you will find that the current gets significantly smaller as we move further away from the resonant frequency. Moreover, E and I are out of phase by $\theta = -63.4°$, and get further out of phase as we move away from the resonant frequency.

EXERCISE

20.4

For the *RLC* circuit, find the inductive reactance X_L, the capacitive reactance X_C, and the impedance $\vec{Z}$:

1. $R = 1$ kΩ, $L = 8.8$ H, $C = 0.8$ μF, and $f = 60$ Hz
2. $R = 12$ kΩ, $L = 44.4$ mH, $C = 57$ pF, and $f = 100$ kHz

Find the resonant frequency, and then find X_L and X_C:

3. $L = 550$ mH and $C = 120$ pF
4. $L = 30$ mH and $C = 0.06$ μF

For the *RLC* circuit, find the current $\vec{I}$ at the resonant frequency:

5. $L = 550$ mH, $C = 120$ pF, $R = 500$ Ω, and $E = 2$ V
   (see Exercise 3)
6. $L = 30$ mH, $C = 0.06$ μF, $R = 12.5$ Ω, and $E = 10$ V
   (see Exercise 4)
7. $L = 150$ mH, $C = 0.075$ μF, $R = 30$ Ω, and $E = 10$ V
8. $L = 3$ mH, $C = 330$ pF, $R = 18$ Ω, and $E = 6$ V

Find the current $\vec{I}$ at the given frequencies:

9. $L = 200$ mH, $C = 500$ pF, $R = 100$ Ω, and $E = 5$ V at
   a. $f = 15$ kHz   b. $f = 17$ kHz
   (see Example 20.18)
10. $L = 200$ mH, $C = 500$ pF, $R = 100$ Ω, and $E = 5$ V at
    a. $f = 10$ kHz   b. $f = 20$ kHz
    (see Example 20.18)
11. $L = 550$ mH, $C = 120$ pF, $R = 500$ Ω, and $E = 2$ V at
    a. $f = 15$ kHz   b. $f = 25$ kHz
    (see Exercises 3 and 5)
12. $L = 30$ mH, $C = 0.06$ μF, $R = 12.5$ Ω, and $E = 10$ V at
    a. $f = 3.5$ kHz   b. $f = 4$ kHz
    (see Exercises 4 and 6)

AC Parallel Circuits

Suppose a resistor and an inductor are placed in parallel with an AC source to form an **AC parallel RL circuit**. At any time t, the voltage is the same across each branch of the AC parallel circuit and equal to the source voltage E. The current i_R in the branch containing the resistor is in phase with the voltage. The current i_L in the branch containing the inductor, however, *lags* the voltage by 90°.

The peak values of i_R and i_L are I_R and I_L. These values can be represented by a phasor diagram. We draw I_R along the positive x-axis. Then, since I_L lags I_R by 90°, we draw I_L along the negative y-axis:

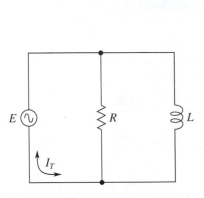

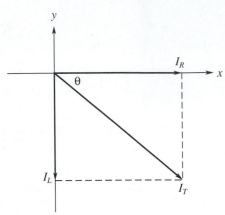

An AC parallel RL circuit.

Phasor diagram showing currents for the parallel RL circuit.

The phasor diagram shows that the total current is given by

$$I_T = \sqrt{I_R^2 + I_L^2},$$

and the phase angle θ between I_T and I_R is given by

$$\theta = -\tan^{-1}\left(\frac{I_L}{I_R}\right),$$

where θ is negative so we take the negative of the inverse tangent.

EXAMPLE 20.19

In a parallel RL circuit, $R = 12\ \Omega$, $X_L = 8\ \Omega$, and $E = 24$ V. Find I_R, I_L, $\vec{I}_T$, and the total impedance $\vec{Z}_T$ as a complex number in polar form and in rectangular form.

SOLUTION

The voltage is the same across each branch of the parallel circuit, and is the same as the source voltage E. Therefore, applying Ohm's law to each branch, we have

$$I_R = \frac{E}{R}$$

$$= \frac{24\text{ V}}{12\ \Omega}$$

$$= 2\text{ A},$$

and

$$I_L = \frac{E}{X_L}$$

$$= \frac{24\text{ V}}{8\ \Omega}$$

$$= 3\text{ A}.$$

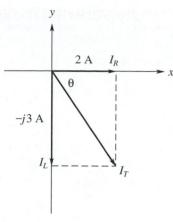

Phasor diagram for Example 20.19.

We draw a phasor diagram showing I_R, I_L, I_T, and θ, remembering that, for the parallel RL circuit, I_L is on the negative y-axis and θ is negative. The total current $\vec{I}_T$ is then represented by the complex number

$$\vec{I}_T = 2 - j3 \text{ A}.$$

We can write this complex number in polar form. The absolute value is given by

$$I_T = \sqrt{I_R^2 + I_L^2}$$
$$= \sqrt{2^2 + 3^2} \text{ A}$$
$$= 3.61 \text{ A}.$$

The argument is given by

$$\theta = -\tan^{-1}\left(\frac{I_L}{I_R}\right)$$
$$= -\tan^{-1}\left(\frac{3}{2}\right)$$
$$= -56.3°.$$

Thus, in polar form the total current $\vec{I}_T$ is

$$\vec{I}_T = 3.61\underline{/-56.3°} \text{ A}.$$

To find the impedance $\vec{Z}_T$, we may apply Ohm's law in the form

$$\vec{Z}_T = \frac{\vec{E}}{\vec{I}_T}.$$

We know that $\vec{E}$ and I_R are in phase, so both are at an angle of 0°. Since $E = 24$ V, we write $\vec{E} = 24\underline{/0°}$ V. Thus, we have

$$\vec{Z}_T = \frac{24\underline{/0°} \text{ V}}{3.61\underline{/-56.3°} \text{ A}}.$$

Then, dividing the absolute values and subtracting the arguments, we have the complex number in polar form

$$\vec{Z}_T = 6.65\underline{/56.3°} \ \Omega.$$

We can also write the complex number in the rectangular form

$$\vec{Z}_T = 3.69 + j5.53 \ \Omega.$$

We observe that this complex number also represents the impedance $\vec{Z}$ of a series RL circuit with $R = 3.69 \ \Omega$ and $X_L = 5.53 \ \Omega$. Thus, the given parallel RL circuit is equivalent to the series RL circuit with that R and X_L.   ▲

We can find total impedance by using a formula similar to the parallel resistance formula derived in Section 8.2. For two parallel impedances $\vec{Z}_1$ and $\vec{Z}_2$, and the total impedance $\vec{Z}_T$ is given by

$$\frac{1}{\vec{Z}_T} = \frac{1}{\vec{Z}_1} + \frac{1}{\vec{Z}_2}.$$

Furthermore, we may follow steps similar to those in Section 9.3 to derive the alternate form

$$\vec{Z}_T = \frac{\vec{Z}_1\vec{Z}_2}{\vec{Z}_1 + \vec{Z}_2}.$$

If a value for E is not given, or if we prefer, we can find the total impedance of the RL circuit in Example 20.19 by using the parallel impedance formula. The first impedance is

$$\vec{Z}_1 = R.$$

The second impedance is

$$\vec{Z}_2 = \frac{\vec{E}}{\vec{I}_L}$$

$$= \frac{E\underline{/0°}}{I_L\underline{/-90°}}$$

$$= X_L\underline{/90°}$$

$$= jX_L.$$

We observe that X_L is on the *positive* y-axis.

EXAMPLE 20.20   For the parallel RL circuit in Example 20.19, $R = 12\ \Omega$ and $X_L = 8\ \Omega$. Find the total impedance $\vec{Z}_T$ as a complex number in polar form.

SOLUTION   We represent the impedances by

$$\vec{Z}_1 = R$$

$$= 12\ \Omega,$$

and

$$\vec{Z}_2 = jX_L$$

$$= j8\ \Omega.$$

Therefore, we write

$$\vec{Z}_T = \frac{\vec{Z}_1\vec{Z}_2}{\vec{Z}_1 + \vec{Z}_2}$$

$$= \frac{(12\ \Omega)(j8\ \Omega)}{(12\ \Omega) + (j8\ \Omega)}$$

$$= \frac{j96}{12 + j8}\ \Omega.$$

Now, we write all the complex numbers in polar form, and divide to obtain

$$\vec{Z}_T = \frac{96\underline{/90°}}{14.4\underline{/33.7°}}\ \Omega$$

$$= 6.67\underline{/56.3°}\ \Omega,$$

which agrees with our result in Example 20.19.   ▲

PARALLEL RC CIRCUITS

Now, suppose a resistor and a capacitor are placed in parallel in an AC circuit to form a **parallel RC circuit**. The current i_R in the branch containing the resistor is in phase with the voltage. The current i_C in the branch containing the capacitor, however, *leads* the voltage by 90°.

The peak values of i_R and i_C are I_R and I_C. These values can be represented by a phasor diagram. We draw I_R along the positive x-axis. Then, since I_C leads I_R by 90°, we draw I_C along the positive y-axis:

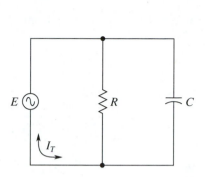

An AC parallel RC circuit.

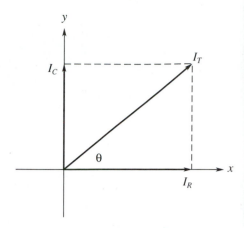

Phasor diagram showing currents for the parallel RC circuit.

The phasor diagram shows that the total current is given by

$$I_T = \sqrt{I_R^2 + I_C^2},$$

and the phase angle θ between I_T and I_R is given by

$$\theta = \tan^{-1}\left(\frac{I_C}{I_R}\right).$$

EXAMPLE 20.21 ▶ In a parallel RC circuit, $R = 12\ \Omega$, $X_C = 6\ \Omega$, and $E = 24$ V. Find I_R, I_C, $\vec{I}_T$, and the total impedance $\vec{Z}_T$ as a complex number in polar form and in rectangular form.

SOLUTION Applying Ohm's law to each branch, we have

$$I_R = 2\ \text{A},$$

as in Example 20.19, and

$$I_C = \frac{E}{X_C}$$

$$= \frac{24\ \text{V}}{6\ \Omega}$$

$$= 4\ \text{A}.$$

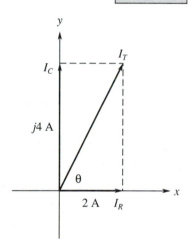

Phasor diagram for Example 20.21.

We draw a phasor diagram showing I_R, I_C, I_T, and θ, remembering that, for the parallel RC circuit, I_C is on the positive y-axis. The total current $\vec{I}_T$ is then represented by the complex number

$$\vec{I}_T = 2 + j4\ \text{A}.$$

We can write this complex number in polar form. The absolute value is given by

$$\vec{I}_T = \sqrt{I_R^2 + I_C^2}$$

$$= \sqrt{2^2 + 4^2}$$

$$= 4.47\ \text{A}.$$

The phase angle θ is given by

$$\theta = \tan^{-1}\left(\frac{I_C}{I_R}\right)$$

$$= \tan^{-1}\left(\frac{4}{2}\right)$$

$$= 63.4°.$$

Thus, in polar form the total current $\vec{I}_T$ is

$$\vec{I}_T = 4.47\underline{/63.4°} \text{ A.}$$

To find the total impedance $\vec{Z}_T$, we apply Ohm's law in the form

$$\vec{Z}_T = \frac{\vec{E}}{\vec{I}_T},$$

where $\vec{E}$ is at an angle of 0°. Therefore, we write

$$\vec{Z}_T = \frac{24\underline{/0°} \text{ V}}{4.47\underline{/63.4°} \text{ A}}$$

$$= 5.37\underline{/-63.4°} \text{ } \Omega.$$

We can write the complex number in the rectangular form

$$\vec{Z}_T = 2.4 - j4.8 \text{ } \Omega.$$

We observe that this complex number also represents the impedance $\vec{Z}$ of a series RC circuit with $R = 2.4 \text{ } \Omega$ and $X_C = 4.8 \text{ } \Omega$. Thus, the given parallel RC circuit is equivalent to the series RC circuit with that R and X_C. ▲

We can also find the total impedance of the RC circuit in Example 20.21 by using the parallel impedance formula

$$\vec{Z}_T = \frac{\vec{Z}_1\vec{Z}_2}{\vec{Z}_1 + \vec{Z}_2}.$$

As before, the first impedance is

$$\vec{Z}_1 = R.$$

The second impedance is

$$\vec{Z}_2 = \frac{\vec{E}}{\vec{I}_C}$$

$$= \frac{E\underline{/0°}}{I_C\underline{/90°}}$$

$$= X_C\underline{/-90°}$$

$$= -jX_C.$$

We observe that X_C is on the *negative* y-axis.

EXAMPLE 20.22 For the parallel RC circuit in Example 20.21, $R = 12 \text{ } \Omega$ and $X_C = 6 \text{ } \Omega$. Find the total impedance $\vec{Z}_T$ as a complex number in polar form.

SOLUTION We represent the impedances by

$$\vec{Z}_1 = R$$

$$= 12 \text{ } \Omega,$$

and

$$\vec{Z}_2 = -jX_C$$

$$= -j6 \text{ } \Omega.$$

Therefore, we write

$$\vec{Z}_T = \frac{\vec{Z}_1\vec{Z}_2}{\vec{Z}_1 + \vec{Z}_2}$$

$$= \frac{(12\ \Omega)(-j6\ \Omega)}{(12\ \Omega) + (-j6\ \Omega)}$$

$$= \frac{-j72}{12 - j6}\ \Omega.$$

Now, we write all the complex numbers in polar form, and divide to obtain

$$\vec{Z}_T = \frac{72\underline{/-90°}}{13.4\underline{/-26.6°}}\ \Omega$$

$$= 5.37\underline{/-63.4°}\ \Omega$$

which agrees with our previous result. ▲

PARALLEL RLC CIRCUITS

A **parallel RLC circuit** contains a resistor, an inductor, and a capacitor, each in parallel with the others:

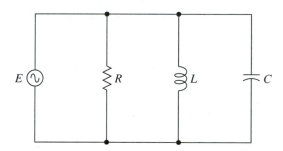

By using complex numbers, we can reduce parallel *RLC* circuits to either parallel *RL* or parallel *RC* circuits and then to either series *RL* or series *RC* circuits.

EXAMPLE 20.23 ▶ In a parallel *RLC* circuit, $R = 12\ \Omega$, $X_L = 8\ \Omega$, $X_C = 6\ \Omega$, and $E = 24$ V. Find I_R, I_L, I_C, $\vec{I}_T$, and the total impedance $\vec{Z}_T$ as a complex number in polar form and in rectangular form.

SOLUTION Applying Ohm's law to each branch, we have

$$I_R = 2\ \text{A},$$

$$I_L = 3\ \text{A},$$

and

$$I_C = 4\ \text{A},$$

as in Examples 20.19 and 20.21. Using imaginary numbers, and remembering that I_L is on the negative *y*-axis while I_C is on the positive *y*-axis, we represent I_L as $-j3$ A and I_C as $j4$ A. The phasor diagram is shown at the left on page 474. We can combine I_L and I_C by adding the imaginary numbers

$$I_L + I_C = -j3\ \text{A} + j4\ \text{A}$$

$$= j1\ \text{A}.$$

Thus, I_L and I_C combined act like a current of 1 A in the branch containing the capacitor. The circuit is equivalent to the parallel *RC* circuit with the phasor diagram shown at the right on page 474:

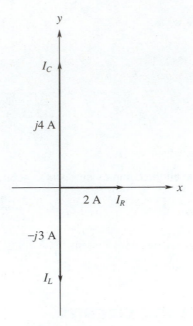

Phasor diagram for Example 20.23.

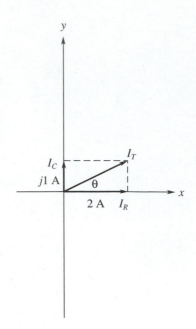

Phasor diagram for an equivalent
parallel *RC* circuit.

The total current $\vec{I}_T$ is then represented by the complex number

$$\vec{I}_T = 2 + j1 \text{ A.}$$

We can also write the complex number in the polar form

$$\vec{I}_T = 2.24\underline{/26.6°} \text{ A.}$$

To find the total impedance $\vec{Z}_T$, we apply Ohm's law in the form

$$\vec{Z}_T = \frac{\vec{E}}{\vec{I}_T},$$

where $\vec{E}$ is at an angle 0°. Therefore, we write

$$\vec{Z}_T = \frac{24\underline{/0°} \text{ V}}{2.24\underline{/26.6°} \text{ A}}$$

$$= 10.7\underline{/-26.6°} \text{ Ω.}$$

We can return the complex number to the rectangular form

$$\vec{Z}_T = 9.57 - j4.79 \text{ Ω.}$$

Thus, the given parallel *RLC* circuit is equivalent to the series *RC* circuit with
$R = 9.57$ Ω and $X_C = 4.79$ Ω. ▲

The form of the parallel impedance formula

$$\vec{Z}_T = \frac{\vec{Z}_1\vec{Z}_2}{\vec{Z}_1 + \vec{Z}_2}$$

can only be used for two impedances. To find $\vec{Z}_T$ for Example 20.23 by using this formula,
we must use the formula twice.

EXAMPLE 20.24 ▶

For the parallel *RC* circuit in Example 20.23, $R = 12$ Ω, $X_L = 8$ Ω, and $X_C = 6$ Ω.
Find the total impedance $\vec{Z}_T$ as a complex number in polar form.

There are three impedances. The first impedance is

$$\vec{Z}_1 = R$$
$$= 12\ \Omega,$$

the second impedance is

$$\vec{Z}_2 = jX_L$$
$$= j8\ \Omega,$$

and the third impedance is

$$\vec{Z}_3 = -jX_C$$
$$= -j6\ \Omega.$$

We used the parallel impedance formula in Example 20.20 to find the total impedance for $\vec{Z}_1$ and $\vec{Z}_2$. Thus, we have the partial result

$$\vec{Z}_X = 6.67\underline{/56.3°}\ \Omega,$$

or in rectangular form,

$$\vec{Z}_X = 3.7 + j5.55\ \Omega.$$

Now, we use the parallel impedance formula a second time, with $\vec{Z}_X$ and $\vec{Z}_3$, to obtain

$$\vec{Z}_T = \frac{\vec{Z}_X\vec{Z}_3}{\vec{Z}_X + \vec{Z}_3}$$

$$= \frac{(3.7 + j5.55)(-j6)}{(3.7 + j5.55) + (-j6)}\ \Omega$$

$$= \frac{(3.7 + j5.55)(-j6)}{3.7 - j0.45}\ \Omega.$$

We can multiply and divide by writing all the complex numbers in polar form:

$$\vec{Z}_T = \frac{(6.67\underline{/56.3°})(6\underline{/-90°})}{3.73\underline{/-6.93°}}\ \Omega$$

$$= \frac{40.02\underline{/-33.7°}}{3.73\underline{/-6.93°}}\ \Omega$$

$$= 10.7\underline{/-26.8°}\ \Omega,$$

which agrees with our previous result. ▲

SERIES-PARALLEL CIRCUITS

We conclude by finding the total impedance for some AC series-parallel circuits.

EXAMPLE 20.25 ▶ Find the total impedance $\vec{Z}_T$ for this circuit:

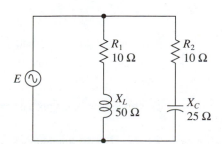

SOLUTION Since R_1 and X_L are in series, we know from Section 20.3 that the impedance of the first branch can be written as the complex number

$$\vec{Z}_1 = 10 + j50 \ \Omega.$$

Since R_2 and X_C are in series, the impedance of the second branch can be written as the complex number

$$\vec{Z}_2 = 10 - j25 \ \Omega.$$

Therefore, by using the parallel impedance formula, we have

$$\vec{Z}_T = \frac{\vec{Z}_1 \vec{Z}_2}{\vec{Z}_1 + \vec{Z}_2}$$

$$= \frac{(10 + j50)(10 - j25)}{(10 + j50) + (10 - j25)} \ \Omega$$

$$= \frac{(10 + j50)(10 - j25)}{20 + j25} \ \Omega.$$

Now, we write all the complex numbers in polar form to obtain

$$\vec{Z}_T = \frac{(51\underline{/78.7°})(26.9\underline{/-68.2°})}{32\underline{/51.3°}} \ \Omega$$

$$= \frac{1372\underline{/10.5°}}{32\underline{/51.3°}} \ \Omega$$

$$= 42.9\underline{/-40.8°} \ \Omega,$$

or in rectangular form,

$$\vec{Z}_T = 32.5 - j28. \qquad \blacktriangle$$

EXAMPLE 20.26   Find the total impedance $\vec{Z}_T$ for this circuit:

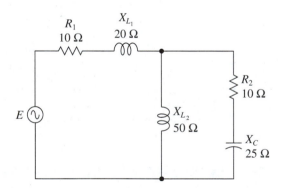

SOLUTION We have three impedances. The impedance $\vec{Z}_1$, consisting of R_1 and X_{L_1}, can be written as the complex number

$$\vec{Z}_1 = 10 + j20 \ \Omega.$$

The impedance $\vec{Z}_2$, consisting of X_{L_2}, can be written as the complex number

$$\vec{Z}_2 = j50 \ \Omega.$$

The impedance $\vec{Z}_3$, consisting of R_2 and X_C, can be written as the complex number

$$\vec{Z}_3 = 10 - j25 \ \Omega.$$

First, we can use the parallel resistance formula to find

$$\vec{Z}_X = \frac{\vec{Z}_2 \vec{Z}_3}{\vec{Z}_2 + \vec{Z}_3}$$

$$= \frac{(j50)(10 - j25)}{(j50) + (10 - j25)} \ \Omega$$

$$= \frac{(j50)(10 - j25)}{10 + j25} \ \Omega.$$

Now, we write all the complex numbers in polar form to obtain

$$\vec{Z}_X = \frac{(50\underline{/90°})(26.9\underline{/-68.2°})}{26.9\underline{/68.2°}} \ \Omega$$

$$= \frac{1345\underline{/21.8°}}{26.9\underline{/68.2°}} \ \Omega$$

$$= 50\underline{/-46.4°} \ \Omega,$$

or in rectangular form,

$$\vec{Z}_X = 34.5 - j36.2 \ \Omega.$$

Then, the impedance $\vec{Z}_1$ is in series with $\vec{Z}_X$, so we add to obtain

$$\vec{Z}_T = \vec{Z}_1 + \vec{Z}_X$$

$$= (10 + j20) + (34.5 - j36.2) \ \Omega$$

$$= 44.5 - j16.2 \ \Omega.$$

▲

EXERCISE

20.5

For the parallel RL circuit, find I_R, I_L, $\vec{I}_T$, and $\vec{Z}_T$:

1. $E = 100$ V, $R = 20 \ \Omega$, and $X_L = 25 \ \Omega$
2. $E = 10$ V, $R = 1.5$ kΩ, and $X_L = 7.5$ kΩ

3. and 4.   Find $\vec{Z}_T$ for Exercises 1 and 2 by using the parallel impedance formula.

For the parallel RC circuit, find I_R, I_C, $\vec{I}_T$, and $\vec{Z}_T$:

5. $E = 20$ V, $R = 10 \ \Omega$, and $X_C = 12.5 \ \Omega$
6. $E = 120$ V, $R = 4$ kΩ, and $X_C = 3$ kΩ

7. and 8.   Find $\vec{Z}_T$ for Exercises 5 and 6 by using the parallel impedance formula.

For the parallel RLC circuit, find I_R, I_L, I_C, $\vec{I}_T$, and $\vec{Z}_T$:

9. $E = 120$ V, $R = 60 \ \Omega$, $X_L = 50 \ \Omega$, and $X_C = 10 \ \Omega$
10. $E = 10$ V, $R = 22 \ \Omega$, $X_L = 44 \ \Omega$, and $X_C = 20 \ \Omega$
11. $E = 100$ V, $R = 25 \ \Omega$, $X_L = 10 \ \Omega$, and $X_C = 40 \ \Omega$
12. $E = 750$ mV, $R = 15 \ \Omega$, $X_L = 25 \ \Omega$, and $X_C = 100 \ \Omega$

13.–16.   Find $\vec{Z}_T$ for Exercises 9–12 by using the parallel impedance formula twice.

17.   Find Z_T for this circuit:

18.   Repeat Exercise 17 for $R_1 = 10 \ \Omega$, $R_2 = 20 \ \Omega$, $X_L = 40 \ \Omega$, and $X_C = 50 \ \Omega$.

19.–22. Find $\vec{Z}_T$ for the circuit shown:

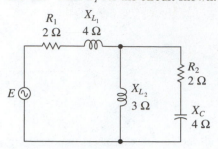

Circuit for Exercise 19.

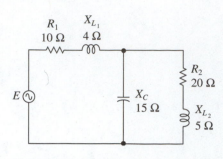

Circuit for Exercise 20.

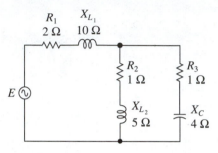

Circuit for Exercise 21.

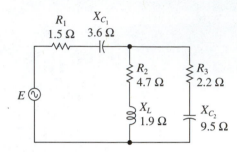

Circuit for Exercise 22.

23. For the circuit shown, find $\vec{Z}_T$ at frequency $f = 10$ kHz.

24. For the circuit shown, find $\vec{Z}_T$ at frequency $f = 50$ Hz.

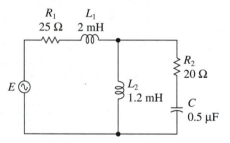

Circuit for Exercise 23.

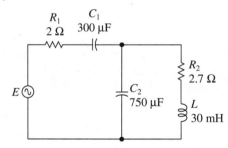

Circuit for Exercise 24.

Self-Test

1. An AC series *RL* circuit has $E = 9$ V and $\theta = 34°$. Find V_R and V_L.

    1. _____

2. An AC series *RC* circuit has frequency $f = 60$ Hz, $C = 0.06$ μF, and $R = 33$ kΩ. Find X_C, Z, and θ.

    2. _____

3. An AC series *RLC* circuit has $R = 39$ kΩ, $X_L = 55$ kΩ, and $X_C = 25$ kΩ. Find $\vec{Z}$ in rectangular form and in polar form.

    3. _____

4. Find the resonant frequency for the AC series *RLC* circuit with $L = 6$ mH and $C = 80$ pF.

    4. _____

5. An AC parallel *RC* circuit has $R = 40$ Ω, $X_C = 75$ Ω, and $E = 300$ mV. Find $\vec{Z}_T$.

    5. _____

UNIT

21

Cumulative Review

Introduction

In Units 1 though 15, you learned about real numbers, and about algebraic expressions, equations, and formulas. You applied Kirchhoff's and Ohm's laws to solving DC circuit problems, and you learned how to solve and apply rational equations. You learned to draw and interpret graphs of basic types of equations in two variables. Then, you used both graphical and algebraic methods to solve systems of equations in two variables and in three variables. You learned how to solve general quadratic equations, and how to draw and interpret graphs of quadratic functions. Finally, you learned about logarithmic and exponential functions, and how to solve and apply exponential equations.

In Units 17 through 20, you learned about the trigonometric ratios and how to use them to solve right triangles. You used right triangle trigonometry to study vectors and vector arithmetic. You learned about complex numbers, and applied right triangle trigonometry to complex number arithmetic. Then, you learned about sine waves and their properties, and about angular motion. You learned how to apply sine waves and angular motion to alternating current and voltage. Finally, you learned how to use right triangle trigonometry to find AC voltages and currents in AC series RL and RC circuits. You learned how complex numbers may be applied to AC series circuits, and you learned about series resonance. You also learned how to apply right triangle trigonometry and complex numbers to AC parallel circuits, and how to find the total impedance in AC series-parallel circuits.

Now, you should review the material in all of the preceding units.

OBJECTIVE

When you have finished this unit you should be able to fulfill every objective of each of Units 1 through 20.

SECTION
21.1

Review Method

For this unit, you should review the Self-Tests for Units 1 through 20. Do each problem in each Self-Test over again. If you cannot do a problem, or if you have the slightest doubt or difficulty, find the appropriate material to review:

1. You will find an objective number next to the answer to every Self-Test problem. Find the objective number of the problem you are working on (this is not necessarily the problem number). Go to the first page of the unit and reread the objective. For example, if you have difficulty with a Self-Test problem in Unit 2, and the objective number next to the answer is Objective 3, reread Objective 3 in the objectives list at the beginning of Unit 2 to find out what concept you need to review.

2. The material you should study to review the concept is in the section that has the same number as the objective. Reread the material in that section, rework the examples, and redo some of the exercises for the section. For example, if you are reviewing Unit 2, Objective 3, refer to Unit 2, Section 3; that is, Section 2.3. Reread Section 2.3, rework examples in that section, and redo problems in Exercise 2.3.

3. Try the Self-Test for the unit again to find out if there is any other objective you need to review in the unit. For example, if you have been reviewing Section 2.3 and Exercise 2.3, try the Self-Test for Unit 2 again.

Review each unit from Unit 1 through Unit 20 in this way. For some units, you may find that you can still do the Self-Test easily. For other units, you might need to review one or more of the objectives in more detail. As a final check, and only after you have reviewed every preceding unit, try the Self-Test for this Cumulative Review Unit 21.

□ Self-Test □

1. Solve $\dfrac{3}{\sqrt{x}} - \dfrac{1}{5} = \dfrac{1}{10}$ and check the solution.

1. _____

2. For the formula $Q = \dfrac{1}{R}\sqrt{\dfrac{L}{C}}$, find C if $Q = 200$, $R = 2\ \mathrm{k\Omega}$, and $L = 4\ \mathrm{H}$.

2. _____

3. For the formula $A = P(1 + R)^N$, find N if $A = \$4000$, $P = \$2000$, and $R = 6\%$.

3. _____

4. Find the slope and y-intercept of the line given by $2x - 3y - 6 = 0$.

4. _____

5. Draw the graphs of the equations and solve the system:

   $y = x + 2$

   $y = x^2$

5. _____

6. Find V_1, V_2, and V_3 for this circuit:

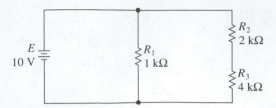

E 10 V, R_1 1 kΩ, R_2 2 kΩ, R_3 4 kΩ

6. _____

7. Find the resultant vector of the horizontal component $x = -22.5$ and vertical component $y = 15.8$.

7. _____

8. Divide the complex numbers $\dfrac{3 - j}{1 + j}$.

8. _____

9. A current and voltage have frequency $f = 100$ Hz. Their peak values are $I_M = 15$ mA and $V_M = 20$ V. The current leads the voltage by 45°. Find their instantaneous values at $t = 5$ ms.

9. _____

10. An AC series RLC circuit has $R = 33$ kΩ, $X_L = 97$ kΩ, and $X_C = 54$ kΩ. Find Z in rectangular form and in polar form.

10. _____

UNIT 22

Analytic Trigonometry

Introduction

In the last several units, you learned topics in trigonometry needed for electronics, and some of their applications to electronics. Other topics in trigonometry are needed for further work in mathematics and, in particular, for the study of calculus. In Unit 19, you interpreted the sine ratio as a function. In this unit, you will learn how to interpret the other trigonometric ratios as functions. You will learn about three more trigonometric functions, and about inverses of the trigonometric functions. Then, you will learn how to simplify trigonometric expressions by using identities. You will learn two formulas called the laws of sines and cosines. Finally, you will simplify trigonometric expressions by using identities derived from these laws. Several proofs of laws and identities are given at the end of the unit.

OBJECTIVES

When you have finished this unit you should be able to:

1. Draw graphs of the trigonometric functions.
2. Find principal values of the inverse trigonometric functions.
3. Use identities to simplify trigonometric expressions.
4. Use the laws of sines and cosines to solve oblique triangles and to find sums of vectors.
5. Use the sum and difference identities and the double-angle identities to simplify trigonometric expressions.

SECTION 22.1

The Trigonometric Functions

In Unit 17, we defined three trigonometric ratios, the sine, cosine, and tangent, in terms of the sides of a triangle. In Unit 19, we also defined the sine *function* in terms of the vertical component of a radius vector in the unit circle. We drew graphs of sine waves by using the x-axis as a θ-axis. In analytic trigonometry, it is common to define the sine function as a relation in x and y by the equation

$$y = \sin x,$$

where x is an angle in radians. One cycle of the graph of the sine function is shown on page 486. We saw in Unit 19 that the cycles of the sine function continue indefinitely to both the right and the left.

For any function in x and y, we define the **domain** and **range** of the function.

Domain of a Function: The **domain** is the set of all x-values for which a y-value is defined.

Since the cycles of the sine function continue indefinitely to both the right and the left, the domain of the sine function is all real numbers.

> **Range of a Function:**   The **range** is the set of all y-values of the function.

The y-values of the sine function are always between 1 and -1. Therefore, the range of the sine function is

$$-1 \leq y \leq 1.$$

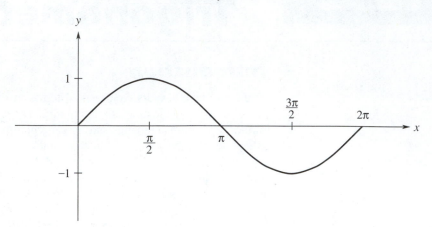

One cycle of the graph of $y = \sin x$.

THE COSINE FUNCTION

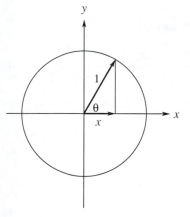

Horizontal component of a radius vector in the unit circle.

We can define the **cosine function** in a similar way. We concentrate on the *horizontal* component of a radius vector in the unit circle. The horizontal component is given by

$$\cos \theta = \frac{x}{1}$$

$$\cos \theta = x,$$

or

$$x = \cos \theta.$$

Values of x for some angles θ in radians between 0 and $\frac{\pi}{2}$ are shown in this chart:

θ	0	$\frac{\pi}{6}$	$\frac{\pi}{4}$	$\frac{\pi}{3}$	$\frac{\pi}{2}$
x	1	0.866	0.707	0.5	0

You should use your calculator to check these values. Observe that the values are the same as those for the sine function, but in reverse order.

To draw the graph of the cosine function, we can follow the length of the x-component of the radius vector. For angles from $\theta = 0$ to $\theta = \frac{\pi}{2}$, this length decreases from 1 to 0 as in the chart. For $\theta = \frac{\pi}{2}$ to $\theta = \pi$, the length decreases through negatives of the same x-values from 0 to -1. For $\theta = \pi$ to $\theta = \frac{3\pi}{2}$, the length increases through negatives of the x-values from -1 to 0. Finally, for $\theta = \frac{3\pi}{2}$ to $\theta = 2\pi$, the length increases through the x-values from 0 to 1.

When we draw the graph of the cosine function, the length of the x-component is drawn on the y-axis. Therefore, the cosine function as a relation in x and y is

$$y = \cos x.$$

You should be careful not to confuse x and y in this relation with the x and y components of the radius vector. One cycle of the graph of the cosine function is shown at the top of page 487.

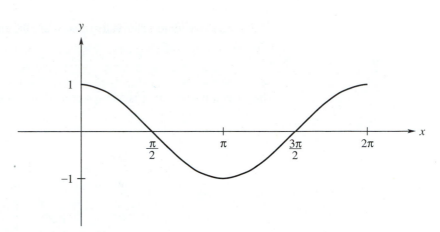

Like the sine function, the cosine function repeats in cycles indefinitely to both the right and the left. Therefore, the domain of the cosine function is all real numbers. Also, the y-values of the cosine function are always between 1 and -1. Therefore, the range of the sine function is

$$-1 \le y \le 1.$$

EXAMPLE 22.1 ▶ Draw two cycles of the graph of $y = \cos x$, starting from $x = -\dfrac{\pi}{2}$.

SOLUTION We draw one-quarter of a cycle to the left of the original cycle, and three-quarters of a cycle to the right:

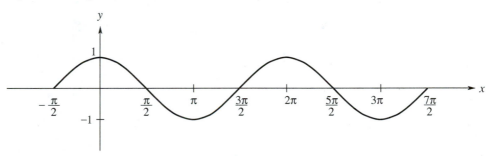

We observe that the cosine function is the same as a sine function with a leading phase angle $x = \frac{\pi}{2}$; that is,

$$\cos x = \sin\left(x + \frac{\pi}{2}\right).$$

▲

THE TANGENT FUNCTION

To define the tangent function, we use a vector on the unit circle that has the horizontal component $x = 1$. The y-component, representing the tangent, is part of a tangent line to the circle. In this diagram, we have

$$\tan \theta = \frac{y}{1}$$

$$\tan \theta = y,$$

or

$$y = \tan \theta.$$

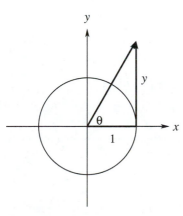

A vector with horizontal component $x = 1$.

As θ increases from $\theta = 0$ toward $\theta = \frac{\pi}{2}$, the height of the y-component starts from 0 and increases indefinitely. When $\theta = \frac{\pi}{2}$, there is no triangle that includes the y-component, and $\tan \frac{\pi}{2}$ is undefined.

We can calculate other values and draw the first part of the graph. In particular, we note that

$$\tan \frac{\pi}{4} = 1.$$

The dashed line at $x = \frac{\pi}{2}$ is an asymptote of the graph:

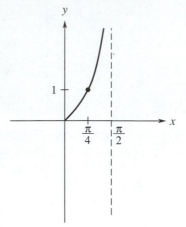

Now, we consider negative angles from $\theta = 0$ to $\theta = -\frac{\pi}{2}$. We use the negative half of the tangent line. As θ decreases from $\theta = 0$ toward $\theta = -\frac{\pi}{2}$, the y-component starts from 0 and decreases until $\tan\left(-\frac{\pi}{2}\right)$ is undefined. We draw the second part of the graph to correspond with the first part. The dashed line at $x = -\frac{\pi}{2}$ is also an asymptote of the graph:

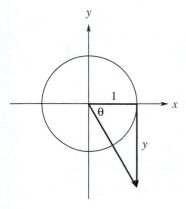

A vector with horizontal component $x = 1$ and negative y-component.

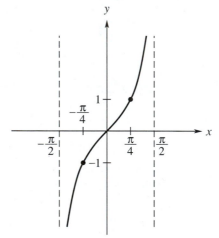

The graph of the tangent function repeats this pattern indefinitely to both the right and the left:

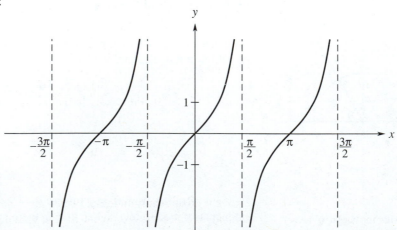

The graph approaches the asymptotes but the tangent function is not defined at the asymptotes. Therefore, the domain of the tangent function is all real numbers x such that

$$x \neq \pm\frac{\pi}{2},\ \pm\frac{3\pi}{2},\ \pm\frac{5\pi}{2},\ \ldots$$

The graph continues indefinitely upward and downward between each pair of asymptotes. Therefore, the range of the tangent function is all real numbers. Finally, we observe that the graph repeats between each pair of asymptotes. Therefore, the period of the tangent function is π.

EXAMPLE 22.2 ▶

Draw the graph of $y = \tan x$ from $x = 0$ to $x = 2\pi$.

SOLUTION

We draw one-half of a period of the original graph, and one and one-half periods to the right:

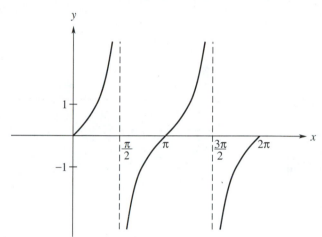

COTANGENT, SECANT, AND COSECANT

Three more trigonometric functions are defined, which are the reciprocals of the tangent, cosine, and sine.

The **cotangent** is the reciprocal of the tangent:

$$\cot x = \frac{1}{\tan x}\ \text{and}\ \tan x = \frac{1}{\cot x}.$$

The **secant** is the reciprocal of the cosine:

$$\sec x = \frac{1}{\cos x}\ \text{and}\ \cos x = \frac{1}{\sec x}.$$

The **cosecant** is the reciprocal of the sine:

$$\csc x = \frac{1}{\sin x}\ \text{and}\ \sin x = \frac{1}{\csc x}.$$

(The abbreviation of *cosecant* is "csc" to distinguish it from *cosine*.)

To draw the graph of the cotangent function, we can take the reciprocals of several values of the tangent function. In particular, when $\tan x = 0$,

$$\cot x = \frac{1}{\tan x}$$

$$= \frac{1}{0}.$$

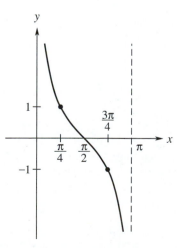

One period of the graph of the cotangent function.

Thus, the cotangent is undefined when $\tan x = 0$. Similarly, $\cot x = 0$ when the tangent is undefined. Since the reciprocal of 1 is 1, and the reciprocal of -1 is -1, the points with those y-values are unchanged.

The domain of the cotangent function is all real numbers x such that

$$x \neq 0, \ \pm\pi, \ \pm 2\pi, \ \dots$$

Like the tangent function, the range of the cotangent function is the set of all real numbers, and the period is π.

EXAMPLE 22.3 ► Draw the graph of $y = \cot x$ from $x = -\dfrac{\pi}{2}$ to $x = \dfrac{3\pi}{2}$.

SOLUTION   We draw one-half of a period to the left of the original graph, and one-half of a period to the right:

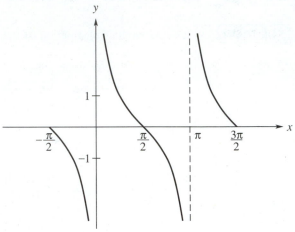

▲

To draw the graph of the secant function, we can take the reciprocals of several values of the cosine function. In particular, the secant is undefined when $\cos x = 0$. Also, $\sec x = 1$ when $\cos x = 1$, and $\sec x = -1$ when $\cos x = -1$. One period of the graph of $y = \sec x$ is shown with $y = \cos x$:

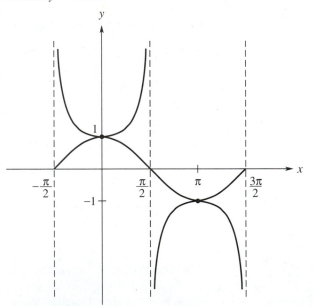

The domain of the secant function is all real numbers x such that

$$x \neq \pm\frac{\pi}{2}, \ \pm\frac{3\pi}{2}, \ \pm\frac{5\pi}{2}, \ \dots$$

The range of the secant function is

$$y \geq 1 \ \text{and} \ y \leq -1.$$

Like the cosine function, the period is 2π.

EXAMPLE 22.4 ▶ On one set of axes, draw the graphs of $y = \cos x$ and $y = \sec x$ from $x = 0$ to $x = 2\pi$.

SOLUTION   We draw the graph of $y = \cos x$ from $x = 0$ to $x = 2\pi$. Then, we can draw the asymptotes and fill in the graph of $y = \sec x$:

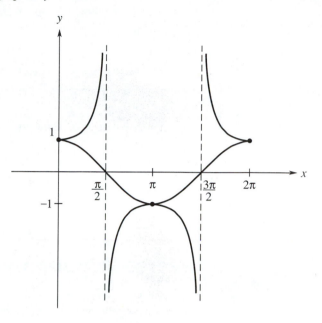

▲

To draw the graph of the cosecant function, we can take the reciprocals of several values of the sine function. In particular, the cosecant is undefined when $\sin x = 0$. One period of the graph of $y = \csc x$ is shown with $y = \sin x$:

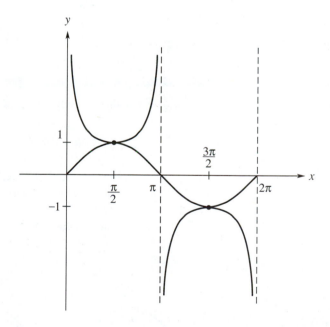

The domain of the cosecant function is all real numbers x such that

$$x \neq 0, \ \pm\pi, \ \pm 2\pi, \ \ldots$$

The range of the cosecant function is

$$y \geq 1 \ \text{and} \ y \leq -1.$$

Like the sine function, the period is 2π.

1. Draw two cycles of the graph of $y = \sin x$, starting from $x = 0$.

2. Draw two cycles of the graph of $y = \sin x$, starting from $x = -4\pi$.

3. Draw two cycles of the graph of $y = \cos x$, starting from $x = 0$.

4. Draw two cycles of the graph of $y = \cos x$, starting from $x = -4\pi$.

5. Draw the graph of $y = \tan x$ from $x = -\frac{\pi}{2}$ to $x = \frac{5\pi}{2}$.

6. Draw the graph of $y = \tan x$ from $x = -\pi$ to $x = \pi$.

7. Draw the graph of $y = \cot x$ from $x = -\pi$ to $x = \pi$.

8. Draw the graph of $y = \cot x$ from $x = -\frac{\pi}{2}$ to $x = \frac{5\pi}{2}$.

9. On one set of axes, draw the graphs of $y = \cos x$ and $y = \sec x$ from $x = -\frac{3\pi}{2}$ to $x = \frac{5\pi}{2}$.

10. On one set of axes, draw the graphs of $y = \cos x$ and $y = \sec x$ from $x = 0$ to $x = 4\pi$.

11. On one set of axes, draw the graphs of $y = \sin x$ and $y = \csc x$ from $x = 0$ to $x = 4\pi$.

12. On one set of axes, draw the graphs of $y = \sin x$ and $y = \csc x$ from $x = -\frac{3\pi}{2}$ to $x = \frac{5\pi}{2}$.

Inverse Trigonometric Functions

In Section 10.4, we defined the inverse of a function. We recall that the inverse of a function is the relation obtained by reversing the coordinates of the ordered pairs of the function. The inverse may be obtained by exchanging the variables of the function. The resulting relation might not be a function.

We can apply this concept to the trigonometric functions. To write the inverse of $y = \sin x$, we exchange x and y to obtain

$$x = \sin y.$$

There is no algebraic method by which this inverse relation can be solved for y. Therefore, we define an equation that is the solution for y.

Inverse Sine Relation: The **inverse sine** relation
$$y = \sin^{-1} x,$$
where $-1 \leq x \leq 1$, is the solution for y of the relation $x = \sin y$.

The equation $y = \sin^{-1} x$ is read "y is the inverse sine of x," or "y is the angle whose sine is x." We recall that $\sin^{-1} x$ *does not mean* $\frac{1}{\sin x}$. The inverse sine is also written $y = \arcsin x$, read "y is the arc sine of x."

To draw the graph of the inverse sine relation, we start with the graph of the sine function. Two cycles of the graph, starting from $x = -2\pi$, are

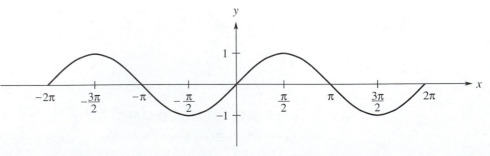

Then, we reverse the ordered pairs of this graph to obtain ordered pairs of the inverse sine relation. For example, the ordered pair

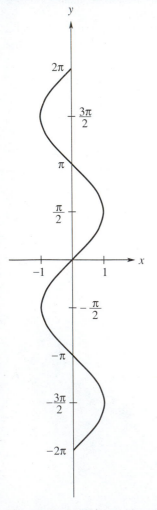

The graph of the inverse sine relation $y = \sin^{-1} x$.

$$\left(\frac{\pi}{2}, 1\right) \text{ becomes } \left(1, \frac{\pi}{2}\right).$$

Two cycles of the graph of $y = \sin^{-1} x$ are shown at the left on page 492.
 We observe that the x-values are between −1 and 1. Thus, the domain of the relation is

$$-1 \leq x \leq 1.$$

The cycles continue indefinitely both upward and downward, so the range of the relation is all real numbers. Finally, we observe that there are many y-values for each x in the domain. Therefore, the relation is *not a function*.
 We can restrict the range of the inverse sine relation so that it is a function. When the y-values are between $-\frac{\pi}{2}$ and $\frac{\pi}{2}$, there is exactly one y for each x. The restricted range is called the **range of principal values**. When we restrict the inverse sine to the range of principal values, we write the equation

$$y = \text{Sin}^{-1} x,$$

with a capital S. This relation is a function with the range

$$-\frac{\pi}{2} \leq y \leq \frac{\pi}{2}.$$

When x is positive, the principal value of the inverse sine function is in the first quadrant. When x is negative, the principal value of the inverse sine function is in the *fourth quadrant*.
 You should be careful not to confuse the quadrant of the angle with the quadrant of the graph. Your calculator gives values in the range of principal values when you press

 followed by .

The graph of the inverse sine function $y = \text{Sin}^{-1} x$, for the range of principal values.

EXAMPLE 22.5 ▶ Find the value of $\text{Sin}^{-1}(-0.5)$ in

a. radians   **b.** degrees

SOLUTIONS Since x is negative, the principal value is in the fourth quadrant.

a. With a calculator in radian mode, we find

$$\text{Sin}^{-1}(-0.5) = -0.524 \text{ rad.}$$

This result is an angle in the fourth quadrant, a little beyond one-half a radian.

b. With the calculator in degree mode, we find

$$\text{Sin}^{-1}(-0.5) = -30°.$$

We can see that these results agree by using the ratio $\frac{180°}{\pi}$:

$$-0.524 = -0.524\left(\frac{180°}{\pi}\right)$$

$$= -30°. \qquad ▲$$

THE INVERSE COSINE

To write the inverse of $y = \cos x$, we exchange x and y to obtain

$$x = \cos y.$$

Then, we define an equation that is the solution for y.

Inverse Cosine Relation: The **inverse cosine** relation

$$y = \cos^{-1} x,$$

where $-1 \leq x \leq 1$, is the solution for y of the relation $x = \cos y$.

To draw the graph of the inverse cosine relation, we start with the graph of the cosine function. Two cycles of the graph, starting from $x = -2\pi$, are

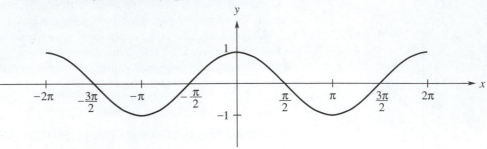

Then, we reverse the ordered pairs of this graph to obtain ordered pairs of the inverse cosine relation. For example, the ordered pair

$$(0, 1) \quad \text{becomes} \quad (1, 0).$$

Two cycles of the graph of $y = \cos^{-1} x$ are shown at the left.

We observe that the x-values are between -1 and 1. Thus, the domain of the relation is

$$-1 \leq x \leq 1.$$

The cycles continue indefinitely both upward and downward, so the range of the relation is all real numbers. Again, there are many y-values for each x in the domain, so the relation is *not a function*.

To restrict the range of the inverse cosine relation, we use the y-values between 0 and π. In this range, there is exactly one y for each x. The function

$$y = \text{Cos}^{-1} x$$

has the range of principal values

$$0 \leq y \leq \pi.$$

When x is positive, the principal value of the inverse cosine function is in the first quadrant. When x is negative, the principal value of the inverse cosine function is in the *second quadrant*. There are no negative angles in the range of principal values for the inverse cosine function:

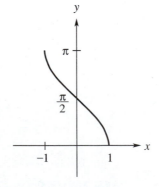

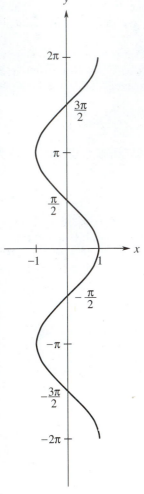

The graph of the inverse cosine relation $y = \cos^{-1} x$.

EXAMPLE 22.6 ▶ Find the value of $\text{Cos}^{-1} (-0.5)$ in

   a. radians   **b.** degrees

SOLUTIONS   Since x is negative, the principal value is in the second quadrant.

a. With a calculator in radian mode, we find

$$\text{Cos}^{-1}(-0.5) = 2.09 \text{ rad.}$$

This result is in the second quadrant, one-third of the way through.

b. With the calculator in degree mode, we find

$$\text{Cos}^{-1}(-0.5) = 120°.$$

You can see that these results agree by using the ratio $\frac{180°}{\pi}$. ▲

THE INVERSE TANGENT

To write the inverse of $y = \tan x$, we exchange x and y to obtain

$$x = \tan y.$$

Then, we define an equation that is the solution for y.

Inverse Tangent Relation: The **inverse tangent** relation

$$y = \tan^{-1} x$$

is the solution for y of the relation $x = \tan y$.

To draw the graph of the inverse tangent relation, we start with the graph of the tangent function. A natural restriction is to one cycle between two asymptotes. Then, we reverse the ordered pairs of this graph to obtain ordered pairs of the inverse tangent relation. We also reverse the asymptotes:

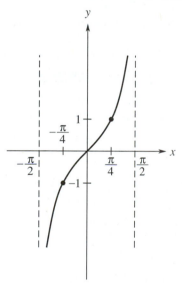

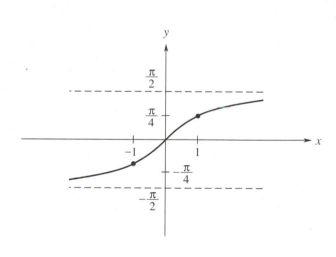

One period of the graph of $y = \tan x$.

The graph of the inverse tangent function $y = \text{Tan}^{-1} x$, for the range of principal values.

There is exactly one y for each x, so the relation is the function

$$y = \text{Tan}^{-1} x,$$

with the range of principal values

$$-\frac{\pi}{2} < y < \frac{\pi}{2}.$$

When x is positive, the principal value of the inverse tangent function is in the first quadrant. When x is negative, the principal value of the inverse tangent function is in the fourth quadrant.

EXAMPLE 22.7 　Find the value of $\text{Tan}^{-1}(-1)$ in

　　a. radians　　　　　　　　　　　　　　　　**b.** degrees

SOLUTIONS　Since x is negative, the principal value is in the fourth quadrant.

　　a. With a calculator in radian mode, we find

$$\text{Tan}^{-1}(-1) = -0.785 \text{ rad.}$$

This result is in the fourth quadrant, one-half of the way through.

　　b. With the calculator in degree mode, we find

$$\text{Tan}^{-1}(-1) = -45°.$$

You can see that these results agree by using the ratio $\frac{180°}{\pi}$.　　　　▲

EXAMPLE 22.8　Find the value of $\text{Tan}^{-1}(-2)$ in degrees.

SOLUTION　We can use a calculator to find

$$\text{Tan}^{-1}(-2) = -63.4°.$$　　　　▲

　　The inverse tangent function is defined for all real values of x. However, recall that the functions

$$y = \text{Sin}^{-1} x \text{ and } y = \text{Cos}^{-1} x$$

are defined only for

$$-1 \le x \le 1.$$

If you attempt to find $\text{Sin}^{-1} 2$ or $\text{Sin}^{-1}(-2)$ by using a calculator, you will get an error indication.

INVERSE COTANGENT, SECANT, AND COSECANT

Inverses of the cotangent, secant, and cosecant functions can be defined by using their relationships to the tangent, cosine, and sine functions. For example, we define

$$y = \csc^{-1} x$$

as the solution of

$$x = \csc y.$$

Then we use the relationship $\csc y = \dfrac{1}{\sin y}$ to write

$$x = \frac{1}{\sin y}.$$

Taking the reciprocal of each side, we have

$$\frac{1}{x} = \sin y$$

or

$$y = \sin^{-1} \frac{1}{x}.$$

Therefore, the inverse cosecant is related to the inverse sine by using the reciprocal of x to write

$$\csc^{-1} x = \sin^{-1} \frac{1}{x}.$$

Similarly, we can show that

$$\sec^{-1} x = \cos^{-1} \frac{1}{x},$$

and

$$\cot^{-1} x = \tan^{-1} \frac{1}{x}.$$

EXAMPLE 22.9 ▶ Find the value of $\text{Csc}^{-1}(-2)$ in degrees.

SOLUTION We relate the inverse cosecant to the inverse sine by using the reciprocal of x:

$$\text{Csc}^{-1}(-2) = \text{Sin}^{-1}\left(\frac{1}{-2}\right)$$

$$= \text{Sin}^{-1}(-0.5)$$

$$= -30°.$$ ▲

Since the inverse sine and cosine are defined only for x between -1 and 1, the inverse cosecant and secant are defined only for $x \geq 1$ and $x \leq -1$. If you attempt to find $\text{Csc}^{-1} 0.5$ or $\text{Csc}^{-1}(-0.5)$ by using a calculator, you will get an error indication. The inverse cotangent is defined for all real numbers.

EXERCISE
22.2

Find the value in

a. radians   **b.** degrees

1. $\text{Sin}^{-1} 0.5$ **2.** $\text{Cos}^{-1} 0.5$

3. $\text{Sin}^{-1}(-0.866)$ **4.** $\text{Sin}^{-1}(-0.8)$

5. $\text{Cos}^{-1}(-0.866)$ **6.** $\text{Cos}^{-1}(-0.8)$

7. $\text{Cos}^{-1} 1$ **8.** $\text{Sin}^{-1} 1$

9. $\text{Tan}^{-1}(-0.5)$ **10.** $\text{Tan}^{-1}(-0.866)$

11. $\text{Tan}^{-1}(-1.33)$ **12.** $\text{Tan}^{-1} 10$

Find the value in degrees:

13. $\text{Csc}^{-1}(-1.25)$ **14.** $\text{Csc}^{-1} 10$

15. $\text{Sec}^{-1}(-2)$ **16.** $\text{Sec}^{-1}(-1)$

17. $\text{Cot}^{-1}(-1)$ **18.** $\text{Cot}^{-1} 10$

SECTION
22.3

Trigonometric Identities

Identities are equations that are true for all allowable values of the variables. For example, the commutative property for addition

$$x + y = y + x$$

is an identity.

There are many useful trigonometric identities. The definitions relating the cotangent, secant, and cosecant to the tangent, cosine, and sine are identities called the **reciprocal identities**.

The Reciprocal Identities:

$$\cot x = \frac{1}{\tan x} \quad \text{and} \quad \tan x = \frac{1}{\cot x}$$

$$\sec x = \frac{1}{\cos x} \quad \text{and} \quad \cos x = \frac{1}{\sec x}$$

$$\csc x = \frac{1}{\sin x} \quad \text{and} \quad \sin x = \frac{1}{\csc x}$$

EXAMPLE 22.10 ▶   Write $\dfrac{\sec x}{\csc x}$ in terms of $\sin x$ and $\cos x$.

SOLUTION   Using the reciprocal identities, we replace $\sec x$ by $\dfrac{1}{\cos x}$ and $\csc x$ by $\dfrac{1}{\sin x}$:

$$\frac{\sec x}{\csc x} = \frac{\dfrac{1}{\cos x}}{\dfrac{1}{\sin x}}.$$

Then, we divide the fractions to obtain

$$\frac{\dfrac{1}{\cos x}}{\dfrac{1}{\sin x}} = \frac{1}{\cos x} \cdot \frac{\sin x}{1}$$

$$= \frac{\sin x}{\cos x}.$$ ▲

THE RATIO IDENTITIES

The **ratio identities** relate the tangent and cotangent to the sine and cosine. We recall the right triangle with legs a and b, and hypotenuse c. We know that

$$\sin \theta = \frac{a}{c}, \quad \cos \theta = \frac{b}{c}, \quad \text{and } \tan \theta = \frac{a}{b}.$$

Therefore, we can write

A right triangle with legs a and b and hypotenuse c.

$$\frac{\sin \theta}{\cos \theta} = \frac{\dfrac{a}{c}}{\dfrac{b}{c}}$$

$$= \frac{a}{c} \cdot \frac{c}{b}$$

$$= \frac{a}{b}$$

$$= \tan \theta.$$

Thus, we have

$$\tan \theta = \frac{\sin \theta}{\cos \theta}.$$

Similarly, since $\cot \theta = \dfrac{1}{\tan \theta}$, we can write

$$\cot \theta = \frac{\cos \theta}{\sin \theta}.$$

Replacing θ by x, we have the ratio identities.

> **The Ratio Identities:**
>
> $$\tan x = \frac{\sin x}{\cos x}$$
>
> $$\cot x = \frac{\cos x}{\sin x}$$

EXAMPLE 22.11   Write $\sec x \; \cot x$ in terms of $\sin x$ and $\cos x$.

SOLUTION   We use the reciprocal identity

$$\sec x = \frac{1}{\cos x}$$

and the ratio identity

$$\cot x = \frac{\cos x}{\sin x}.$$

Then, we can write

$$\sec x \; \cot x = \frac{1}{\cos x} \cdot \frac{\cos x}{\sin x}$$

$$= \frac{1}{\sin x}. \qquad \blacktriangle$$

THE PYTHAGOREAN IDENTITY

The **Pythagorean identity** relates the sine and the cosine. Returning to the right triangle with legs a and b, and hypotenuse c, we recall the Pythagorean theorem,

$$a^2 + b^2 = c^2.$$

Dividing both sides by c^2, we have

$$\frac{a^2 + b^2}{c^2} = \frac{c^2}{c^2}$$

$$\frac{a^2}{c^2} + \frac{b^2}{c^2} = 1$$

$$\left(\frac{a}{c}\right)^2 + \left(\frac{b}{c}\right)^2 = 1$$

$$(\sin \theta)^2 + (\cos \theta)^2 = 1.$$

To avoid excessive use of parentheses, it is common to write $\sin^2 \theta$ in place of $(\sin \theta)^2$, and $\cos^2 \theta$ in place of $(\cos \theta)^2$. Thus, we write

$$\sin^2 \theta + \cos^2 \theta = 1.$$

This result is often useful in forms where it is solved for the sine or the cosine:

$$\sin^2 \theta = 1 - \cos^2 \theta$$

and

$$\cos^2 \theta = 1 - \sin^2 \theta.$$

Replacing θ by x, we have the Pythagorean identity and its alternate forms.

> **The Pythagorean Identity:**
>
> $$\sin^2 x + \cos^2 x = 1$$
>
> Alternate Forms:
>
> $$\sin^2 x = 1 - \cos^2 x$$
> $$\cos^2 x = 1 - \sin^2 x$$

EXAMPLE 22.12 Simplify $\sec x - \sin x \, \tan x$ by writing the expression in terms of $\sin x$ and $\cos x$.

SOLUTION First, we use a reciprocal identity and a ratio identity to write the expression in terms of $\sin x$ and $\cos x$:

$$\sec x - \sin x \, \tan x = \frac{1}{\cos x} - \sin x \cdot \frac{\sin x}{\cos x}$$

$$= \frac{1}{\cos x} - \frac{\sin^2 x}{\cos x}$$

$$= \frac{1 - \sin^2 x}{\cos x}.$$

We can use an alternate form of the Pythagorean identity to write

$$\frac{1 - \sin^2 x}{\cos x} = \frac{\cos^2 x}{\cos x}$$

$$= \cos x. \qquad \blacktriangle$$

EXAMPLE 22.13 Simplify $\tan x + \cot x$ by writing the expression in terms of $\sin x$ and $\cos x$.

SOLUTION We use the ratio identities to write

$$\tan x + \cot x = \frac{\sin x}{\cos x} + \frac{\cos x}{\sin x}.$$

We can add these fractions by using the common denominator $\sin x \cos x$. We multiply both the numerator and the denominator of the first fraction by $\sin x$, and we multiply both the numerator and the denominator of the second fraction by $\cos x$:

$$\frac{\sin x}{\cos x} + \frac{\cos x}{\sin x} = \frac{\sin x \sin x}{\sin x \cos x} + \frac{\cos x \cos x}{\sin x \cos x}$$

$$= \frac{\sin^2 x}{\sin x \cos x} + \frac{\cos^2 x}{\sin x \cos x}$$

$$= \frac{\sin^2 x + \cos^2 x}{\sin x \cos x}.$$

Then, by using the Pythagorean identity, we have

$$\frac{\sin^2 x + \cos^2 x}{\sin x \cos x} = \frac{1}{\sin x \cos x}. \qquad \blacktriangle$$

We can avoid some complex fractions by deriving two additional forms of the Pythagorean identity. By dividing both sides of the Pythagorean identity by $\cos^2 x$, we have

$$\sin^2 x + \cos^2 x = 1$$

$$\frac{\sin^2 x + \cos^2 x}{\cos^2 x} = \frac{1}{\cos^2 x}$$

$$\frac{\sin^2 x}{\cos^2 x} + \frac{\cos^2 x}{\cos^2 x} = \frac{1}{\cos^2 x}$$

$$\tan^2 x + 1 = \sec^2 x.$$

Similarly, by dividing both sides of the Pythagorean identity by $\sin^2 x$, we can show that

$$\cot^2 x + 1 = \csc^2 x.$$

Other Pythagorean Identities:

$$\tan^2 x + 1 = \sec^2 x$$
$$\cot^2 x + 1 = \csc^2 x$$

Alternate Forms:

$$\tan^2 x = \sec^2 x - 1$$
$$\cot^2 x = \csc^2 x - 1$$

EXAMPLE 22.14 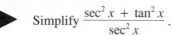 Simplify $\dfrac{\sec^2 x + \tan^2 x}{\sec^2 x}$.

SOLUTION   We can use a Pythagorean identity to write

$$\frac{\sec^2 x + \tan^2 x}{\sec^2 x} = \frac{\sec^2 x + (\sec^2 x - 1)}{\sec^2 x}$$

$$= \frac{2\sec^2 x - 1}{\sec^2 x}$$

$$= \frac{2\sec^2 x}{\sec^2 x} - \frac{1}{\sec^2 x}$$

$$= 2 - \frac{1}{\sec^2 x}$$

$$= 2 - \cos^2 x. \qquad \blacktriangle$$

OTHER ALGEBRAIC TECHNIQUES

We have used several techniques of algebra in the preceding examples. Many other algebraic techniques are useful in simplifying trigonometric expressions. Some trigonometric expressions can be simplified by using factoring.

EXAMPLE 22.15  Simplify $\dfrac{\cos^2 x}{1 - \sin x}$.

SOLUTION   First, we use a Pythagorean identity to write

$$\frac{\cos^2 x}{1 - \sin x} = \frac{1 - \sin^2 x}{1 - \sin x}.$$

Now, by using factoring techniques from Unit 14, we can factor the numerator to obtain

$$\frac{1 - \sin^2 x}{1 - \sin x} = \frac{(1 + \sin x)(1 - \sin x)}{1 - \sin x}.$$

Thus, dividing out the common factor $1 - \sin x$, we have

$$\frac{(1 + \sin x)(1 - \sin x)}{1 - \sin x} = 1 + \sin x. \qquad \blacktriangle$$

Another useful technique for simplifying trigonometric expressions is multiplication by a conjugate.

EXAMPLE 22.16   Simplify $\dfrac{\cos x}{1 - \sin x}$.

SOLUTION   We recall that conjugates were used to divide complex numbers in Unit 18. The conjugate of $1 - \sin x$ is $1 + \sin x$. We multiply both the numerator and the denominator by the conjugate to obtain

$$\frac{\cos x}{1 - \sin x} = \frac{\cos x \ (1 + \sin x)}{(1 - \sin x)(1 + \sin x)}.$$

Then, multiplying the factors in the denominator and using an alternate form of the Pythagorean identity, we have

$$\frac{\cos x \ (1 + \sin x)}{(1 - \sin x)(1 + \sin x)} = \frac{\cos x \ (1 + \sin x)}{1 - \sin^2 x}$$

$$= \frac{\cos x \ (1 + \sin x)}{\cos^2 x}$$

$$= \frac{1 + \sin x}{\cos x}.$$

The expression is in simpler form when there is just one term in the denominator. We can also write this result in the form

$$\frac{1 + \sin x}{\cos x} = \frac{1}{\cos x} + \frac{\sin x}{\cos x}$$

$$= \sec x + \tan x. \qquad \blacktriangle$$

EXERCISE 22.3

Simplify by writing in terms of $\sin x$ and $\cos x$:

1. $\sin x \sec x$

2. $\cos x \csc x$

3. $\dfrac{\csc x}{\sec x}$

4. $\dfrac{\cos x}{\csc x}$

5. $\tan x \csc x$

6. $\sin x \cot x$

7. $\sin x \cot x \sec x$

8. $\cos x \tan x \cot x \sec x$

9. $\dfrac{\sec x}{\tan x}$

10. $\dfrac{\cot x}{\csc x}$

11. $\csc x - \cos x \cot x$

12. $\sec^2 x - \tan^2 x$

13. $\sin x \ \sec^2 x - \tan^2 x$

14. $\cos x \csc^2 x + \cot^2 x$

15. $\sec^2 x + \csc^2 x$

16. $\dfrac{\tan x}{\csc x} + \dfrac{\sin x}{\tan x}$

Simplify:

17. $\dfrac{\cot^2 x + \csc^2 x}{\csc^2 x}$

18. $\dfrac{\sec^2 x - \tan^2 x}{\sec^2 x}$

19. $\dfrac{\sin^2 x}{1 - \cos x}$

20. $\dfrac{\tan^2 x}{\sec x - 1}$

21. $\dfrac{\tan^2 x}{\sec^2 x - \sec x}$

22. $\dfrac{\cot^2 x}{\csc^2 x + \csc x}$

23. $\dfrac{\sin x}{1 + \cos x}$

24. $\dfrac{\tan x}{\sec x - 1}$

Laws of Sines and Cosines

Triangles that are not right triangles are called **oblique** triangles. The trigonometric ratios cannot be applied directly to oblique triangles. There are two other rules that can be used to solve oblique triangles.

If two angles and one side are given, an oblique triangle can be solved by using the **law of sines**.

Law of Sines:

$$\frac{a}{\sin A} = \frac{b}{\sin B} = \frac{c}{\sin C}$$

The proof of the first equality is given at the end of this unit.

EXAMPLE 22.17 ▶ Solve the triangle with $A = 48°$, $B = 75°$, and $c = 112$.

SOLUTION We draw the triangle with side a across from angle A, side b across from angle B, and side c across from angle C. Since the sum of the angles of a triangle is 180°, we can find angle C by using

$$C = 180° - (A + B)$$
$$= 180° - 123°$$
$$= 57°.$$

Triangle for Example 22.17.

To find sides a and b, we use the law of sines. By using the first and third parts, we can write

$$\frac{a}{\sin A} = \frac{c}{\sin C}$$
$$a = \frac{c \sin A}{\sin C}$$
$$= \frac{112 \sin 48°}{\sin 57°}$$
$$= 99.$$

Similarly, by using the second and third parts, we can write

$$\frac{b}{\sin B} = \frac{c}{\sin C}$$
$$b = \frac{c \sin B}{\sin C}$$
$$= \frac{112 \sin 75°}{\sin 57°}$$
$$= 129.$$ ▲

TWO SIDES AND INCLUDED ANGLE

The **included angle** for two sides is the angle formed by the two sides. If two sides and the included angle are given, an oblique triangle can be solved by using the **law of cosines**.

Law of Cosines:

$$a^2 = b^2 + c^2 - 2bc \cos A$$
$$b^2 = a^2 + c^2 - 2ac \cos B$$
$$c^2 = a^2 + b^2 - 2ab \cos C$$

The proof of the first form is given at the end of this unit.

EXAMPLE 22.18 Solve the triangle with $a = 22$, $b = 30$, and $C = 55°$.

SOLUTION We draw the triangle and observe that C is the included angle for sides a and b. We can find side c by using the law of cosines. Using the third form, we write

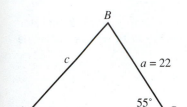

Triangle for Example 22.18.

$$c^2 = a^2 + b^2 - 2ab \cos C$$
$$= 22^2 + 30^2 - 2(22)(30) \cos 55°.$$

Calculating the expression, and then its square root, we have

$$c = 25.$$

Now, we can find the remaining angles by using the law of sines. To find A, we use the first and third parts:

$$\frac{a}{\sin A} = \frac{c}{\sin C}.$$

Taking the reciprocal of each side, we have

$$\frac{\sin A}{a} = \frac{\sin C}{c}$$

$$\sin A = \frac{a \sin C}{c}.$$

Then, we use the inverse sine to find

$$A = \text{Sin}^{-1}\left(\frac{a \sin C}{c}\right)$$

$$= \text{Sin}^{-1}\left(\frac{22 \sin 55°}{25}\right)$$

$$= 46.1°.$$

Similarly, we can find B by using the reciprocals of the second and third parts:

$$\frac{\sin B}{b} = \frac{\sin C}{c}$$

$$\sin B = \frac{b \sin C}{c}.$$

Thus, we have

$$B = \text{Sin}^{-1}\left(\frac{b \sin C}{c}\right)$$

$$= \text{Sin}^{-1}\left(\frac{30 \sin 55°}{25}\right)$$

$$= 79.4°.$$

We observe that

$$A + B + C = 46.1° + 79.4° + 55°$$

$$= 180.5°.$$

In solving oblique triangles, we often use previous results, which can cause rounding errors.   ▲

Angles that are less than 90° are called **acute** angles. Angles that are greater than 90° are called **obtuse** angles. Obtuse angles cannot be found directly by using the law of sines.

EXAMPLE 22.19 ▶ Solve the triangle with $b = 120$, $c = 64$, and $A = 53°$.

SOLUTION   We draw the triangle and observe that A is the included angle for sides b and c. To find side a, we use the law of cosines:

$$a^2 = b^2 + c^2 - 2bc \cos A$$

$$a^2 = 120^2 + 64^2 - 2(120)(64) \cos 53°$$

$$a = 96.2.$$

If we use the law of sines to find B, it appears that

$$\frac{\sin B}{b} = \frac{\sin A}{a}$$

$$\sin B = \frac{b \sin A}{a},$$

and so we might have

$$B = \text{Sin}^{-1}\left(\frac{b \sin A}{a}\right)$$

$$= \text{Sin}^{-1}\left(\frac{120 \sin 53°}{96.2}\right)$$

$$= 85°.$$

We recall, however, that the range of principal values for $y = \text{Sin}^{-1} x$ is

$$-\frac{\pi}{2} \le y \le \frac{\pi}{2},$$

or, in degrees,

$$-90° \le y \le 90°.$$

Obtuse angles are not included in this range. However, the range of principal values for $y = \text{Cos}^{-1} x$ is

$$0 \le y \le \pi,$$

or, in degrees,

$$0° \le y \le 180°.$$

Thus, we may determine whether B is an obtuse angle by using the inverse cosine. We solve the second form of the law of cosines for $\cos B$ to obtain

$$b^2 = a^2 + c^2 - 2ac \cos B$$

$$2ac \cos B = a^2 + c^2 - b^2$$

$$\cos B = \frac{a^2 + c^2 - b^2}{2ac}.$$

B

$c = 64$ a

$53°$

A $b = 120$ C

Triangle for Example 22.19.

Then, we find

$$B = \text{Cos}^{-1}\left(\frac{a^2 + c^2 - b^2}{2ac}\right)$$

$$= \text{Cos}^{-1}\left(\frac{96.2^2 + 64^2 - 120^2}{2(96.2)(64)}\right)$$

$$= 94.9°.$$

Since $a^2 + c^2 - b^2$ is negative, B is an obtuse angle.

The remaining angle C must then be an acute angle, so we may use either the law of sines or the law of cosines to find

$$C = 32.1°.$$ ▲

THREE SIDES

We can use the law of cosines to find the angles when three sides of an oblique triangle are given.

EXAMPLE 22.20   Solve the triangle with sides $a = 9.24$, $b = 15.5$, and $c = 11.8$.

SOLUTION To find angle A in the triangle shown, we solve the first form of the law of cosines for $\cos A$:

$$a^2 = b^2 + c^2 - 2bc \cos A$$

$$\cos A = \frac{b^2 + c^2 - a^2}{2bc}.$$

Thus, we have

$$A = \text{Cos}^{-1}\left(\frac{b^2 + c^2 - a^2}{2bc}\right)$$

$$= \text{Cos}^{-1}\left(\frac{15.5^2 + 11.8^2 - 9.24^2}{2(15.5)(11.8)}\right)$$

$$= 36.5°.$$

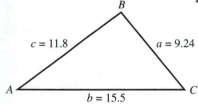

Triangle for Example 22.20.

To find angle B, we use the second form solved for B as in the preceding example:

$$B = \text{Cos}^{-1}\left(\frac{a^2 + c^2 - b^2}{2ac}\right)$$

$$= \text{Cos}^{-1}\left(\frac{9.24^2 + 11.8^2 - 15.5^2}{2(9.24)(11.8)}\right)$$

$$= 94.1°.$$

The expression $a^2 + c^2 - b^2$ is negative, so B is an obtuse angle. We can use the third form to find

$$C = 49.4°.$$

We observe that

$$A + B + C = 36.5° + 94.1° + 49.4°$$

$$= 180°.$$ ▲

A NONINCLUDED ANGLE

An oblique triangle is uniquely determined by any two angles and one side, by any two sides and their *included* angle, or by the three sides. While it is possible to solve some triangles given two sides and a nonincluded angle, some such triangles are not uniquely determined, and the results can be unpredictable.

EXAMPLE 22.21

A ship sails from a harbor 1.4 nautical miles in a northeasterly direction. It then sails due south 1.6 nautical miles. The captain calculates that the angle to the harbor is 70°. Find angle θ in the diagram shown.

SOLUTION

It appears that we can use the law of sines to write

$$\frac{\sin \theta}{1.6} = \frac{\sin 70°}{1.4}$$

$$\theta = \text{Sin}^{-1}\left(\frac{1.6 \sin 70°}{1.4}\right).$$

However, when we do the calculation, we find that

$$\theta = \text{Sin}^{-1} 1.074.$$

The calculation leads to an error. There is no triangle with the two sides and nonincluded angle shown. ▲

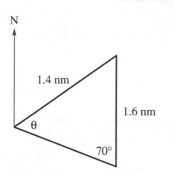

N

1.4 nm

1.6 nm

θ

70°

Triangle for Example 22.21.

VECTOR ADDITION

In Section 17.4, we added vectors by finding their horizontal and vertical components, adding the respective components, and then returning the result to vector form. We can also add vectors by using oblique triangles. In the following example, we add the vectors from Example 17.17.

EXAMPLE 22.22

Find the sum of the vectors $V_1 = 6\underline{/30°}$ and $V_2 = 8\underline{/75°}$.

SOLUTION

We draw a diagram of the vectors, place V_2 at the tip of V_1, and draw the sum from the base of V_1 to the tip of V_2, as in Example 17.17:

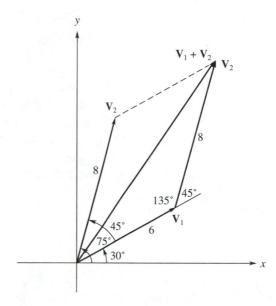

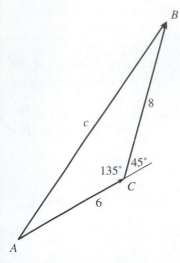

The two vectors and their sum as an oblique triangle.

The angle between the vectors is

$$75° - 30° = 45°.$$

Then, when $\mathbf{V}_2$ is placed at the tip of $\mathbf{V}_1$, the angle between $\mathbf{V}_1$ and $\mathbf{V}_2$ is

$$180° - 45° = 135°.$$

The magnitude of the sum is side c of an oblique triangle. We can find c by using the law of cosines:

$$c^2 = a^2 + b^2 - 2ab \cos C$$

$$c^2 = 8^2 + 6^2 - 2(8)(6) \cos 135°$$

$$c = 13.$$

The angle between $\mathbf{V}_1$ and the sum is angle A. We can find A by using the law of sines:

$$\frac{\sin A}{a} = \frac{\sin C}{c}$$

$$\sin A = \frac{a \sin C}{c}$$

$$\sin A = \frac{8 \sin 135°}{13}$$

$$A = 25.8°.$$

Therefore, the direction of the sum is

$$25.8° + 30° = 55.8°,$$

and the sum is the vector 13/55.8°. A small difference between this result and our result in Section 17.4 is due to rounding errors.   ▲

EXERCISE 22.4

Solve the triangle:

1. $A = 55°, B = 65°, c = 20$
2. $A = 35°, C = 40°, b = 15$
3. $A = 50°, B = 85°, a = 10$
4. $A = 82°, C = 28°, c = 105$
5. $a = 5.5, b = 7.5, C = 78°$
6. $b = 20, c = 22, A = 40°$
7. $b = 22, c = 64, A = 45°$
8. $a = 12.5, b = 9.2, C = 36°$
9. $a = 10, b = 15, c = 12$
10. $a = 16, b = 22, c = 13$

11. Technicians must wire point A to point B, but there is an obstruction between them. They wire 9 feet to point C, and then 6 feet to point B. The angle at C is 70°. What is the distance between A and B, and what is the angle θ where they start?

12. Technicians must wire point A to point B as in Exercise 11 (A and B are not necessarily the same distance apart as before). They start at an angle of 45° and lay 9 feet of wire to point C, and then 6 feet to point B. What is the angle θ at point B?

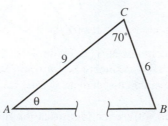

Triangle for Exercise 11.

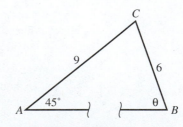

Triangle for Exercise 12.

13. A ship sails from a harbor 4 nautical miles in a generally southeasterly direction, and then 3 nautical miles in a generally southwesterly direction. The angle between the two directions is 100°. How far is the ship from the harbor, and what is the angle θ?

14. A ship sails at an angle of 45° southeast from a harbor for 5.2 nautical miles, and then 3.4 nautical miles in a generally southwesterly direction. What is the angle θ where the ship finishes?

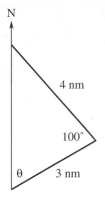

Triangle for Exercise 13.   Triangle for Exercise 14.

15. and **16.** Find the sum of the vectors in Exercise 17.4, #9 and #10, by using oblique triangle methods.

17. and **18.** Find the sum of the vectors in Exercise 17.4, #13 and #14, by using oblique triangle methods.

Sum and Difference Identities

SECTION

22.5

The law of cosines can be used to prove another group of useful identities. First, we consider the cosine of the sum of two angles. If the angles are α and β, then the cosine of their sum is $\cos(\alpha + \beta)$.

EXAMPLE 22.23

Use $\alpha = 60°$ and $\beta = 30°$ to show that $\cos(\alpha + \beta)$ is not the same as $\cos\alpha + \cos\beta$.

SOLUTION Using $\alpha = 60°$ and $\beta = 30°$, we find that

$$\cos(\alpha + \beta) = \cos(60° + 30°)$$

$$= \cos 90°$$

$$= 0.$$

However, we find that

$$\cos\alpha + \cos\beta = \cos 60° + \cos 30°$$

$$= 0.5 + 0.866$$

$$= 1.366.$$

Therefore, we conclude that $\cos(\alpha + \beta) \neq \cos\alpha + \cos\beta$. ▲

The type of demonstration in Example 22.23 is a *proof by counterexample*. A counterexample shows that a statement is *not true*. We cannot prove that a statement is true by examples. To prove that a statement is true, we must prove the statement for all allowable values.

The **sum and difference identities for the cosine** provide the correct identities for $\cos(\alpha + \beta)$ and $\cos(\alpha - \beta)$.

> **Sum and Difference Identities for the Cosine:**
> $$\cos(\alpha + \beta) = \cos \alpha \cos \beta - \sin \alpha \sin \beta$$
> $$\cos(\alpha - \beta) = \cos \alpha \cos \beta + \sin \alpha \sin \beta$$

To prove the sum and difference identities for the cosine, we must show that they are true for *all* values of α and β. The proof is given at the end of this unit.

EXAMPLE 22.24   Use $\alpha = 60°$ and $\beta = 30°$ to *illustrate* the sum identity for the cosine,

$$\cos(\alpha + \beta) = \cos \alpha \cos \beta - \sin \alpha \sin \beta.$$

SOLUTION In Example 22.23, we found that

$$\cos(60° + 30°) = 0.$$

Now, using $\alpha = 60°$ and $\beta = 30°$, we find that

$$\cos \alpha \cos \beta - \sin \alpha \sin \beta = \cos 60° \cos 30° - \sin 60° \sin 30°$$
$$= (0.5)(0.866) - (0.866)(0.5)$$
$$= 0.$$

Therefore, we conclude that the identity is true for $\alpha = 60°$ and $\beta = 30°$. ▲

We can use the sum and difference identities for the cosine to find other useful relationships.

EXAMPLE 22.25   Use the difference identity for the cosine to simplify $\cos(-x)$.

SOLUTION We can write

$$\cos(-x) = \cos(0 - x).$$

Therefore, we use the identity

$$\cos(\alpha - \beta) = \cos \alpha \cos \beta + \sin \alpha \sin \beta,$$

with $\alpha = 0$ and $\beta = x$, to write

$$\cos(0 - x) = \cos 0 \cos x + \sin 0 \sin x$$
$$= 1 \cos x + 0 \sin x$$
$$= \cos x.$$

Thus, we have

$$\cos(-x) = \cos x.$$ ▲

EXAMPLE 22.26   Use the sum identity for the cosine to simplify $\cos\left(x + \dfrac{\pi}{2}\right)$.

SOLUTION We use the identity

$$\cos(\alpha + \beta) = \cos \alpha \cos \beta - \sin \alpha \sin \beta,$$

with $\alpha = x$ and $\beta = \frac{\pi}{2}$. Thus, we have

$$\cos\left(x + \frac{\pi}{2}\right) = \cos x \cos \frac{\pi}{2} - \sin x \sin \frac{\pi}{2}$$
$$= (\cos x)(0) - (\sin x)(1)$$
$$= -\sin x.$$

The derivation of the similar relationship

$$\cos\left(x - \frac{\pi}{2}\right) = \sin x$$

is in Exercise 22.5.   ▲

The **sum and difference identities for the sine** provide identities for $\sin(\alpha + \beta)$ and $\sin(\alpha - \beta)$.

Sum and Difference Identities for the Sine:

$$\sin(\alpha + \beta) = \sin\alpha\cos\beta + \cos\alpha\sin\beta$$
$$\sin(\alpha - \beta) = \sin\alpha\cos\beta - \cos\alpha\sin\beta$$

The sum and difference identities for the sine are derived from those for the cosine at the end of this unit.

EXAMPLE 22.27   Use the difference identity for the sine to simplify $\sin(\pi - x)$.

SOLUTION   We use the identity

$$\sin(\alpha - \beta) = \sin\alpha\cos\beta - \cos\alpha\sin\beta,$$

with $\alpha = \pi$ and $\beta = x$. Thus, we have

$$\sin(\pi - x) = \sin\pi\cos x - \cos\pi\sin x$$
$$= 0\cos x - (-1)\sin x$$
$$= 0 + \sin x$$
$$= \sin x.$$ ▲

THE TANGENT IDENTITIES

We can derive similar identities for the tangent by using

$$\tan x = \frac{\sin x}{\cos x}.$$

We replace x by $\alpha + \beta$ to obtain

$$\tan(\alpha + \beta) = \frac{\sin(\alpha + \beta)}{\cos(\alpha + \beta)}$$

$$= \frac{\sin\alpha\cos\beta + \cos\alpha\sin\beta}{\cos\alpha\cos\beta - \sin\alpha\sin\beta}.$$

We divide the numerator and the denominator by $\cos\alpha\cos\beta$, which is the same as dividing each term by $\cos\alpha\cos\beta$:

$$\tan(\alpha + \beta) = \frac{\dfrac{\sin\alpha\cos\beta}{\cos\alpha\cos\beta} + \dfrac{\cos\alpha\sin\beta}{\cos\alpha\cos\beta}}{\dfrac{\cos\alpha\cos\beta}{\cos\alpha\cos\beta} - \dfrac{\sin\alpha\sin\beta}{\cos\alpha\cos\beta}}$$

$$= \frac{\dfrac{\sin\alpha}{\cos\alpha} + \dfrac{\sin\beta}{\cos\beta}}{1 - \dfrac{\sin\alpha\sin\beta}{\cos\alpha\cos\beta}}$$

$$= \frac{\tan\alpha + \tan\beta}{1 - \tan\alpha\,\tan\beta}.$$

Proof of the corresponding difference identity

$$\tan(\alpha - \beta) = \frac{\tan\alpha - \tan\beta}{1 + \tan\alpha\,\tan\beta}$$

is in Exercise 22.5.

EXAMPLE 22.28 Use the difference identity for the tangent to simplify $\tan(\pi - x)$.

SOLUTION We use the identity

$$\tan(\alpha - \beta) = \frac{\tan\alpha - \tan\beta}{1 + \tan\alpha\tan\beta},$$

with $\alpha = \pi$ and $\beta = x$. Thus, we have

$$\tan(\pi - x) = \frac{\tan\pi - \tan x}{1 + \tan\pi\,\tan x}$$

$$= \frac{0 - \tan x}{1 + 0\,\tan x}$$

$$= \frac{-\tan x}{1}$$

$$= -\tan x. \qquad \blacktriangle$$

We recall that the tangent function is not defined at the asymptotes. Therefore, we cannot use the tangent identities when α or β has a value such as $\frac{\pi}{2}$ or $\frac{3\pi}{2}$. We can avoid this situation by using

$$\tan(\alpha + \beta) = \frac{\sin(\alpha + \beta)}{\cos(\alpha + \beta)}$$

and

$$\tan(\alpha - \beta) = \frac{\sin(\alpha - \beta)}{\cos(\alpha - \beta)}.$$

EXAMPLE 22.29 Simplify $\tan\left(\frac{\pi}{2} + x\right)$.

SOLUTION We use

$$\tan(\alpha + \beta) = \frac{\sin(\alpha + \beta)}{\cos(\alpha + \beta)},$$

with $\alpha = \frac{\pi}{2}$ and $\beta = x$. Thus, we have

$$\tan\left(\frac{\pi}{2} + x\right) = \frac{\sin\left(\frac{\pi}{2} + x\right)}{\cos\left(\frac{\pi}{2} + x\right)}$$

$$= \frac{\sin\frac{\pi}{2}\cos x + \cos\frac{\pi}{2}\sin x}{\cos\frac{\pi}{2}\cos x - \sin\frac{\pi}{2}\sin x}$$

$$= \frac{1\cos x + 0\sin x}{0\cos x - 1\sin x}$$

$$= \frac{\cos x}{-\sin x}$$

$$= -\cot x. \qquad \blacktriangle$$

DOUBLE-ANGLE IDENTITIES

Many other identities are derived from the sum and difference identities for the sine and cosine. Two that are easy to derive are double-angle identities for the sine and cosine. To derive the double-angle identity for the sine, we use

$$\sin(\alpha + \beta) = \sin\alpha\cos\beta + \cos\alpha\sin\beta,$$

with $\alpha = x$ and $\beta = x$. Then, we have

$$\sin(x + x) = \sin x\cos x + \cos x\sin x$$
$$\sin 2x = 2\sin x\cos x.$$

Similarly, to derive the double-angle identity for the cosine, we use

$$\cos(\alpha + \beta) = \cos\alpha\cos\beta - \sin\alpha\sin\beta$$

with $\alpha = x$ and $\beta = x$. Then, we have

$$\cos(x + x) = \cos x\cos x - \sin x\sin x$$
$$\cos 2x = \cos^2 x - \sin^2 x.$$

EXAMPLE 22.30 Simplify $\dfrac{\cos 2x}{\cos^2 x}$.

SOLUTION We write

$$\frac{\cos 2x}{\cos^2 x} = \frac{\cos^2 x - \sin^2 x}{\cos^2 x}$$

$$= \frac{\cos^2 x}{\cos^2 x} - \frac{\sin^2 x}{\cos^2 x}$$

$$= 1 - \tan^2 x. \qquad \blacktriangle$$

EXERCISE 22.5

Use $\alpha = 60°$ and $\beta = 30°$ to show by counterexample:

1. $\sin(\alpha + \beta) \neq \sin\alpha + \sin\beta$
2. $\sin(\alpha - \beta) \neq \sin\alpha - \sin\beta$
3. $\cos(\alpha - \beta) \neq \cos\alpha - \cos\beta$
4. $\tan(\alpha + \beta) \neq \tan\alpha + \tan\beta$

Use $\alpha = 60°$ and $\beta = 30°$ to illustrate:

5. $\sin(\alpha + \beta) = \sin\alpha\cos\beta + \cos\alpha\sin\beta$
6. $\sin(\alpha - \beta) = \sin\alpha\cos\beta - \cos\alpha\sin\beta$
7. $\cos(\alpha - \beta) = \cos\alpha\cos\beta + \sin\alpha\sin\beta$
8. $\tan(\alpha + \beta) = \dfrac{\tan\alpha + \tan\beta}{1 - \tan\alpha\tan\beta}$

Simplify:

9. $\cos(\pi - x)$
10. $\cos\left(x - \dfrac{\pi}{2}\right)$

11. $\cos(\pi + x)$
12. $\cos\left(\dfrac{3\pi}{2} + x\right)$

13. $\sin(\pi + x)$
14. $\sin\left(\dfrac{3\pi}{2} - x\right)$

15. $\tan (\pi + x)$

16. $\tan \left(\dfrac{\pi}{2} - x \right)$

17. $\cot (\pi - x)$

18. $\cot \left(\dfrac{\pi}{2} + x \right)$

19. $\dfrac{\cos 2x}{\sin^2 x}$

20. $\dfrac{\cos 2x}{\cos x \sin x}$

21. $\dfrac{\sin 2x}{2 \sin x}$

22. $\sin 2x \csc x$

23. $\dfrac{2}{1 + \cos 2x}$

24. $\dfrac{2}{1 - \cos 2x}$

25. $\dfrac{\sin 2x}{\cos 2x + 1}$

26. $\dfrac{\sin 2x}{\cos 2x - 1}$

Prove the identity:

27. $\tan (\alpha - \beta) = \dfrac{\tan \alpha - \tan \beta}{1 + \tan \alpha \, \tan \beta}$

28. $\cot (\alpha + \beta) = \dfrac{\cot \alpha \, \cot \beta - 1}{\cot \beta + \cot \alpha}$

29. $\tan 2x = \dfrac{2 \tan x}{1 - \tan^2 x}$

30. $\cot 2x = \dfrac{\cot^2 x - 1}{2 \cot x}$

Proofs of Laws and Identities

Proof of the Law of Sines

The height h in the triangle shown forms two triangles. In the left-hand triangle,

$$\sin A = \frac{h}{b}$$

$$h = b \sin A.$$

In the right-hand triangle,

$$\sin B = \frac{h}{a}$$

$$h = a \sin B.$$

Therefore,

$$a \sin B = b \sin A,$$

and dividing both sides by $\sin A \sin B$,

$$\frac{a \sin B}{\sin A \sin B} = \frac{b \sin A}{\sin A \sin B}$$

$$\frac{a}{\sin A} = \frac{b}{\sin B}.$$

The other parts are proved similarly.

Diagram for proofs of the laws of sines and cosines.

Proof of the Law of Cosines

Using the same diagram as for the law of sines, and applying the Pythagorean theorem to the left-hand triangle,

$$x^2 + h^2 = b^2$$

$$h^2 = b^2 - x^2.$$

Applying the Pythagorean theorem to the right-hand triangle,

$$(c - x)^2 + h^2 = a^2$$

$$h^2 = a^2 - (c - x)^2.$$

Therefore,

$$a^2 - (c - x)^2 = b^2 - x^2$$

$$a^2 - (c^2 - 2cx + x^2) = b^2 - x^2$$

$$a^2 - c^2 + 2cx - x^2 = b^2 - x^2$$

$$a^2 - c^2 + 2cx = b^2$$

$$a^2 = b^2 + c^2 - 2cx.$$

In the left-hand triangle, $x = b \cos A$; therefore,

$$a^2 = b^2 + c^2 - 2c \, (b \cos A)$$

$$= b^2 + c^2 - 2bc \cos A.$$

The other forms are proved similarly.

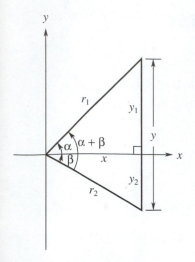

Diagram for proof of the sum
identity for the cosine.

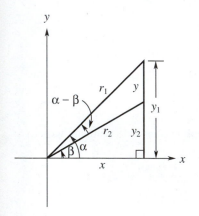

Diagram for proof of the difference
identity for the cosine.

Proofs of the Sum and Difference Identities for the Cosine

(For simplicity, assume β in the diagram shown is counterclockwise so is positive, and y_2 is directed upward so is positive.) Applying the law of cosines to $\alpha + \beta$,

$$y^2 = r_1^2 + r_2^2 - 2r_1r_2 \cos(\alpha + \beta).$$

Solving for $\cos(\alpha + \beta)$,

$$\cos(\alpha + \beta) = \frac{r_1^2 + r_2^2 - y^2}{2r_1r_2}$$

$$= \frac{r_1^2 + r_2^2 - (y_1 + y_2)^2}{2r_1r_2}$$

$$= \frac{r_1^2 + r_2^2 - (y_1^2 + 2y_1y_2 + y_2^2)}{2r_1r_2}$$

$$= \frac{r_1^2 + r_2^2 - y_1^2 - 2y_1y_2 - y_2^2}{2r_1r_2}$$

$$= \frac{(r_1^2 - y_1^2) + (r_2^2 - y_2^2) - 2y_1y_2}{2r_1r_2}.$$

Applying the Pythagorean theorem to the separate triangles, $r_1^2 - y_1^2 = x^2$ and $r_2^2 - y_2^2 = x^2$; therefore,

$$\cos(\alpha + \beta) = \frac{x^2 + x^2 - 2y_1y_2}{2r_1r_2}$$

$$= \frac{2x^2 - 2y_1y_2}{2r_1r_2}$$

$$= \frac{x^2}{r_1r_2} - \frac{y_1y_2}{r_1r_2}$$

$$= \frac{x}{r_1} \cdot \frac{x}{r_2} - \frac{y_1}{r_1} \cdot \frac{y_2}{r_2}$$

$$= \cos\alpha\cos\beta - \sin\alpha\sin\beta.$$

The difference identity for the cosine is proved similarly by using the diagram shown.

Proofs of the Sum and Difference Identities for the Sine

Recall from Example 22.26 that

$$\cos\left(x + \frac{\pi}{2}\right) = -\sin x.$$

Therefore, replacing x by $\alpha + \beta$ and using the sum identity for the cosine,

$$-\sin x = \cos\left(x + \frac{\pi}{2}\right)$$

$$-\sin(\alpha + \beta) = \cos\left[(\alpha + \beta) + \frac{\pi}{2}\right]$$

$$= \cos\left[\alpha + \left(\beta + \frac{\pi}{2}\right)\right]$$

$$= \cos\alpha\cos\left(\beta + \frac{\pi}{2}\right) - \sin\alpha\sin\left(\beta + \frac{\pi}{2}\right).$$

Also, recall from Section 22.1 that the cosine function is the same as a sine function with leading phase angle $\frac{\pi}{2}$, so

$$\sin\left(x + \frac{\pi}{2}\right) = \cos x.$$

Then, $\cos(\beta + \frac{\pi}{2}) = -\sin\beta$ and $\sin(\beta + \frac{\pi}{2}) = \cos\beta$; therefore,

$$-\sin(\alpha + \beta) = (\cos\alpha)(-\sin\beta) - \sin\alpha\cos\beta$$
$$= -\cos\alpha\sin\beta - \sin\alpha\cos\beta,$$

and so

$$\sin(\alpha + \beta) = \sin\alpha\cos\beta + \cos\alpha\sin\beta.$$

Furthermore, recall from Example 22.25 that

$$\cos(-x) = \cos x,$$

and from an observation at the end of Section 19.2 that

$$\sin(-x) = -\sin x.$$

Therefore,

$$\sin(\alpha - \beta) = \sin[\alpha + (-\beta)]$$
$$= \sin\alpha\cos(-\beta) + \cos\alpha\sin(-\beta)$$
$$= \sin\alpha\cos\beta + (\cos\alpha)(-\sin\beta)$$
$$= \sin\alpha\cos\beta - \cos\alpha\sin\beta.$$

Proofs of Formulas for Multiplication and Division of Complex Numbers in Polar Form

Recall from Section 18.2 that

$$r\underline{/\theta} = (r\cos\theta) + j(r\sin\theta)$$
$$= r(\cos\theta + j\sin\theta).$$

Therefore,

$$(r_1\underline{/\theta_1})(r_2\underline{/\theta_2}) = [r_1(\cos\theta_1 + j\sin\theta_1)][r_2(\cos\theta_2 + j\sin\theta_2)]$$
$$= r_1r_2(\cos\theta_1 + j\sin\theta_1)(\cos\theta_2 + j\sin\theta_2)$$
$$= r_1r_2(\cos\theta_1\cos\theta_2 + j\cos\theta_1\sin\theta_2 + j\sin\theta_1\cos\theta_2 + j^2\sin\theta_1\sin\theta_2)$$
$$= r_1r_2[(\cos\theta_1\cos\theta_2 - \sin\theta_1\sin\theta_2) + j(\sin\theta_1\cos\theta_2 + \cos\theta_1\sin\theta_2)]$$
$$= r_1r_2[\cos(\theta_1 + \theta_2) + j\sin(\theta_1 + \theta_2)]$$
$$= r_1r_2\underline{/\theta_1 + \theta_2}.$$

Similarly, dividing by using the conjugate of the denominator,

$$\frac{r_1 \underline{/\theta_1}}{r_2 \underline{/\theta_2}} = \frac{r_1(\cos\theta_1 + j\sin\theta_1)}{r_2(\cos\theta_2 + j\sin\theta_2)}$$

$$= \frac{r_1}{r_2}\left[\frac{(\cos\theta_1 + j\sin\theta_1)(\cos\theta_2 - j\sin\theta_2)}{(\cos\theta_2 + j\sin\theta_2)(\cos\theta_2 - j\sin\theta_2)}\right]$$

$$= \frac{r_1}{r_2}\left[\frac{(\cos\theta_1\cos\theta_2 - j\cos\theta_1\sin\theta_2 + j\sin\theta_1\cos\theta_2 - j^2\sin\theta_1\sin\theta_2}{\cos^2\theta_2 - j\cos\theta_2\sin\theta_2 + j\sin\theta_2\cos\theta_2 - j^2\sin^2\theta_2}\right]$$

$$= \frac{r_1}{r_2}\left[\frac{\cos\theta_1\cos\theta_2 + \sin\theta_1\sin\theta_2 + j(\sin\theta_1\cos\theta_2 - \cos\theta_1\sin\theta_2)}{\cos^2\theta_2 + \sin^2\theta_2}\right]$$

$$= \frac{r_1}{r_2}\left[\frac{\cos(\theta_1 - \theta_2) + j\sin(\theta_1 - \theta_2)}{1}\right]$$

$$= \frac{r_1}{r_2}\underline{/\theta_1 - \theta_2}\,.$$

Self-Test

1. On one set of axes, draw the graphs of $y = \sin x$ and $y = \csc x$ from $x = -2\pi$ to $x = 2\pi$.

2. Find the value of $\text{Tan}^{-1}(-1.73)$ in

    a. radians   b. degrees

    2a. _____

    2b. _____

3. Simplify $\cot^2 x - \dfrac{\cos^2 x}{\tan^2 x}$.

    3. _____

4. Simplify $\cot\left(\dfrac{3\pi}{2} + x\right)$.

    4. _____

5. Find angle C for the triangle with sides $a = 7$, $b = 8$, and $c = 10$.

    5. _____

UNIT 23

Mathematics for Computers

Introduction

The mathematics you have studied so far applies to *analog* devices. Analog devices work in a continuous manner, like a sweep second hand on a clock. The common types of computers are *digital* in nature. Digital devices have separate or discrete states, like a light switch that is either off or on. Computers have two states that are represented by 0 and 1. Thus, computer math is done entirely in zeros and ones. In this unit, you will learn about the binary number system, in which the only digits are 0 and 1, and about two related number systems used with computers, the octal and hexadecimal number systems. You will also learn how computers do arithmetic in the binary number system. Then, you will learn about Boolean algebra, a logic system that uses only 0 and 1. Finally, you will learn how Boolean algebra is applied to digital electronics.

OBJECTIVES

When you have finished this unit you should be able to:

1. Transfer whole numbers between the decimal system and the binary, octal, and hexadecimal systems, and between the binary system and the octal and hexadecimal systems.

2. Add whole numbers in binary, use 1's and 2's complements to subtract whole numbers in binary, multiply whole numbers in binary, and transfer fractional numbers between decimal and binary.

3. Make truth tables for Boolean expressions, and use truth tables or properties of Boolean algebra to prove Boolean identities.

4. Compute outputs of digital circuits, draw digital circuits for Boolean expressions, and use Boolean algebra to simplify digital circuits.

SECTION 23.1

Number Systems

In Unit R, we reviewed the **decimal number system**. The decimal system uses **base 10**, and the ten **digits**

$$0, 1, 2, 3, 4, 5, 6, 7, 8, 9.$$

Decimal numbers are constructed from powers of ten. For example, the number 222 means

$$222 = 2 \times 100 + 2 \times 10 + 2.$$

In Unit 2, we studied powers of ten. We may write the place values for decimal numbers as powers of ten, such as

$$100 = 10^2$$

$$10 = 10^1$$

$$1 = 10^0.$$

Thus, we may write decimal numbers in terms of powers of 10. For example, the number 222 can be written

$$222 = 2 \times 100 + 2 \times 10 + 2 \times 1$$
$$= 2 \times 10^2 + 2 \times 10^1 + 2 \times 10^0.$$

THE BINARY NUMBER SYSTEM

Computers use the **binary number system**. The binary system uses **base 2**, and has only two **binary digits**, 0 and 1. Some examples of binary numbers are

$$111_2, \ 1101_2, \ 101011101_2.$$

The subscript 2 means the number is a binary number, not the decimal number with the same digits. Binary numbers are constructed from powers of two. To transfer a binary number to decimal we write the number in terms of binary place values, which are powers of two.

EXAMPLE 23.1   Write 111_2 as a decimal number.

SOLUTION   We write the number in terms of powers of two:

$$111_2 = 1 \times 2^2 + 1 \times 2^1 + 1 \times 2^0$$
$$= 1 \times 4 + 1 \times 2 + 1 \times 1$$
$$= 4 + 2 + 1$$
$$= 7.$$ ▲

EXAMPLE 23.2   Write 1101_2 as a decimal number.

SOLUTION   We write the number in terms of powers of two:

$$1101_2 = 1 \times 2^3 + 1 \times 2^2 + 0 \times 2^1 + 1 \times 2^0$$
$$= 1 \times 8 + 1 \times 4 + 0 \times 2 + 1 \times 1$$
$$= 8 + 4 + 1$$
$$= 13.$$ ▲

To write larger binary numbers as decimal numbers, we must continue the list of powers of two:

$2^4 = 16$	$2^7 = 128$	$2^{10} = 1024$
$2^5 = 32$	$2^8 = 256$	•
		•
$2^6 = 64$	$2^9 = 512$	•

Each power of two can be obtained by multiplying the preceding power by 2.

EXAMPLE 23.3   Write 101011101_2 as a decimal number.

SOLUTION   We write the number in terms of powers of two:

$$101011101_2 = 1 \times 2^8 + 0 \times 2^7 + 1 \times 2^6 + 0 \times 2^5 + 1 \times 2^4 + 1 \times 2^3 + 1 \times 2^2 + 0 \times 2^1 + 1 \times 2^0$$

$$= 1 \times 256 + 0 \times 128 + 1 \times 64 + 0 \times 32 + 1 \times 16 + 1 \times 8 + 1 \times 4 + 0 \times 2 + 1 \times 1$$

$$= 256 + 64 + 16 + 8 + 4 + 1$$

$$= 349.$$

▲

To convert decimal numbers to binary numbers, we may subtract binary place values, which are powers of two.

EXAMPLE 23.4   Write 13 as a binary number.

SOLUTION   We subtract powers of two beginning with the largest power less than or equal to 13, which is $2^3 = 8$:

$$13 - 8 = 5.$$

Now, we subtract the largest power less than or equal to 5, which is $2^2 = 4$:

$$5 - 4 = 1.$$

The largest power less than or equal to 1 is $2^0 = 1$:

$$1 - 1 = 0.$$

The powers of two contained in the binary number are those we subtracted. Thus, the binary number is

$$1 \times 8 + 1 \times 4 + 0 \times 2 + 1 \times 1 = 1101_2.$$

We found that the binary number 1101_2 and the decimal number 13 are equivalent in Example 23.2.   ▲

EXAMPLE 23.5   Write 89 as a binary number.

SOLUTION   We subtract powers of two beginning with the largest power less than or equal to 89, which is $2^6 = 64$:

$$89 - 64 = 25.$$

Then, continuing to subtract the largest power less than or equal to each result, we have

$$25 - 16 = 9$$

$$9 - 8 = 1$$

$$1 - 1 = 0.$$

The powers of two contained in the binary number are those we subtracted. Thus, the binary number is

$$1 \times 64 + 0 \times 32 + 1 \times 16 + 1 \times 8 + 0 \times 4 + 0 \times 2 + 1 \times 1 = 1011001_2.$$

You should check this result by returning the binary number 1011001_2 to decimal form. ▲

We may also find the binary form by using a method called **repeated division**. To use repeated division, we divide the decimal number by the base 2, and then divide each resulting quotient by the base 2. The remainder of each division is either a 0 or a 1. These remainders are the binary digits of the binary number. The binary digits are obtained *in reverse order*.

EXAMPLE 23.6   Write 89 as a binary number by using repeated division.

SOLUTION   We divide 89 by 2, writing the result as a quotient and a remainder R:

$$89 \div 2 = 44, \quad R = 1.$$

Then, continuing to divide each quotient by 2, we obtain the binary digits as the remainders R:

$$44 \div 2 = 22, \quad R = 0$$
$$22 \div 2 = 11, \quad R = 0$$
$$11 \div 2 = 5, \quad R = 1$$
$$5 \div 2 = 2, \quad R = 1$$
$$2 \div 2 = 1, \quad R = 0$$
$$1 \div 2 = 0, \quad R = 1.$$

These remainders are the binary digits *in reverse order*. Thus, the binary number is

$$1011001_2.$$ ▲

THE OCTAL NUMBER SYSTEM

Because converting between decimal numbers and binary numbers is often a tedious process, the **octal number system** can be used in association with computers in place of the decimal system. The octal number system uses **base 8**, and the eight **octal digits**

$$0, 1, 2, 3, 4, 5, 6, 7.$$

Octal numbers are constructed from powers of eight. To transfer an octal number to decimal, we write the number in terms of octal place values, which are powers of eight.

EXAMPLE 23.7   Write 702_8 as a decimal number.

SOLUTION   We write the number in terms of powers of eight:

$$702_8 = 7 \times 8^2 + 0 \times 8^1 + 2 \times 8^0$$
$$= 7 \times 64 + 0 \times 8 + 2 \times 1$$
$$= 448 + 2$$
$$= 450.$$ ▲

To convert decimal numbers to octal numbers, we use repeated division, dividing by eight.

EXAMPLE 23.8   Write 199 as an octal number.

SOLUTION   We divide 199 by 8, writing the result as a quotient and a remainder R:

$$199 \div 8 = 24, \quad R = 7.$$

Then, we divide the quotient 24 by 8, writing the result as a quotient and a remainder R:

$$24 \div 8 = 3, \quad R = 0.$$

Finally, we divide the quotient 3 by 8, observing that the result is 0, with a remainder 3:

$$3 \div 8 = 0, \quad R = 3.$$

These remainders are the octal digits *in reverse order*. Thus, the octal number is 307_8. ▲

Octal numbers are useful in working with computers because it is very easy to transfer between binary numbers and octal numbers. Each octal digit corresponds to a group of three binary digits:

Octal	Binary		Octal	Binary
0	000		4	100
1	001		5	101
2	010		6	110
3	011		7	111

EXAMPLE 23.9   Write 361_8 as a binary number.

SOLUTION   We write each octal digit as the corresponding group of binary digits, using left-hand zeros when necessary to make groups of three binary digits:

$$361_8 = 11,110,001_2.$$

The group at the left does not need to have three binary digits. Removing the grouping commas, we have

$$361_8 = 11110001_2.$$

You should check this result by writing both the octal and binary numbers as decimal numbers. ▲

EXAMPLE 23.10   Write 10011101_2 as an octal number.

SOLUTION   We separate the binary digits into groups of three starting from the right. The last group at the left does not need to have three binary digits:

$$10011101_2 = 10,011,101_2.$$

Then, we write each group of binary digits as the corresponding octal digit:

$$10,011,101_2 = 235_8.$$

You should check this result by writing both the binary and octal numbers as decimal numbers. ▲

THE HEXADECIMAL NUMBER SYSTEM

The number system that is most often used in association with computers is the **hexadecimal number system**. The hexadecimal number system uses **base 16**, and has sixteen **hexadecimal digits**. Since we have only the ten decimal digits available, the letters A through F are used as the final six digits. Thus, the hexadecimal digits are

$$0, 1, 2, 3, 4, 5, 6, 7, 8, 9, A, B, C, D, E, F,$$

where the digit A corresponds to the decimal number 10, B to 11, C to 12, D to 13, E to 14, and F to 15.

Hexadecimal numbers are constructed from powers of sixteen. To transfer a hexadecimal number to decimal we write the number in terms of hexadecimal place values, which are powers of sixteen.

EXAMPLE 23.11   Write $1C0_{16}$ as a decimal number.

SOLUTION   We write the number in terms of powers of sixteen:

$$1C0_{16} = 1 \times 16^2 + C \times 16^1 + 0 \times 16^0.$$

The hexadecimal digit C corresponds to the decimal number 12; therefore, we write

$$1C0_{16} = 1 \times 16^2 + 12 \times 16^1 + 0 \times 16^0$$

$$= 1 \times 256 + 12 \times 16 + 0 \times 1$$

$$= 256 + 192 + 0$$

$$= 448.$$

▲

Hexadecimal numbers are often written with an H at the end rather than the subscript 16. The hexadecimal number 1C0H is the same as the hexadecimal number $1C0_{16}$. The letter H is *not* a hexadecimal digit, but simply indicates that the number is a hexadecimal number.

To convert decimal numbers to hexadecimal numbers, we use repeated division, dividing by sixteen.

EXAMPLE 23.12   Use repeated division to write 1249 as a hexadecimal number.

SOLUTION   We divide 1249 by 16, writing the result as a quotient and a remainder R:

$$1249 \div 16 = 78, \quad R = 1.$$

Then, we divide the quotient 78 by 16, writing the result as a quotient and a remainder R:

$$78 \div 16 = 4, \quad R = 14 = E.$$

The remainder 14 is not a hexadecimal digit, but corresponds to the hexadecimal digit E. Finally, we divide the quotient 4 by 16, observing that the quotient is 0 with a remainder $R = 4$:

$$4 \div 16 = 0, \quad R = 4.$$

These remainders are the hexadecimal digits *in reverse order*. Thus, the hexadecimal number is 4E1H.   ▲

It is easy to transfer between binary numbers and hexadecimal numbers. Each hexadecimal digit corresponds to a group of four binary digits:

Hexadecimal	Binary	Hexadecimal	Binary
0	0000	8	1000
1	0001	9	1001
2	0010	A	1010
3	0011	B	1011
4	0100	C	1100
5	0101	D	1101
6	0110	E	1110
7	0111	F	1111

EXAMPLE 23.13   Write 4E1H as a binary number.

SOLUTION   There are three hexadecimal digits. Recall that the H indicates a hexadecimal number, but it is not a hexadecimal digit. We write each hexadecimal digit as the corresponding group of binary digits, using left-hand zeros when necessary to make groups of four binary digits:

$$4E1H = 100{,}1110{,}0001_2.$$

The group at the left does not need to have four binary digits. Removing the grouping commas, we have

$$4E1H = 10011100001_2.$$

You should check this result by writing both the binary and the hexadecimal numbers as decimal numbers.   ▲

EXAMPLE 23.14   Write 111010101_2 as a hexadecimal number.

SOLUTION   We separate the binary digits into groups of four, starting from the right. The last group at the left does not need to have four binary digits:

$$111010101_2 = 1{,}1101{,}0101_2.$$

Then, we write each group of binary digits as the corresponding hexadecimal digit:

$$1{,}1101{,}0101_2 = 1D5H.$$

You should check this result by writing both the binary and the hexadecimal numbers as decimal numbers.   ▲

Write each binary number as a decimal number:

1. 1011_2 **2.** 11011_2 **3.** 111000_2 **4.** 1110001_2

5. 10100000_2 **6.** 100000000_2

Write each decimal number as a binary number:

7. 10   **8.** 17   **9.** 40   **10.** 77

11. 127   **12.** 128

Write each octal number as a decimal number:

13. 74_8 **14.** 206_8 **15.** 377_8 **16.** 1307_8

Write each decimal number as an octal number:

17. 28   **18.** 66   **19.** 288   **20.** 575

Write each octal number as a binary number:

21. 27_8 **22.** 41_8 **23.** 506_8 **24.** 1017_8

Write each binary number as an octal number:

25. 1010011_2 **26.** 100001101_2 **27.** 111000010_2 **28.** 1110000100_2

Write each hexadecimal number as a decimal number:

29. 4BH   **30.** AFH   **31.** 1E0H   **32.** AF1H

Write each decimal number as a hexadecimal number:

33. 199   **34.** 208   **35.** 510   **36.** 1024

Write each hexadecimal number as a binary number:

37. 4BH   **38.** F5H   **39.** D0CH   **40.** F3EH

Write each binary number as a hexadecimal number:

41. 111010_2 **42.** 11100111_2

43. 11111110000_2 **44.** 1010000001100_2

Binary Arithmetic

There are just four basic addition facts in the binary number system:

$$0 + 0 = 0$$

$$0 + 1 = 1$$

$$1 + 0 = 1$$

$$1 + 1 = 10.$$

(In this section, when we have stated that the numbers are binary numbers, we will not always write the subscript 2.)

The first three addition facts are the same as for addition of decimal numbers. For the fourth, we observe that, *in decimal*,

$$1 + 1 = 2.$$

But, we know from Section 23.1 that

$$2 = 10_2;$$

therefore, we have

$$1 + 1 = 10_2.$$

To add two binary numbers, we add corresponding **weighted digits**; that is, digits with the same place values, exactly as in addition of decimal numbers.

EXAMPLE 23.15 Add 10 + 101.

SOLUTION We add the corresponding weighted digits:

$$
\begin{array}{r}
10 \\
+\ 101 \\
\hline
111
\end{array}
$$

We may check by converting the addition to decimal:

$$
\begin{array}{rcr}
10 &=& 2 \\
+\ 101 &=& +5 \\
\hline
111 &=& 7
\end{array}
$$
▲

When the sum of two weighted digits is more than 1, we *carry* into the next place to the left.

EXAMPLE 23.16 Add 11 + 1.

SOLUTION We add the corresponding weighted digits. We may show the process of carrying by writing the carried digit above the column into which it is carried. Recalling that $1 + 1 = 10$, we have

$$
\begin{array}{r}
1 \\
11 \\
+\ \ 1 \\
\hline
100
\end{array}
$$

In decimal, this addition is simply $3 + 1 = 4$. ▲

EXAMPLE 23.17 Add 1011 + 101.

SOLUTION We add the corresponding weighted digits, carrying ones into the next column:

$$
\begin{array}{r}
111 \\
1011 \\
+\ \ \ 101 \\
\hline
10000
\end{array}
$$

Checking by converting to decimal, we have

$$
\begin{array}{rcr}
1011 &=& 11 \\
+\ \ \ 101 &=& +\ 5 \\
\hline
10000 &=& 16
\end{array}
$$
▲

Additions with carries often lead to the sum of three ones. We observe that, in binary,

$$
1 + 1 + 1 = 10 + 1
$$
$$
= 11.
$$

EXAMPLE 23.18   Add 10111 + 11011.

SOLUTION   We add the corresponding weighted digits, carrying ones into the next column:

$$
\begin{array}{r}
1111 \\
10111 \\
+\ 11011 \\
\hline
110010
\end{array}
$$

Checking by converting to decimal, we have

$$
\begin{array}{rcr}
10111 & = & 23 \\
+\ 11011 & = & +27 \\
\hline
110010 & = & 50
\end{array}
$$

▲

SUBTRACTION

To subtract binary numbers, we subtract corresponding weighted digits, as in subtraction of decimal numbers.

EXAMPLE 23.19   Subtract 111 − 10.

SOLUTION   We subtract the corresponding weighted digits:

$$
\begin{array}{r}
111 \\
-\ 10 \\
\hline
101
\end{array}
$$

We may check by converting the subtraction to decimal:

$$
\begin{array}{rcr}
111 & = & 7 \\
-\ 10 & = & -2 \\
\hline
101 & = & 5
\end{array}
$$

We could also check by adding in binary:

$$
\begin{array}{r}
10 \\
+\ 101 \\
\hline
111
\end{array}
$$

▲

When a subtraction in binary will not result in a 0 or a 1, we may *borrow* from the next place in the first number. We know that

$$1 + 1 = 10,$$

and therefore, we have

$$10 - 1 = 1.$$

In decimal, this subtraction is simply 2 − 1 = 1.

EXAMPLE 23.20   Subtract 110 − 11.

SOLUTION   We subtract the corresponding weighted digits, borrowing when necessary from the next column:

$$0\ 10\ 10$$
$$\cancel{1}\ \cancel{1}\ 0$$
$$-\quad 1\ \ 1$$
$$\overline{\quad\quad 1\ \ 1}$$

In decimal, this subtraction is 6 − 3 = 3. ▲

Computers do subtraction in binary by using addition. One way subtractions can be written as additions is by using the **1's complement**. To form the 1's complement of a binary number, we change each 0 to a 1 and each 1 to a 0.

EXAMPLE 23.21 ▶ Subtract 110 − 11 by using the 1's complement.

SOLUTION   We must write the binary numbers with the same number of digits:

$$110$$
$$-\ 011$$

Then, we form the 1's complement of 011 by changing the 0's to 1's, and the 1's to 0's to obtain 100. Writing the subtraction as an addition of the 1's complement, we have

$$110$$
$$+\ 100$$
$$\overline{1010}$$

This result, which is larger than either of the given numbers, clearly can not result from the subtraction of the two original numbers. To complete the process, we transfer the final carry, which is the left-most 1, to an addition of 1:

$$110$$
$$+\ 100$$
$$\overline{\cancel{1}010}$$
$$+\quad 1$$
$$\overline{\quad\quad 11}$$

We know from Example 23.20 that this result is correct. The transfer of the final carry to an addition is called an *end-around carry*. ▲

EXAMPLE 23.22 ▶ Subtract 1011 − 101 by using the 1's complement.

SOLUTION   We write each number with four digits:

$$1011$$
$$-\ 0101$$

Then, we form the 1's complement of the second number, write the subtraction as an addition of this 1's complement, and perform an end-around carry:

$$1011$$
$$+\ 1010$$
$$\overline{\cancel{1}0101}$$
$$+\quad\quad 1$$
$$\overline{\quad\quad 110}$$

We may check by converting the subtraction to decimal:

$$
\begin{array}{rcr}
1011 &=& 11 \\
-\ \ 101 &=& -\ 5 \\
\hline
110 &=& 6
\end{array}
$$

We may also check by addition in binary:

$$
\begin{array}{r}
101 \\
+\ \ 110 \\
\hline
1011
\end{array}
$$

▲

EXAMPLE 23.23 ▶ Subtract $1011 - 1101$ by using the 1's complement.

SOLUTION Each number has four digits. We form the 1's complement of the second number, and add the result to the first number:

$$
\begin{array}{r}
1011 \\
+\ 0010 \\
\hline
1101
\end{array}
$$

We observe that there is *no final carry*. We also observe that we have subtracted a larger number from a smaller number. In this case, the answer is *negative*, and this result is the *1's complement* of the answer. We take the 1's complement of the result to obtain 10. Therefore, the result is the negative binary number –10. Checking by converting to decimal, we have

$$
\begin{array}{rcr}
1011 &=& 11 \\
-\ 1101 &=& -\ 13 \\
\hline
-\ \ \ 10 &=& -\ 2
\end{array}
$$

▲

Subtractions can also be written as additions by using the **2's complement**. To form the 2's complement of a binary number, we first form the 1's complement by changing each 0 to a 1 and each 1 to a 0. Then, we form the 2's complement by adding 1 to the 1's complement.

EXAMPLE 23.24 ▶ Subtract $110 - 11$ by using the 2's complement.

SOLUTION We write the binary numbers with the same number of digits:

$$
\begin{array}{r}
110 \\
-\ 011 \\
\hline
\end{array}
$$

Then, we form the 2's complement of 011 by forming the 1's complement and then adding 1:

$$
\begin{array}{r}
100 \\
+\ \ \ 1 \\
\hline
101
\end{array}
$$

Writing the subtraction as an addition of this 2's complement, we have

$$
\begin{array}{r}
110 \\
+\ 101 \\
\hline
1011
\end{array}
$$

When we have used the 2's complement, we simply drop the final carry, which is the left-most 1:

$$\begin{array}{r} 110 \\ + \ 101 \\ \hline \cancel{1}011 \end{array}$$

Thus, the resulting binary number is 11. We know from Examples 23.20 and 23.21 that this result is correct.   ▲

EXAMPLE 23.25   Subtract 1011 − 101 by using the 2's complement.

SOLUTION   We write each number with four digits:

$$\begin{array}{r} 1011 \\ - \ 0101 \\ \hline \end{array}$$

Then, we form the 2's complement of 0101 by forming the 1's complement and adding 1:

$$\begin{array}{r} 1010 \\ + \quad 1 \\ \hline 1011 \end{array}$$

We write the subtraction as an addition of this 2's complement, and drop the final carry:

$$\begin{array}{r} 1011 \\ + \ 1011 \\ \hline \cancel{1}0110 \end{array}$$

The resulting binary number is 110. We checked this result in Example 23.22.   ▲

EXAMPLE 23.26   Subtract 1011 − 1101 by using the 2's complement.

SOLUTION   Each number has four digits. We form the 2's complement of the second number:

$$\begin{array}{r} 0010 \\ + \quad 1 \\ \hline 0011 \end{array}$$

Then, we add the 2's complement:

$$\begin{array}{r} 1011 \\ + \ 0011 \\ \hline 1110 \end{array}$$

There is *no final carry*, and we have subtracted a larger number from a smaller number. In this case, the answer is *negative*, and this result is the *2's complement* of the answer. We take the 2's complement of the result to obtain

$$\begin{array}{r} 0001 \\ + \quad 1 \\ \hline 0010 \end{array}$$

Thus, the result is the negative binary number −10. We checked this result in Example 23.23.   ▲

MULTIPLICATION

There are four basic multiplication facts in the binary number system:

$$0 \times 0 = 0$$
$$0 \times 1 = 0$$
$$1 \times 0 = 0$$
$$1 \times 1 = 1.$$

These multiplication facts are the same as for multiplication of decimal numbers. To multiply binary numbers, we follow the pattern for multiplication of decimal numbers. Multiplication by the binary number 10 puts a zero on the end of a binary number.

EXAMPLE 23.27 Multiply 101 × 10.

SOLUTION We follow the pattern for multiplication of decimal numbers:

$$
\begin{array}{r}
101 \\
\times \quad 10 \\
\hline
000 \\
101 \quad\ \\
\hline
1010 \\
\end{array}
$$

We may check by converting the multiplication to decimal:

$$
\begin{array}{rcr}
101 & = & 5 \\
\times \quad 10 & = & \times\ 2 \\
\hline
1010 & = & 10 \\
\end{array}
$$
▲

We observe that there are no carries in multiplication of the binary numbers. There may, however, be carries in the addition of partial products.

EXAMPLE 23.28 Multiply 1011 × 111.

SOLUTION We follow the pattern for multiplication of decimal numbers:

$$
\begin{array}{r}
1011 \\
\times \quad 111 \\
\hline
1011 \\
1011\ \ \\
1011\ \ \ \\
\hline
1001101 \\
\end{array}
$$

Checking by converting to decimal, we have

$$
\begin{array}{rcr}
1011 & = & 11 \\
\times \quad 111 & = & \times\ 7 \\
\hline
1001101 & = & 77 \\
\end{array}
$$
▲

Multiplication can also be done as repeated addition. For example, because 10_2 means 2, the multiplication in Example 23.27,

$$
\begin{array}{r}
101 \\
\times \quad 10 \\
\hline
1010 \\
\end{array}
$$

is the same as the addition

$$
\begin{array}{r}
101 \\
+\ \ 101 \\
\hline
1010
\end{array}
$$

Similarly, because 11_2 means 3, the multiplication

$$
\begin{array}{r}
101 \\
\times\ \ \ 11 \\
\hline
101 \\
101 \\
\hline
1111
\end{array}
$$

is the same as the addition

$$
\begin{array}{r}
101 \\
101 \\
+\ \ 101 \\
\hline
1111
\end{array}
$$

In decimal, this multiplication is $5 \times 3 = 15$.

DIVISION

We may do division of binary numbers by following the pattern for division of decimal numbers, or by using repeated subtraction, which is how computers divide. For example, to divide 110 by 011 by using repeated subtraction, we form the 2's complement of 011:

$$
\begin{array}{r}
100 \\
+\ \ \ 1 \\
\hline
101
\end{array}
$$

Adding the 2's complement, we have

$$
\begin{array}{r}
110 \\
+\ \ 101 \\
\hline
\cancel{1}011
\end{array}
$$

or 011, dropping the final carry. If we subtract by adding the 2's complement a second time, we have

$$
\begin{array}{r}
011 \\
+\ \ 101 \\
\hline
\cancel{1}000
\end{array}
$$

or 0. The quotient is the number of subtractions, which is 2 or 10_2. In decimal, this division is $6 \div 3 = 2$, or as a repeated subtraction, 3 subtracted twice:

$$
\begin{array}{r}
6 \\
-\ 3 \\
\hline
3 \\
-\ 3 \\
\hline
0
\end{array}
$$

BINARY FRACTIONS

Finally, we look at the representation of fractional numbers in binary arithmetic. Decimal numbers that contain a decimal point and a fractional part are **fractional decimal numbers** or **decimal fractions**. The fractional part of a decimal number is constructed from reciprocals of powers of ten. For example, the number 2.22 means

$$2.22 = 2 \times 1 + 2 \times \frac{1}{10} + 2 \times \frac{1}{100}$$

$$= 2 \times 10^0 + 2 \times \frac{1}{10^1} + 2 \times \frac{1}{10^2}.$$

In Unit 2, we used powers of ten with negative exponents, such as

$$\frac{1}{10^1} = 10^{-1}$$

$$\frac{1}{10^2} = 10^{-2}.$$

Thus, the number 2.22 can be written

$$2.22 = 2 \times 10^0 + 2 \times 10^{-1} + 2 \times 10^{-2}.$$

Similarly, binary numbers may contain a **binary point** and a fractional part. The binary point plays the same role as a decimal point. Binary numbers that contain a binary point and a fractional part are **fractional binary numbers** or **binary fractions**. The fractional part of a binary number is constructed from powers of two with negative exponents. To transfer a binary fraction to a decimal fraction, we write the binary fraction in terms of powers of two.

EXAMPLE 23.29 Write 0.11_2 as a decimal fraction.

SOLUTION We write the binary fraction in terms of powers of two:

$$0.11_2 = 1 \times 2^{-1} + 1 \times 2^{-2}$$

$$= 1 \times \frac{1}{2} + 1 \times \frac{1}{2^2}$$

$$= 1 \times \frac{1}{2} + 1 \times \frac{1}{4}$$

$$= 0.5 + 0.25$$

$$= 0.75. \qquad \blacktriangle$$

EXAMPLE 23.30 Write 0.0101_2 as a decimal fraction.

SOLUTION We write the binary fraction in terms of powers of two:

$$0.0101_2 = 0 \times 2^{-1} + 1 \times 2^{-2} + 0 \times 2^{-3} + 1 \times 2^{-4}$$

$$= 1 \times \frac{1}{2^2} + 1 \times \frac{1}{2^4}$$

$$= 1 \times \frac{1}{4} + 1 \times \frac{1}{16}$$

$$= 0.25 + 0.0625$$

$$= 0.3125. \qquad \blacktriangle$$

To convert decimal fractions to binary fractions, we use **repeated multiplication**. Repeated multiplication is the opposite of repeated division. We multiply the decimal fraction by 2, and then multiply each resulting fractional part by 2. The whole number part of each multiplication is either a 0 or a 1. These whole number parts are the binary digits of the binary number. The binary digits appear in the binary fraction in the order in which they are obtained.

EXAMPLE 23.31   Write 0.375 as a binary fraction.

SOLUTION   We multiply 0.375 by 2:

$$0.375 \times 2 = 0.75, \quad C = 0.$$

The whole number part, or carry, is $C = 0$. Then, we multiply 0.75 by 2:

$$0.75 \times 2 = 1.5, \quad C = 1.$$

The carry is $C = 1$. We multiply the fractional part .5 by 2:

$$0.5 \times 2 = 1.0, \quad C = 1.$$

The carry is $C = 1$, and the fractional part is 0. The carries are the binary digits, in order. Thus, the binary fraction is 0.011_2. You should check this result by returning the binary fraction 0.011_2 to decimal form.   ▲

EXAMPLE 23.32   Write 0.1 as a binary fraction.

SOLUTION   The number 0.1 is the *decimal* fraction $\frac{1}{10}$. To write the *binary* fraction, we multiply 0.1 by 2, and then multiply each resulting fractional part by 2:

$$0.1 \times 2 = 0.2, \quad C = 0$$
$$0.2 \times 2 = 0.4, \quad C = 0$$
$$0.4 \times 2 = 0.8, \quad C = 0$$
$$0.8 \times 2 = 1.6, \quad C = 1.$$

So far, the binary fraction is 0.0001_2. We continue, multiplying the fractional parts only:

$$0.6 \times 2 = 1.2, \quad C = 1$$
$$0.2 \times 2 = 0.4, \quad C = 0$$
$$0.4 \times 2 = 0.8, \quad C = 0$$
$$0.8 \times 2 = 1.6, \quad C = 1$$
$$0.6 \times 2 = 1.2, \quad C = 1$$
$$0.2 \times 2 = 0.4, \quad C = 0$$
$$0.4 \times 2 = 0.8, \quad C = 0$$
$$0.8 \times 2 = 1.6, \quad C = 1.$$

The fractional parts will not reach zero, but will continue in the pattern 0.6, 0.2, 0.4, 0.8, . . . , with the carries 1, 0, 0, 1. Thus, the result is a nonterminating repeating binary fraction

$$0.000110011001 \ldots {}_2.$$

You can check this result by using a calculator to evaluate the powers of two. You will get closer to the decimal fraction 0.1 as you continue to find more places.   ▲

EXERCISE 23.2   Add the binary numbers:

1. $100 + 11$
2. $1001 + 110$
3. $10 + 10$
4. $101 + 11$
5. $1101 + 110$
6. $10110 + 1011$
7. $10111 + 10110$
8. $111011 + 110111$

Subtract by using borrowing when necessary:

9. 111 − 101 **10.** 1101 − 1000

11. 110 − 101 **12.** 1110 − 111

Subtract by using the 1's complement:

13. 111 − 101 **14.** 1101 − 1000

15. 1101 − 111 **16.** 11010 − 1011

17. 100001 − 10001 **18.** 111110 − 100010

19. 1001 − 1110 **20.** 10110 − 11001

Subtract by using the 2's complement:

21. 111 − 101 **22.** 1101 − 1000

23. 1001 − 110 **24.** 10111 − 1101

25. 101100 − 11100 **26.** 111110 − 100001

27. 1011 − 1110 **28.** 10010 − 11010

Multiply the binary numbers:

29. 111 × 10 **30.** 110 × 100 **31.** 111 × 101 **32.** 1101 × 111

Write the binary fraction as a decimal fraction:

33. 0.1_2 **34.** 0.101_2 **35.** 0.01_2 **36.** 0.011_2

37. 0.00011_2 **38.** 0.010101_2

Write the decimal fraction as a binary fraction:

39. 0.125 **40.** 0.875 **41.** 0.1875 **42.** 0.078125

43. 0.4 **44.** 0.9

SECTION 23.3

Boolean Algebra

As we saw in the preceding sections, computers do arithmetic by using only the digits 0 and 1. These digits correspond to the electronic states off and on, or more accurately, low and high voltages. Computer circuits operate by using a type of logic that also uses just 0 and 1. This logic is based on a field of mathematics called **Boolean algebra**, named for George Boole (1815–1864), an English mathematician and logician.

Boolean algebra in its mathematical form is sometimes called *propositional calculus*, because it deals with mathematical *propositions*, or statements. Each statement has the value *true* or the value *false*. The basic properties of propositional calculus can be demonstrated by using English statements. For example, the statement "This book has 24 units" is true. The statements "This book *does not* have 24 units" and "*It is false that* this book has 24 units" are false.

For any statement A, the statement "*it is false that* A" is called the **complement of A**, or the **negation of A**, or **NOT A**. The complement of A is denoted by the symbol $\overline{A}$, read "not A" or "A bar." If A is true then $\overline{A}$ is false. If A is false then $\overline{A}$ is true. In computer logic, 0 represents false and 1 represents true. Thus, if $A = 1$ then $\overline{A} = 0$, and if $A = 0$ then $\overline{A} = 1$. We summarize these results by using a **truth table** as shown at the left.

The **disjunction of A and B** combines two statements A and B by A **OR** B, which is the **inclusive OR**. In computer logic, A OR B is represented by the plus sign: $A + B$. If A and B are both true then $A + B$ is true. If A is true but B is false, or if A is false but B is true, then $A + B$ is true. If A and B are both false then $A + B$ is false. The truth values for $A + B$ are given by the truth table at the left. We observe that, in Boolean algebra, we have the unusual fact that

$$1 + 1 = 1.$$

In Boolean algebra there is neither a 2 nor a 10_2.

A	$\overline{A}$
0	1
1	0

Complementation Truth Table

A	B	$A + B$
0	0	0
0	1	1
1	0	1
1	1	1

Disjunction Truth Table

A	B	$A \cdot B$
0	0	0
0	1	0
1	0	0
1	1	1

Conjunction Truth Table

The **conjunction of A and B** is A AND B. In computer logic, A AND B is represented by the multiplication dot: $A \cdot B$. If A and B are both true then $A \cdot B$ is true. If A is true but B is false, or if A is false but B is true, then $A \cdot B$ is false. If A and B are both false then $A \cdot B$ is false. The truth values for $A \cdot B$ are given by the truth table at the left.

Complementation, disjunction, and conjunction, are **logical operators**. The logical operators have many properties, which are the basis of Boolean algebra. Some of the properties are similar to properties of ordinary arithmetic and algebra. For example, the **commutative properties** are

$$A + B = B + A$$
$$A \cdot B = B \cdot A.$$

The **associative properties** are

$$(A + B) + C = A + (B + C)$$
$$(A \cdot B) \cdot C = A \cdot (B \cdot C).$$

The **distributive property** of Boolean algebra is

$$A \cdot (B + C) = A \cdot B + A \cdot C.$$

There also are properties for 0 and 1. The **identity properties** are

$$A + 0 = A$$
$$A \cdot 1 = A$$

and a special property of 0 is

$$A \cdot 0 = 0.$$

Other properties are different from properties of ordinary arithmetic and algebra. For example, in Boolean algebra a special property of 1 is

$$A + 1 = 1,$$

and the **redundancy properties** are

$$A + A = A$$
$$A \cdot A = A.$$

We prove such properties by using truth tables.

EXAMPLE 23.33 Use truth tables to prove the property $A + 1 = 1$.

SOLUTION Since A can be 0 or 1, we start the truth table with columns for A and 1. Then, combining the first two columns by using disjunction truth values, we have

A	1	$A + 1$
0	1	1
1	1	1

We observe that the result is always 1.   ▲

EXAMPLE 23.34 Use truth tables to prove the property $A + A = A$.

SOLUTION We start with two identical columns for A and combine by using disjunction truth values:

A	A	$A + A$
0	0	0
1	1	1

We observe that the resulting values are the same as the original values of A.   ▲

Some **complementation properties** are

$$A + \overline{A} = 1$$

$$A \cdot \overline{A} = 0,$$

and the **double complementation property** is

$$\overline{\overline{A}} = A.$$

EXAMPLE 23.35 Use truth tables to prove the property $A \cdot \overline{A} = 0$.

SOLUTION We complement A and then combine A and $\overline{A}$ by using conjunction truth values:

A	$\overline{A}$	$A \cdot \overline{A}$
0	1	0
1	0	0

We observe that the result is always 0. ▲

PROPERTIES INVOLVING TWO STATEMENTS

To prove properties involving one statement A, we have used truth tables with *two* lines. To prove properties involving two statements A and B, we must use truth tables with *four* lines like those defining $A + B$ and $A \cdot B$. Some properties involving two statements A and B are the **absorption properties**

$$A + A \cdot B = A$$

$$A + \overline{A} \cdot B = A + B.$$

For properties involving more than one logical operator, such as the absorption properties, the precedence order for logical operations is

1. complementation, $\overline{A}$
2. conjunction, $A \cdot B$
3. disjunction, $A + B$.

EXAMPLE 23.36 Use truth tables to prove $A + \overline{A} \cdot B = A + B$.

SOLUTION We start with the four possible combinations of truth values for A and B, as in the truth tables for $A + B$ and $A \cdot B$. Then, we construct several columns to evaluate the complement, conjunction, and disjunction, following the precedence order:

A	B	$\overline{A}$	$\overline{A} \cdot B$	$A + \overline{A} \cdot B$
0	0	1	0	0
0	1	1	1	1
1	0	0	0	1
1	1	0	0	1

The column under $\overline{A}$ complements the first column. The column under $\overline{A} \cdot B$ combines the second and third columns by using conjunction truth values. The column under $A + \overline{A} \cdot B$ combines the first and fourth columns by using disjunction truth values. We observe that the resulting values are the same as the truth values for $A + B$. ▲

Two properties involving complements of A and B are called **DeMorgan's laws**, named for Augustus DeMorgan (1806–1871), an English mathematician and friend of Boole. DeMorgan's laws relate complementation with disjunction and conjunction:

$$\overline{A + B} = \overline{A} \cdot \overline{B}$$

$$\overline{A \cdot B} = \overline{A} + \overline{B}.$$

EXAMPLE 23.37   Use truth tables to prove $\overline{A + B} = \overline{A} \cdot \overline{B}$.

SOLUTION   For the left-hand side, we construct the truth table for $A + B$, and then complement each value:

A	B	$A + B$	$\overline{A + B}$
0	0	0	1
0	1	1	0
1	0	1	0
1	1	1	0

For the right-hand side, we construct a truth table in which we complement A, complement B, and then combine the complements by using conjunction truth values:

A	B	$\overline{A}$	$\overline{B}$	$\overline{A} \cdot \overline{B}$
0	0	1	1	1
0	1	1	0	0
1	0	0	1	0
1	1	0	0	0

When the resulting values in each table are the same, the Boolean expressions are equal. ▲

PROPERTIES INVOLVING THREE STATEMENTS

To prove properties involving three statements A, B, and C, we must use truth tables with *eight* lines. The eight possible combinations of truth values for A, B, and C, are shown in the following examples.

EXAMPLE 23.38   Use truth tables to prove the associative property $(A + B) + C = A + (B + C)$.

SOLUTION   For the left-hand side, we combine A and B by using disjunction truth values, and then combine the resulting values with C. For the right-hand side, we combine B and C by using disjunction truth values, and then combine A with the resulting values:

A	B	C	$A + B$	$(A + B) + C$		A	B	C	$B + C$	$A + (B + C)$
0	0	0	0	0		0	0	0	0	0
0	0	1	0	1		0	0	1	1	1
0	1	0	1	1		0	1	0	1	1
0	1	1	1	1		0	1	1	1	1
1	0	0	1	1		1	0	0	0	1
1	0	1	1	1		1	0	1	1	1
1	1	0	1	1		1	1	0	1	1
1	1	1	1	1		1	1	1	1	1

Left-hand side   Right-hand side

The resulting values in each table are the same so the Boolean expressions are equal. ▲

EXAMPLE 23.39   Use truth tables to prove the distributive property $A \cdot (B + C) = A \cdot B + A \cdot C$.

SOLUTION   As in ordinary arithmetic and algebra, parentheses have priority over all logical operators. Therefore, for the left-hand side, we combine B and C by using disjunction truth values, and then combine A with the resulting values by using conjunction truth values. For the right-hand side, we combine A and B by using conjunction truth values, and A and C by using conjunction truth values, and then combine these results by using disjunction truth values:

A	B	C	$B + C$	$A \cdot (B + C)$
0	0	0	0	0
0	0	1	1	0
0	1	0	1	0
0	1	1	1	0
1	0	0	0	0
1	0	1	1	1
1	1	0	1	1
1	1	1	1	1

A	B	C	$A \cdot B$	$A \cdot C$	$A \cdot B + A \cdot C$
0	0	0	0	0	0
0	0	1	0	0	0
0	1	0	0	0	0
0	1	1	0	0	0
1	0	0	0	0	0
1	0	1	0	1	1
1	1	0	1	0	1
1	1	1	1	1	1

Left-hand side   Right-hand side

The resulting values in each table are the same so the Boolean expressions are equal. ▲

PROPERTIES OF BOOLEAN ALGEBRA

We have used truth tables to prove several properties of Boolean algebra. Because of the similarity of many of these properties to algebraic properties, the logical conjunction $A \cdot B$ of is often indicated by writing AB, as in algebraic multiplication. Thus, for example, the distributive property

$$A \cdot (B + C) = A \cdot B + A \cdot C$$

may be written

$$A(B + C) = AB + AC.$$

It should be understood, however, that the logical operation conjunction is *not the same as* the arithmetic operation multiplication, just as the logical operation disjunction is *not the same as* the arithmetic operation addition. The properties of Boolean algebra are listed in the following table.

Properties of Boolean Algebra:

Commutative Properties:
1. $A + B = B + A$
2. $AB = BA$

Associative Properties:
1. $(A + B) + C = A + (B + C)$
2. $(AB)C = A(BC)$

Distributive Property:
$A(B + C) = AB + AC$

Identity Properties:
1. $A + 0 = A$
2. $A \cdot 1 = A$

Special Properties for 0 and 1:
1. $A + 1 = 1$
2. $A \cdot 0 = 0$

Redundancy Properties:
1. $A + A = A$
2. $AA = A$

Complementation Properties:
1. $A + \bar{A} = 1$
2. $A\bar{A} = 0$

continued

Properties of Boolean Algebra—*continued*

Double Complementation Property:
$$\overline{\overline{A}} = A$$

Absorption Properties:
1. $A + AB = A$
2. $A + \overline{A}B = A + B$

DeMorgan's Laws:
1. $\overline{A + B} = \overline{A} \cdot \overline{B}$
2. $\overline{AB} = \overline{A} + \overline{B}$

We may use the properties of Boolean algebra to simplify Boolean expressions and to prove other equalities.

EXAMPLE 23.40   Use properties of Boolean algebra to prove the absorption property $A + AB = A$.

SOLUTION   The distributive property may be used to factor expressions as in ordinary algebra. Thus, using the second identity property and then factoring, we write

$$A + AB = A \cdot 1 + AB$$
$$= A(1 + B).$$

We use the first commutative property and then the special property for 1 to write $1 + B = B + 1$ and $B + 1 = 1$. Thus, we have

$$A(1 + B) = A(1).$$

Finally, we use the second identity property again to obtain

$$A(1) = A. \qquad \blacktriangle$$

EXAMPLE 23.41   Use properties of Boolean algebra to prove $A(\overline{A}B + \overline{A}C) = 0$.

SOLUTION   We can proceed in either of two ways.

Method 1.   We may use the distributive property to write

$$A(\overline{A}B + \overline{A}C) = A(\overline{A}B) + A(\overline{A}C).$$

Then, we use the second associative and complementation properties to write

$$A(\overline{A}B) + A(\overline{A}C) = (A\overline{A})B + (A\overline{A})C$$
$$= 0(B) + 0(C).$$

From the second commutative property, we know that $0 \cdot B = B \cdot 0$ and $0 \cdot C = C \cdot 0$, so we can use the special property for 0 to write

$$0(B) + 0(C) = 0 + 0.$$

Then, using the first identity property, we have

$$0 + 0 = 0.$$

Method 2.   We may use the distributive property to factor:

$$A(\overline{A}B + \overline{A}C) = A[\overline{A}(B + C)].$$

Then, we use the second associative and complementation properties to write

$$A[\overline{A}(B + C)] = (A\overline{A})(B + C)$$
$$= 0(B + C).$$

Using the special property for 0, we have

$$0(B + C) = 0. \qquad \blacktriangle$$

The commutative and associative properties are used so frequently that we will not always point out their use.

There is a second distributive property that reverses the roles of disjunction and conjunction. The first distributive property is for conjunction with respect to disjunction:

$$A(B + C) = AB + AC.$$

The second distributive property is for disjunction with respect to conjunction:

$$A + BC = (A + B)(A + C).$$

This second distributive property has no counterpart in ordinary arithmetic or algebra.

EXAMPLE 23.42   Use properties of Boolean algebra to prove $A + BC = (A + B)(A + C)$.

SOLUTION We start with the more complicated side, and recall the method for multiplying binomials in Section 14.1. This method is a double application of the first distributive property, so we may use it to write

$$(A + B)(A + C) = AA + AC + BA + BC.$$

Then, from the second redundancy property, we have

$$AA + AC + BA + BC = A + AC + BA + BC.$$

Now, we may use the method in Example 23.40, or the first absorption property applied to A and C, to write $A + AC = A$. Therefore, we have

$$A + AC + BA + BC = A + BA + BC.$$

Similarly, we know that $A + BA = A$, and so

$$A + BA + BC = A + BC.$$ ▲

EXAMPLE 23.43   Use properties of Boolean algebra to prove $\overline{A} + B = A\overline{B}$.

SOLUTION We use DeMorgan's first law applied to $\overline{A}$ and B to write

$$\overline{\overline{A} + B} = \overline{\overline{A}} \cdot \overline{B}$$

Then, by using the double complementation property, we have

$$\overline{\overline{A}} \cdot \overline{B} = A \cdot \overline{B}.$$ ▲

**EXERCISE
23.3**

Prove by using truth tables:

1. $A + 0 = A$ 2. $A \cdot 1 = A$
3. $A \cdot 0 = 0$ 4. $A \cdot A = A$
5. $A + \overline{A} = 1$ 6. $\overline{\overline{A}} = A$ (Use columns for A, $\overline{A}$, and $\overline{\overline{A}}$.)
7. $A + (A \cdot B) = A$ 8. $A \cdot (A + B) = A$
9. $\overline{A \cdot B} = \overline{A} + \overline{B}$ 10. $A \cdot B \cdot (\overline{A} + \overline{B}) = 0$
11. $\overline{\overline{A} \cdot B} = A + \overline{B}$ 12. $\overline{\overline{A} + B} = A \cdot \overline{B}$
13. $(A \cdot B) \cdot C = A \cdot (B \cdot C)$ 14. $(A \cdot B) \cdot (B \cdot C) = A \cdot B \cdot C$
15. $A + B \cdot C = (A + B) \cdot (A + C)$ 16. $A + B + A \cdot B + A \cdot C = A + B$
17. $\overline{A + B + C} = \overline{A} \cdot \overline{B} \cdot \overline{C}$ 18. $\overline{A \cdot B \cdot C} = \overline{A} + \overline{B} + \overline{C}$
19. $\overline{\overline{A} + \overline{B} \cdot \overline{C}} = A \cdot B \cdot C$ 20. $\overline{\overline{A} + \overline{B} \cdot \overline{C}} = A \cdot (B + C)$

Prove by using properties of Boolean algebra:

21. $A\overline{B} + AB = A$ 22. $A + AB + \overline{A}B = A + B$
23. $A(A + B) = A$ 24. $A(AB + B) = AB$

25. $A(\overline{B} + AB) = A$ **26.** $AB(\overline{A} + \overline{B}) = 0$

27. $A + ABC = A$ **28.** $ABC + A\overline{BC} = A$

29. $ABC + AB\overline{C} = AB$ **30.** $ABC(A\overline{B} + A\overline{C}) = 0$

31. $(AB + BC)(AC + BC) = BC$ **32.** $(\overline{A} + B)(A + \overline{B}) = AB + \overline{A}{\cdot}\overline{B}$

33. $\overline{\overline{A} + B} = A\overline{B}$ **34.** $\overline{\overline{A} + \overline{B}} = AB$

35. $\overline{\overline{A} + \overline{B}{\cdot}\overline{C}} = A(B + C)$ **36.** $\overline{A(\overline{B} + \overline{C})} = A + BC$

<table>
<tr><td>

SECTION

23.4

</td></tr>
</table>

Digital Circuits

Digital circuits operate at two voltage levels, low and high, or off and on, or in terms of Boolean algebra, the logic values 0 and 1. Any device that has just two states, such as a switch that is either off or on, can be the basis of a digital circuit. Digital circuits constructed from relay switches, a switching device used in telephone networks, were used in electromechanical calculating machines built in the late 1930s and early 1940s. These relay switch machines were the forerunners of today's electronic computers. Digital circuits constructed from vacuum tubes were used in the Electronic Numerical Integrator and Computer (ENIAC). This machine, usually considered to have been the first electronic computer, was built during the years 1943–1946 at the University of Pennsylvania. Later, digital circuits were constructed by using the transistor, which was invented in 1947 at AT&T's Bell Laboratories. Several transistors were incorporated into a single unit by the invention of the integrated circuit in the late 1950s at Texas Instruments. Today's digital circuits are integrated circuits that can incorporate millions of electronic devices etched on silicon chips.

A vacuum tube and a 1 million bit chip.
Courtesy of *International Business Machines Corporation*.

Digital circuits are built from **logic gates**. A logic gate is an integrated circuit that performs a logical operation such as NOT, OR, or AND. The **NOT gate** is also called an **inverter**. Each logic gate has a **logic symbol** that is used to represent the gate in circuit diagrams. The logic symbol for the inverter is shown at the left below. The inverter takes an input signal A and converts it to an output signal $X = \overline{A}$. If the input signal is low then the output signal is high, and if the input signal is high then the output signal is low. When we use 0 for low and 1 for high, the truth table for the inverter is the same as the complement or NOT truth table of Boolean algebra:

$A \longrightarrow\!\!\!\!\triangleright\!\!\circ\!\!\longrightarrow X = \overline{A}$

Logic symbol and truth table for the inverter.

Input	Output
A	X
0	1
1	0

The **OR gate** has the logic symbol shown at the left below. The OR gate takes two input signals A and B and converts them to an output signal $X = A + B$. The truth table for the OR gate is the same as the disjunction or inclusive-OR truth table of Boolean algebra:

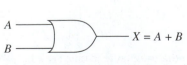

A ——
B —— $X = A + B$

Logic symbol and truth table for the OR gate.

Inputs		Output
A	B	X
0	0	0
0	1	1
1	0	1
1	1	1

The **AND gate** has the logic symbol shown at the left on the top of page 545. The AND gate takes two input signals A and B and converts them to an output signal $X = AB$. The truth table for the AND gate is the same as the conjunction or AND truth table of Boolean algebra:

Inputs		Output
A	B	X
0	0	0
0	1	0
1	0	0
1	1	1

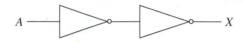

A —[ ]— $X = AB$
B —

Logic symbol and truth table for
the AND gate

EXAMPLE 23.44 ▶ Determine the output X of this circuit if

   a. $A = 0$ **b.** $A = 1$

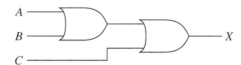

A ——▷○——▷○—— X

SOLUTIONS   We use the truth table for the inverter.

a. If $A = 0$, the first inverter converts A to 1, and the second converts the 1 back to 0.
Therefore, the output is $X = 0$.

b. If $A = 1$, the first inverter converts A to 0, and the second converts the 0 back to 1.
Therefore, the output is $X = 1$. We observe that this circuit illustrates the double complementation property, $\overline{\overline{A}} = A$.   ▲

EXAMPLE 23.45 ▶ Determine the output X of this circuit if $A = 0$, $B = 1$, and $C = 1$:

A —
B —[OR]—[OR]— X
C —

SOLUTION   We use the truth table for the OR gate. If $A = 0$ and $B = 1$, then

$$A + B = 0 + 1 = 1.$$

A —
B —[ ]— X
C —

A three-input OR gate.

If $A + B = 1$ and $C = 1$, then

$$(A + B) + C = 1 + 1 = 1.$$

Therefore, the output is $X = 1$. This circuit illustrates the left-hand side of the associative
property, $(A + B) + C$.   ▲

   The circuit for $(A + B) + C$ or $A + (B + C)$ is usually represented by the *three-
input* OR gate shown at the left above. Similarly, the circuit for $(AB)C$ or $A(BC)$ is usually
represented by the *three-input* AND gate shown at the left.

A —
B —[ ]— X
C —

A three-input AND gate.

EXAMPLE 23.46 ▶ Determine the output X of this circuit if $A = 0$ and $B = 1$:

A —[OR]—▷○—— X
B —

SOLUTION   First, we use the truth table for the OR gate. If $A = 0$ and $B = 1$, then

$$A + B = 0 + 1 = 1.$$

Then, we use the truth table for the inverter. If $A + B = 1$ then

$$\overline{A + B} = 0.$$

Therefore, the output is $X = 0$.   ▲

NOR AND NAND GATES

A gate called the **NOR gate** combines the two gates in Example 23.46. The NOR gate has the logic symbol shown at the left below. The NOR gate takes two input signals A and B and converts them to an output signal $\overline{A + B}$. The truth table for the NOR gate is

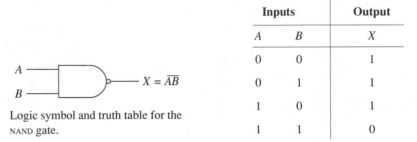

Inputs		Output
A	B	X
0	0	1
0	1	0
1	0	0
1	1	0

$X = \overline{A + B}$

Logic symbol and truth table for the NOR gate.

The **NAND gate** combines an AND gate with an inverter. The NAND gate has the logic symbol shown at the left below. The NAND gate takes two input signals A and B and converts them to an output signal $\overline{AB}$. The truth table for the NAND gate is

Inputs		Output
A	B	X
0	0	1
0	1	1
1	0	1
1	1	0

$X = \overline{AB}$

Logic symbol and truth table for the NAND gate.

EXAMPLE 23.47 ▶ Determine the output X of this circuit if $A = 0$ and $B = 1$:

SOLUTION First, we use the truth table for the inverter. If $A = 0$ and $B = 1$, then

$$\overline{A} = 1$$

and

$$\overline{B} = 0.$$

Then, using the truth table for the AND gate, we have

$$\overline{A} \cdot \overline{B} = (1)(0) = 0.$$

Therefore, the output is $X = 0$.   ▲

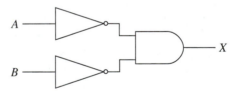

$X = \overline{A} \cdot \overline{B}$

A negative-AND gate.

$X = \overline{A} + \overline{B}$

A negative-OR gate.

A negative-AND gate combines the gates in Example 23.47. The negative-AND gate has the logic symbol shown at the left. The *NOR* gate and the *negative-AND* gate are equivalent, and illustrate DeMorgan's law

$$\overline{A + B} = \overline{A} \cdot \overline{B}.$$

Similarly, a negative-OR gate combines two inverters with an OR gate. The negative-OR gate has the logic symbol shown at the left. The *NAND* gate and the *negative-OR* gate are equivalent, and illustrate DeMorgan's law

$$\overline{AB} = \overline{A} + \overline{B}.$$

CIRCUITS AND BOOLEAN EXPRESSIONS

Given a Boolean expression, we can draw a circuit that represents the expression by using logic gates.

EXAMPLE 23.48

Draw the circuit that represents $X = AB + C$.

SOLUTION

The inputs A and B are combined by an AND gate. The input C is then combined with the *output* of the AND gate by an OR gate. Thus, the output of this circuit is $X = AB + C$:

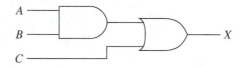

▲

EXAMPLE 23.49

Draw the circuit that represents $X = (\overline{A} + B)C$.

SOLUTION

The input A is inverted *before* being combined with B by an OR gate. The input C is then combined with the *output* of the OR gate by an AND gate. Thus, the output of this circuit is $X = (\overline{A} + B)C$:

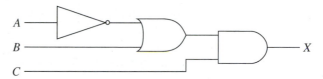

▲

Similarly, given a diagram of a logic circuit, we can write a Boolean expression represented by the output of the circuit.

EXAMPLE 23.50

Write the Boolean expression represented by the output X of this circuit:

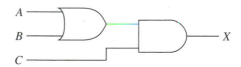

SOLUTION

The OR gate combining the inputs A and B represents $A + B$. Then, the AND gate combining the output of the OR gate with C represents $(A + B)C$. Thus, the Boolean expression represented by the output of the circuit is

$$X = (A + B)C.$$

▲

EXAMPLE 23.51

Write the Boolean expression represented by the output X of this circuit:

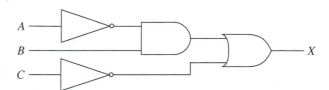

SOLUTION

The input A is inverted, and then the AND gate combining $\overline{A}$ with B represents $\overline{A}B$. The input C is inverted, and then the OR gate combining the output of the AND gate with $\overline{C}$ represents $\overline{A}B + \overline{C}$. Thus, the Boolean expression represented by the output of the circuit is

$$X = \overline{A}B + \overline{C}.$$

▲

SIMPLIFYING CIRCUITS

Often, a circuit can be simplified by writing its Boolean expression and then simplifying the Boolean expression. The circuit represented by the simplified expression has the same output as the original circuit.

EXAMPLE 23.52

Write the Boolean expression represented by the output X of the circuit, simplify the expression, and draw the circuit represented by the resulting expression:

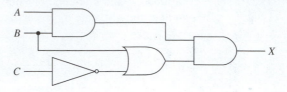

SOLUTION

The AND gate combining the inputs A and B represents AB. The dot connecting B to the OR gate means that B is also an input for the OR gate. The input C is inverted, so the OR gate combining B and $\overline{C}$ represents $B + \overline{C}$. Finally, the outputs of the AND and OR gates are combined by another AND gate. Thus, the Boolean expression represented by the output of the circuit is

$$X = AB(B + \overline{C}).$$

We can simplify this expression by using the distributive property:

$$AB(B + \overline{C}) = ABB + AB\overline{C}.$$

Using the second redundancy property, the second identity property, and then factoring, we have

$$ABB + AB\overline{C} = AB + AB\overline{C}$$

$$= AB{\cdot}1 \ AB\overline{C}$$

$$= AB(1 + \overline{C}).$$

Then, the special property for 1 and second identity property give

$$AB(1 + \overline{C}) = AB(1)$$

$$= AB.$$

Therefore, the original circuit can simply be replaced by an AND gate with inputs A and B.

We can check this result by using truth tables. The truth table for the original expression is

A	B	C	$\overline{C}$	AB	$B + \overline{C}$	$AB(B + \overline{C})$
0	0	0	1	0	1	0
0	0	1	0	0	0	0
0	1	0	1	0	1	0
0	1	1	0	0	1	0
1	0	0	1	0	1	0
1	0	1	0	0	0	0
1	1	0	1	1	1	1
1	1	1	0	1	1	1

Comparing the final column to the column for AB, we see that the truth values are the same.   ▲

EXAMPLE 23.53 ▶ Write the Boolean expression represented by the output X of the circuit, simplify the expression, and draw the circuit represented by the resulting expression:

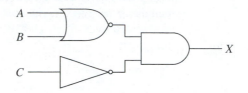

SOLUTION The NOR gate combining the inputs A and B represents $\overline{A + B}$. The input C is inverted, and then the AND gate combining the output of the NOR gate with $\overline{C}$ represents $(\overline{A + B})\overline{C}$. Thus, the Boolean expression represented by the output of the circuit is

$$X = (\overline{A + B})\overline{C}.$$

We can simplify this expression by using DeMorgan's second law, $\overline{A + B} = \overline{A}\cdot\overline{B}$, used from right to left to read $\overline{A}\cdot\overline{B} = \overline{A + B}$, and applied with $A + B$ replacing A and C replacing B:

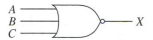

$$(\overline{A + B})\overline{C} = \overline{(A + B) + C}.$$

A three-input NOR gate.

The circuit for $\overline{(A + B) + C}$ can be represented by a *three-input* NOR gate, as shown at the left. ▲

EXAMPLE 23.54 ▶ Write the Boolean expression represented by the output X of the circuit, simplify the expression, and draw the circuit represented by the resulting expression:

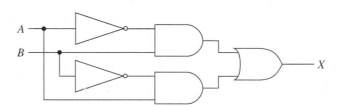

SOLUTION The input A is inverted, so the first AND gate combining $\overline{A}$ with B represents $\overline{A}B$. The dot connecting A to the second AND gate means that A is also an input for the second AND gate. The dot connecting B to the second inverter means that $\overline{B}$ is an input for the second AND gate. Therefore, the second AND gate combining A with $\overline{B}$ represents $A\overline{B}$. Finally, the outputs of the two AND gates are combined by an OR gate. Thus, the Boolean expression represented by the output of the circuit is

$$X = \overline{A}B + A\overline{B}.$$

This expression cannot be simplified by using properties of Boolean algebra. The truth table for the expression is

A	B	$\overline{A}$	$\overline{B}$	$\overline{A}B$	$A\overline{B}$	$\overline{A}B + A\overline{B}$
0	0	1	1	0	0	0
0	1	1	0	1	0	1
1	0	0	1	0	1	1
1	1	0	0	0	0	0

▲

THE XOR GATE

The truth table in Example 23.54 represents the **exclusive-OR**, or **XOR** logic gate. The OR truth values are different from the XOR truth values by one value. If A and B are both true then the *inclusive*-OR is true, but the *exclusive*-OR, XOR, is *false*. The exclusive-OR, XOR, is represented by $A \oplus B$. The XOR gate has the logic symbol shown. The truth table for the XOR gate is

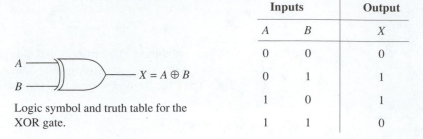

Logic symbol and truth table for the XOR gate.

Inputs		Output
A	B	X
0	0	0
0	1	1
1	0	1
1	1	0

EXAMPLE 23.55 ▶ This circuit is called a half-adder:

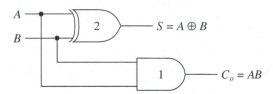

The output S of gate 2 is the sum of two binary digits, and the output C_o of gate 1 is the carry. Find the outputs S and C_o for all inputs A and B.

SOLUTION We find S and C_o by using the truth values for $A \oplus B$ and AB. We can show the results in the form of a truth table:

		C_o Gate 1	S Gate 2
A	B	AB	$A \oplus B$
0	0	0	0
0	1	0	1
1	0	0	1
1	1	1	0

These results show the binary arithmetic facts

$$0 + 0 = 0$$

$$0 + 1 = 1$$

$$1 + 0 = 1$$

$$1 + 1 = 10_2.$$

The last result is the sum $S = 0$ and the carry $C_o = 1$. ▲

A half-adder has only a carry-out C_o. A more complicated circuit called a full-adder also has a carry-in C_i, which represents a carry from the previous sum. Thus, a full adder can also do the sum

$$1 + 1 + 1 = 11_2.$$

The full-adder is shown in Exercise 23.4, #30.

EXERCISE

23.4

1–4. Determine the output X of the circuit shown if

a. $A = 0$ **b.** $A = 1$

(The connected inputs mean that both inputs are A.)

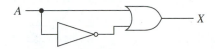

Circuit for Exercise 1.

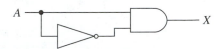

Circuit for Exercise 2.

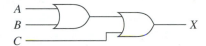

Circuit for Exercise 3.

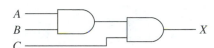

Circuit for Exercise 4.

Determine the output X of the circuit shown below for the given inputs:

5. **a.** $A = 0, B = 0, C = 0$ **b.** $A = 0, B = 0, C = 1$

   c. $A = 0, B = 1, C = 0$ **d.** $A = 1, B = 1, C = 0$

   e. $A = 1, B = 1, C = 1$

6. **a.** $A = 0, B = 1, C = 1$ **b.** $A = 0, B = 0, C = 1$

   c. $A = 0, B = 1, C = 0$ **d.** $A = 1, B = 1, C = 0$

   e. $A = 1, B = 1, C = 1$

Circuit for Exercise 5.

Circuit for Exercise 6.

7–10. Determine the output X of the circuit shown below for the given inputs:

   a. $A = 0, B = 0$ **b.** $A = 1, B = 0$ **c.** $A = 1, B = 1$

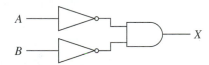

Circuit for Exercise 7.

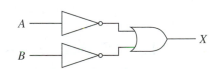

Circuit for Exercise 8.

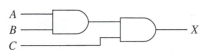

Circuit for Exercise 9.

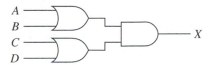

Circuit for Exercise 10.

Draw the circuit that represents the Boolean expression:

11. $(A + B)C$ **12.** $AB + AC$

13. $\overline{A}B + C$ **14.** $(A\overline{B}) + \overline{D}$

15–18. Write the Boolean expression represented by the output X of the circuit:

Circuit for Exercise 15.

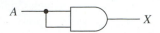

Circuit for Exercise 16.

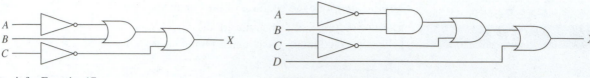

Circuit for Exercise 17.   Circuit for Exercise 18.

19–28.   Write the Boolean expression represented by the output X of the circuit, simplify the expression, and draw the circuit represented by the resulting expression:

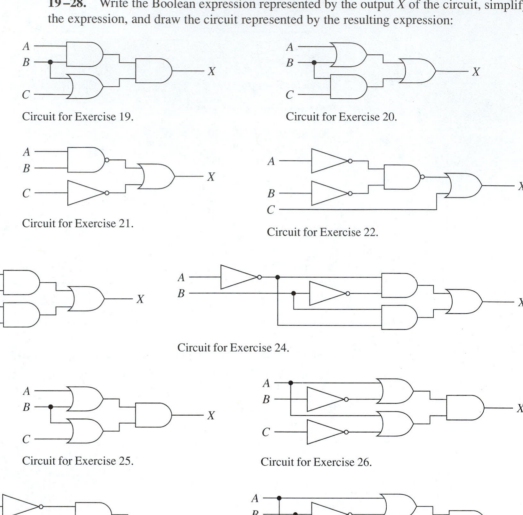

Circuit for Exercise 19.   Circuit for Exercise 20.

Circuit for Exercise 21.   Circuit for Exercise 22.

Circuit for Exercise 23.   Circuit for Exercise 24.

Circuit for Exercise 25.   Circuit for Exercise 26.

Circuit for Exercise 27.   Circuit for Exercise 28.

29.   The negative exclusive-OR or XNOR gate has the logic symbol and truth table

$X = \overline{A \oplus B}$

Logic symbol and truth table for the XNOR gate.

Inputs		Output
A	B	X
0	0	1
0	1	0
1	0	0
1	1	1

(See Exercise 23.27.)

continued

29. *continued*

Find the output X of this circuit for all inputs A, B, and C:

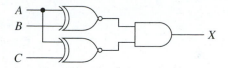

Show the results in this truth table:

A	B	C	$\overline{A \oplus B}$	$\overline{A \oplus C}$	$\overline{(A \oplus B)} \cdot \overline{(A \oplus C)}$
0	0	0			
0	0	1			
0	1	0			
0	1	1			
1	0	0			
1	0	1			
1	1	0			
1	1	1			

30. Find the outputs S and C_o for all inputs A, B, and C_i of this circuit, the full-adder:

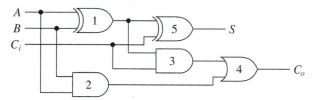

Show the results in this truth table:

A	B	C_i	Gate 1 $A \oplus B$	Gate 2 AB	Gate 3 Gate 1·C_i	C_o Gate 4 Gate 2 + Gate 3	S Gate 5 Gate 1 $\oplus$ C_i
0	0	0					
0	0	1					
0	1	0					
0	1	1					
1	0	0					
1	0	1					
1	1	0					
1	1	1					

Self-Test

1. Subtract $11011_2 - 1100_2$ by using the 1's complement or the 2's complement.

1. _____

2. Write 11110011_2 as

    a. a decimal number

    b. a hexadecimal number

2a. _____

2b. _____

3. Use truth tables to prove $\overline{A} \cdot \overline{B} + \overline{A}B + A\overline{B} = \overline{A} + \overline{B}$.

4. Use properties of Boolean algebra to prove $AB(A + C) = AB$.

5. Write the Boolean expression represented by the output X of the circuit, simplify the expression, and draw the circuit represented by the resulting expression:

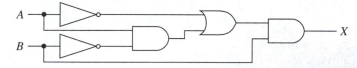

APPENDIX A: Names and Symbols

GREEK ALPHABET

Name	Capital	Lowercase
alpha	A	α
beta	B	β
gamma	Γ	γ
delta	Δ	δ
epsilon	E	ε
zeta	Z	ζ
eta	H	η
theta	Θ	θ
iota	I	ι
kappa	K	κ
lambda	Λ	λ
mu	M	μ
nu	N	ν
xi	Ξ	ξ
omicron	O	o
pi	Π	π
rho	P	ρ
sigma	Σ	σ
tau	T	τ
upsilon	Y	υ
phi	Φ	φ
chi	X	χ
psi	Ψ	ψ
omega	Ω	ω

UNITS

Name	Symbol	Quantity
ampere	A	current (I or i)
coulomb	C	charge (Q or q)
farad	F	capacitance (C)
henry	H	inductance (L)
hertz	Hz	frequency (f)
ohm	Ω	resistance (R) reactance (X) impedance (Z)
second	s	time (t) period (T)
siemens	S	conductance (G)
volt	V	voltage (V or v) electromotive force (E or e)
watt	W	power (P)

PREFIXES

Name	Symbol	Power of Ten
atto-	a	10^{-18}
femto-	f	10^{-15}
giga-	G	10^{9}
kilo-	k	10^{3}
mega-	M	10^{6}
micro-	μ	10^{-6}
milli-	m	10^{-3}
nano-	n	10^{-9}
peta-	P	10^{15}
pico-	p	10^{-12}
tera-	T	10^{12}

CIRCUIT SYMBOLS

Device	Symbol
AC voltage source	
battery (2 cells); DC voltage source	
capacitor	
inductor	
resistor	

LOGIC SYMBOLS

Gate	Symbol
AND	
exclusive-OR	
inverter	
NAND	
negative-AND	
negative-OR	
NOR	
OR	

CALCULATOR KEYS

Function	Symbol
addition	$+$
change sign	$+/-$
common logarithm	log
cosine	cos
degree-rad-grad	DRG or mode
division	$\div$
equals	$=$
exponent entry	EXP or EE
inverse	INV or 2nd
mode	MODE or degree-rad-grad
multiplication	$\times$
natural logarithm	lnx or ln
parentheses	(and)
pi	π
power	y^x or x^y
recall	RCL or other variation
reciprocal	1/x
second function	2nd or INV
sine	sin
square	x^2
square root	$\sqrt{x}$ or $\sqrt{}$
store	STO or other variation
subtraction	$-$
tangent	tan

APPENDIX B: Glossary

absolute value   Unsigned form of a number: $|A| = A$ and $|-A| = A$ where A is positive or 0; magnitude of a vector: r in the polar form r/θ.

absorption property   Properties of Boolean algebra: $A + A \cdot B = A$ and $A + \overline{A} \cdot B = A + B$.

AC resistance   See dynamic resistance.

acute angle   An angle that is greater than 0° and less than 90°.

adjacent side   The leg of a right triangle that with the hypotenuse forms a specified acute angle.

algebraic expression   Numbers and letters combined by arithmetic operations, representing a number.

algorithm   A sequence of steps to solve a problem.

alternating current (AC)   Current that reverses direction.

ampere (A)   Unit for current.

amplitude   Height of a sine wave: $|A|$ for $y = A \sin \theta$.

AND gate   Logic gate that produces the truth values for conjunction.

angle   An amount of rotation: $\theta = 2\pi ft$ or $\theta = 360°ft$.

angular velocity (ω)   Rate of change of an angle: $\omega = 2\pi f$ or $\omega = 360°f$.

argument   Direction of a vector: θ in the polar form r/θ.

arithmetic operation   Combines two numbers to produce a number: addition, subtraction, multiplication, and division.

associative property   Properties of algebra and Boolean algebra: $(A + B) + C = A + (B + C)$ and $(AB)C = A(BC)$.

asymptote   A straight line approached by a graph as its distance from the origin increases.

atto- (a)   Prefix representing 10^{-18}.

axis   A number line used to form the rectangular coordinate system (plural: axes).

binary fraction   A binary number containing a binary point and a nonzero fractional part.

binary number   A number written by using the base 2 and the digits 0 and 1.

binary point   A dot separating the whole number and fractional parts of a binary fraction.

binomial   A polynomial with two terms.

Boolean algebra   A field of mathematics that uses only 0 and 1, and is the source of the logic used by computer circuits.

branch   A current path in a parallel or series-parallel circuit.

branch point   See node.

capacitance (C)   Ability of a capacitor to store charge.

capacitive reactance (X_C)   Opposition to current provided by a capacitor.

capacitor   A device that stores opposite charges on conducting plates separated by insulating material.

Cartesian coordinate system   See rectangular coordinate system.

circle   The set of all points a fixed distance from a given point; graph of an equation of the form $x^2 + y^2 = a^2$.

circumference   Distance around a circle: $c = 2\pi r$.

closed loop   A current path that begins and ends at the same point in a circuit.

coefficient   See numerical coefficient.

cofactor   Signed minor of an element of a determinant.

coil   See inductor.

common logarithm   Logarithm to the base 10.

commutative property   Properties of algebra and Boolean algebra: $A + B = B + A$ and $AB = BA$.

complement of A ($\overline{A}$)   Operation of Boolean algebra that changes 0 to 1 and 1 to 0.

complementation property   Properties of Boolean algebra: $A + \overline{A} = 1$ and $A \cdot \overline{A} = 0$.

complex conjugate   Complex numbers of the form $a + jb$ and $a - jb$, which are the same except for the sign of their imaginary parts.

complex number   A number of the form $a + jb$, where a is a real number and jb is an imaginary number.

conductance (G)   The reciprocal of resistance: $G = \frac{1}{R}$.

conjunction of A and B ($A \cdot B$)   Operation of Boolean algebra that gives 1 only when both A and B are 1.

constant of variation   The constant k in the variation formulas $y = kx$ and $y = \frac{k}{x}$.

constant term   A term that contains no variables.

conventional current   Direction of current flow taken from the positive terminal of a DC source to the negative terminal.

coordinate   Number associated with a point on the number line; a number in an ordered pair associated with a point in the rectangular coordinate system.

cosecant   The reciprocal of the sine.

cosine   A trigonometric function; in a right triangle, the ratio of the adjacent side to the hypotenuse.

cotangent   The reciprocal of the tangent.

coulomb (C)   Unit for electrical charge.

counting number   A number in the set 1, 2, 3, 4, 5, . . .

current (I)   Electron flow or rate of charge movement in a circuit.

cycle   One rotation of a phasor or one repetition of a waveform such as a sine wave.

DC resistance   See static resistance.

decimal fraction   A decimal number containing a decimal point and a nonzero fractional part.

decimal number   A number written by using the base 10 and the ten digits 0, 1, 2, 3, 4, 5, 6, 7, 8, 9.

decimal point   A dot separating the whole number and fractional parts of a decimal fraction.

degree   An angle measure; 1/360th of a circle.

DeMorgan's laws   Properties of Boolean algebra: $\overline{A + B} = \overline{A} \cdot \overline{B}$ and $\overline{A \cdot B} = \overline{A} + \overline{B}$.

denominator   The part $b \neq 0$ of a fraction $\frac{a}{b}$.

determinant   A square array of numbers, representing a number.

digit   For decimal numbers, 0, 1, 2, 3, 4, 5, 6, 7, 8, and 9. See also binary, hexadecimal, and octal numbers.

digital circuit   A circuit that operates at only two voltage levels, which represent the logic values 0 and 1.

direct current (DC)   Current that flows in only one direction.

direct variation   Relation of the form $y = kx$ in which the dependent variable increases or decreases as the independent variable increases or decreases.

disjunction of A and B ($A + B$)   Operation of Boolean algebra that gives 1 whenever $A = 1$ or $B = 1$ or both.

distributive property   Property of algebra and Boolean algebra: $A(B + C) = AB + AC$.

domain   The set of all x-values of a function for which a y-value is defined.

double complementation property   Property of Boolean algebra: $\overline{\overline{A}} = A$.

dynamic resistance   The reciprocal of the slope of the tangent line at any point of an *I-V* characteristic curve.

electromotive force (EMF)   See voltage rise.

electron flow current   Direction of current flow taken from the negative terminal of a DC source to the positive terminal.

exclusive-OR ($A \oplus B$)   Operation of Boolean algebra that gives 1 when $A = 1$ or when $B = 1$, but not both.

exponent   The part x of a power b^x.

exponential equation   An equation with a variable in an exponent.

extraneous solution   An apparent solution of an equation that does not check in the equation, or does not fit the original problem.

farad (F)   Unit for capacitance.

femto- (f)   Prefix representing 10^{-15}.

fraction   A number of the form $\frac{a}{b}$, where $b \neq 0$.

fractional binary number   See binary fraction.

fractional decimal number   See decimal fraction.

fractional part   The part of a decimal or binary fraction following the decimal or binary point.

frequency (f)   Rate of rotation of a phasor or rate of repetition of a waveform such as a sine wave.

function   A relation in x and y where for each allowable value of x there is exactly one value of y.

giga- (G)   Prefix representing 10^9.

henry (H)   Unit for inductance.

hertz (Hz)   Unit for frequency; one hertz is one cycle per second.

hexadecimal number   A number written by using the base 16 and the sixteen digits 0, 1, 2, 3, 4, 5, 6, 7, 8, 9, A, B, C, D, E, F.

horizontal component   A vector given by the x-coordinate of the tip of a vector in the rectangular coordinate system.

hyperbola   Graph of an equation of the form $y = \frac{a}{x}$, or $x^2 - y^2 = a^2$, or $y^2 - x^2 = a^2$.

hypotenuse   The side of a right triangle across from the right angle.

ideal transformer   A transformer with 100% efficiency: output power equals input power.

identity   An equation that is true for all allowable values of the variables.

identity property   Properties of algebra and Boolean algebra: $A + 0 = A$ and $A \cdot 1 = A$.

imaginary axis   The vertical axis when complex numbers are represented in the rectangular coordinate system.

imaginary number   A number of the form bi where $i = \sqrt{-1}$, or jb where $j = \sqrt{-1}$.

imaginary part   The part jb of a complex number written in the form $a + jb$.

impedance (Z)   Total opposition to current in an AC circuit; resultant of resistance and reactance.

included angle   The angle formed by two sides of a triangle.

inclusive-OR   See disjunction of A and B.

inductance (L)   The ability of an inductor to produce an opposing voltage.

inductive reactance (X_L)   Opposition to current provided by an inductor.

inductor   A device in which a change in current produces an opposing voltage.

instantaneous value   Value of an alternating voltage or current at a specified time.

integer   A number that is a counting number, zero, or the negative of a counting number.

intercept   A point where the graph of an equation crosses an axis in the rectangular coordinate system.

inverse relation   The relation obtained by reversing the coordinates in the ordered pairs of a function.

inverse variation   Relation of the form $y = \frac{k}{x}$ in which the dependent variable decreases as the independent variable increases and increases as the independent variable decreases.

inverter   Logic gate that produces the truth values for the complement.

irrational number   A real number that cannot be written as a ratio of two integers.

***I-V* characteristic curve**   A graph representing current I as a function of voltage V for a circuit component.

***j*-operator**   The symbol for an imaginary number $j = \sqrt{-1}$; rotates a phasor by 90°.

kilo- (k)   Prefix representing 10^3.

Kirchhoff's current law   At any node of a circuit, the sum of the currents into the node is equal to the sum of the currents out of the node.

Kirchhoff's voltage law   In any closed loop of a circuit, the sum of the voltage rises is equal to the sum of the voltage drops.

law of cosines   Relates the sides of an oblique triangle with an angle: $a^2 = b^2 + c^2 - 2bc \cos A$, and related forms.

law of sines   Relates sides and angles of an oblique triangle: $\dfrac{a}{\sin A} = \dfrac{b}{\sin B} = \dfrac{c}{\sin C}$.

leg   A side of a right triangle that is not the hypotenuse.

like terms   Terms with all of their literal parts, the variables and their exponents, identical.

linear equation   An equation containing only linear expressions.

linear expression   An expression in which every term is a constant, a variable, or a product of a constant and a variable.

logarithm   The solution for y of $x = b^y$: $y = \log_b x$, where $b > 0$ and $b \neq 1$.

logic gate   An integrated circuit that performs a logical operation such as NOT, OR, or AND.

logic symbol   A symbol used to represent a logic gate (see Appendix A).

logical operator   An operation of Boolean algebra such as complement, disjunction, and conjunction.

major diagonal   See principal diagonal.

mega- (M)   Prefix representing 10^6.

micro- (μ)   Prefix representing 10^{-6}.

milli- (m)   Prefix representing 10^{-3}.

minor   A determinant formed by deleting the row and column of a specified element of a determinant.

minor diagonal   See secondary diagonal.

monomial   An expression with just one term.

NAND gate Logic gate that produces the complement of the AND gate.

nano- (n) Prefix representing 10^{-9}.

natural logarithm Logarithm to the base e where $e \approx 2.718$.

natural number See counting number.

negation of A See complement of A.

negative number A number that is less than zero or to the left of zero on the number line.

node A point in a circuit connecting two or more branches.

NOR gate Logic gate that produces the complement of the OR gate.

NOT gate See inverter.

number line A line that associates every real number with a point on the line.

numerator The part a of a fraction $\frac{a}{b}$.

numerical coefficient Numerical part of a term of an algebraic expression.

oblique triangle A triangle that is not a right triangle.

obtuse angle An angle that is greater than 90° and less than 180°.

octal number A number written by using the base 8 and the eight digits 0, 1, 2, 3, 4, 5, 6, 7.

ohm (Ω) Unit for resistance, reactance, and impedance.

Ohm's law Current is equal to the ratio of voltage to resistance: $I = \frac{V}{R}$.

1's complement A number formed from a binary number by changing each 0 to a 1 and each 1 to a 0.

opposite side The leg of a right triangle across from a specified acute angle.

OR gate Logic gate that produces the truth values for disjunction.

order relation Compares two numbers: equals (=), less than (<), and greater than (>).

ordered pair Two numbers with a specified first and second, written (x, y).

origin The point on a number line with the coordinate 0; the point of the rectangular coordinate system with coordinates (0, 0) where the axes cross.

parabola Graph of an equation of the form $y = ax^2 + bx + c$.

parallel circuit A circuit that has components in two or more closed loops.

peak value Maximum value of an alternating voltage or current.

percent A number divided by 100.

percentage Result of finding a percent of a number.

period The reciprocal of frequency; time for a phasor to complete one cycle; length of one cycle of a sine wave.

peta- (P) Prefix representing 10^{15}.

phase angle Angle at which rotation of a phasor starts, or an angle between two phasors; the amount by which a sine wave is shifted.

phasor A vector for which the direction is defined in terms of time.

pico- (p) Prefix representing 10^{-12}.

place value Amount by which a digit is multiplied because of its position in a number.

polar form A way of representing vectors and complex numbers: $r\underline{/\theta}$.

polynomial An algebraic expression in which every term is a constant, a variable, or a product of constants and variables with positive integer exponents.

positive number A number that is greater than zero or to the right of zero on the number line.

potential difference See voltage.

power An expression of the form b^x; rate at which energy is used: $P = VI$.

prefix A modifier that changes a unit to a smaller or a larger unit (see Appendix A).

primary winding Input coil of a transformer.

principal diagonal The elements in the line from top left to bottom right of a determinant.

proportion Two equal ratios.

purely resistive circuit A circuit that contains only an energy source and resistors.

Pythagorean identities Identities relating the sine and cosine, and related identities: $\sin^2 x + \cos^2 x = 1$, $\tan^2 x + 1 = \sec^2 x$, and $\cot^2 x + 1 = \csc^2 x$.

Pythagorean theorem The square of the hypotenuse of a right triangle is equal to the sum of the squares of the legs: $c^2 = a^2 + b^2$.

quadrant A section of the rectangular coordinate system formed by the axes, numbered I, II, III, IV counterclockwise from the upper right.

quadrantal angle An angle with one side on the positive x-axis and its other side also on an axis: 0°, 90°, 180°, 270°, and 360°.

quadratic equation An equation involving the square of a variable or the product of two variables.

quadratic formula A formula that gives the solutions of a quadratic equation in standard form: $x = \frac{-b \pm \sqrt{b^2 - 4ac}}{2a}$.

quadratic function An equation in two variables of the form $y = ax^2 + bx + c$.

radian An angle measure: π radians $= 180°$.

radical equation An equation that includes the square root of a variable.

radius The distance of a circle from its center.

radius vector A radius rotated around the unit circle.

range The set of all y-values of a function.

range of principal values A range that restricts an inverse trigonometric relation so that it is a function.

ratio A comparison of two quantities in the form of a fraction.

ratio identities Identities relating the tangent and cotangent to the sine and cosine: $\tan x = \dfrac{\sin x}{\cos x}$ and $\cot x = \dfrac{\cos x}{\sin x}$.

rational equation An equation that includes a variable in a denominator.

rational number A real number that can be written as a ratio of two integers.

RC circuit A circuit containing an energy source, resistors, and capacitors.

reactance (X) Opposition to current provided by a capacitor or inductor.

real axis The horizontal axis when complex numbers are represented in the rectangular coordinate system.

real number A positive, zero, or negative number; a number that is associated with a point on the number line.

real part The part a of a complex number written in the form $a + jb$.

reciprocal One divided by a number or expression: $\frac{1}{A}$, where $A \neq 0$.

reciprocal identities The definitions relating the cotangent, secant, and cosecant to the tangent, cosine, and sine:

$$\cot x = \frac{1}{\tan x}, \quad \sec x = \frac{1}{\cos x}, \quad \text{and} \quad \csc x = \frac{1}{\sin x}.$$

rectangular coordinate system Two number lines placed at right angles to one another with a common origin.

rectangular form A way of representing complex numbers: $a + jb$.

redundancy property Properties of Boolean algebra: $A + A = A$ and $A \cdot A = A$.

reference angle The positive acute angle between a vector and its horizontal component.

relation in x and y An equation in two variables x and y.

resistance (R) Opposition to current provided by a resistor.

resistor A circuit component providing opposition to current.

resonant frequency (f_r) The frequency at which inductive reactance and capacitive reactance in an RLC circuit are equal.

resultant The vector sum of a horizontal vector and a vertical vector.

right angle A 90° angle.

right triangle A triangle with one right angle.

RL circuit A circuit containing an energy source, resistors, and inductors.

RLC circuit A circuit containing an energy source, resistors, inductors, and capacitors.

scalar A quantity that has magnitude only.

secant The reciprocal of the cosine.

secondary diagonal The elements in a line from top right to bottom left of a determinant.

secondary winding Output coil of a transformer.

series circuit A circuit that has all its components in one closed loop.

siemens (S) Unit for conductance.

significant digits Nonzero digits, zeros between two nonzero digits, and trailing zeros in some cases.

similar triangles Triangles with corresponding angles equal.

sine A trigonometric function; in a right triangle, the ratio of the opposite side to the hypotenuse.

sine wave Graph of a function of the form $y = A \sin(B\theta + C)$.

slope The measure of how steeply a line rises or falls: $m = \frac{\Delta y}{\Delta x}$, where $\Delta y = y_2 - y_1$ and $\Delta x = x_2 - x_1$.

slope-intercept form The equation of a line in the form $y = mx + b$, where m is the slope and b is the y-intercept.

standard position Location of an angle that is formed by a rotation starting on the positive x-axis.

static resistance V divided by I at any point of an I-V characteristic curve.

straight angle A 180° angle.

sum and difference identities Identities for the sine and cosine of the sum of two angles:

$\sin(\alpha \pm \beta) = \sin\alpha\cos\beta \pm \cos\alpha\sin\beta$ and
$\cos(\alpha \pm \beta) = \cos\alpha\cos\beta \mp \sin\alpha\sin\beta$.

symmetry with respect to a line A property in which graphs are mirror images of one another with the line as the mirror.

system of equations Two or more equations in two or more variables.

tangent A trigonometric function; in a right triangle, the ratio of the opposite side to the adjacent side.

tera- (T) Prefix representing 10^{12}.

term A part of an algebraic expression consisting of products and quotients of numbers and letters.

time constant (τ) The value given by $\tau = RC$ for an RC circuit or $\tau = \frac{L}{R}$ for an RL circuit.

transformer A device that steps up or steps down AC voltage.

trigonometric form See polar form.

trigonometric ratio Ratios that relate an acute angle of a right triangle to the sides of the triangle; see sine, cosine, and tangent.

trinomial A polynomial with three terms.

2's complement A number formed from a binary number by forming the 1's complement and then adding 1.

unit Distance on the number line from the origin to the point with coordinate 1; standard amounts by which quantities are measured (see Appendix A).

unit circle A circle in the rectangular coordinate system with its center at the origin and radius 1.

universal time-constant curves Graphs that express the voltage across the resistor and the voltage across the capacitor (or inductor) in a RC (or RL) circuit as percents of the DC source voltage.

variable Quantity represented by a letter, which can take on different values.

variation formula Formulas that relate two or more variables by direct or inverse variation: $y = kx$ or $y = \frac{k}{x}$.

vector A quantity that has both magnitude and direction.

vertex The minimum or maximum point on a parabola.

vertical component A vector given by the y-coordinate of the tip of a vector in the rectangular coordinate system.

volt (V) Unit for potential difference or voltage.

voltage (E, V) Difference in energy or energy required to move charge between two points of a circuit.

voltage drop Loss of energy across a component that provides opposition to current.

voltage rise Gain of energy across a voltage source.

watt (W) Unit for power.

weight See place value.

weighted digit A digit with its place value.

whole number A number in the set 0, 1, 2, 3, 4, 5, ...

whole number part The part of a decimal or binary fraction before the decimal or binary point.

x-component See horizontal component.

XOR gate Logic gate that gives the truth values for the exclusive-OR.

y-component See vertical component.

zero product rule If $ab = 0$ then $a = 0$ or $b = 0$.

APPENDIX C: Reverse Polish Notation

Polish notation, which uses no parentheses, was developed by Polish logician Jan Lukasiewicz (1878–1956) in 1929. In ordinary arithmetic, an operator is placed between two numbers, for example,

$$2 + 3.$$

In Polish notation, Lukasiewicz placed the operator in front of the numbers.

In the arrangement called *reverse Polish notation* (RPN), the operator is placed *after* the numbers, for example,

$$2 \quad 3 \quad +.$$

Some calculators, such as some Hewlett-Packard (HP) models, use RPN. The HP model 32SII has been used for the RPN algorithms in this appendix.

Polish and reverse Polish notations avoid ambiguities that require parentheses. For example, in ordinary notation, the expressions

$$2 \times 3 + 4$$

and

$$2 \times (3 + 4)$$

are distinguished by the parentheses. In RPN, the first expression is given by

$$2 \quad 3 \quad \times \quad 4 \quad +$$

and the second expression is given by

$$2 \quad 3 \quad 4 \quad + \quad \times.$$

No parentheses are needed.

When using RPN calculators, numbers are stored in a set of registers called the stack. The HP 32SII has four registers, called X, Y, Z, and T. It is important to understand that the display is the current register, which is the X-register. Thus a number entered on the display is in the stack. You press

$$\boxed{\text{ENTER}}$$

to move the number to the second register, which is the Y-register, and then you can enter another number on the display.

A 32SII reverse Polish notation scientific calculator.
Courtesy of *Hewlett Packard*.

EXAMPLE C.1 Add 2 + 3.

SOLUTION You enter 2 on the display, move it to the Y-register by pressing ENTER, and then enter 3 and add:

		display:
Enter 2		2_
Press $\boxed{\text{ENTER}}$		2
Enter 3		3_
Press $\boxed{+}$		5

Observe that you do not press an equals key. The sum 5 is in the X-register, which is the display. If you get 6, it is because you pressed ENTER after entering 3, thereby storing 3 in both the X- and Y-registers and adding 3 to itself. ▲

You can view the contents of the stack registers by pressing the roll-down key marked

Recall that the display *is* the current or X-register. You can view the contents of the other registers by pressing the roll-down key once for each register. Pressing the roll-down key four times views each resister and returns to the X-register.

To demonstrate the roll-down key, enter 1 and press ENTER, 2 and press ENTER, 3 and press ENTER, and 4 but *do not* press ENTER because the display *is* the remaining register. Press the roll-down key once to display 3, again to display 2, again to display 1, and the fourth time to return to 4.

You can clear the register that is currently displayed by pressing the C key. You can clear any register by pressing the roll-down key until you reach the register, and then pressing C.

For the displays in this appendix, the calculator is in the ALL display mode. You can show the display menu by pressing the orange left-shift key.

and then the E key, which has DISP marked in orange above it. The arrow under ALL in the display points to the key marked Σ+. Press that key to activate the ALL display.

You can subtract, multiply, and divide by using steps similar to those in Example C.1. Expressions with more than one operation can be evaluated without using parentheses.

EXAMPLE C.2 ▶ Evaluate $2 \times 3 + 4$.

SOLUTION You enter 2, move it to the Y-register, then enter 3 and multiply, and then enter 4 and add:

	display:
Enter 2	2_
Press ENTER	2
Enter 3	3_
Press ×	6
Enter 4	4_
Press +	10 ▲

EXAMPLE C.3 ▶ Evaluate $2 \times (3 + 4)$.

SOLUTION You enter 2 and 3, then enter 4 and add, and then multiply:

	display:
Enter 2	2_
Press ENTER	2
Enter 3	3_
Press ENTER	3
Enter 4	4_
Press +	7
Press ×	14 ▲

Expressions containing other grouping symbols can be evaluated without using parentheses.

EXAMPLE C.4 ▶ Evaluate $\dfrac{3.3 \times 8.2}{3.3 + 8.2}$ (Example 1.31).

SOLUTION You multiply 3.3 and 8.2, add 3.3 and 8.2, and then divide:

		display:
Enter	3.3	3.3_
Press	ENTER	3.3
Enter	8.2	8.2_
Press	$\times$	27.06
Enter	3.3	3.3_
Press	ENTER	3.3
Enter	8.2	8.2_
Press	$+$	11.5
Press	$\div$	2.35304347826 ▲

The square, square root, and reciprocal are functions. The square is labelled in orange above the square-root key. To square a number, you enter the number and then press

$$\boxed{x^2}$$

by pressing the orange left-shift key and then the square-root key. To find a square root, you enter the number and then press the square-root key marked

$$\boxed{\sqrt{x}}\,.$$

To find the reciprocal of a number, you enter the number and then press the reciprocal key marked

$$\boxed{1/x}\,.$$

EXAMPLE C.5 ▶ Evaluate the formula $\dfrac{1}{R_T} = \dfrac{1}{R_1} + \dfrac{1}{R_2}$ for $R_1 = 1.2\ \Omega$ and $R_2 = 1.8\ \Omega$ (Example 5.7).

SOLUTION To calculate $\dfrac{1}{R_T} = \dfrac{1}{1.2} + \dfrac{1}{1.8}$, you find both reciprocals and then add:

		display:
Enter	1.2	1.2_
Press	1/x	8.33333333E–1
Enter	1.8	1.8_
Press	1/x	5.55555556E–1
Press	$+$	1.38888888889

To find R_T, press the reciprocal key one more time:

Press	1/x	7.2E–1

Observe that decimals such as 0.72 are sometimes shown in scientific notation. ▲

The number π is labelled in blue above the sine key. You press the blue right-shift key

$$\boxed{\longrightarrow}$$

and then the sine key to press

$$\boxed{\pi}\,.$$

EXAMPLE C.6 ▶ Evaluate the formula $X_L = 2\pi fL$ for $f = 60$ MHz and $L = 4\ \mu$H (Example 5.8).

SOLUTION To calculate $X_L = 2\pi(60)(4)$, you enter 2 and multiply by π, then by 60, and then by 4:

	display:
Enter 2	2_
Press ENTER	2
Press π	3.14159265359
Press ×	6.28318530718
Enter 60	60_
Press ×	376.991118431
Enter 4	4_
Press ×	1,507.96447372

Alternatively, since there are four registers in the stack and four factors in the product, you can enter all the factors and then multiply. The number π is treated as a function so you *do not* press ENTER to separate π from the number that follows:

	display:
Enter 2	2_
Press ENTER	2
Press π	3.14159265359
Enter 60	60_
Press ENTER	60
Enter 4	4_
Press ×	240
Press ×	753.982236862
Press ×	1,507.96447372

You can use the exponent entry key marked E to calculate powers of ten, as on an ordinary scientific calculator:

	display:
Enter 2	2_
Press ENTER	2
Press π	3.14159265359
Press ×	6.28318530718
Enter 60	60_
Press E	60E_
Enter 6	60E6_
Press ×	376,991,118.431
Enter 4	4_
Press E	4E_
Enter 6	4E6_
Press +/−	4E–6_
Press ×	1,507.96447372 ▲

Common and natural logarithms are also functions. For the common log, you must press the orange left-shift key and then the LN key to press log .

EXAMPLE C.7 Evaluate $\dfrac{\log 25}{\log 15}$ (Example 15.11).

SOLUTION You find both logs and then divide:

	display:
Enter 25	25_
Press ☐ log	1.39794000867
Enter 15	15_
Press ☐ log	1.17609125906
Press ☐ ÷	1.18863225783

To check this example, you can evaluate $15^{1.19}$ by using the power key
☐ y^x .

You enter the base 15, move it to the Y-register by pressing ENTER, then enter the exponent 1.19. The power key is treated as an operator, or two-number function, and is pressed last:

	display:
Enter 15	15_
Press ENTER	15
Enter 1.19	1.19_
Press ☐ y^x	25.0927695603 ▲

Calculations with the natural logarithm, 10^x, and e^x are similar to common logarithm calculations. Calculations with the sine, cosine, and tangent are also similar.

EXAMPLE C.8 Evaluate 10 sin 30° (Example 17.5).

SOLUTION You enter 10, move it to the Y-register by pressing ENTER, then find sin 30° and multiply:

	display:
Enter 10	10_
Press ENTER	10
Enter 30	30_
Press ☐ sin	0.5
Press ☐ ×	5 ▲

The inverse sine, cosine, and tangent are labelled ASIN, ACOS, and ATAN, for arcsine, arccosine, and arctangent. They are found by using the orange left-shift key followed by the sine, cosine, and tangent.

EXAMPLE C.9 Evaluate $\sin^{-1}\left(\dfrac{560}{820}\right)$ (Example 17.8).

SOLUTION You divide 560 by 820, and then use the orange left-shift key to find ASIN:

	display:
Enter 560	560_
Press ENTER	560
Enter 820	820_
Press ☐ ÷	6.82926829E−1
Press ASIN	43.0727813467 ▲

You change to radian mode by pressing the orange left-shift key and then the change-sign key, which has MODES marked in orange above it. Press the LN key as indicated by the arrow under RD in the display. The letters RAD appear on the display.

EXAMPLE C.10   Evaluate $\sin \dfrac{\pi}{6}$.

SOLUTION Change the calculator to radian mode. Then, you can divide π by 6 and find the sine. Recall that π is treated as a function so you do not press ENTER after π:

		display:
Press	π	3.14159265359
Enter 6		6_
Press	$\div$	5.23598776E–1
Press	SIN	0.5 ▲

To change back to degree mode, press MODES by using the orange left-shift key and then the change-sign key. Press the square-root key as indicated by the arrow under DG in the display.

APPENDIX D: ANSWERS

UNIT R

Exercise R.1

1. $20 + 6$

2. $800 + 90 + 2$

3. $4000 + 700 + 60 + 9$

4. $80{,}000 + 5000 + 500 + 30 + 1$

5. $600 + 8$

6. $5000 + 40$

7. $5000 + 5$

8. $60{,}000 + 10 + 8$

9. 89   **10.** 886   **11.** 72   **12.** 164

13. 793   **14.** 1136   **15.** 41   **16.** 245

17. 9   **18.** 46   **19.** 21   **20.** 707

21. 86   **22.** 52   **23.** 140   **24.** 469

25. 1376   **26.** 2898   **27.** 10,800   **28.** 17,510

29. 42   **30.** 56   **31.** 65   **32.** 43

33. 22 Remainder 1   **34.** 99 Remainder 3   **35.** 21   **36.** 61 Remainder 1

Exercise R.2

1. $\dfrac{1}{3}$ **2.** $\dfrac{1}{5}$ **3.** $\dfrac{2}{9}$ **4.** $\dfrac{3}{4}$ **5.** $\dfrac{1}{6}$ **6.** $\dfrac{2}{15}$

7. $\dfrac{4}{5}$ **8.** $\dfrac{3}{20}$ **9.** $\dfrac{1}{6}$ **10.** $\dfrac{1}{15}$ **11.** $\dfrac{6}{25}$ **12.** $\dfrac{4}{9}$

13. $\dfrac{2}{3}$ **14.** $\dfrac{8}{3}$ **15.** $\dfrac{5}{3}$ **16.** $\dfrac{4}{3}$ **17.** $\dfrac{1}{10}$ **18.** 5

19. $\dfrac{8}{5}$ **20.** $\dfrac{2}{9}$ **21.** $\dfrac{1}{2}$ **22.** 1   **23.** $\dfrac{3}{8}$ **24.** $\dfrac{3}{4}$

25. $\dfrac{7}{12}$ **26.** $\dfrac{13}{40}$ **27.** $\dfrac{23}{20}$ **28.** $\dfrac{43}{30}$ **29.** $\dfrac{29}{24}$ **30.** $\dfrac{25}{24}$

31. $\dfrac{15}{4}$ **32.** $\dfrac{35}{6}$ **33.** $\dfrac{1}{2}$ **34.** $\dfrac{1}{24}$ **35.** $\dfrac{3}{4}$ **36.** $\dfrac{17}{20}$

Exercise R.3

1. $3 + \dfrac{1}{10} + \dfrac{5}{100}$

2. $70 + 4 + \dfrac{9}{10} + \dfrac{1}{100}$

3. $\dfrac{2}{10} + \dfrac{6}{1000}$

4. $\dfrac{4}{100} + \dfrac{5}{1000}$

5. $\dfrac{3}{1000} + \dfrac{2}{100{,}000}$

6. $\dfrac{1}{1{,}000{,}000}$

7. 7.8   **8.** 8.9   **9.** 9.3   **10.** 11.7   **11.** 9.73   **12.** 4.009

13. 1.2   **14.** 7   **15.** 7.5   **16.** 0.4   **17.** 3.14   **18.** 3.67

19. 26   **20.** 31.2   **21.** 2.7   **22.** 11.96   **23.** 0.364   **24.** 2.6499

25. 1.57   **26.** 0.131   **27.** 8.1   **28.** 65   **29.** 75.7   **30.** 0.143

31. 0.5   **32.** 0.625   **33.** 0.16   **34.** 3.75   **35.** $0.333\ldots$ **36.** $0.666\ldots$

37. $0.2727\ldots$ **38.** $0.714285714285\ldots$

Exercise R.4

1. $\dfrac{1}{5}$ 2. $\dfrac{3}{2}$ 3. $\dfrac{3}{200}$ 4. $\dfrac{21}{200}$ 5. 0.33 6. 0.202

7. 0.045 8. 1.21 9. 125 10. 24 11. 4.4 12. 37.5

13. 18 14. 330 15. 0.3 16. 0.6875

17. **a.** 20.9 to 23.1 ohms **b.** 19.8 to 24.2 ohms **c.** 17.6 to 26.4 ohms

18. **a.** 6.46 to 7.14 ohms **b.** 6.12 to 7.48 ohms **c.** 5.44 to 8.16 ohms

Exercise R.5

1. 50.05 2. 97.5 3. 2.431 4. 0.8492 5. 53.06 6. 4.61

7. 3.136 8. 0.2008 9. 16.33 10. 0.663 11. 0.2852 12. 0.0231

13. 3.92 14. 2.35 15. 3.162 16. 2.718 17. 0.1592 18. 0.4343

19. 0.301 20. 1.661 21. 4.47 22. 1.51 23. 0.0224 24. 0.000318

25. 7390 26. 98,700 27. 26.9 28. 3520 29. 0.39 30. 0.000126

31. 2.54 32. 0.64 33. 0.195 34. 335 35. 3.06 36. 0.32

37. 3520 38. 726,000 39. 2.24 40. 3.16 41. 8.62 42. 13.7

Self-Test

1. **a.** 600 **b.** 105 (Objective 1) 2. **a.** $\dfrac{1}{4}$ **b.** $\dfrac{5}{6}$ (Objective 2)

3. **a.** 0.18 **b.** 2.3 (Objective 3) 4. 0.96 to 1.44 ohms (Objective 4)

5. **a.** 0.92 **b.** 0.596 (Objective 5)

UNIT 1

Exercise 1.1

1.

e. $\sqrt{-6}$ is not a real number.

2.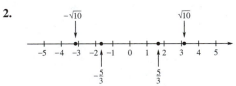

e. $\sqrt{-10}$ is not a real number.

3.
$\sqrt{17} > 4$

4.
$\dfrac{17}{4} > 4$

5.
$\sqrt{8} > 0$

6.
$-\sqrt{8} < 0$

7.
$-4 > -5$

8.
$-\dfrac{7}{2} < -\sqrt{7}$

9.
$-\dfrac{10}{3} < -\dfrac{11}{4}$

10.
$-\dfrac{4}{3} > -\dfrac{7}{5}$

Exercise 1.2

1. 6	**2.** 6	**3.** −6	**4.** −6	**5.** 9	**6.** 12
7. 14	**8.** 16	**9.** 0	**10.** 0	**11.** 18	**12.** 80
13. 10	**14.** 1	**15.** 1	**16.** 9	**17.** 2	**18.** $\frac{3}{2}$
19. 8	**20.** 17	**21.** 9	**22.** 0	**23.** $\frac{1}{5}$	**24.** 4
25. 0.504	**26.** 0	**27.** 4.18	**28.** 3.46		

Exercise 1.3

1. −11	**2.** −35	**3.** −26	**4.** −48	**5.** 2	**6.** 6
7. 6	**8.** 16	**9.** −7	**10.** −14	**11.** 15	**12.** −11
13. −10	**14.** −32	**15.** 0	**16.** 0	**17.** −21	**18.** −50
19. 26	**20.** 42	**21.** 3	**22.** 12	**23.** −9	**24.** −18
25. 10	**26.** −20	**27.** 0	**28.** 0	**29.** −8	**30.** 9
31. −12	**32.** −5	**33.** 5	**34.** 1	**35.** 8	**36.** 0
37. −1	**38.** 1				

Exercise 1.4

1. −40	**2.** −132	**3.** −55	**4.** −96	**5.** 48	**6.** 2
7. −6	**8.** −20	**9.** −10	**10.** −0.25	**11.** 6	**12.** 2.5
13. −0.667	**14.** −0.444	**15.** 1.36	**16.** 1.17	**17.** 6.93	**18.** −1.73
19. −3.54	**20.** 1.58	**21.** 0	**22.** 0	**23.** 0	**24.** 0
25. 0	**26.** Undefined	**27.** Undefined	**28.** 0		

Exercise 1.5

1. −14	**2.** 5	**3.** 11	**4.** −10	**5.** 6	**6.** −4
7. 1	**8.** 2	**9.** 16	**10.** 3.5	**11.** 25	**12.** 4
13. −12	**14.** 8	**15.** 0.25	**16.** 8	**17.** 18	**18.** −2
19. −50	**20.** −24	**21.** −1	**22.** −3	**23.** 1.5	**24.** 0.012
25. −100	**26.** 36	**27.** 13	**28.** 8	**29.** 9.88	**30.** 171
31. 0.053	**32.** 0.0411				

Self-Test

1.

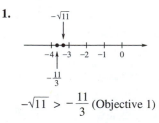

$-\sqrt{11} > -\frac{11}{3}$ (Objective 1)

2. a. 12   **b.** 12 (Objective 2)

3. a. −20   **b.** −9 (Objective 3)

4. a. −11   **b.** 1.22 (Objective 4)

5. 8 (Objective 5)

UNIT 2

Exercise 2.1

1. 10^9 2. 10^6 3. 10^4 4. 1 5. $\dfrac{1}{10^2}$ 6. $\dfrac{1}{10^3}$

7. 10^{12} 8. 10^{10} 9. 10^{12} 10. 10^{14} 11. 10^6 12. $\dfrac{1}{10^8}$

13. 10^4 14. $\dfrac{1}{10^9}$ 15. 10^5 16. $\dfrac{1}{10^6}$

Exercise 2.2

1. 10^2 2. $\dfrac{1}{10^3}$ 3. 1 4. 1 5. 10^9 6. 10^6

7. $\dfrac{1}{10^4}$ 8. $\dfrac{1}{10^6}$ 9. $\dfrac{1}{10}$ 10. 10^6 11. 10^2 12. 10^9

13. 10^{10} 14. 1 15. $\dfrac{1}{10^6}$ 16. 10^8 17. 10^3 18. 10^{-6}

19. 10^5 20. 10 21. 10^6 22. 10^3 23. $10^5 \times \sqrt{10}$ 24. $10^7 \times \sqrt{10}$

25. $10^{-8} \times \sqrt{10}$ 26. $10^{-5} \times \sqrt{10}$ 27. 5×10^4 28. 12×10^{-5} 29. 11×10^4 30. 2.55×10^3

31. 1.58×10^{-3} 32. 5×10^{-5}

Exercise 2.3

1. 25.4   2. 6190   3. 807,000,000   4. 10,500,000,000
5. 0.464   6. 0.00704   7. 0.0000302   8. 0.0000000416
9. 1.29   10. 5.02   11. 5.75×10^2 12. 8.74×10^4
13. 2.01×10^6 14. 6.03×10^{11} 15. 6.43×10^{-1} 16. 8.02×10^{-4}
17. 4.04×10^{-6} 18. 2.69×10^{-10} 19. 2.34×10^0 20. 1.00×10^0
21. 1.91×10^8 22. 2.52×10^{15} 23. 4.07×10^4 24. 2.04×10^5
25. 1.60×10^{-12} 26. 6.01×10^{-16} 27. 8.81×10^8 28. 4.24×10^{11}
29. 4.14×10^{12} 30. 2.94×10^{-1} 31. 3.55×10^{-10} 32. 5.28×10^{-11}
33. 2.70×10^6 miles   34. 1.98×10^{30} kilograms   35. 3.95×10^{-22} grams   36. 1.47×10^{-17} coulombs
37. 500 seconds   38. 88 electrons

Exercise 2.4

1. 16,000,000 Hz   2. 2500 Hz   3. 180,000,000,000 Hz   4. 702,000 Hz
5. 0.15 s   6. 0.000000693 s   7. 0.0000122 s   8. 0.00000000102 s
9. 23,800 W   10. 0.0745 g   11. 0.00000000000173 g   12. 6,050,000,000 W
13. 272 kHz   14. 9.42 kHz   15. 69 GHz   16. 350 kHz
17. 492 ms   18. 189 µs   19. 62.2 µs   20. 820 ps
21. 5.85 ns   22. 4.65 GHz   23. 7.96 MW   24. 1.88 mW
25. 685 kHz   26. 76.7 GHz   27. 250 µs   28. 15.5 ns
29. 38 kW   30. 893 MW   31. 54.4 ps   32. 90.8 ms
33. 0.466 MW   34. 7.3 GW   35. 0.00708 ms   36. 0.0304 µs
37. 965,000 kHz   38. 7710 MHz   39. 47.4 µs   40. 8.08 nsZZ

41. 785,000 ps   **42.** 18,000 kW   **43.** 0.000722 ms   **44.** 0.00628 GHz

45. 150,000,000 km, 150 Gm   **46.** 7500 Tm, 7.5 Pm

47. 0.00000464 pC, 0.00464 fC, 4.64 aC   **48.** 0.0158 fC, 15.8 aC

Self-Test

1. a. 78.7 µs   **b.** 15 GHz (Objective 4)   **2. a.** 2.97×10^7 **b.** 5.99×10^{-5} (Objective 3)

3. $\dfrac{1}{10}$ (Objective 1)   **4.** 10^7 (Objective 2)   **5.** 273 µs (Objective 4)

UNIT 3

Exercise 3.1

1. 13   **2.** -4 **3.** 6   **4.** 15

5. 8   **6.** -11 **7.** 3.6   **8.** 2.25

9. 54   **10.** 0.032   **11.** -12.96 **12.** 1.53

13. $31.9 \dfrac{m}{h}$ **14.** $0.05 \dfrac{mi}{min}$ **15.** $3.33 \times 10^{-3} \dfrac{V}{\Omega}$ **16.** $6.67 \times 10^{-6} \dfrac{V}{\Omega}$

17. 8.5 mi   **18.** 132 ft   **19.** 24×10^{-3} V•A   **20.** 6.75×10^{-3} V•A

21. $25 \times 10^3 \dfrac{V}{A}$ **22.** $0.4 \dfrac{V}{A}$ **23.** 14.1 A•Ω **24.** 49.5 A•Ω

25. 56×10^3 A²•Ω **26.** 1.18 A²•Ω **27.** $192 \times 10^{-3} \dfrac{V^2}{\Omega}$ **28.** $7.2 \times 10^{-3} \dfrac{V^2}{\Omega}$

29. $55 \times 10^{-6} \sqrt{\dfrac{W}{\Omega}}$ **30.** $91.3 \times 10^{-3} \sqrt{\dfrac{W}{\Omega}}$ **31.** $260 \sqrt{W•\Omega}$ **32.** $62.9 \times 10^3 \sqrt{W•\Omega}$

Exercise 3.2

1. $10x + 10y$ **2.** $0.5x + 0.5y$ **3.** $5x - 5y$ **4.** $3.3y - 3.3z$

5. $9x - 9y - 9z$ **6.** $1.2z - 1.2y - 1.2x$ **7.** $-x + y$ **8.** $-x - y + z$

9. $-2x + 2y$ **10.** $-3y + 3x$ **11.** $-10x + 15y + 10z$ **12.** $-2x - 3y + 4z$

13. $10(x + y)$ **14.** $5(x - y)$ **15.** $4(x + 2y)$ **16.** $5(2x - 3y)$

17. $6x$ **18.** $12x$ **19.** x **20.** $-3x$

21. $-x$ **22.** $-2x$ **23.** $-2x$ **24.** $-4x$

25. $-x - y$ **26.** $3x + 5y$ **27.** $x - y$ **28.** $2x - 9y$

29. $x - 2y$ **30.** $-2x + 6z$ **31.** $-x + y + 6z$ **32.** $-x - y - z$

33. $2x + 2xy + 8y$ **34.** $2x - xy - 3y$ **35.** $-2x - 5xy + 7y$ **36.** $-3x - 4xy + y$

Exercise 3.3

1. $-18x + 9$ **2.** $3x + 6$ **3.** $-4x + 8y + 1$ **4.** $-2x - 9y + 3$

5. $-9x - 3$ **6.** $-4y + 4$ **7.** $3x + 13$ **8.** $-11x + 2$

9. $11x + y$ **10.** $7x - 23y$ **11.** 6   **12.** $28x$

13. $-12x + 2y$ **14.** $3x - 9y$ **15.** $-10x + 15$ **16.** $15x$

Exercise 3.4

1. $15x^8$
2. $56x^7$
3. $5x^2 + 15x^3$
4. $3x^4 + 3x^3$
5. $6x^3 + 16x^{-1}$
6. $35x^5 - 28x^{-1}$
7. $x^{-4} - 1$
8. $10x^4 + 5$
9. $2x^2y$
10. $4x^3y^6$
11. $4.5x^5y^{-2}$
12. $0.333x^{-6}y^4$
13. $3x^4 + 2x^3$
14. $xy^3 + y^2$
15. $4xy^{-1} + 2x^{-1}y$
16. $3.5xy^{-1} + 4.5x^{-1}y$
17. $x^{-1} - x^{-2} + 1$
18. $xy^{-2} + x^{-2}y - 1$
19. x^2y^6
20. $x^{-4}y^6$
21. x^6y^4
22. x^4y^{-8}
23. $x^3y^{-6}z^9$
24. $x^8y^4z^{-12}$
25. x^2y^{-1}
26. $x^{-1}y$
27. $xy\sqrt{x}$
28. $x^2y\sqrt{xy}$
29. $1.73\,x^4y^{-2}\sqrt{x}$
30. $2.24y\sqrt{y}$

Self-Test

1. $9x$ (Objective 2)
2. $-12x - 5y$ (Objective 3)
3. $1.25xy^2 + 0.75y^3$ (Objective 4)
4. $2.45xy^2\sqrt{xy}$ (Objective 4)
5. $151 \times 10^{-3}\sqrt{\dfrac{W}{\Omega}}$ (Objective 1)

UNIT 4

Exercise 4.1

1. 7
2. -1
3. -1.9
4. $\dfrac{3}{2}$
5. 18
6. 10.1
7. 15
8. 1.17
9. -6
10. $\dfrac{1}{10}$
11. 36
12. 19
13. 6
14. $\dfrac{5}{8}$
15. $\dfrac{3}{8}$
16. $\dfrac{14}{3}$

Exercise 4.2

1. 3
2. 6
3. $\dfrac{9}{2}$
4. $\dfrac{3}{2}$
5. -3
6. $-\dfrac{10}{3}$
7. 10
8. 30
9. 3
10. 4
11. 1
12. -4
13. 6
14. 2
15. 1
16. 2
17. $\dfrac{1}{2}$
18. $\dfrac{3}{4}$
19. $\dfrac{2}{3}$
20. $\dfrac{1}{3}$

Exercise 4.3

1. $\dfrac{2}{3}$
2. $\dfrac{1}{4}$
3. 1
4. -1
5. $\dfrac{1}{3}$
6. $\dfrac{3}{4}$
7. 4
8. 3
9. 9
10. 4
11. 10
12. 4
13. 12
14. 10
15. 4
16. 5
17. 1
18. 3
19. No solution
20. No solution

Exercise 4.4

1. ± 5
2. ± 11
3. ± 3.61
4. ± 6.32
5. ± 4
6. ± 3
7. ± 3
8. ± 10
9. ± 2.35
10. ± 0.894
11. No solution
12. No solution
13. ± 2.83
14. ± 11.2
15. ± 1.41
16. ± 2.19
17. 4
18. 6.25
19. 9
20. 1
21. No solution
22. No solution
23. 14.1
24. 1.44

Exercise 4.5

1. 20 A	**2.** 4.55 A	**3.** 16.7 mA	**4.** 4 μA	**5.** 5.4 V	**6.** 5.6 V
7. 15 V	**8.** 60 mV	**9.** 10 kΩ	**10.** 4.8 MΩ	**11.** 2.5 kΩ	**12.** 500 Ω
13. 1 W	**14.** 1.44 mW	**15.** 1.8 mW	**16.** 594 μW	**17.** 750 mW	**18.** 4.32 W
19. 65.5 mW	**20.** 6.98 mW	**21.** 15.3 mA	**22.** 51.8 μA	**23.** 115 V	**24.** 99.5 V

Self-Test

1. $\frac{4}{3}$ (Objective 2) **2.** 12 (Objective 1) **3.** −3 (Objective 3)

4. ±1.06 (Objective 4) **5.** 1.12 μW (Objective 5)

UNIT 5

Exercise 5.1

1. 1.71 μA	**2.** 110 μA	**3.** −5	**4.** 8	**5.** 2	**6.** −8
7. 2	**8.** −2	**9.** $50	**10.** $150	**11.** $3	**12.** $52.50
13. $322.50	**14.** $618.75	**15.** 600 Ω	**16.** 1.38 MΩ	**17.** 11 Ω	**18.** 5.42 Ω
19. 15.8 Ω	**20.** 2.89 kΩ	**21.** 0.962	**22.** 0.995	**23.** 2.67 kΩ	**24.** 78.5 Ω
25. 265 Ω	**26.** 1.77 kΩ	**27.** 10.3 kHz	**28.** 2.42 MHz	**29.** 200	**30.** 122

Exercise 5.2

1. $m = \dfrac{y}{x}$ **2.** $T = \dfrac{I}{PR}$ **3.** $t = \dfrac{Q}{I}$ **4.** $m = \dfrac{F}{a}$

5. $a = \dfrac{2s}{t^2}$ **6.** $Q_2 = \dfrac{Fr^2}{kQ_1}$ **7.** $I_{CO} = I_C - \alpha I_E$ **8.** $I_E = \dfrac{I_C - I_{CO}}{\alpha}$

9. $I_E = \dfrac{V_C - V_{CE}}{R_E}$ **10.** $I_C = \dfrac{V_{CC} - V_{CE} - I_E R_E}{R_C}$ **11.** $I = \dfrac{E}{R_1 + R_2}$ **12.** $I_C = \dfrac{V_{CC} - V_{CE}}{R_C + R_E}$

13. $R_1 = \dfrac{E - IR_2}{I}$ **14.** $R_C = \dfrac{V_{CC} - V_{CE} - I_C R_E}{I_C}$ **15.** $t = \sqrt{\dfrac{2s}{a}}$ **16.** $r = \sqrt{\dfrac{kQ_1 Q_2}{F}}$

17. $s = \dfrac{v^2}{2a}$ **18.** $m = \dfrac{2E}{v^2}$ **19.** $X_C = \sqrt{Z^2 - R^2}$ **20.** $V_R = \sqrt{E^2 - V_C^2}$

21. $C = \dfrac{1}{(2\pi f_r)^2 L}$ **22.** $L = (QR)^2 C$ **23.** $C = \dfrac{L}{(QR)^2}$ **24.** $g = \dfrac{(2\pi)^2 l}{T^2}$

25. 14.5 μH **26.** 35 MHz **27.** 22.1 nF **28.** 17.7 MHz

29. 25 Ω **30.** 0.00393 per °C **31.** 1.55 Ω **32.** 661 mV

33. 15 μH **34.** 422 pF **35.** 3.5 H **36.** 3 pF

Exercise 5.3

1. $P = kV$ **2.** $\omega = kf$ **3.** $P = \dfrac{k}{R}$ **4.** $\omega = \dfrac{k}{T}$

5. $W = kFd$ **6.** $F = kQ_1 Q_2$ **7.** $P = \dfrac{kF}{t}$ **8.** $H = \dfrac{kI}{r}$

9. $P = \left(5.5 \, \dfrac{\text{W}}{\text{V}}\right) V$ **10.** $V = \left(9.8 \, \dfrac{\text{m}}{\text{s}^2}\right) t$ **11.** $P = \dfrac{222 \, \text{W} \cdot \Omega}{R}$ **12.** $\lambda = \dfrac{3 \times 10^8 \, \text{m} \cdot \text{Hz}}{f}$

13. $V_1 = \left(\dfrac{0.13}{\Omega}\right) ER_1$ **14.** $\theta = \left(6.28 \, \dfrac{\text{radians}}{\text{Hz} \cdot \text{s}}\right) ft$ **15.** $R = \dfrac{\left(10.4 \, \dfrac{\Omega \cdot \text{c.mils}}{\text{ft}}\right) l}{A}$ **16.** $X_C = \dfrac{159 \, \Omega \cdot \text{Hz} \cdot \text{F}}{fC}$

17. 1 W
18. 61.1 mW
19. 0.754 radians
20. 1.31 Ω
21. 5.55 kW
22. 200 ft
23. 810 lb
24. 1 mN
25. $87.6 \dfrac{\text{ft}}{\text{s}}$
26. 0.11 mH

Exercise 5.4

1. 2.8
2. 4
3. 68.2
4. 10.0
5. 167 min, or 2.78 h
6. 262.5 mi
7. 304.8 mm
8. 54 mi
9. 2500 operations
10. 400 s, or 6.67 min
11. 8 V
12. 12.5 V
13. 800 Ω
14. 24 V
15. 10 mA
16. 27.5 kΩ
17. 5 V
18. 1.41 A
19. 13.5 W
20. 3 W

Exercise 5.5

1. 5.95 V
2. 180 turns
3. 2.03 V
4. 224 turns
5. 13.3 V
6. 1.29 V
7. 1 : 8.5
8. 3.1 : 1
9. 90 mA
10. 10.8 mA
11. 330 Ω
12. 1200 turns
13. 300 kΩ
14. 28.3 V
15. 1.15 mA
16. 7.78 mA

Self-Test

1. $R_E = \dfrac{V_{CC} - I_C R_C}{I_C}$ (Objective 2)
2. 252 kHz (Objective 1)
3. 2.5 s (Objective 4)
4. 15.6 kΩ (Objective 5)
5. 66.2 Ω (Objective 3)

UNIT 6

Self-Test

1. −13 (Unit 1)
2. 4.51×10^6 (Unit 2)
3. **a.** 1220 ps (Unit 2)
   b. 0.00036 GHz (Unit 2)
4. $9x - 11$ (Unit 3)
5. $2xy^2$ (Unit 3)
6. 9 (Unit 4)
7. ±2 (Unit 4)
8. 2.2 kΩ (Unit 4)
9. 9.65 nF (Unit 5)
10. 30 cu ft (Unit 5)

UNIT 7

Exercise 7.1

1. 11 V
2. 325 mV
3. 7.4 V
4. 150 mV
5. 50 V
6. 44 V
7. 1.5 V
8. 980 mV
9. $V_1 = 1.5\,V, V_3 = 2.1\,V$
10. $E = 60\,V, V_2 = 15\,V$
11. $V_2 = 9\,V, V_5 = 1\,V$
12. $E = 13\,V, V_5 = 3.5\,V$
13. $V_1 = 7.3\,V, V_4 = 2.3\,V, V_5 = 2.3\,V$
14. $E = 47\,V, V_3 = 34\,V, V_4 = 13\,V$
15. $V_2 = 6.1\,V, V_5 = 5.2\,V$
16. $V_3 = 1.05\,V, V_4 = 850\,mV$
17. $V_1 = 10\,V, V_3 = 3.4\,V, V_4 = 3.4\,V$
18. $E = 40\,V, V_1 = 40\,V, V_4 = 25\,V$
19. $V_1 = 30\,V, V_4 = 30\,V, V_5 = 60\,V, V_6 = 100\,V$
20. $E = 30\,V, V_1 = 10\,V, V_2 = 10\,V, V_3 = 10\,V$

Exercise 7.2

1. 210 μA
2. 1.7 mA
3. 800 mA
4. 70 μA
5. 48.1 mA
6. 2.5 mA
7. 2.81 A
8. 1.59 mA

9. $I_1 = 3.45$ mA, $I_3 = 1.43$ mA<ZZ>

10. $I_1 = 120$ μA, $I_2 = 75$ μA

11. $I_2 = 6.8$ mA, $I_3 = 6.8$ mA

12. $I_1 = 375$ μA, $I_3 = 825$ μA

13. $I_3 = 7.3$ mA, $I_4 = 11.6$ mA

14. $I_1 = 85.6$ μA, $I_4 = 25.4$ μA

15. $I_2 = 2.9$ A, $I_3 = 1.3$ A

16. $I_1 = 52$ mA, $I_2 = 32$ mA

17. $I_4 = 10$ mA, $I_5 = 20$ mA, $I_6 = 30$ mA

18. $I_T = 20$ mA, $I_2 = 10$ mA, $I_3 = 15$ mA

19. $I_3 = 10$ A, $I_4 = 9$ A, $I_5 = 1$ A left to right

20. $I_1 = 1$ A, $I_3 = 1$ A, $I_5 = 4$ A right to left

Self-Test

1. 310 mA (Objective 2)

2. $V_1 = V_2 = V_3 = 9.2$ V (Objective 1)

3. 690 mV (Objective 1)

4. $V_1 = 5.7$ V, $V_3 = 10$ V (Objective 1)

5. $I_1 = 21.8$ mA, $I_4 = 17.2$ mA (Objective 2)

UNIT 8

Exercise 8.1

1. 510 Ω

2. 90 kΩ

3. 5.6 kΩ

4. 5.1 MΩ

5. 2.77 kΩ

6. 3.26 MΩ

7. 470 Ω

8. 510 kΩ

9. $I = 0.125$ mA, $R_T = 40$ kΩ, $V_1 = 2.25$ V, $V_2 = 2.75$ V

10. $I = 1.16$ μA, $R_T = 1.29$ MΩ, $V_1 = 0.951$ V, $V_2 = 0.545$ V

11. $E = 9$ V, $R_T = 300$ Ω, $R_1 = 180$ Ω, $V_2 = 3.6$ V

12. $R_T = 13.2$ kΩ, $R_1 = 7.6$ kΩ, $V_1 = 14.4$ V, $V_2 = 10.6$ V

13. $I = 0.5$ mA, $R_T = 80$ kΩ, $R_1 = 33$ kΩ, $V_2 = 23.5$ V

14. $I = 12$ μA, $R_1 = 600$ kΩ, $R_2 = 400$ kΩ, $V_1 = 7.2$ V

15. $I = 11.1$ mA, $R_T = 450$ Ω, $V_1 = 1.33$ V, $V_2 = 1.67$ V, $V_3 = 2$ V

16. $I = 0.667$ mA, $R_T = 21$ kΩ, $V_1 = 4.53$ V, $V_2 = 3.4$ V, $V_3 = 6.07$ V

17. $I = 28.3$ μA, $R_T = 1.41$ MΩ, $R_3 = 380$ kΩ, $V_1 = 15.8$ V, $V_3 = 10.9$ V

18. $I = 10.1$ μA, $R_T = 2.48$ MΩ, $R_2 = 910$ kΩ, $V_2 = 9.19$ V, $V_3 = 7.58$ V

19. $E = 24$ V, $R_1 = 470$ Ω, $R_3 = 400$ Ω, $V_1 = 9.4$ V, $V_2 = 6.6$ V

20. $E = 60$ V, $R_T = 4$ MΩ, $R_3 = 980$ kΩ, $V_1 = 12.3$ V, $V_2 = 33$ V

Exercise 8.2

1. 114 Ω

2. 3.72 kΩ

3. 220 kΩ

4. 12 MΩ

5. 72.6 Ω

6. 410 Ω

7. 511 Ω

8. 810 kΩ

9. $R_T = 77.6$ Ω, $I_T = 64.4$ mA, $I_1 = 41.7$ mA, $I_2 = 22.7$ mA

10. $R_T = 3.33$ kΩ, $I_T = 1.8$ mA, $I_1 = 1.07$ mA, $I_2 = 0.732$ mA

11. $E = 145$ V, $R_T = 1.93$ kΩ, $I_1 = 53.7$ mA, $I_2 = 21.3$ mA

12. $E = 41.8$ V, $R_T = 17.3$ MΩ, $R_1 = 21.9$ MΩ, $I_2 = 0.51$ μA

13. $R_T = 1.79$ MΩ, $R_2 = 3.29$ MΩ, $I_1 = 6.41$ μA, $I_2 = 7.59$ μA

14. $R_T = 289$ Ω, $R_2 = 750$ Ω, $I_1 = 16$ μA, $I_2 = 10$ μA

15. $R_T = 13.5$ kΩ, $I_T = 1.11$ mA, $I_1 = 0.27$ mA, $I_2 = 0.385$ mA, $I_3 = 0.455$ mA

16. $R_T = 470$ kΩ, $I_T = 63.8$ μA, $I_1 = 13.6$ μA, $I_2 = 13.6$ μA, $I_3 = 36.5$ μA

17. $R_T = 66.7$ kΩ, $R_1 = 150$ kΩ, $R_3 = 360$ kΩ, $I_2 = 667$ μA, $I_3 = 333$ μA

18. $R_T = 429$ kΩ, $R_2 = 1.5$ MΩ, $R_3 = 1$ MΩ, $I_1 = 8$ μA, $I_3 = 12$ μA

19. $E = 60.2$ V, $R_T = 5.68$ kΩ, $R_3 = 9.15$ kΩ, $I_2 = 2.74$ mA, $I_3 = 6.58$ mA

20. $E = 120$ mV, $R_1 = 750$ Ω, $R_2 = 822$ Ω, $R_3 = 1.79$ kΩ, $I_3 = 67$ μA

Exercise 8.3

1. $R_T = 14.2$ kΩ, $I_T = 3.52$ mA, $V_1 = E$, $V_2 = 20$ V, $V_3 = 30$ V, $I_1 = 1.52$ mA, $I_2 = I_3 = 2$ mA
2. $R_T = 0.9$ MΩ, $I_T = 11.1$ μA, $V_1 = E$, $V_2 = 1.67$ V, $V_3 = 8.33$ V, $I_1 = 10$ μA, $I_2 = I_3 = 1.11$ μA
3. $R_T = 39$ kΩ, $I_T = I_1 = 1.28$ mA, $V_1 = 42.3$ V, $V_2 = V_3 = 7.69$ V, $I_2 = 0.768$ mA, $I_3 = 0.512$ mA
4. $R_T = 2.25$ MΩ, $I_T = I_1 = 4.44$ μA, $V_1 = 4.44$ V, $V_2 = V_3 = 5.56$ V, $I_2 = 3.71$ μA, $I_3 = 0.741$ μA
5. $R_T = 2.91$ Ω, $I_T = 3.43$ A, $V_1 = V_2 = 4.12$ V, $V_3 = V_4 = 5.88$ V, $I_1 = 2.06$ A, $I_2 = 1.37$ A, $I_3 = 1.96$ A, $I_4 = 1.47$ A
6. $R_T = 16.7$ kΩ, $I_T = 0.3$ mA, $V_1 = V_2 = 2.67$ V, $V_3 = V_4 = 2.33$ V, $I_1 = 0.178$ mA, $I_2 = 0.122$ mA, $I_3 = 0.106$ mA, $I_4 = 0.194$ mA
7. $R_T = 1.67$ Ω, $I_T = 6$ A, $V_1 = 10$ V, $V_2 = V_3 = 3$ V, $V_4 = 4$ V, $I_1 = 5$ A, $I_2 = I_3 = I_4 = 1$ A
8. $R_T = 1.72$ kΩ, $I_T = 34.9$ μA, $V_1 = 60$ mV, $V_2 = V_3 = V_4 = 20$ mV, $I_1 = 18.2$ μA, $I_2 = I_3 = I_4 = 16.7$ μA
9. $R_T = 8.33$ Ω, $I_T = I_1 = I_3 = 1.2$ A, $V_1 = 2.4$ V, $V_2 = V_4 = 4$ V, $V_3 = 3.6$ V, $I_2 = 0.8$ A, $I_4 = 0.4$ A
10. $R_T = 4.45$ kΩ, $I_T = I_1 = I_3 = 13.5$ μA, $V_1 = 29.7$ mV, $V_2 = V_4 = 10.1$ mV, $V_3 = 20.2$ mV, $I_2 = I_4 = 6.75$ μA
11. $R_T = 2.3$ Ω, $I_T = 4.35$ A, $V_1 = E$, $V_2 = 5.55$ V, $V_3 = V_L = 4.45$ V, $I_1 = 2.5$ A, $I_2 = 1.85$ A, $I_3 = 1.11$ A, $I_L = 0.742$ A
12. $R_T = 6.5$ kΩ, $I_T = 18.5$ mA, $V_1 = E$, $V_2 = 65$ V, $V_3 = V_L = 55$ V, $I_1 = 12$ mA, $I_2 = 6.5$ mA, $I_3 = 3.67$ mA, $I_L = 2.75$ mA
13. $R_T = 1.87$ Ω, $I_T = 5.36$ A, $V_1 = 5.71$ V, $V_2 = V_3 = 4.29$ V, $V_4 = E$, $I_1 = 2.86$ A, $I_2 = I_3 = 1.43$ A, $I_4 = 2.5$ A
14. $R_T = 5.05$ kΩ, $I_T = 11.9$ mA, $V_1 = 38.8$ V, $V_2 = V_3 = 21.2$ V, $V_4 = E$, $I_1 = 3.88$ mA, $I_2 = 2.12$ mA, $I_3 = 1.76$ mA, $I_4 = 8$ mA
15. $R_1 = 4$ kΩ, $R_2 = 667$ Ω, $R_3 = 2$ kΩ, $R_L = 1$ kΩ
16. $R_1 = 167$ Ω, $R_2 = 333$ Ω, $R_3 = 2$ kΩ, $R_{L1} = 800$ Ω, $R_{L2} = 1$ kΩ, $R_{L3} = 1$ kΩ

Self-Test

1. 1.32 kΩ (Objective 2)
2. 1.05 kΩ (Objective 1)
3. 6.5 Ω, 1.54 A (Objective 3)
4. 5.4 V (Objective 3)
5. 0.31 A (Objective 3)

UNIT 9

Exercise 9.1

1. $6(x + y)$
2. $12(x - y)$
3. $6(x - 2y)$
4. $2(3x + 4y)$
5. $3x(x + 3)$
6. $5x(2x - 3)$
7. $3xy(x - 4y)$
8. $5xy(3x - y)$
9. $2x(x^2 - 2x + 2)$
10. $3x(4x^2 + 5x - 3)$
11. $6x^2(x + 1)$
12. $4x(2x + 1)$
13. $8x(1 - 2y)$
14. $3y(1 + 3x)$
15. $5x(2x^2 + 2x + 1)$
16. $4xy(1 - 2x - 2y)$

Exercise 9.2

1. 3
2. 2
3. $\dfrac{3}{4}$
4. $\dfrac{5}{3}$
5. $x = \dfrac{3y}{y - 4}$
6. $x = \dfrac{3y}{3 - y}$
7. $y = \dfrac{4x}{x - 3}$
8. $y = \dfrac{3x}{x + 3}$
9. $x = \dfrac{y}{3y - 2}$, $y = \dfrac{2x}{3x - 1}$
10. $x = \dfrac{2y}{y + 4}$, $y = \dfrac{4x}{2 - x}$
11. $x = \dfrac{2y - 2}{y}$, $y = \dfrac{2}{2 - x}$
12. $x = \dfrac{3y + 6}{2y}$, $y = \dfrac{6}{2x - 3}$
13. $R_1 = \dfrac{I_T R_2 - I_1 R_2}{I_1}$
14. $r_e = \dfrac{\beta R_{BB} - A_i R_{BB}}{\beta A_i}$
15. $R_2 = \dfrac{I_1 R_1}{I_T - I_1}$
16. $R_{BB} = \dfrac{\beta A_i r_e}{\beta - A_i}$

17. $R_1 = \dfrac{R_T R_2}{R_2 - R_T}$

18. $R_B = \dfrac{A_i Z_b}{h_{fe} - A_i}$

19. $R_B = \dfrac{V_{BB} - V_{BE} - (\beta + 1)I_B R_E}{I_B}$

20. $Z_b = \dfrac{(1 + h_{fe})R_B - A_i R_B}{A_i}$

21. $R_E = \dfrac{\beta R_B - \beta A_i r_e - A_i R_B}{\beta A_i}$

22. $R_B = \dfrac{\beta A_i (r_e + R_E)}{\beta - A_i}$

Exercise 9.3

1. $0.72\ \Omega$ **2.** $9.9\ \text{k}\Omega$ **3.** $25.6\ \text{k}\Omega$ **4.** $0.975\ \text{M}\Omega$ **5.** $39\ \Omega$ **6.** $56\ \Omega$

7. $3.93\ \text{k}\Omega$ **8.** $676\ \Omega$ **9.** $4\ \Omega$ **10.** $1.8\ \Omega$

Exercise 9.4

1. $V_1 = 16.5\ \text{V}, V_2 = 23.5\ \text{V}$

2. $V_1 = 1.51\ \text{V}, V_2 = 3.49\ \text{V}$

3. $V_1 = V_2 = 20\ \text{V}$

4. $V_1 = V_2 = 2.5\ \text{V}$

5. $V_1 = 2.18\ \text{V}, V_2 = 3.2\ \text{V}, V_3 = 2.62\ \text{V}$

6. $V_1 = 28\ \text{mV}, V_2 = 23.5\ \text{mV}, V_3 = 13.5\ \text{mV}$

7. $V_1 = V_2 = V_3 = 4\ \text{V}$

8. $V_1 = 6\ \text{V}, V_2 = V_3 = 3\ \text{V}$

9. $I_1 = 3.57\ \text{mA}, I_2 = 1.43\ \text{mA}$

10. $I_1 = 20.5\ \mu\text{A}, I_2 = 29.5\ \mu\text{A}$

11. $I_1 = I_2 = 2.5\ \text{mA}$

12. $I_1 = I_2 = 25\ \mu\text{A}$

Self-Test

1. $3x(2 - 2x + 3x^2)$ (Objective 1)

2. $y = \dfrac{9}{3 + x}$ (Objective 2)

3. $R_S = \dfrac{A_v}{g_m - g_m A_v}$ (Objective 2)

4. $248\ \Omega$ (Objective 3)

5. $I_1 = 14.3\ \text{mA}, I_2 = 10.7\ \text{mA}$ (Objective 4)

UNIT 10

Exercise 10.1

1

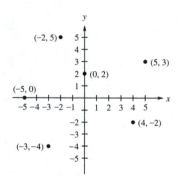

2.

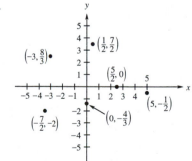

3.

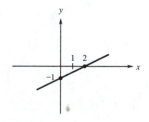

4.

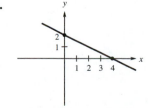

5.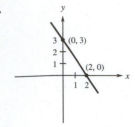

6.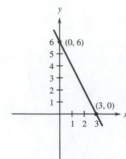

7.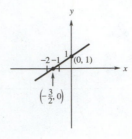

8.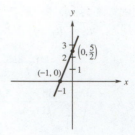

9.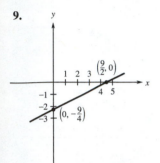

10.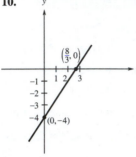

11.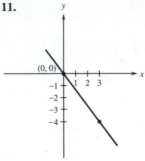

12.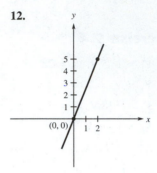

13.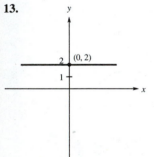

14.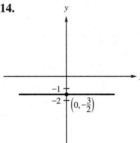

15.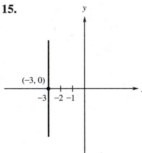

16.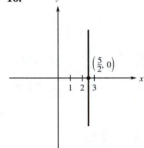

Exercise 10.2

1. $\dfrac{2}{3}$ **2.** 2 **3.** -3 **4.** $-\dfrac{3}{2}$ **5.** -3 **6.** $-\dfrac{1}{2}$

7. $\dfrac{1}{2}$ **8.** $-\dfrac{1}{2}$ **9.** $-\dfrac{3}{4}$ **10.** 1 **11.** 2 **12.** $-\dfrac{1}{2}$

13. 0 **14.** Undefined **15.** Undefined **16.** 0

Exercise 10.3

1. $2, (0, 4)$ **2.** $1, \left(0, -\dfrac{3}{2}\right)$ **3.** $-\dfrac{5}{2}, (0, 5)$ **4.** $-\dfrac{3}{4}, (0, -2)$ **5.** $-\dfrac{1}{4}, \left(0, \dfrac{1}{2}\right)$ **6.** $\dfrac{1}{2}, \left(0, \dfrac{3}{4}\right)$

7. $\dfrac{1}{2}, \left(0, -\dfrac{2}{3}\right)$ **8.** $-\dfrac{1}{3}, \left(0, -\dfrac{3}{2}\right)$ **9.** $-\dfrac{2}{5}, (0, 0)$ **10.** $\dfrac{4}{3}, (0, 0)$ **11.** $0, \left(0, -\dfrac{2}{3}\right)$ **12.** $0, \left(0, \dfrac{5}{2}\right)$

Exercise 10.4

1.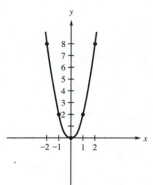

2.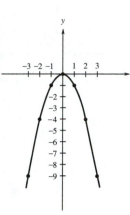

3.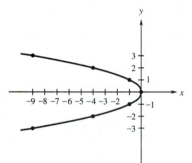

4.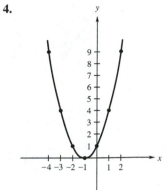

5.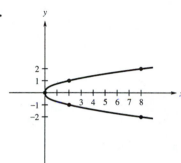

6.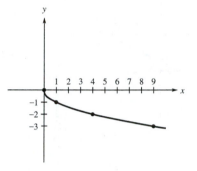

7.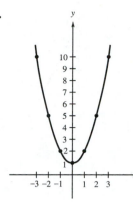

8.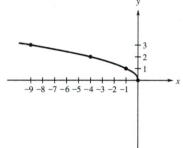

9.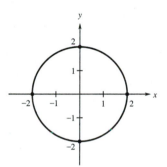

10.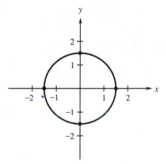

11.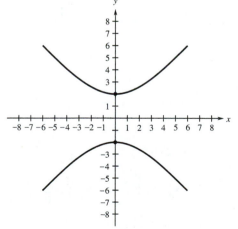

12.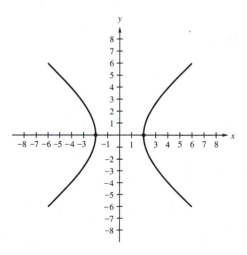

13.

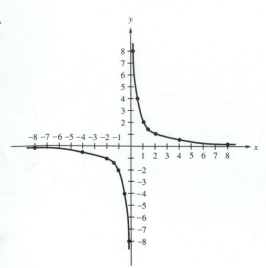

14.

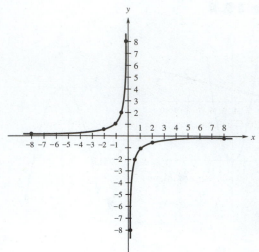

15.

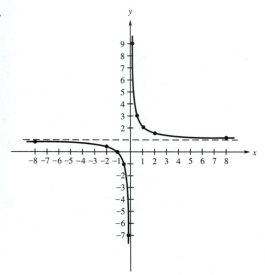

16.

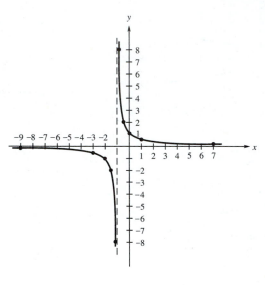

17.

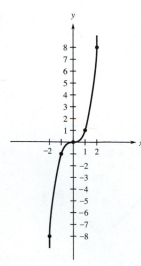

18.

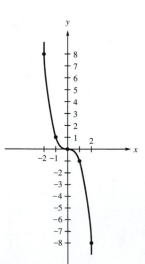

19.

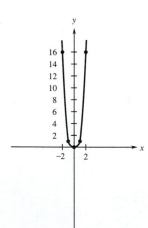

20.

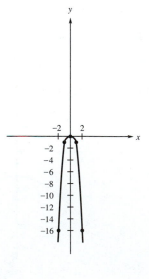

Exercise 10.5

1.

$m = 0.26$ (The chart was made using a 3.9 Ω resistor.)

2.

$m = 0.12$ (The chart was made using an 8.2 Ω resistor.)

3. 470 Ω   **4.** 160 Ω   **5.** 200 Ω   **6.** 37.5 Ω

7. ~ 60 Ω   **8.** ~ 20 Ω   **9.** ~ 40 Ω   **10.** ~ 6 Ω

(The symbol ~ means "approximately.")

Self-Test

1.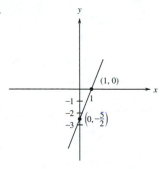

(Objective 1)

2.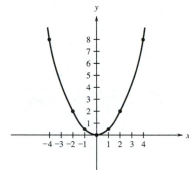

(Objective 4)

3. $-\dfrac{5}{4}, \left(0, \dfrac{5}{2}\right)$ (Objective 3)

4. $-\dfrac{1}{2}$ (Objective 2)

5. ~ 50 Ω (Objective 5)

UNIT 11

Self-Test

1. 100 (Unit 1)

2. 10^9 (Unit 2)

3. $6x - 2$ (Unit 3)

4. 9 (Unit 4)

5. 3.8×10^8 m (Unit 5)

6. $y = \dfrac{12x}{3 + 2x}$ (Unit 9)

7. $V_2 = 5$ V, $V_5 = 10$ V (Unit 7)

8. $R_T = 7.56$ Ω, $I_T = 1.32$ A (Unit 8)

9.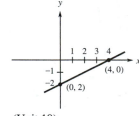

(Unit 10)

10. $m = 20$ (Unit 10)

UNIT 12

Exercise 12.1

1.

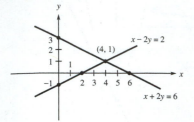

2.

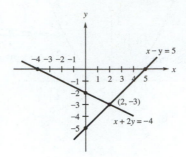

3.

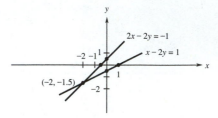

4.

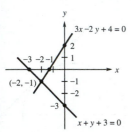

5.

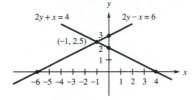

6.

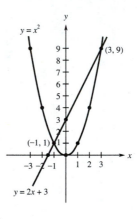

7.

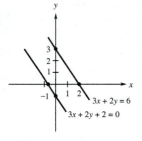

8.

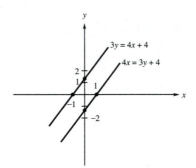

9.

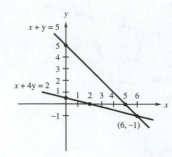

10.

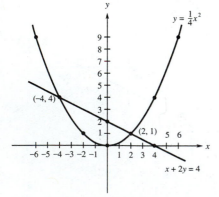

11.

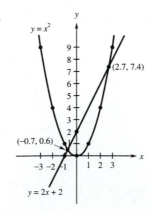

12.

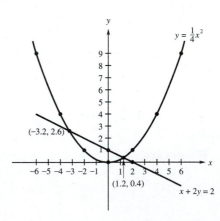

13.

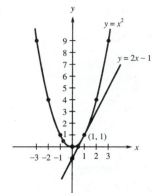

14.

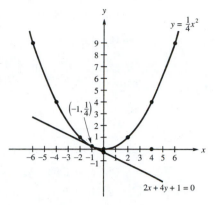

15.

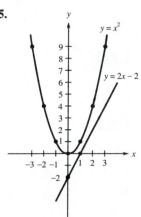

16.

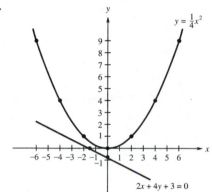

17.

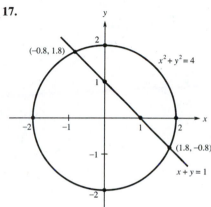

18.

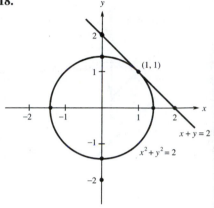

19.

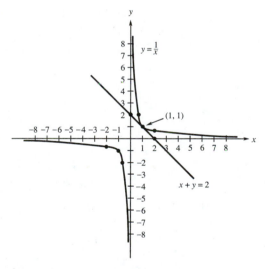

20.

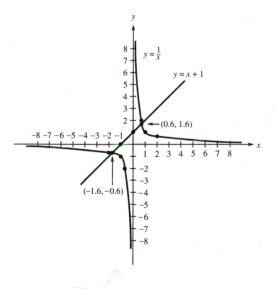

Exercise 12.2

1. $(3, 1)$

2. $\left(\dfrac{1}{2}, -\dfrac{3}{2}\right)$

3. $(-1, 3)$

4. $(5, -2)$

5. $\left(\dfrac{3}{2}, \dfrac{5}{2}\right)$

6. $\left(\dfrac{1}{3}, \dfrac{2}{3}\right)$

7. $(1, 1)$

8. $(2, -1)$

9. $\left(2, \dfrac{1}{2}\right)$

10. $\left(\dfrac{5}{2}, -1\right)$

11. $(2, 6)$

12. $(5, 4)$

13. Inconsistent   **14.** Inconsistent   **15.** $(0, 0), (2, 4)$ **16.** $(0, 0), (-2, 1)$

17. $(1, 1), (-1, -1)$ **18.** $\left(2, \dfrac{1}{2}\right), \left(-2, -\dfrac{1}{2}\right)$ **19.** Inconsistent   **20.** Inconsistent

21. $(\sqrt{5}, 2), (-\sqrt{5}, 2), (\sqrt{5}, -2), (-\sqrt{5}, -2)$ **22.** $(2, 0), (-2, 0)$

23. Inconsistent   **24.** Inconsistent

Exercise 12.3

1. $I_1 = 2$ mA, $I_2 = 0.8$ mA   **2.** $I_1 = 2.78$ mA, $I_2 = 2$ mA

3. $I_1 = 1.25$ mA, $I_2 = 0.5$ mA, $I_3 = 0.75$ mA   **4.** $I_1 = 2.08$ mA, $I_2 = 1.25$ mA, $I_3 = 0.833$ mA

5. $I_1 = 4.74$ A, $I_2 = 5.79$ A, $I_3 = 1.05$ A   **6.** $I_1 = 7$ A, $I_2 = 6.5$ A, $I_3 = 0.5$ A to the left

7. $I_1 = 2.42$ A, $I_2 = 0.31$ A upward, $I_3 = 2.11$ A   **8.** $I_1 = 2.47$ A, $I_2 = 1.18$ A downward, $I_3 = 3.65$ A

9. $I_1 = 11.8$ A, $I_2 = 9.09$ A, $I_3 = 2.73$ A upward   **10.** $I_1 = 6.37$ A, $I_2 = 1.82$ A upward, $I_3 = 4.55$ A upward

Self-Test

1.

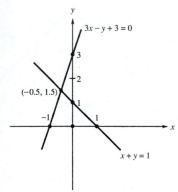

(Objective 1)

2.

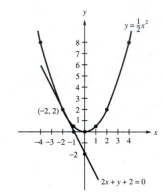

(Objective 1)

3. Inconsistent (Objective 2)   **4.** $\left(1, -\dfrac{1}{2}\right)$ (Objective 2)

5. $I_1 = 3.3$ A, $I_2 = 1.2$ A, $I_3 = 2.1$ A (Objective 3)

UNIT 13

Exercise 13.1

1. 14   **2.** 20   **3.** 10   **4.** 144   **5.** 12   **6.** −16

7. 10.6   **8.** −2.715   **9.** $(2, 1)$ **10.** $\left(3, \dfrac{1}{2}\right)$ **11.** $(-1, 3)$ **12.** $(5, -3)$

13. $\left(1, -\dfrac{1}{2}\right)$ **14.** $\left(\dfrac{3}{4}, \dfrac{1}{2}\right)$ **15.** $(-1, 2)$ **16.** $(-3, -5)$ **17.** $\left(\dfrac{1}{2}, \dfrac{1}{3}\right)$ **18.** $(2, 6)$

19. $\left(\dfrac{3}{2}, -1\right)$ **20.** $\left(-\dfrac{3}{2}, -\dfrac{2}{3}\right)$ **21.** $(3.67, 5.78)$ **22.** $(-3.5, 1)$ **23.** $(-1.18, 7.2)$ **24.** $(2.16, 3.04)$

Exercise 13.2

1. 5   **2.** −10   **3.** −15   **4.** −7   **5.** 11   **6.** −12

7. −12   **8.** 8   **9.** −18   **10.** −24   **11.** 0   **12.** 0

13. $(2, 1, -1)$ **14.** $(-3, -1, 2)$ **15.** $(-1, -2, -3)$ **16.** $\left(\dfrac{3}{2}, -\dfrac{3}{2}, \dfrac{1}{2}\right)$ **17.** $(2, 1, 4)$ **18.** $\left(2, -1, \dfrac{1}{2}\right)$

19. $\left(\dfrac{1}{2}, \dfrac{1}{3}, 1\right)$ **20.** $(-2, -4, -3)$

Exercise 13.3

13. 5 **14.** −24 **15.** −12 **16.** 1680

25. $x = 1, y = 2, z = -1, u = -2$ **26.** $x = 2, y = \frac{1}{2}, z = \frac{1}{2}, u = 1$

27. $x = 4, y = -3, z = 2, u = -1$ **28.** $x = 1, y = -1, z = 1, u = 0$

Exercise 13.4

1. $I_1 = 3.53$ mA, $I_2 = 1.57$ mA, $I_3 = 1.96$ mA **2.** $I_1 = 0.199$ mA, $I_2 = 0.118$ mA, $I_3 = 0.0808$ mA

3. $I_1 = 0.378$ A, $I_2 = 0.433$ A, $I_3 = 0.0556$ A **4.** $I_1 = 0.0514$ A to the right, $I_2 = 0.114$ A, $I_3 = 0.166$ A

5. $I_1 = 3.33$ A, $I_2 = 1.53$ A upward, $I_3 = 1.8$ A **6.** $I_1 = 2.75$ A, $I_2 = 3.52$ A downward, $I_3 = 6.27$ A

7. $I_5 = 0.769$ A **8.** $I_5 = 0$ A (The Wheatstone bridge is said to be balanced.)

9. $I_1 = 3.95$ A, $I_2 = 3.02$ A, $I_3 = 0.93$ A to the left, $I_4 = 2.56$ A, $I_5 = 1.63$ A

10. $I_1 = 2.5$ A, $I_2 = 3.5$ A, $I_3 = 1$ A to the right, $I_4 = 4.5$ A, $I_5 = 5.5$ A

Self-Test

1. 92 (Objective 1) **2.** −38 (Objective 2 or 3)

3. (1, −2) (Objective 1) **4.** $(2, 2, -\frac{1}{2})$ (Objective 2 or 3)

5. $I_1 = 6.13$ A, $I_2 = 2.65$ A upward, $I_3 = 3.48$ A (Objective 4)

UNIT 14

Exercise 14.1

1. 0, 3 **2.** $0, \frac{1}{4}$ **3.** 0, 2 **4.** 0, −2

5. $x^2 + 6x + 8$ **6.** $x^2 - 7x + 12$ **7.** $x^2 - x - 12$ **8.** $x^2 + 3x - 18$

9. $2x^2 - x - 10$ **10.** $3x^2 - 10x - 8$ **11.** $4x^2 - 8x + 3$ **12.** $8x^2 + 18x - 5$

13. $(x + 1)(x + 3)$ **14.** $(x + 2)(x + 9)$ **15.** $(x - 3)(x - 8)$ **16.** $(x - 1)(x - 12)$

17. $(x + 3)(x - 4)$ **18.** $(x - 4)(x + 9)$ **19.** $(x + 4)(2x - 3)$ **20.** $(x - 3)(3x + 2)$

21. $(2x + 3)(2x - 1)$ **22.** $(x - 2)(4x + 3)$ **23.** $3(x - 1)(2x + 1)$ **24.** $2(x - 1)(4x - 1)$

25. 3, 4 **26.** −3, 5 **27.** $-1, \frac{1}{3}$ **28.** $2, -\frac{3}{2}$

29. $-2, \frac{1}{4}$ **30.** $\frac{3}{2}, -\frac{1}{2}$ **31.** −3, −4 **32.** $\frac{1}{2}, \frac{1}{3}$

Exercise 14.2

1. 3, 2 **2.** $\frac{1}{2}, -2$ **3.** 0.414, −2.41 **4.** 3.24, −1.24

5. 0.618, −1.62 **6.** 3.45, −1.45 **7.** 1.71, 0.293 **8.** 2.19, −0.686

9. 0.291, −2.29 **10.** 1.17, −0.171 **11.** 1.28, −0.781 **12.** 2.37, 0.634

13. 0.448, −1.12 **14.** 1.88, 0.319 **15.** No real number solutions **16.** No real number solutions

Exercise 14.3

1.

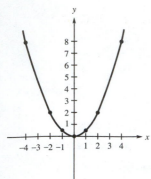

2.

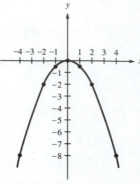

3.

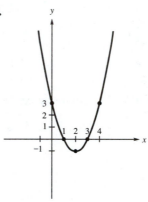

4.

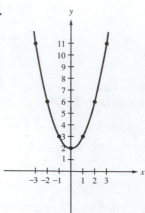

5.

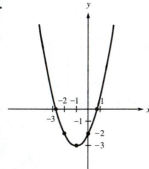

6.

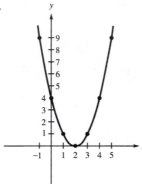

7.

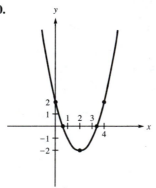

8.

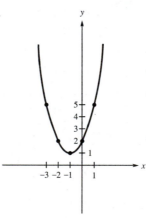

9.

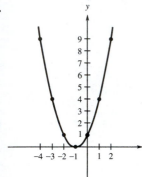

10.

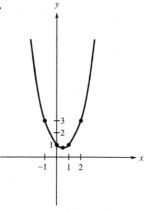

11.

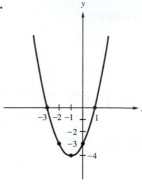

12.

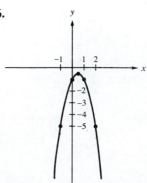

13.

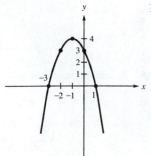

14.

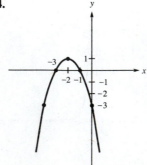

15.

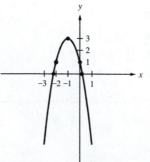

16.

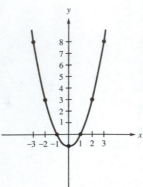

Exercise 14.4

1. $(1, 1), (-2, 4)$ **2.** $(1, 2), \left(-\dfrac{1}{2}, \dfrac{1}{2}\right)$ **11.** $(0, -1), (1, 0)$ **12.** $(-1, 2), (2, -1)$

14. No real number solutions   **16.** $(2, -2)$

Self-Test

1. $0.581, -2.58$ (Objective 2)   **2.** No real number solutions (Objective 2)

3. $\left(-\dfrac{1}{2}, \dfrac{3}{2}\right)$ (Objective 1)   **4.**

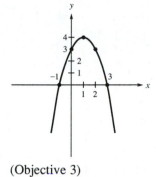

5. $(0.618, 0.191), (-1.62, 1.31)$ (Objective 4)

(Objective 3)

UNIT 15

Exercise 15.1

1.

2.

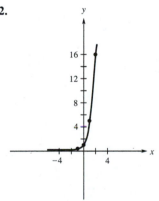

3.

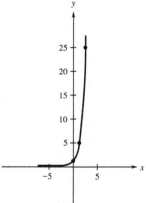

4.

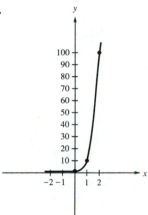

5.

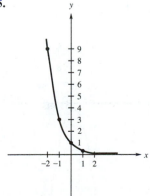

6.

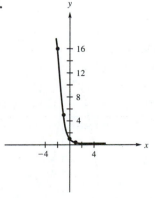

7.

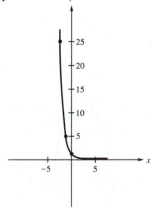

8.

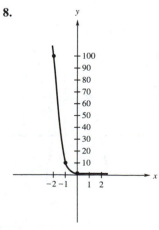

Exercise 15.2

1. $\log_2 8 = 3$

2. $\log_3 9 = 2$

3. $\log_8 2 = \dfrac{1}{3}$

4. $\log_9 3 = \dfrac{1}{2}$

5. $\log_2 \dfrac{1}{8} = -3$

6. $\log_3 \dfrac{1}{9} = -2$

7. $\log_{10} 0.01 = -2$

8. $\log_{100} 10 = \dfrac{1}{2}$

9. $3^4 = 81$

10. $4^3 = 64$

11. $81^{\frac{1}{4}} = 3$

12. $64^{\frac{1}{3}} = 4$

13. $3^{-4} = \dfrac{1}{81}$

14. $4^{-3} = \dfrac{1}{64}$

15. $10^3 = 1000$

16. $10^{-4} = 0.0001$

17. 2   18. 3   19. 6   20. 3   21. $\dfrac{1}{2}$ 22. $\dfrac{1}{3}$ 23. $\dfrac{1}{3}$ 24. $\dfrac{1}{2}$

25. −2   26. −1   27. −3   28. −2   29. 5   30. −5   31. 1   32. 0

Exercise 15.3

1.
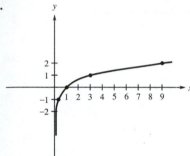

2.
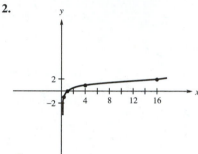

3.
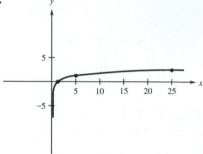

4.
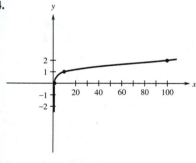

5.
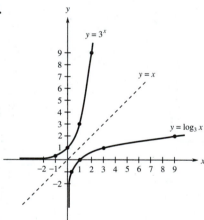

6.
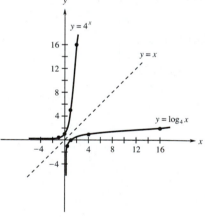

Exercise 15.4

1. 1.18   2. 0.544   3. 1.57   4. 0.852   5. −0.4   6. −1.5

7. 3.68   8. 0.646   9. 4.7   10. 2.21   11. 0.48   12. 1.98

13. $2300   14. $154   15. 12.6 yr   16. 14.6 yr   17. 2.28 yr   18. 0.751 yr

19. 45.2 V   20. 6.77 V   21. 22.3 ms   22. 2.77 s   23. 4.76 V   24. 161 ms

25. 5570 yr   26. 8820 yr   27. 16.8 mA   28. **a.** 118mA   **b.** 824 mA   **c.** 5.77 A

Exercise 15.5

1. $v_C = 9.52$ V, $v_R = 90.5$ V, $i = 1.81$ mA
2. $v_C = 8.24$ V, $v_R = 16.8$ V, $i = 11.2$ μA
3. 55.8 ms
4. 6.69 s
5. 6.27 ms
6. 131 μs
7. 19.1 ms
8. 49 μs
9. 13.5 V
10. 22.2 V
11. 10.2 ms
12. 26 ms
13. 69.3 ms
14. 0.693 τ
15. 2.11 mA
16. 173 mA
17. 51.1 μs
18. 16.1 ms
19. 7.79 mA
20. 10.9 ms

Self-Test

1. $\frac{1}{2}$ (Objective 2)
2. 0.882 (Objective 4)
3. 146 μs (Objective 5)

4.

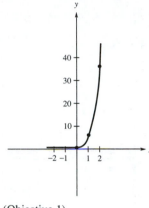

(Objective 1)

5.

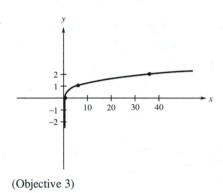

(Objective 3)

UNIT 16

Self-Test

1. 5 (Unit 4)
2. 1.58, −7.58 (Unit 14)
3. $Z_1 = \dfrac{Z_T Z_2}{Z_2 - Z_T}$ (Unit 9)
4. 0.7 V (Unit 5)

5.

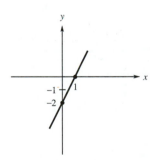

$m = 2, (0, -2)$ (Unit 10)

6.

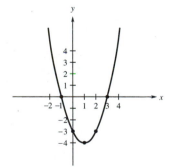

Vertex $(1, -4)$, y-intercept $(0, -3)$,
x-intercepts $(-1, 0)$ and $(3, 0)$ (Unit 14)

7. $\left(\dfrac{3}{2}, -2\right)$ (Unit 12 or 13)

8. $I_1 = 5.4$ mA, $I_2 = 2.3$ mA, $I_3 = 3.1$ mA (Unit 8 or 12; the third digit will vary depending on method used)

9. 3.45 (Unit 15)

10. $v_C = 3.3$ V, $v_R = 6.7$ V (Unit 15)

UNIT 17

Exercise 17.1

1. 10
2. 13
3. 8
4. 2
5. 3.39
6. 8.98
7. 1.7
8. 22.4
9. 3.32 m
10. 5.66 in
11. $2\sqrt{3}$ or 3.46 m
12. $10\sqrt{3}$ or 17.3 in
13. 1.32 ft
14. 11.3 ft
15. 2.24 m
16. 12 in

Exercise 17.2

1. If $A = 30°$ and $c = 150$, then $a = 75$, $b = 130$, $B = 60°$
2. If $A = 30°$ and $c = 2.5$, then $a = 1.25$, $b = 2.17$, $B = 60°$
3. If $A = 42°$ and $a = 75$, then $b = 83.3$, $c = 112$, $B = 48°$
4. If $A = 69°$ and $a = 35.6$, then $b = 13.7$, $c = 38.1$, $B = 21°$
5. If $A = 64°$ and $b = 7.8$, then $a = 16$, $c = 17.8$, $B = 26°$
6. If $A = 18.4°$ and $b = 460$, then $a = 153$, $c = 485$, $B = 71.6°$
7. If $A = 17°$ and $c = 4.8$, then $a = 1.4$, $b = 4.59$, $B = 73°$
8. If $A = 67°$ and $c = 26$, then $a = 23.9$, $b = 10.2$, $B = 23°$
9. If $a = 10$ and $b = 12$, then $A = 39.8°$, $B = 50.2°$, $c = 15.6$
10. If $a = 7.2$ and $b = 3.3$, then $A = 65.4°$, $B = 24.6°$, $c = 7.92$
11. If $a = 22$ and $c = 47$, then $A = 27.9°$, $B = 62.1°$, $b = 41.5$
12. If $a = 3.29$ and $c = 5.89$, then $A = 34°$, $B = 56°$, $b = 4.89$
13. $a = 18$, $c = 29.2$, $A = 38°$
14. $a = 8.02$, $c = 8.85$, $A = 65°$
15. $b = 495$, $c = 803$, $A = 52°$
16. $b = 132$, $c = 137$, $A = 15°$
17. $a = 1.95$, $b = 1.84$, $A = 46.6°$
18. $a = 52.9$, $b = 66.5$, $A = 38.5°$
19. $a = 29$, $A = 69.2°$, $B = 20.8°$
20. $a = 42.7$, $A = 37.5°$, $B = 52.5°$

Exercise 17.3

1. $x = 10.9$, $y = 5.07$
2. $x = 2.12$, $y = 2.91$
3. $x = -4.33$, $y = 2.5$
4. $x = -54.9$, $y = 60.9$
5. $x = -1.15$, $y = -1.99$
6. $x = -35.4$, $y = -19.6$
7. $x = 2.6$, $y = -2.18$
8. $x = 310$, $y = -583$
9. $26.1\underline{/57.5°}$
10. $368\underline{/18.2°}$
11. $26.6\underline{/146°}$
12. $9.6\underline{/135°}$
13. $9.45\underline{/210°}$
14. $2.25\underline{/238°}$
15. $882\underline{/-60.8°}$
16. $14.8\underline{/-39°}$

Exercise 17.4

1. $13\underline{/90°}$
2. $8.3\underline{/30°}$
3. $2\underline{/180°}$
4. $3\underline{/0°}$
5. $3.2\underline{/51.3°}$
6. $16\underline{/39.2°}$
7. $3.35\underline{/153°}$
8. $3.97\underline{/236°}$
9. $7.81\underline{/36.3°}$
10. $9.9\underline{/56°}$
11. $14\underline{/76.2°}$
12. $1.09\underline{/149°}$
13. $5.02\underline{/33.7°}$
14. $13.2\underline{/-33.4°}$
15. $3\underline{/270°}$
16. $1.1\underline{/30°}$
17. $5\underline{/-36.9°}$
18. $2.51\underline{/119°}$
19. $3.61\underline{/-13.9°}$
20. $5.66\underline{/38°}$

Self-Test

1. 8.7 (Objective 1)
2. If $a = 390$ and $b = 140$, then $A = 70.3°$, $B = 19.7°$, $c = 414$ (Objective 2)
3. If $A = 40°$ and $b = 29$, then $a = 24.3$, $c = 37.9$, $B = 50°$ (Objective 2)
4. $x = 44.2$, $y = -8.28$ (Objective 3)
5. $4.91\underline{/96.9°}$ (Objective 4)

UNIT 18

Exercise 18.1

1. $j2$
2. $j6$
3. $j4.47$
4. $j8.49$
5. $-j5$
6. $-j3.16$
7. -2
8. -10
9. -4
10. -10
11. $j8$
12. $j5.2$

Exercise 18.2

1.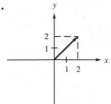
2.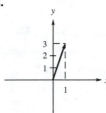
3.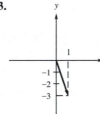
4.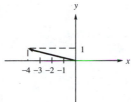
5.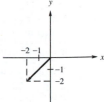

6.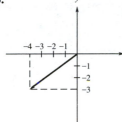
7.
8.
9.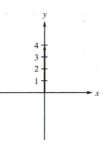
10.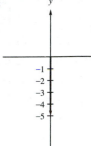

11. $1.41\underline{/45°}$
12. $5\underline{/53.1°}$
13. $3.61\underline{/146°}$
14. $3.61\underline{/-56.3°}$
15. $5.66\underline{/225°}$
16. $2.92\underline{/211°}$
17. $10\underline{/0°}$
18. $1.5\underline{/180°}$
19. $1\underline{/90°}$
20. $2.2\underline{/270°}$
21. $1 + j1.73$
22. $0.707 + j0.707$
23. $-1 + j$
24. $4.33 - j2.5$
25. $-5 - j8.66$
26. $1.91 - j1.1$
27. -3
28. $-j4$
29. $j1.5$
30. 1

Exercise 18.3

1. $9 + j6$
2. $11 + j14$
3. $16 + j3$
4. $8 - j9$
5. $12 - j10$
6. $5 - j$
7. 10
8. j
9. $2 + j8$
10. $-4.7 + j2.6$
11. $j7.2$
12. 2
13. $2.24\underline{/71.6°}$
14. $6.08\underline{/94.7°}$
15. $5.41\underline{/43.9°}$
16. $2\underline{/-90°}$
17. $10.8\underline{/38.2°}$
18. $3.15\underline{/-22.1°}$
19. $4\underline{/-90°}$
20. $7.07\underline{/0°}$

Exercise 18.4

1. $1 + j7$
2. $-14 + j23$
3. $14 + j3$
4. $19 - j9$
5. $-22 - j34$
6. $36 + j24$
7. 17
8. 40
9. -53
10. -10
11. $1 - j$
12. $0.2 + j0.4$
13. $0.412 + j0.647$
14. $1.38 + j1.08$
15. $0.75 - j0.5$
16. $3.5 + j3$
17. $7.08\underline{/81.8°}$ (see Exercise #1)
18. $40.4\underline{/-123°}$ or $40.4\underline{/237°}$ (see Exercise #5)
19. $1.41\underline{/-45°}$ (see Exercise #11)
20. $0.767\underline{/57.6°}$ (see Exercise #13)
21. $0.2\underline{/-36.9°}$
22. $1.41\underline{/45°}$
23. $1\underline{/143°}$
24. $2\underline{/-180°}$ or $2\underline{/180°}$

Self-Test

1. $j10$ (Objective 1)

2. $2.5\underline{/-36.9°}$ (Objective 2)

3. $5.6\underline{/59.6°}$ (Objective 3)

4. $-0.5 + j2.5$ (Objective 4)

5. $3\underline{/135°}$ (Objective 4)

UNIT 19

Exercise 19.1

1. About one-third of first quadrant; 28.6°

2. Near end of fourth quadrant; 344°

3. Middle of third quadrant; 229°

4. Near beginning of fourth quadrant; 286°

5. Near end of first quadrant; 85.9°

6. Near end of third quadrant; 258°

7. First quadrant, second cycle; 401°

8. Third quadrant, second cycle; 573°

9. 1.05 rad **10.** 1.75 rad **11.** 3.93 rad **12.** 6.11 rad **13.** 6.98 rad **14.** 8.73 rad

Exercise 19.2

1.

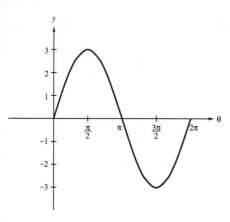

2.

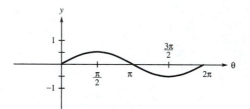

3.

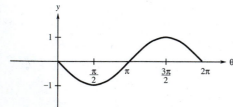

4.

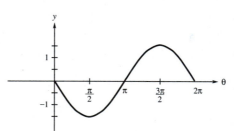

5.

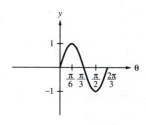

6.

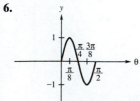

7.

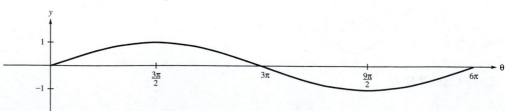

8.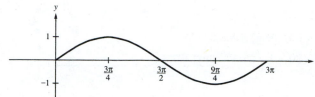

9.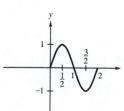

10.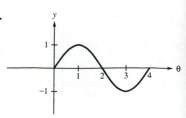

11.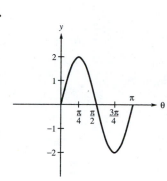

12.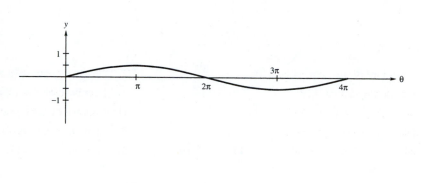

Exercise 19.3

1.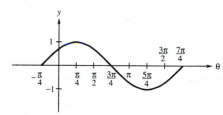

2.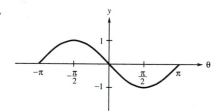

3.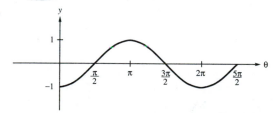

4.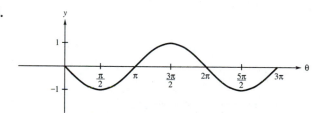

5.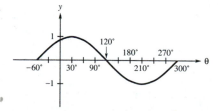

6.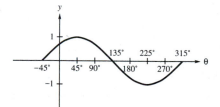

7.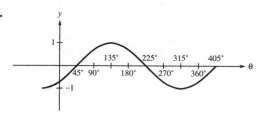

8.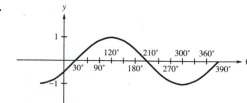

9.

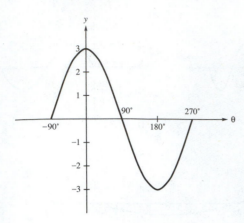

10.

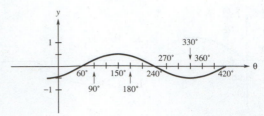

11.

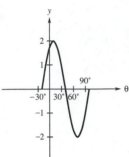

12.

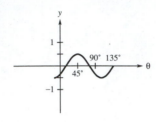

13.

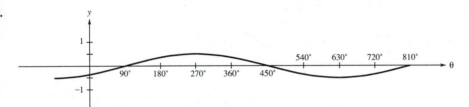

14.

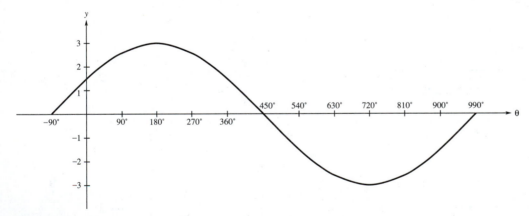

15.

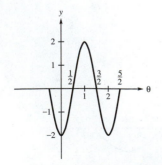

16.

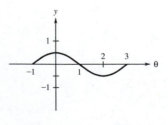

Exercise 19.4

1. 20 Hz	**2.** 10 kHz	**3.** 40 kHz	**4.** 0.5 Hz
5. 100 μs	**6.** 1.25 s	**7.** 83.3 ms	**8.** 62.5 ns

9. a. $25.1 \frac{\text{rad}}{\text{ms}}$ **b.** $1.44 \frac{\text{deg}}{\text{μs}}$ **10. a.** $314 \frac{\text{rad}}{\text{ms}}$ **b.** $18 \frac{\text{deg}}{\text{μs}}$

11. a. $62.8 \frac{\text{rad}}{\text{s}}$ **b.** $3.6 \frac{\text{deg}}{\text{ms}}$ **12. a.** $75.4 \frac{\text{rad}}{\text{μs}}$ **b.** $4.32 \frac{\text{deg}}{\text{ns}}$

13. a. 0.628 rad	**b.** 36°	**14. a.** 62.8 rad	**b.** 3600°
15. a. 7.85 rad	**b.** 450°	**16. a.** 0.163 rad	**b.** 9.36°
17. 2.5 kHz	**18.** 0.75 Hz	**19.** 3.18 kHz	**20.** 10 Hz
21. 100 μs	**22.** 741 ns	**23.** 1 μs	**24.** 26.5 ms

Exercise 19.5

1. 14.1 mA	**2.** 20 A	**3.** 100 V	**4.** 0 V
5. −144 V	**6.** 19.4 V	**7.** −10.6 μA	**8.** 4.11 V

9. $i = 2.94$ mA, $v = 20.2$ V   **10.** $i = 5$ mA, $v = 12.5$ V

11. $i = 2.94$ mA, $v = 24.9$ V   **12.** $i = 0$ mA, $v = 25$ V

13. $i = 1.25$ mA, $v = -61.3$ V   **14.** $i = -5.88$ mA, $v = 58.8$ V

15. $v = 11.8$ V, $i = 8.09$ mA   **16.** $v = 0$ V, $i = -7.07$ mA

Self-Test

1. About one-third of third quadrant; 201° (Objective 1)   **2.** 125 kHz (Objective 4)

3.

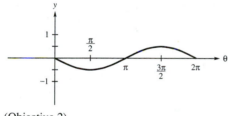

(Objective 2)

4.

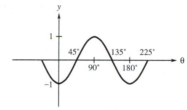

(Objective 3)

5. $i = -5$ mA, $v = -8.66$ V (Objective 5)

UNIT 20

Exercise 20.1

1. $E = 50$ V, $\theta = 36.9°$ **2.** $E = 12.8$ V, $\theta = 51.3°$

3. $V_R = 86.6$ V, $V_L = 50$ V   **4.** $V_R = 13.5$ V, $V_L = 29$ V

5. $Z = 3.55\ \Omega$, $\theta = 40.4°$ **6.** $Z = 11.3\ \Omega$, $\theta = 7.58°$

7. $R = 330\ \Omega$, $X_L = 440\ \Omega$ **8.** $R = 8.37$ kΩ, $X_L = 1.48$ kΩ

9. $X_L = 628\ \Omega$, $Z = 803\ \Omega$, $\theta = 51.5°$ **10.** $X_L = 0.88$ kΩ, $Z = 1.74$ kΩ, $\theta = 30.4°$

11. $X_L = 12.6\ \Omega$, $Z = 14.9\ \Omega$, $\theta = 57.6°$, $I = 805$ mA   **12.** $X_L = 31.4$ kΩ, $Z = 32.9$ kΩ, $\theta = 72.5°$, $I = 152$ μA

13. $I = 66.7$ mA, $Z = 16.5\ \Omega$, $\theta = 43.3°$ **14.** $I = 8$ A, $Z = 2.12\ \Omega$, $\theta = 45°$

Exercise 20.2

1. $E = 25$ V, $\theta = -53.1°$
2. $E = 165$ V, $\theta = -14°$
3. $V_R = 7.07$ V, $V_C = 7.07$ V
4. $V_R = 67.1$ V, $V_C = 99.5$ V
5. $Z = 28$ kΩ, $\theta = -34.8°$
6. $Z = 1.5$ kΩ, $\theta = -53.1°$
7. $R = 170\ \Omega$, $X_C = 1.21$ kΩ
8. $R = 830\ \Omega$, $X_C = 722\ \Omega$
9. $X_C = 15.9\ \Omega$, $Z = 31.3\ \Omega$, $\theta = -30.5°$
10. $X_C = 24.1$ kΩ, $Z = 26.1$ kΩ, $\theta = -67.5°$
11. $X_C = 79.6\ \Omega$, $Z = 128\ \Omega$, $\theta = -38.5°$, $I = 234$ mA
12. $X_C = 265$ kΩ, $Z = 344$ kΩ, $\theta = -50.3°$, $I = 419\ \mu$A
13. $X_C = 1.27$ kΩ, $Z = 1.62$ kΩ, $\theta = -51.8°$, $E = 16.2$ V
14. $X_C = 424$ kΩ, $Z = 656$ kΩ, $\theta = -40.3°$, $E = 31.5$ V
15. $I = 25$ mA, $Z = 4$ kΩ, $\theta = -60°$
16. $I = 1.82$ mA, $Z = 2.75$ kΩ, $\theta = -36.9°$

Exercise 20.3

1. $100 + j60$ V, $117\underline{/31°}$ V
2. $5 + j24$ V, $24.5\underline{/78.2°}$ V
3. $20 - j21$ V, $29\underline{/-46.4°}$ V
4. $9.9 - j3$ V, $10.3\underline{/-16.9°}$ V
5. $18.2 + j52\ \Omega$, $55.1\underline{/70.7°}\ \Omega$
6. $6.7 + j\,4.5$ kΩ, $8.07\underline{/33.9°}$ kΩ
7. $9.1 - j3.13$ kΩ, $9.62\underline{/-19°}$ kΩ
8. $100 - j500\ \Omega$, $510\underline{/-78.7°}\ \Omega$
9. $5 + j21.9\ \Omega$, $22.5\underline{/77.1°}\ \Omega$
10. $1.5 - j0.601$ kΩ, $1.62\underline{/-21.8°}$ kΩ
11. $15 + j5.65\ \Omega$, $16\underline{/20.6°}\ \Omega$
12. $1.2 + j0.801$ kΩ, $1.44\underline{/33.7°}$ kΩ
13. $1 - j0.849\ \Omega$, $1.31\underline{/-40.3°}\ \Omega$
14. $1 - j0.181$ kΩ, $1.02\underline{/-10.3°}$ kΩ

Exercise 20.4

1. $X_L = X_C = 3.32$ kΩ, $\vec{Z} = 1\underline{/0°}$ kΩ
2. $X_L = X_C = 27.9$ kΩ, $\vec{Z} = 12\underline{/0°}$ kΩ
3. $f_r = 19.6$ kHz, $X_L = X_C = 67.7$ kΩ
4. $f_r = 3.75$ kHz, $X_L = X_C = 707\ \Omega$
5. $4\underline{/0°}$ mA
6. $800\underline{/0°}$ mA
7. $333\underline{/0°}$ mA
8. $333\underline{/0°}$ mA
9. a. $2.11\underline{/87.6°}$ mA
   b. $1.89\underline{/-87.8°}$ mA
10. a. $259\underline{/89.7°}\ \mu$A
    b. $543\underline{/-89.4°}\ \mu$A
11. a. $54.6\underline{/89.2°}\ \mu$A
    b. $60.1\underline{/-89.1°}\ \mu$A
12. a. $101\underline{/82.7°}$ mA
    b. $109\underline{/-82.2°}$ mA

Exercise 20.5

1. $I_R = 5$ A, $I_L = 4$ A, $\vec{I}_T = 6.4\underline{/-38.7°}$ A, $\vec{Z}_T = 15.6\underline{/38.7°}\ \Omega$
2. $I_R = 6.67$ mA, $I_L = 1.33$ mA, $\vec{I}_T = 6.8\underline{/-11.3°}$ mA, $\vec{Z}_T = 1.47\underline{/11.3°}$ kΩ
5. $I_R = 2$ A, $I_C = 1.6$ A, $\vec{I}_T = 2.56\underline{/38.7°}$ A, $\vec{Z}_T = 7.81\underline{/-38.7°}\ \Omega$
6. $I_R = 30$ mA, $I_C = 40$ mA, $\vec{I}_T = 50\underline{/53.1°}$ mA, $\vec{Z}_T = 2.4\underline{/-53.1°}$ kΩ
9. $I_R = 2$ A, $I_L = 2.4$ A, $I_C = 12$ A, $\vec{I}_T = 9.81\underline{/78.2°}$ A, $\vec{Z}_T = 12.2\underline{/-78.2°}\ \Omega$
10. $I_R = 455$ mA, $I_L = 227$ mA, $I_C = 500$ mA, $\vec{I}_T = 531\underline{/31°}$ mA, $\vec{Z}_T = 18.8\underline{/-31°}\ \Omega$
11. $I_R = 4$ A, $I_L = 10$ A, $I_C = 2.5$ A, $\vec{I}_T = 8.5\underline{/-61.9°}$ A, $\vec{Z}_T = 11.8\underline{/61.9°}\ \Omega$
12. $I_R = 50$ mA, $I_L = 30$ mA, $I_C = 7.5$ mA, $\vec{I}_T = 54.8\underline{/-24.2°}$ mA, $\vec{Z}_T = 13.7\underline{/24.2°}\ \Omega$
17. $59.6\underline{/-41.1°}\ \Omega$
18. $70.3\underline{/26.2°}\ \Omega$
19. $5.59 + j8.8\ \Omega$
20. $19 - j6.48\ \Omega$
21. $10.6 + j6.2\ \Omega$
22. $6.26 - j4.2\ \Omega$
23. $74.5 + j93.6\ \Omega$
24. $3.42 - j17.6\ \Omega$

Self-Test

1. $V_R = 7.46$ V, $V_L = 5.03$ V (Objective 1)
2. $X_C = 44.2$ kΩ, $Z = 55.2$ kΩ, $\theta = -53.3°$ (Objective 2)
3. $39 + j30$ kΩ, $49.2\underline{/37.6°}$ kΩ (Objective 3)
4. 230 kHz (Objective 4)
5. $35.3\underline{/-28.1°}\ \Omega$ (Objective 5)

UNIT 21

Self-Test

1. 100 (Unit 4) **2.** 25 pF (Unit 5) **3.** 11.9 yr (Unit 15) **4.** $\frac{2}{3}$, $(0, -2)$ (Unit 10)

5.

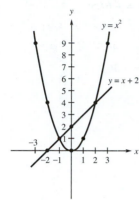

6. $V_1 = 10$ V, $V_2 = 3.33$ V, $V_3 = 6.67$ V (Unit 8 or 9)

7. $27.5\underline{/145°}$ (Unit 17)

8. $1 - j2$ (Unit 18)

9. $v = 0$ V, $i = -10.6$ mA (Unit 19)

10. $33 + j43$ kΩ, $54.2\underline{/52.5°}$ kΩ (Unit 20)

$(-1, 1), (2, 4)$ (Unit 12 or 14)

UNIT 22

Exercise 22.1

1.

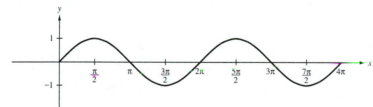

2.

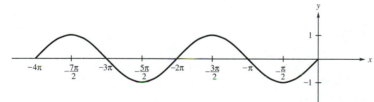

3.

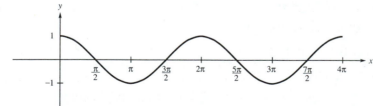

4.

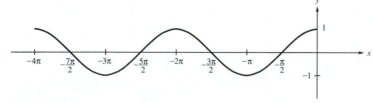

5.

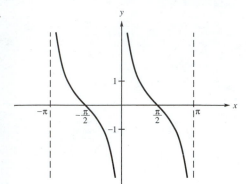

6.

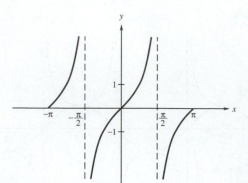

7.

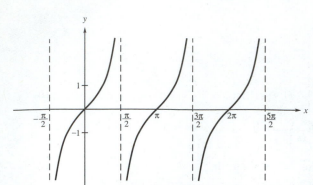

8.

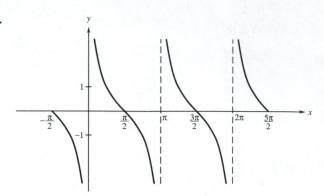

9.

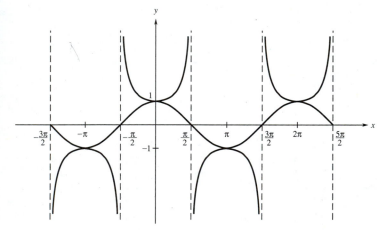

10.

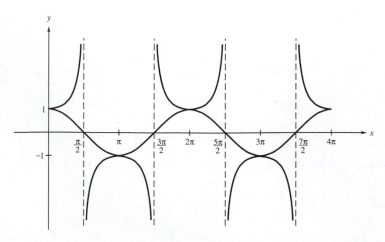

11.

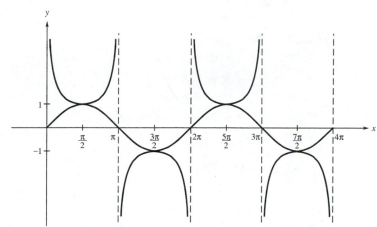

12.

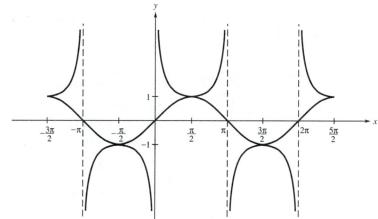

Exercise 22.2

1.	**a.** 0.524 rad	**b.** 30°		**2.**	**a.** 1.05 rad	**b.** 60°
3.	**a.** −1.05 rad	**b.** −60°		**4.**	**a.** −0.927 rad	**b.** −53.1°
5.	**a.** 2.62 rad	**b.** 150°		**6.**	**a.** 2.5 rad	**b.** 143°
7.	**a.** 0 rad	**b.** 0°		**8.**	**a.** 1.57 rad	**b.** 90°
9.	**a.** −0.464 rad	**b.** −26.6°		**10.**	**a.** −0.714 rad	**b.** −40.9°
11.	**a.** −0.926 rad	**b.** −53.1°		**12.**	**a.** 1.47 rad	**b.** 84.3°

13. −53.1° **14.** 5.74° **15.** 120° **16.** 180° **17.** −45° **18.** 5.71°

Exercise 22.3

1. $\dfrac{\sin x}{\cos x}$ **2.** $\dfrac{\cos x}{\sin x}$ **3.** $\dfrac{\cos x}{\sin x}$ **4.** $\sin x \cos x$ **5.** $\dfrac{1}{\cos x}$ **6.** $\cos x$

7. 1 **8.** 1 **9.** $\dfrac{1}{\sin x}$ **10.** $\cos x$ **11.** $\sin x$ **12.** 1

13. $\dfrac{\sin x}{1 + \sin x}$ **14.** $\dfrac{\cos x}{1 - \cos x}$ **15.** $\dfrac{1}{\sin^2 x \cos^2 x}$ **16.** $\dfrac{1}{\cos x}$ **17.** $1 + \cos^2 x$ **18.** $\cos^2 x$

19. $1 + \cos x$ **20.** $\sec x + 1$ **21.** $1 + \cos x$ **22.** $1 - \sin x$ **23.** $\csc x - \cot x$ **24.** $\csc x + \cot x$

Exercise 22.4

1. $C = 60°, a = 18.9, b = 20.9$
2. $B = 105°, a = 8.91, c = 9.98$
3. $C = 45°, b = 13, c = 9.23$
4. $B = 70°, a = 221, b = 210$
5. $c = 8.33, A = 40.2°, B = 61.7°$
6. $a = 14.5, B = 62.5°, C = 77.4°$
7. $a = 50.9, B = 17.8°, C = 117°$
8. $c = 7.4, A = 97°, B = 47°$
9. $A = 41.6°, B = 85.5°, C = 52.9°$
10. $A = 46°, B = 98.2°, C = 35.8°$
11. 8.95 ft, 39°
12. No solution
13. 5.4 nm, 46.8°
14. No solution

Exercise 22.5

(Counterexamples, illustrations, and proofs are not included.)

9. $-\cos x$
10. $\sin x$
11. $-\cos x$
12. $\sin x$
13. $-\sin x$
14. $-\cos x$
15. $\tan x$
16. $\cot x$
17. $-\cot x$
18. $-\tan x$
19. $\cot^2 x - 1$
20. $\cot x - \tan x$
21. $\cos x$
22. $2\cos x$
23. $\sec^2 x$
24. $\csc^2 x$
25. $\tan x$
26. $-\cot x$

Self-Test

1.

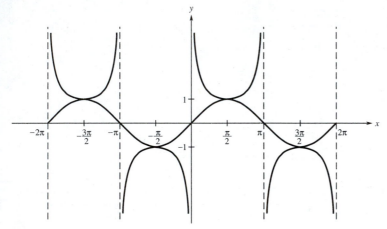

(Objective 1)

2. a. -1.05 rad
   b. $-60°$ (Objective 2)
3. $\cos^2 x$ (Objective 3)
4. $-\tan x$ (Objective 5)
5. $83.3°$ (Objective 4)

UNIT 23

Exercise 23.1

1. 11
2. 27
3. 56
4. 113
5. 160
6. 256
7. 1010_2
8. 10001_2
9. 101000_2
10. 1001101_2
11. 1111111_2
12. 10000000_2
13. 60
14. 134
15. 255
16. 711
17. 34_8
18. 102_8
19. 440_8
20. 1077_8
21. 10111_2
22. 100001_2
23. 101000110_2
24. 1000001111_2
25. 123_8
26. 415_8
27. 702_8
28. 1604_8
29. 75
30. 175
31. 480
32. 2801
33. C7H
34. D0H
35. 1FEH
36. 400H
37. 1001011_2
38. 11110101_2
39. 110100001100_2
40. 111100111110_2
41. 3AH
42. E7H
43. 7F0H
44. 140CH

Exercise 23.2

1. 111	**2.** 1111	**3.** 100	**4.** 1000	**5.** 10011	**6.** 100001
7. 101101	**8.** 1110010	**9.** 10	**10.** 101	**11.** 1	**12.** 111
13. 10	**14.** 101	**15.** 110	**16.** 1111	**17.** 10000	**18.** 11100
19. −101	**20.** −11	**21.** 10	**22.** 101	**23.** 11	**24.** 1010
25. 10000	**26.** 11101	**27.** −11	**28.** −1000	**29.** 1110	**30.** 11000
31. 100011	**32.** 1011011	**33.** 0.5	**34.** 0.625	**35.** 0.25	**36.** 0.375
37. 0.09375	**38.** 0.328125	**39.** 0.001_2	**40.** 0.111_2	**41.** 0.0011_2	**42.** 0.000101_2

43. $0.011001100110\ldots_2$ **44.** $0.11100110011001\ldots_2$

Exercise 23.3

(Proofs are not included.)

Exercise 23.4

1. **a.** 0	**b.** 1	
3. **a.** 1	**b.** 1	
5. **a.** 0	**b.** 1	**c.** 1
6. **a.** 0	**b.** 0	**c.** 0
7. **a.** 1	**b.** 0	**c.** 0
9. **a.** 1	**b.** 0	**c.** 0

2. **a.** 0	**b.** 1	
4. **a.** 0	**b.** 0	
d. 1	**e.** 1	
d. 0	**e.** 1	
8. **a.** 1	**b.** 1	**c.** 0
10. **a.** 1	**b.** 1	**c.** 0

11.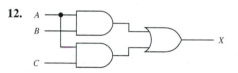

12.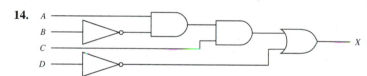

13.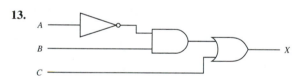

14.

15. $(AB)C$

16. $(A + B)(C + D)$

17. $(\overline{A} + B) + \overline{C}$

18. $\overline{A}B + \overline{C} + D$

19. A ——⊐ X = AB
B
$AB(B + C) = AB$

20. A ——⊐ X = A + B
B
$A + B + BC = A + B$

21. A B C ——⊐ X
$\overline{AB} + \overline{C} = \overline{ABC}$

22. A B C ——⊐ X
$(\overline{A \cdot B}) + C = A + B + C$

23. $\overline{A}B + AB = B$

24. $\overline{A \cdot B} + \overline{A}B = \overline{A}$

25. A B C ——⊐ X
$(A + B)(B + C) = AC + B$

26. A B C ——⊐ X
$(A + \overline{B})(A + \overline{C}) = A + \overline{B + C}$

27. A B ——⊐ X = $\overline{A \oplus B}$
$\overline{A}B + A\overline{B} = \overline{A \oplus B}$, the exclusive-NOR, or XNOR

28. A B ——⊐ X = $A \oplus B$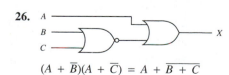
$(A + \overline{B})(\overline{A} + B) = \overline{A}B + A\overline{B}$, the exclusive-OR, or XOR

29.

A	B	C	$\overline{A \oplus B}$	$\overline{A \oplus C}$	$\overline{(A \oplus B)} \cdot \overline{(A \oplus C)}$
0	0	0	1	1	1
0	0	1	1	0	0
0	1	0	0	1	0
0	1	1	0	0	0
1	0	0	0	0	0
1	0	1	0	1	0
1	1	0	1	0	0
1	1	1	1	1	1

30.

A	B	C_i	Gate 1 $A \oplus B$	Gate 2 AB	Gate 3 Gate1·C_i	C_o Gate 4 Gate2 + Gate3	S Gate 5 Gate1 $\oplus$ C_i
0	0	0	0	0	0	0	0
0	0	1	0	0	0	0	1
0	1	0	1	0	0	0	1
0	1	1	1	0	1	1	0
1	0	0	1	0	0	0	1
1	0	1	1	0	1	1	0
1	1	0	0	1	0	1	0
1	1	1	0	1	0	1	1

Self-Test

1. 1111 (Objective 2)

2. a. 243   **b.** F3H (Objective 1)

3, 4. (Objective 3; Proofs are not included)

5.

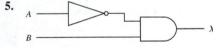

$B(A\overline{B} + \overline{A}) = \overline{A}B$ (Objective 4)

Index